P. Johnson

Functional Avian Endocrinology

Functional Avian Endocrinology

Editors
Alistair Dawson
Peter J Sharp

Narosa Publishing House
New Delhi Chennai Mumbai Kolkata

Editors
Alistair Dawson
Centre for Ecology and Hydrology
Monks Wood, Abbots Ripton, Huntingdon
Cambs. PE28 2LS, U.K.

Peter J Sharp
Division of Integrative Biology
Roslin Institute (Edinburgh)
Roslin, Midlothian EH25 9PS
Scotland, U.K.

NAROSA PUBLISHING HOUSE PVT. LTD.

22 Daryaganj, Delhi Medical Association Road, New Delhi 110 002
35-36 Greams Road, Thousand Lights, Chennai 600 006
306 Shiv Centre, D.B.C. Sector 17, K.U. Bazar P.O., Navi Mumbai 400 703
2F-2G Shivam Chambers, 53 Syed Amir Ali Avenue, Kolkata 700 019

www.narosa.com

Printed from the camera-ready copy provided by the Editors.

ISBN 81-7319-568-4

Cover Photograph: Alex Badyaev

Published by N. K. Mehra for Narosa Publishing House Pvt. Ltd., 22 Daryaganj, Delhi Medical Association Road, New Delhi 110 002 and printed at Rajkamal Electric Press Delhi 110 033, India.

Preface

This book contains selected presentations given at the 8th International Symposium on Avian Endocrinology, held in Scottsdale, Arizona, June 6-11th 2004. The Symposia are held every four years to provide an opportunity for scientists with an interest in avian endocrinology to meet with like-minded colleagues to discuss and present their recent research.

Significant progress has been made in areas of research traditionally presented represented at ISAE meetings. As at the first meeting in Calcutta in 1976, research on neuroendocrine mechanisms underlying seasonal breeding, on the endocrinology of gonadal, pineal, adrenal and thyroid function, and on developmental endocrinology continued to be strongly represented at the Scottsdale meeting. Some areas of research covered by the Calcutta meeting, notably intermediary metabolism and renal endocrinology, appear to have declined, while others are new or have become more prominent. Amongst these are endocrine disruptors and the biology of thyroid hormones. Curiously, no papers were presented at the Calcutta meeting on behavioural endocrinology, which has been a feature of subsequent ISAE meetings, including Scottsdale. Many of the advances in avian endocrinology during the last 25 years have been possible through the development of research tools, the most important of which include radioimmunoassay, immunocytochemistry and techniques in molecular biology, including situ hybridisation and PCR. More new research tools have become available though further advances in molecular biology. One of these is cDNA microarray technology which makes it possible to identify novel genes underlying endocrine function. The first section of this book describes this technique and illustrates the way it being used to open up new opportunities for avian endocrinologists.

Financial support for the meeting is gratefully acknowledged from Arizona State University, in particular, the Department of Biology, the Office of the Dean, the College of Liberal Arts and Sciences, the Office of the Vice-Provost for Research and Administration.

Thanks go to Pierre Deviche and the students and administrative staff of the School of Life Sciences, Arizona State University for the organization of all practical aspects of the conference and the preparation of the programme and conference schedule. Members of the International Scientific Programme Committee are thanked for their help with the selection of plenary lecturers and symposia. And finally, a special thanks to Doreen Wade who provided editorial assistance with the preparation of the manuscripts for publication.

Alistair Dawson

Peter J Sharp

Eberhard Gwinner (1938-2004)

In September 2004, the field of avian endocrinology has lost an eminent scientist and invaluable colleague with the unexpected passing of Eberhard Gwinner. Ebo was an intuitive naturalist and rigorous experimental scientist, who greatly advanced avian endocrinology and numerous other fields with his integrative organismal research. His infectious enthusiasm and charismatic personality will be deeply missed by colleagues, collaborators, and students around the world.

From the beginning Ebo was an ardent ornithologist, who published 13 manuscripts on bird behavior and natural history even before he obtained a formal academic degree. His Ph.D. on social behavior of ravens was mentored by two of Germany's finest ethologists, Gustav Kramer and Konrad Lorenz from the Max-Planck Institute for Behavioral Physiology. A short while later in Jürgen Aschoff's department at the same institute, he pioneered the study of endogenous circannual rhythms. This jump-started his long-term career studying circannual rhythms, the subject for which he has become most famous. During this formative period, he spent half a year in Congo studying migration in Phylloscopus warblers, which sparked in him a life-long fascination with the tropics, and Africa in particular. After a two-year stint in the New World working with Donald Farner at the University of Washington and Colin Pittendrigh at Stanford University, he returned to Germany and the Max-Planck Institute, becoming the Head of the Vogelwarte Radolfzell in 1979, and eventually a director of the institute in 1991.

Ebo was an excellent observer of bird behavior and was fascinated by nature in general. Owing to his solid ethological background, he was able to consider mechanistic questions without losing sight of the organisms he studied. Intriguing natural phenomena attracted his curiosity and his thorough experimental investigations pushed many frontiers. The result of his prolific career is almost 250 scientific publications; more are still forthcoming.

Ebo conducted stunning long-term studies on birds, most notably on the circannual system of his main avian muse, the stonechat (Saxicola torquata). He succeeded in the long-term captive breeding of wild birds, eventually establishing hybrids between different subspecies of stonechat to understand the mechanisms underlying ecological variation in circannual rhythms. His landmark demonstration of highly persistent circannual rhythms in equatorial stonechats directed new attention to seasonal phenomena in the tropics. He determined that circannual rhythms are synchronized by photoperiod and showed beautifully how photoperiodic responses of birds are adapted to the ecological requirements of each population. Among his highly influential findings was the demonstration that migration is controlled by the circannual system and that both the timing and the direction of migration are endogenously programmed. The question how birds find their wintering and breeding grounds continued to fascinate him. In recent years, his research led to the discovery that migratory species have a better long-term memory than non-migratory species, and that the hippocampus may be the area in the brain in which such spatial information is stored.

Another field in which Ebo's work was highly influential is that of circadian rhythms. He was one of the first to demonstrate that social cues could act as a zeitgeber. Later, he worked extensively on the role of melatonin in the circadian system of birds. He conceptualized the model of internal resonance, suggesting that the two main oscillators of the avian circadian system, the pineal gland and the suprachiasmatic nucleus (SCN), amplify each other via their respective output. In support of this paradigm, he began documenting how variations in the output of the pineal gland, the pattern of nocturnal melatonin secretion, affect overall circadian function in birds from across the globe, and suggested adaptive reasons for that variation. In recent years, he successfully ventured into cellular aspects of circadian rhythmicity such as the storage of photoperiodic information in the pineal gland, and contributed to the search for the location of the SCN in the avian brain.

Ebo was an ingenious researcher, who loved and lived his life to the fullest and spread his enthusiasm for science wherever he went. He was a spirited and engaged mentor, who thoroughly enjoyed interacting with students. Ebo will be sorely missed as an inspiring teacher, warm colleague and good friend.

Michaela Hau
(Photo courtesy Laura Spinney)

Contents

Part one

Functional genomics

Northern Mockingbird
Pierre Deviche

Functional Avian Endocrinology
A Dawson & P J Sharp (Eds)

Genomic tools for endocrine research

J Burnside, LA Cogburn, R Talbot[1] and JB Dodgson[2]

Department of Animal and Food Sciences, University of Delaware, Newark DE 19717
[1]Dept. of Genomics & Bioinformatics, Roslin Institute, Roslin, Midlothian EH25 9PS, UK
[2]Microbiology and Molecular Genetics, Michigan State University, East Lansing, MI 48824

Introduction

The recent release of a draft chicken genome sequence (1), as well as the development of a large collection of expressed sequence tags (ESTs), genome-wide-BAC-based physical maps and supporting bioinformatics, has dramatically changed the landscape of avian biology. These new resources re-establish the chicken as a major experimental model organism. Furthermore, it seems certain that sequence and mechanistic similarities to other birds will extend the use of chicken genome information throughout avian biology. In particular, the chicken sequence, along with comparative avian genomics research, offers the long-range potential to connect complex avian phenotypes, such as growth and behaviour, to their fundamental molecular mechanistic causes. This chapter reviews the current status of chicken genomics and the resources newly available to avian researchers.

Chicken genome sequence and resources

Although individual chicken gene sequences have been emerging for over 25 years, sequencing the full genome began following the submission of a white paper to NIH by J.D. McPherson, J.B. Dodgson, R. Krumlauf and O. Pourquié (2). This highlighted the role of the chicken not only as a food animal but also as a major model organism in developmental biology, evolutionary biology, and disease research. The National Human Genome Research Institute (NHGRI) designated this a "high priority" project and sequencing began at the Washington University Genome Sequencing Center (WUGSC) (3). The strategy was to produce 6-fold whole-genome shotgun (WGS) coverage of the genome of a single female (the heterogametic sex) of the inbred UCD001 Red Jungle Fowl line. A single inbred bird was chosen to minimize DNA sequence polymorphisms and UCD001 was chosen based on its previous use in developing linkage maps and BAC libraries.

The Red Jungle Fowl is the nearest wild relative of the domestic chicken and thus represents a "wild type" genome to which subsequent domestic sequences can be compared. Parallel efforts had already begun to generate a full genome BAC contig physical map of the genome (1,4), and these were eventually employed in assembling the 6.6X draft sequence released on March 1, 2004 (5). The term "draft" sequence is used to indicate that a substantial number of gaps and occasional ordering errors remain in the sequence, as is to be expected in a 6.6X WGS assembly, despite the integration of BAC map, EST and cDNA data to improve accuracy. Retrieved sequences may contain short gaps (whose length will often be known based on "end pair" reads from a cloned fragment of known size), and approximately 5% of genes known from cDNA sequences are substantially or completely missing. Furthermore, a portion (12%) of the sequence has yet to be assigned to a specific chromosome location.

The genome sequence, along with a variety of viewing options and analytical tools, can be accessed at three different browsers: the UCSC Chicken Genome Browser Gateway (6); the NCBI Chicken Genome Resources (7); and the EBI's Ensembl Chicken Genome Browser (8). Each browser offers a variety of options and but uses the same genome assembly. The Blat program of the UCSC browser quickly maps a sequence to the genome (similar to BLAST at NCBI and Ensembl, along with SSAHA at Ensembl). The Ensembl and UCSC browsers are highly configurable for displaying (or not) information of interest. In particular, they offer powerful tools to download specific data sets for further analysis (e.g. Ensmart in Ensembl). A more recent addition to the NCBI browser is a presentation the chicken genome in MapViewer including the annotation of partial, complete and predicted genes with extensive documentation and tutorials. Several other options are available at each site, and all browsers have excellent help features to guide the user.

Gene hunters may also access the genome-wide, BAC contig-based physical map that facilitated the assembly of the genome sequence (4). Those BACs that have been aligned with the assembly via their end sequences are available as an optional display section in the UCSC and Ensembl browsers); however, a more complete version of the BAC physical map is available (9) as part of the Wageningen U. ChickAce, which was developed for storage of extensive mapping information, with particular emphasis on linkage, cytogenetic, radiation hybrid and BAC contig maps as well as phenotypic traits (QTL) (10). BAC clones, DNA pools and filters, as well as the BAC libraries themselves are also available (see Resources and/or The BAC Page section of recent Newsletters at http://poultry.mph.msu.edu/ for further details).

In parallel to the WUGSC sequence, the Beijing Genomics Institute in China generated a map of genetic variation by 1/4X shotgun sequence coverage of the genomes of three different domestic chickens (a broiler from the United Kingdom (Roslin Institute), a White Leghorn layer from Sweden (Uppsala University) and a Silkie from China). Nearly three million genetic variation sites, mostly single nucleotide polymorphisms (SNPs) have been identified (11). These data have been integrated into the various genome browsers, and also can be searched at the Chicken Variation Database (12). In addition to being of fundamental biological interest with regard to the domestication and selection of the modern chicken, most of these SNPs are prevalent within domestic lines; therefore, they can be used in high resolution linkage mapping of any phenotypic trait of interest.

A portal to information on the chicken genome and chicken biology was created at the 1st international chicken genome workshop held in Hinxton, UK in 2003 and designated AvianNet (13). This serves as a gateway to a network of WWW sites covering the chicken genome, chicken developmental biology, chicken genetics, biodiversity, chicken immunology, and other species and tools with a shared biological interest. In addition, a mailing list (14) has been established to facilitate rapid sharing of information.

Chicken cDNAs and ESTs

Sequenced cDNAs and ESTs constitute a critical resource for chicken biologists, and the number of chicken ESTs has grown dramatically in the last few years. The BBSRC large scale EST sequencing project produced approximately 340, 000 ESTs from normalized cDNA libraries prepared from 12 adult and 9 embryonic tissues (15). The ESTs can be searched using keywords or the Blast engine at the BBSRC ChickEST website (16). Since these sequences have been isolated from many animals and various strains, mining of the data has identified a large number of coding SNPs which are available on the website. The site also includes an RNAi gene prediction tool and a search engine for non-coding RNAs. The bursal EST project has generated a large (46,595) EST collection from 2-week-old chicken bursa (17). One of the libraries in this project was designed to capture the 5'end and therefore the collection contains a large subset of full-length clones, to aid in obtaining full insert sequence. The EST project at the University of Delaware (18) has focused on generating ESTs (49,000) from tissues of economic importance: liver, muscle, neuroendocrine, reproductive and lymphoid tissues. Clones from the BBSRC, UD and bursal EST projects are available through the web sites or via ARK Genomics (24). The Uppsala project produced 21,869 ESTs from brain and testes. A multi-tissue (brain, ovary, embryos, hypothalamus, skin, and pituitary gland) library approach was taken by the group at INRA, where 12,000 cDNA clones were sequenced from both the 5' and 3' ends (23,273 ESTs total). These and several other projects have made important contributions to the now (November 2004) 527,686 *G. gallus* ESTs in GenBank (Table 1). In other species, full-length cDNA sequences have proven to be essential in accurately annotating genes and their alternative mRNA processing mechanisms through which a single gene may give rise to multiple protein products. Further analysis of all the EST sequences identified a set of unique cDNA clones coding for full length or nearly full length cDNAs. These have been fully sequenced to provide a set of 19,626 fully annotated clones which are searchable through the BBSRC ChickEST website.

The Institute for Genomic Research (TIGR) maintains a gene index of all chicken cDNAs and ESTs in GenBank. The TIGR *G. gallus* Gene Index (GgGI) integrates public data from all *G. gallus* gene sequencing projects with the goal of representing a non-redundant compilation of all chicken genes, including data on their expression patterns, cellular roles, functions, and evolutionary relationships (19). This index now contains 42,988 tentative consensus sequences assembled from all available data. In addition, 72,941 singleton ESTs are indexed. Theses numbers are in excess of the number of genes predicted in the chicken genome (approximately 20-25,000 protein coding genes) (20) and include alternatively-spliced products and non-coding RNAs, as well as lower quality sequences that could not be assembled with the stringent criteria of the TIGR CAP3 assembly program. Along similar lines, NCBI has created a chicken Unigene database that contains 21,000 entries (21). Thus, we are well on our way to having the annotated sequence for every chicken gene.

Gene expression tools

ESTs are invaluable for gene discovery, genome sequence annotation, and identification of orthologous relationships among evolutionarily distinct organisms. In addition, collections of ESTs can provide the basis for the development of microarrays. A number of centers have tackled the production of chicken microarrays (Table 2). A 13,000-element chicken cDNA array is available from the Fred Hutchinson Cancer Research Center (FHCRC) and ARK Genomics, (Roslin Institute, UK) at a cost of $150 per slide. Clones on this array were selected from a total of 363,838 chicken ESTs representing 24 different adult or embryonic tissues. A subset of 11,447 non-redundant ESTs were selected and added to an existing collection of clones (4,162) from immune tissues and a chicken bursal cell line (DT40). The array provides broad coverage of

Table 1 Chicken ESTs in GenBank

Institution	Tissues	# GenBank Entries
University of Delaware	Fat, Liver, Muscle, Pituitary, Oviduct, Ovary, Testis, Lymphoid tissues	49,393
USDA/ARS/BARC	Small intestine from birds infected with coccidia	14,376
Fred Hutchinson Cancer Research Center (FHCRC)	DT40 cells	1,116
BBSRC, UK	Embryo (9 stage or tissue-specific libraries) Adult (12 tissue-specific libraries)	330,096
Roslin Institute, UK	Brain, Embryo, Liver	5,194
Heinrich-Pette-Institute, Germany (Japan)	Bursa of Fabricius	46,595
INRA/Agenae, France	Multi-tissues: brain, ovary, embryos, hypothalamus, skin, pituitary gland	23,273
Uppsala University, Sweden	Brain, Testis (Red Jungle Fowl, Leghorn)	21, 869
University of Sao Paulo ESALQ, Brazil	Embryo (Limb buds, breast muscle)	14,395
All Others		21,379
	Total EST Sequences 10/2004	527,686

mRNAs expressed in many tissues; in addition, clones with expression unique to various tissues can be detected. L. Cogburn and collaborators have produced an 8K metabolic/somatic system array, a 7K neuroendocrine/ reproductive system array, and a 14K integrated systems array (22, 23) (and chapter this volume). These arrays are being used for transcriptional profiling across multiple tissues of divergently selected broiler chicken lines. At ARK genomics (24) four chicken microarrays have been prepared: three targeting specific tissues, neuroendocrine, immune and bone/shell gland, as well as the general chicken 13K array. A range of studies including one on photostimulation in both chicken and quail highlight the ability to use these chicken microarrays in other avian species. Recently, Affymetrix has designed a chicken Gene Chip that contains probe sets representing about 33,000 transcripts. Since the Affymetrix Gene Chips employ short oligonucleotide probes, mostly from 3' untranslated regions of mRNAs, they are less likely to be useful in other avian species. We are rapidly approaching the availability of genome-wide arrays for the chicken, at least some of which should be applicable to other gallinaceous species and probably even more distantly related birds.

Another valuable resource for transcriptional profiling in the chick embryo is the GEISHA gene expression database developed by P. Antin (25). This site collects and displays *in situ* hybridization data at different developmental stages for a large collection of genes. This project is complemented by the Chick-Atlas project where high quality images will serve as references for in-situ hybridisation results (26).

Concluding remarks

Sequencing the chicken genome is not only a significant advance for avian biology but also a milestone in the Human Genome Project. Birds are among the more diverse orders, but their genomes demonstrate a relatively high degree of evolutionary conservation at the karyotype level. From the perspective of vertebrate evolution, the chicken sequence provides a valuable tool in comparative genomics. For example, the chicken genome presently is the nearest "outgroup" with which to analyze mammalian genome evolution. This aids in re-construction of the likely genome arrangement of the common mammalian ancestor and even the common vertebrate ancestor (1). Furthermore, in comparisons made between mammalian genomes (e.g., human vs. mouse), it is often difficult to differentiate evolutionary (i.e., selected for function) conservation from background similarity (especially in non-coding regions). The greater evolutionary distance between birds and mammals reduces the un-selected similarity to near-zero, so any significant conservation observed is almost certainly of functional significance. Third, the chicken genome reveals genes that are unique to, or much more widely used in, mammals, as well as genes that may have been lost in mammals but retained in the chicken and gene families that specifically expanded in avian evolution (1). Continued analysis of the chicken genome will improve our general understanding of how genes evolve, how they function, and the role of conserved non-coding DNA, as well as identifying those genes that give birds their unique character.

Table 2 Chicken Microarray Resources

Institution	Tissues Represented	# Elements
University of Delaware	Lymphoid tissue	5K
University of Delaware	Integrated Systems (Metabolic/Somatic/ Neuroendocrine)	14K
USDA/ARS/BARC	Small Intestine	10K
Fred Hutchinson Cancer Research Center (FHCRC)	Multi-tissue	13K
ARK Genomics, Roslin Institute	Neuroendocrine, Immune, Bone/Shell Gland, Multi-tissue	5K and13K
Uppsala University, Sweden	Brain, Testis	14K
University of Sao Paulo, ESALQ, Brazil	Embryo, muscle, pituitary	5K
China Agricultural University	Liver, muscle, brain, fat, ovary, etc	10K
Affymetrix	Genome-wide Gene Chip®	33K

Finally, the chicken sequence serves as the "model genome" for the ~9600 avian species. Since gene order has been highly conserved during avian evolution, one can expect that genes identified either by molecular analysis or by phenotype in other birds will be located in the analogous position along the sequence to that of their chicken homologs. The location, sequence and function of a chicken homolog provide an entry point for detailed analysis of a trait and/or gene(s) in a bird of interest. Birds, as a group, provide a wealth of species of interest for behavioural, reproductive, nutritional, and physiological traits, and now the chicken sequence now provides a scaffold on which much of this avian-specific biology can be functionally annotated.

References

1 Wallis JW, Aerts J, Groenen M, Crooijmans R, Layman D, Graves T, Scheer D, Kremitzki C, Higgenbotham J, Gaige T, Mead K, Walker J, Albracht D, Divito J, Yang S-P, Leong S, Chinwalla A, Hillier L, Sekhon M, Dodgson JB, Romanov MN, Cheng H, de Jong PJ, Zhang H, McPherson

JD, Krzywinski M, Schein J, Mardis E, Wilson R & Warren WC (2004). A physical map of the chicken genome. *Nature* **432** 761-764
2 www.genome.gov/Page/Research/Sequencing/SeqProposals/Chicken_Genom.pdf
3 http://www.genome.wustl.edu/projects/chicken/index.php
4 Ren C, Lee M-K, Yan B, Ding K, Cox B, Romanov MN, Price JA, Dodgson JB & Zhang H-B (2003). A bac-based physical map of the chicken genome. *Genome Research* **13** 2754-2758
5 http://www.genome.gov/11510730
6 http://genome.ucsc.edu/cgi-bin/hgGateway?org=Chicken&db=0&hgsid=30948908
7 http://www.ncbi.nlm.nih.gov/genome/guide/chicken.
8 http://www.ensembl.org/gallus_gallus.
9 http://hbz.Tamu.Edu/.
10 https://acedb.asg.wur.nl/.
11 International Chicken Polymorphism Map Consortium (2004). A genetic variation map for chicken with 2.8 million single nucleotide polymorphisms. *Nature* **432** 717-722
12 http://chicken.genomics.org.cn/chicken/jsp/about.jsp
13 http://www.chicken-genome.org/.
14 chicken-genome@lists.Bbsrc.Ac.Uk.
15 Boardman E, Sanz-Ezquerro J, Overton IM, Burt DW, Bosch E, Fong WT, Tickle C, Brown WRA, Wilson SA & Hubbard SJ (2002). A comprehensive collection of chicken cDNAs. *Current Biology* **12** 1965-1969
16 http://www.chick.umist.ac.uk/
17 http://swallow.gsf.de/dt40.html
18 http://www.chickest.udel.edu
19 http://www.tigr.org/tdb/tgi/gggi
20 International Chicken Genome Sequencing Consortium (2004). Sequence and comparative analysis of the chicken genome provide unique perspectives on vertebrate evolution. *Nature* 432 695-716
21 ftp://ftp.ncbi.nlm.Nnh.gov/repository/unigene/.
22 Cogburn L, Wang X, Carre W, Rejto L, Porter T, Aggrey S & Simon J (2003). Systems-wide chicken DNA microarrays, gene expression profiling, and discovery of functional genes. *Poultry Science* **82** 939-951
23 Cogburn L, Wang X, Carre W, Rejto L, Aggrey SE, Duclos MJ, Simon J and Porter TE (2004). Functional genomics in chickens: Development of integrated-systems microarrays for transcriptional profiling and discovery of regulatory pathways. *Comparative and Functional Genomics* **2** 253-261
24 www.arkgenomics.org
25 http://geisha.biosci.arizona.edu/
26 http://genex.hgu.mrc.ac.uk/Collaborations/ChickAtlas/intro.html, Duncan Davidsson)

Functional Avian Endocrinology
A Dawson & P J Sharp (Eds)

Functional genomics of the avian pineal gland

V M Cassone, M J Bailey, S P Karaganis, V Kumar[1] and P A Bartell

Center for Research on Biological Clocks, Department of Biology, Texas A&M University, College Station, TX 77843 USA
[1]Department of Zoology, University of Lucknow, Lucknow, UP 226007, India

Introduction

The circadian biological clock is a fundamental property of biological organization in most free-living organisms ranging from cyanobacteria to humans (49). This clock regulates a wide array of biochemical, physiological and behavioral processes with a period of approximately (*circa*) one day (*diem*), synchronizes these to the 24-hr period of the solar day, and orchestrates complex internal processes such that temporal events are coordinated in adaptive efficiency.

Although there is increasing evidence that the capacity to oscillate with a circadian period is a distributed process in multicellular animals (1, 27, 32, 50, 60), in higher vertebrates, regulation of peripheral rhythmicity and overt circadian organization is hierarchically organized by several neuroendocrine structures (27, 50). These are the hypothalamic suprachiasmatic nucleus (SCN), the retina and the pineal gland (11, 12). Here, we will briefly review the role of the pineal gland and retina in the avian circadian clock, discuss the known molecular events involved in the biosynthesis of the pineal hormone melatonin, and explain what is known about the molecular genetics of vertebrate clocks. Then, we will describe our own functional genomics approach to the chick pineal clock, applying the knowledge gained by this approach toward the understanding of avian circadian organization.

The role of the avian pineal gland

The circadian secretion of the indoleamine hormone melatonin by the avian pineal gland is a critical component of the circadian biological clock of many species of birds (10, 13, 38, 41). Surgical removal of the pineal gland of passerine birds, the major source of the hormone in these species, disrupts or abolishes locomotor, body temperature and brain glucose uptake rhythms in birds maintained in constant darkness (DD) (19, 22, 25, 26, 30, 42). Secondly, transplantation of the pineal gland into the anterior chamber of the eye of arrhythmic, pinealectomized birds restores circadian rhythmicity and confers to the recipient birds the circadian phase of the donor birds (63). This observation strongly pointed to the probability that the mechanism by which the pineal gland influences circadian behavior was through the secretion of a hormone, since the effect was rather immediate, before any pineal afferents could be re-established. Of course, this hormone is now known to be the indoleamine hormone melatonin (cf. 10, 41), and several authors have shown that exogenously administered cycles of the hormone can re-establish rhythms of locomotion and brain glucose uptake in pinealectomized house sparrows, European starlings, and pinealectomized/enucleated pigeons (16, 31, 34, 43).

In galliform and columbiform species, the effects of pinealectomy are not as dramatic as they are in passerine birds (16, 23, 45, 54). The surgery has no effect on locomotor rhythms of Japanese quail, *Coturnix coturnix* (54), and only disrupts and/or changes the circadian period of pigeons, *Columba livia* (16, 23). However, in these species, the retina also synthesizes and releases melatonin into the blood stream, augmenting the pineal source of the hormone (23, 54, 55). When pinealectomy is combined with enucleation or retinectomy, quail and pigeons become completely arrhythmic, suggesting that it is the rhythmic presence and absence of melatonin that is the critical factor, rather than the pineal gland itself.

The mechanisms by which melatonin is synthesized are largely understood in a wide variety of vertebrate species (40, 41). In most species, tryptophan is taken up from the bloodstream and is hydroxylated to 5-hydroxytryptophan (5HTP) by tryptophan hydroxylase (TrH; E.C. 1.14.16.4). Aromatic amino acid decarboxylase (AADC; E.C.4.1.1.28) then converts 5HTP to serotonin (5HT). During the night, 5HT is converted to N-acetylserotonin (NAS) by arylalkylamine-N-acetyltransferase (AANAT; E.C. 2.3.1.87). NAS is then converted to melatonin by hydroxyindole-O-methyltransferase (HIOMT; E.C.2.1.1.4) (Fig. 1). However, the specifics of the molecular regulation of rhythmic melatonin biosynthesis vary among species and between tissues in the same species, as we will see below.

The avian pineal gland receives noradrenergic input from the sympathetic nervous system *via* the superior cervical ganglia that can prevent damping of the pineal circadian clock and modify the amplitude of the rhythm of melatonin released by the pineal into the bloodstream (11, 14, 15). Further, the pineal gland is directly photosensitive; light perceived by the retina or the pineal gland can phase shift and acutely inhibit melatonin biosynthesis and, in turn, affects the phase and rhythmicity of pineal coordinated outputs (9, 19, 61).

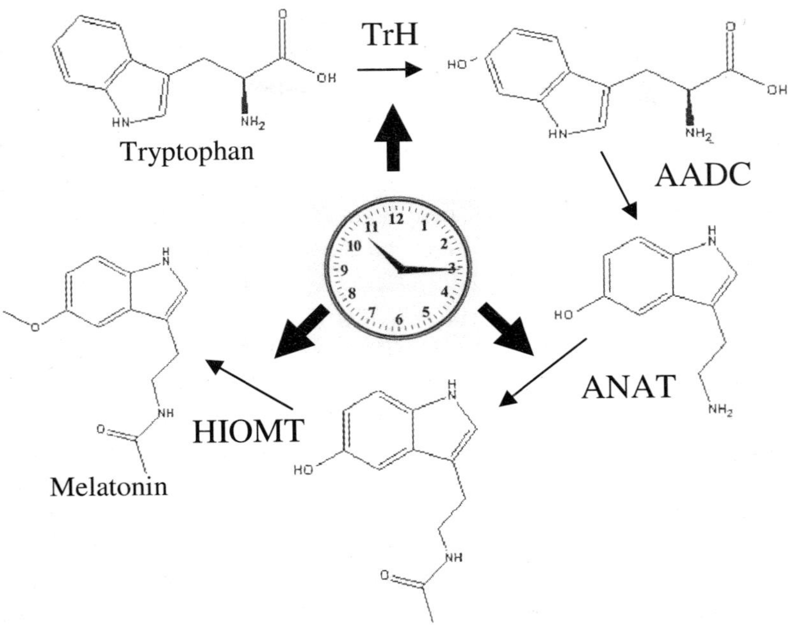

Fig. 1 The biosynthetic pathway for melatonin production. Photoreceptor cells in the pineal gland and retina take up the amino acid tryptophan, which is 5-hydroxylated to 5-hydroxytryptophan and decarboxylated to serotonin. Then during the night, serotonin is converted to N-acetylserotonin, which is subsequently converted to melatonin. The circadian clock regulates at least three enzymes in this pathway.

Cultured chick pineal gland

Several authors have shown that the cultured chick pineal gland expresses daily and circadian rhythms in AANAT activity (9, 19, 39, 56). In these static culture systems, glands from many chicks ranging in age from 1-14 days (depending on the study) were placed in culture media for several days under LD cycles or DD. At intervals of 2-4 hours, 4-5 glands were removed and rapidly frozen. Once the entire experiment was completed, all glands were processed for radioenzymatic assays for AANAT activity. These studies established that cultured chick pineal glands expressed a daily pattern of AANAT activity such that activity was high during the night and low during the day. Secondly, they revealed that the rhythms in AANAT activity persisted for several days even if the glands were placed in DD. Finally, a short pulse of light, or a shift in the LD cycle, were found to phase shift pineal rhythms in AANAT activity. Together, the data from these studies established that the chick pineal gland contained a circadian oscillator, photoreceptors capable of entraining the circadian oscillators to environmental light, and a molecular output, activity of AANAT.

The flow-through pineal culture system developed by Takahashi *et al.* (51) was a significant technical advance in that it allowed much finer temporal resolution of the rhythms expressed by the gland. In this system, pineal glands are placed in a constant flow of oxygenated culture medium collected into a fraction collector. All fractions are assayed for melatonin content by radioimmunoassay. Using this approach, these authors further demonstrated that the chick pineal clock resides in multiple pieces of the gland and that the clock within the gland is temperature compensated.

Zatz and colleagues (44, 61, 62) developed a dispersed pineal cell culture with the significant modification that melatonin production would be assayed in the culture media as in Takahashi *et al.* (51) rather than AANAT activity as in other static culture preparations, which requires harvesting cultured cells. Dispersed culture of pineal cells provided a preparation that could approach the temporal resolution of the flow-through system, and was amenable to pharmacological manipulation. This work demonstrated the role of calcium and cyclic nucleotide metabolism in the regulation of pineal melatonin output and the fact that these second messenger systems were not involved in the regulation of circadian phase in the pineal gland.

Using this system, Zatz *et al.* also showed that light affects pineal melatonin rhythms *via* at least three separable pathways: 1) It acutely decreases melatonin biosynthesis if administered during the night. 2) It decreases the rate of rhythm damping and increases rhythm amplitude, and 3) it phase-shifts the rhythm in a time-dependent fashion. The first two of these pathways appear to be mediated by regulation of cyclic nucleotide regulation and are independent of the pacemaker properties of the gland. This effect is mimicked by application of the sympathetic neurotransmitter norepinephrine (NE), which is normally secreted during the day *in vivo* (11, 15). However, the phase-shifting effect of light is mediated by an independent process, which has not been forthcoming and which presumably involves transcriptional and/or translational effects (44).

Molecular biology of melatonin biosynthesis

The genes encoding the four major enzymes involved in melatonin biosynthesis, TrH, AADC, AANAT and HIOMT, in the domestic fowl, *Gallus gallus*, have been isolated, cloned, sequenced and characterized (cf. 41). Of these, TrH, AANAT and HIOMT have been studied most extensively in the pineal gland and retina, and each have been shown to be regulated by the circadian clock within the gland at the transcriptional, translational and post-translational levels (5, 7, 17, 24, 28, 29, 36). The mRNAs for these enzymes are expressed on a daily basis in both the chicken pineal gland and retina. TrH and AANAT mRNA levels peak primarily at night. In contrast, HIOMT mRNA levels peak at night in the retina, but in the pineal gland it peaks during the day. Further, the rhythms of pineal TrH, AANAT and HIOMT mRNA levels persist in constant darkness (DD) *in vivo* and *in vitro* (8, 28, 29), indicating that a circadian pacemaker within the chick pineal gland controls these genes' transcription. TrH and AANAT, but not HIOMT, mRNA levels also persist in DD in the retina (6, 7, 17), suggesting a circadian pacemaker drives these mRNAs as well, although the locus of this pacemaker is at this stage not known.

The proteins that these mRNAs encode, as well as their enzymatic activities, are also regulated on a daily and circadian basis in chicks, such that protein levels and enzymatic activities are regulated semi-independently of mRNA levels. As an example, acute illumination of cultured chick pineal glands in the middle of the night rapidly decreases both AANAT protein abundance and activity but has no effect on mRNA levels. This particular differential regulation is due to light-induced phosphorylation, ubiquitinylation and proteosomal proteolysis of the enzyme (cf. 37, 41). Since the present review is devoted to functional genomics, we will not further pursue this issue here, although we will discuss it again briefly below.

Molecular biology of vertebrate circadian rhythms

In spite of the fact that we know quite a bit about the molecular biology, biochemistry and cell biology of the melatonin biosynthetic pathway in the chick pineal gland and retina, we know very little about the molecular mechanisms by which the rhythm itself is generated.

Based largely on work on the molecular genetics of circadian rhythms in *Drosophila* (1, 27, 52), genes involved in the regulation of circadian rhythmicity in rodents have been identified by use of reverse and forward genetic approaches (60). These genes, which bear remarkable sequence similarity to their *Drosophila* orthologs, have been further analyzed by *in vitro* expression analyses of their interactions, and studies of knock-in and knockout animals. These studies have been reviewed by many authors (1, 27, 52, 60 and more) and synthesize a common model for the transcriptional/translational regulation of mammalian circadian rhythms, which is remarkably similar to the predicted clock mechanisms in flies. The basic model (Fig. 2) begins with the transcription of "positive elements" *clock, Bmal1* and perhaps *Bmal2*. These are translated, dimerize, and enter the nucleus, where they stimulate the transcription of a wide array of genes whose promoters contain an E-box, to which these transcription factors bind. Among the genes that the "positive elements" stimulate are the so-called "negative elements", *period 1 (per1), period 2 (per2), period 3 (per3), cryptochrome 1 (cry1)* and *cryptochrome 2 (cry2)*. Following transcription and translation of these genes, they oligomerize and enter the nucleus, where they inhibit the action of the "positive elements". This, of course, is a standard feedback/feed-forward regulatory mechanism. However, in this case, a significant delay in the nuclear entry of the "negative elements" is accomplished by the phosphorylation of the PERIOD (PER) proteins by several kinases, particularly casein kinase 1 ε (ck1ε) and casein kinase 1 δ (ck1δ), which targets them for proteosomal proteolysis.

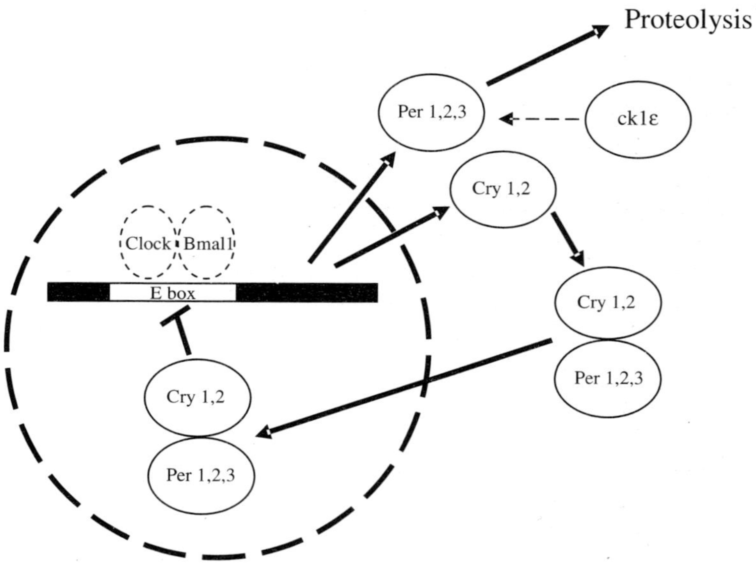

Fig. 2 A very simplified version of the transcription/translation feedback model for the molecular clockworks in mammals. For a complete description, please refer to text and to Refs. 1, 50, 60.

As one might expect, many labs, including ours, have employed PCR-based cloning to identify avian orthologs of these genes (2, 3, 4, 18, 20, 21, 33, 35, 46, 47, 48, 57, 58, 59) and to characterize their distributions, temporal dynamics and phylogenetic relationships to known vertebrate and invertebrate sequences (52). However, the best-case scenario with this type of approach would be to show that birds have the same kinds of mechanisms as in flies and mice. Although this might be satisfying, there is a significant body of literature to suggest that critical components of the molecular clockworks are as yet unknown. Such an approach could never identify these unknown elements nor could it unveil significant input and/or output processes of the circadian clock.

A screen for central clocks components in the chick pineal gland

To approach this system in an objective manner, we set several criteria for identifying the *most likely* mRNA species that are involved in circadian rhythm generation in the chick pineal gland. To do this, we made several assumptions, which may or may not be correct (which we will address below). First, since both the chick retina and pineal gland contain circadian pacemakers capable of generating rhythms of melatonin biosynthesis, we assume that the molecular mechanism underlying circadian rhythms in the chick retina and pineal gland are at least similar, if not identical. Secondly, we assume that the pineal gland itself is a subset of the entire circadian system and that some pineal rhythms *in vivo* derive from rhythmic inputs from the system (4, 5).

Thus, our criteria for most likely clock components in this functional genomics approach were:

1) The mRNA should be expressed rhythmically in the chick pineal gland *in vivo* in both LD and DD.
2) The mRNA should be expressed rhythmically in the chick retina *in vivo* in both LD and DD.
3) The mRNA should be expressed rhythmically in the chick pineal gland *in vitro* in both LD and DD.
4) *Also*, since we know that light directly affects the clock mechanism within the pineal gland, light should affect *in vitro* pineal transcripts.
5) However, since we know that sympathetic norepinephrine (NE) does not *phase* shift the chick pineal clock, but rather affects the pineal by acutely inhibiting melatonin biosynthesis through regulation of cyclic nucleotide metabolism, NE should not affect clock mechanisms at the mRNA level *in vitro*.

Creation of the chick cDNA microarray

Two cDNA libraries were produced from mRNA extracted from pineal glands of 7 day old cockerels maintained in an LD 12:12 cycle. One library was constructed from mRNA extracted at midday (ZT6) and the other was constructed from mRNA extracted at midnight (ZT18). The rationale for this approach was to minimize the possibility that mRNA species that were highly rhythmic and present only at one time of day were not lost in the analysis. Approximately 5000 clones from each of these libraries were isolated and their inserts were sequenced and annotated (n=~9000 cDNAs). Of these, approximately 3500 clones contained known, annotated sequences with highly significant BLAST hits. The remainder was either "no hits" (no close relative in the database) or "no chromats" (sequence was not very reliable).

After amplification, the cDNAs derived from these libraries were spotted on poly-lysine coated glass slides using a GeneMachines Omnigrid microarrayer. Before hybridization, the slides were hydrated, dried, and the DNA was UV-cross-linked. Then, the slides were ready for experimental hybridization (Fig. 3).

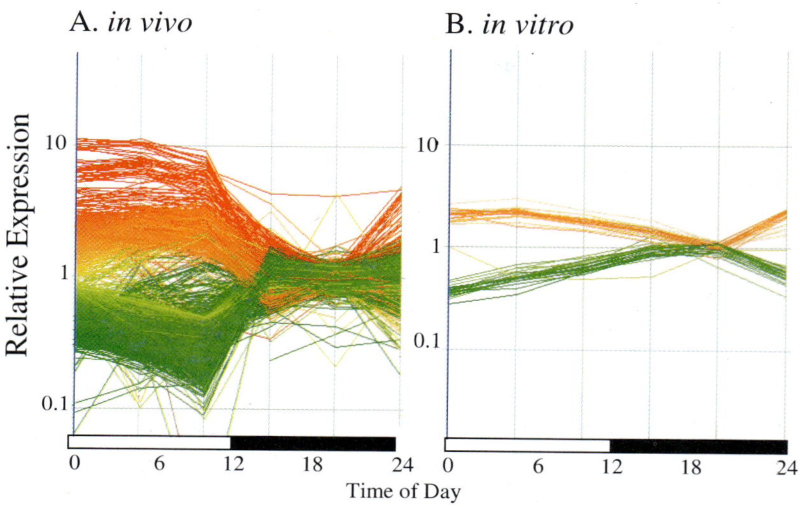

Fig. 3 A DNA microarray analysis of daily changes in gene expression in chicken pineal glands taken from A) chicks at different times over 24hr and from B) pineal glands maintained in culture. Each line represents changes in expression of a gene with time. Note that many genes are expressed rhythmically in LD *in vivo* while the number is drastically reduced *in vitro*.

Transcriptional profile of the chick pineal gland *in vivo*

Because the major biochemical events in the chick pineal gland peak at midnight (ZT18), we semi-arbitrarily assigned ZT18 as our internal control phase. For all time of day experiments, mRNA was extracted at ZT18, amplified to aRNA and labeled with Cy3 fluorescent dye. Then, in all three time of day experiments described below, mRNA was extracted, amplified and labeled with Cy5 fluorescent dye at several different times of day in LD and DD. In LD, the times of day were ZT2 (two hours after lights on in the LD cycle), ZT6, ZT10, ZT14, ZT18, and ZT22. In DD, the times were identical with the exception that samples were taken in complete darkness. Hybridizations were performed with both control (ZT18) and experimental samples simultaneously (N=>4 biological samples / time point), and the relative expression for each gene was expressed relative to its own expression at ZT18.

Of the ~9000 cDNAs on the microarray, 765 classified (known) transcripts oscillated with a 1.5-fold rhythm in LD, and 377 classified transcripts oscillated with a 2-fold or greater amplitude. In DD, the number of transcripts that oscillated with a 1.5 fold rhythm was reduced to 291 and with a 2-fold rhythm to 128. Among these, cDNAs representing all of the melatonin-generating enzymes were represented. As has been previously described, mRNA encoding HIOMT is highest during the day in LD ands subjective day in DD, TrH and AANAT are highest during the night and subjective night respectively, and AADC is not rhythmic. These patterns were corroborated by northern blot analysis. Orthologs of all of the "clock genes" were also represented on the array, and, as with the melatonin generating sequences, many of these were expressed rhythmically in LD and DD. "Positive elements" *Bmal1, Bmal1* and *clock* were expressed primarily during the mid- to late day and subjective day, while "negative elements" *per2* and *per3* were expressed in the late night and subjective night, with *per3* exhibiting the highest amplitude rhythm (No *per1* has been described from any study of avian clock genes.). In contrast, the other "negative elements", *cry1* and *cry2*, are expressed with significantly different phases; *cry1* is expressed with very high amplitude, peaking during the midday and subjective day, while *cry2* exhibits a somewhat bimodal, low amplitude pattern. These were corroborated by northern analysis and with the few published reports on these genes. Both ck1ε and ck1δ were represented on the array, but the mRNAs for these kinases were not rhythmic.

In addition to these seemingly pineal-specific signals, we were surprised to find a broad array of mRNAs encoding a wide range of functions. These included genes involved in phototransduction and the visual cycle, protein dynamics, intermediary metabolism and immune function. This latter category corroborates a view that the chick pineal gland is a lymphopoetic organ. These will be discussed further below.

Transcriptional profile of the chick retina *in vivo*

Using the pineal gland microarray, a similar study was performed in chick retina. In this study, 546 classified (known) transcripts oscillated with a 1.5-fold rhythm in LD, and 175 classified transcripts oscillated with a 2-fold or greater amplitude. In DD, the number of transcripts that oscillated with a 1.5 fold rhythm was reduced to 383 and with a 2-fold rhythm to 49. The retinal transcriptional profile bears some remarkable similarities and equally remarkable differences with the pineal profile. First, the chick retina also expresses rhythmic mRNA involved in melatonin biosynthesis, clock function, phototransductive and visual cycle function, intermediary metabolism, and protein dynamics. However, in many cases, the details of this regulation are quite different.

For example, while HIOMT mRNA in the pineal gland peaks during the subjective day, in the retina, HIOMT mRNA is highest during the night and does not persist in DD. Further, the amplitude of the AANAT mRNA rhythm is much lower than in the pineal gland, such that the trough is significantly higher than in the retina, suggesting that, in the retina, at least, AANAT is not rate-limiting. Several radioenzymatic studies had predicted this to be the case. In contrast, TrH mRNA is robustly rhythmic in the retina, as has been previously shown. Retinal clock genes are also expressed rhythmically. However, only *per3* and *cry1,* both "negative elements", are expressed rhythmically in both LD and DD. Intriguingly, these signals are expressed 180° in antiphase in both the pineal gland and retina, which makes explanation of their role as negative elements in the context of the proposed transcriptional/ translational feedback model somewhat of a stretch.

Although many sequences associated with phototransduction and visual cycle function, intermediary metabolism and protein dynamics are regulated on a circadian basis in these two tissues in the same organisms, relatively few are regulated identically. Of these 9 mRNA species, 2 (cytochrome c oxidase and glyceraldehyde-3-phospharte dehydrogenase) are involved in intermediary metabolism, 1 (carbonic anhydrase) is involved in pH homeostasis, 1 (purpurin) is involved in retinaldehyde transport, 1 (elongation factor 2) is involved in protein dynamics and 1 (cytochrome P450 2A1) is involved in xenobiotic response. As stated above, *cry1* and *per3*, both clock genes, as well as TrH are regulated similarly in the pineal gland and retina *in vivo*.

Transcriptional profile of the chick pineal *in vitro*

Using the chick pineal microarray, a similar study was conducted with dispersed pineal cultures similar to those developed by Zatz and others (see above). This *in vitro* system allowed us to identify rhythmic transcriptional components of the pineal that do not require inputs found only in whole chicks, and to compare those with the *in vivo* rhythmic transcriptome identified in our study (Fig. 3). In our experimental design, pinealocytes were harvested from one-day-old chicks and cultured for 5 days under LD conditions. On day 6, animals were either maintained in LD or switched to DD conditions, and RNA was extracted every 4 hours in a manner consistent with the *in vivo* studies. In addition, melatonin was collected from the media and quantified using radioimmunoasay.

Based on the stringency of our screen, the number of genes exhibiting a 1.5 fold amplitude rhythm in LD ranged from 449 genes (or 5.5% of the arrayed genome) to 90 genes (1.1%). In DD, the number of genes ranged from 339 genes (4.2%) to virtually none using the most stringent analysis. Following this trend of a reduced amplitude in rhythmicity *in vitro* relative to *in vivo*, the number of genes exhibiting a 2 fold amplitude rhythm in LD ranged from 78 genes (0.96%) to 55 genes (0.67 %). In DD, the number of 2-fold rhythmic transcripts is reduced to just 38 genes (0.42%) to none using the most stringent analysis. These data demonstrate that both the number and amplitude of clock controlled genes are reduced *in vitro*, and supports our assumption (stated earlier) that a subset of pineal rhythms are dependent upon physiological inputs from circadian system. However, the functional distribution of this subset of genes is very similar to that of the rhythmic transcriptome *in vivo*, such that genes involved in protein dynamics, intermediary metabolism and immune function comprise the largest portion of rhythmic genes with known functions. Also, as expected, clock genes (especially *cry1* and *per3*) and the three melatonin biosynthesis genes which are rhythmic *in vivo* (TrH, AANAT, and HIOMT), as well as the most abundantly rhythmic transcripts *in vivo* (including cystatin and purpurin) are rhythmic in the same manner *in vitro*.

To address the final two criteria for identifying potential clock components (see criteria four and five, enumerated above), we conducted another experiment in which free-running cultured pinealocytes were given a 6-hour pulse of either light or NE supplemented media (3×10^{-8} M) from CT0-CT6 or from CT12-CT18.

A number of genes expressed in the pineal are regulated (induced or repressed) by light and/or norepinephrine; a significant subset of genes are rhythmic and light driven (i.e., are rhythmic in LD, but not DD, and are regulated by light); a small subset of genes are rhythmic in both LD and DD, and are regulated by light only or by light and norepinephrine. The latter group of genes represents a handful of candidates we have identified which may play an important role as a component of the avian pineal clock mechanism. Further characterization of these genes with respect to clock function is forthcoming.

What does this all mean?

These studies have revealed a dynamic of mRNA abundance that is more complex than one would expect if the sole function of the pineal gland were for the rhythmic production of melatonin. Intermediary metabolism, protein dynamics, and even immune function are regulated within the gland by a circadian clock. In addition, the studies reveal that the regulation of both melatonin biosynthetic and clock gene sequences are not the same in the retina as in the pineal gland, suggesting other processes, perhaps related to the regulation of metabolism and protein dynamics differentially play a role in rhythm generation in these two tissues. Thirdly, the studies call into question the universality of the transcription/translation feedback model for circadian rhythm generation (Fig. 2). Certainly, the fact that two presumed "negative elements" *cry1* and *per3* are expressed in antiphase makes strict adherence to the prevailing model difficult. Perhaps, *cry1* and *per3* encode negative elements regulating two distinct rhythmic processes 180° out of phase. Even so, at the mRNA level, at least, the lack of rhythmic positive elements in the pineal gland *in vitro* and retina *in vivo* suggests that post-transcriptional processes are likely critical for rhythm generation in this system (and probably others as well).

The proof of the pudding, however, will require the ability to manipulate gene expression in an avian system both *in vivo* and *in vitro*. Further, proteomic analysis and analysis of post-translational mechanisms of clock function in this important model system will likely provide clues to the true mechanism of biological clock function. In the coming months, these exciting new vistas will certainly become visible.

Acknowledgements
Supported by NIH P01 NS 39546. The authors thank the Texas A&M Center for Research on Biological Clocks for a dynamic intellectual environment.

References
1 Allada R, Emery P, Takahashi JS & Rosbash M (2001) Stopping time: the genetics of fly and mouse circadian clocks. *Annual Review of Neuroscience* **24** 1091-1119

2 Bailey MJ & Cassone VM (2004) Opsin photoisomerases in the chick retina and pineal gland: characterization, localization, and circadian regulation. *Investigative Ophthalmology and Visual Science* **45** 769-775

3 Baile M.J & Cassone VM (in press) Melanopsin expression in the chick retina and pineal gland. *Molecular Brain Research* doi:10.1016/j.molbrainres.2004.11.003

4 Bailey MJ, Chong NW, Xiong J & Cassone VM (2002) Chickens' Cry2: molecular analysis of an avian cryptochrome in retinal and pineal photoreceptors. *FEBS Letters* **513** 169-174

5 Bailey MJ, Beremand PD, Hammer R, Bell-Pedersen D, Thomas TL & Cassone VM (2003) Transcriptional profiling of the chick pineal gland, a photoreceptive circadian oscillator and pacemaker. *Molecular Endocrinology* **17** 2084-2095

6 Bailey MJ, Beremand PD, Hammer R, Reidel E, Thomas TL & Cassone VM (2004) Transcriptional profiling of circadian patterns of mRNA expression in the chick retina. *Journal of Biological Chemistry* **279** 52247-52254

7 Bernard M, Iuvone PM, Cassone VM, Roseboom PH, Coon SL, & Klein DC (1997) Avian melatonin synthesis: photic and circadian regulation of serotonin N-acetyltransferase mRNA in the chicken pineal gland and retina. *Journal of Neurochemistry* **68** 213-224

8 Bernard M, Klein DC & Zatz M (1997) Chick pineal clock regulates serotonin N-acetyltransferase mRNA rhythm in culture. *Proceedings of the National Academy of Science U S A.* **94** 304-309

9 Binkley S (1979) A timekeeping enzyme in the pineal gland. *Scientific American* **240** 66-71

10 Cassone VM (1998) Melatonin's role in vertebrate circadian rhythms. *Chronobiology International* **15** 457-77

11 Cassone VM & Menaker M (1983) Sympathetic regulation of chicken pineal rhythms. *Brain Research* **272** 311-317

12 Cassone VM & Lu J (1994) The pineal gland and avian circadian organization: the neuroendocrine loop. *Advances in. Pineal Research* **8** 31-40

13 Cassone VM & Natesan AK (1997) Time and time again: the phylogeny of melatonin as a transducer of biological time *Journal of Biological Rhythms* **12** 489-497

14 *Cassone* VM & Menaker M (1984) Is the avian circadian system a neuroendocrine loop? *Journal of Experimental Zoology* **232** 539-549

15 Cassone VM, Forsyth AM & Woodlee GL (1990) Hypothalamic regulation of *circadian* noradrenergic input to the chick pineal gland. *Journal of Comparative Physiology A* **167** 187-192

16 Chabot CC & Menaker M (1992) Effects of physiological cycles of infused melatonin on circadian rhythmicity in pigeons *Journal of Comparative Physiology A* **170** 615-622

17 Chong NW, Cassone VM, Bernard M, Klein DC & Iuvone PM (1998) Circadian expression of tryptophan hydroxylase mRNA in the chicken retina. *Molecular Brain Research* **61** 243-50

18 Chong NW, Bernard M & Klein DC (2000) Characterization of the chicken serotonin N-acetyltransferase gene. Activation via clock gene heterodimer/E box interaction. *Journal of Biological Chemistry* **275** 32991-32998

19 Deguchi TA (1979) A circadian oscillator in cultured cells of chicken pineal gland. *Nature* **282** 94-96

20 Doi M, Okano T, Yujnovsky I, Sassone-Corsi P & Fukada Y (2004) Negative control of circadian clock regulator E4BP4 by casein kinase I epsilon-mediated phosphorylation. *Current Biology* **14** 975-980

21 Doi M, Nakajima Y, Okano T, & Fukada Y (2001). Light-induced phase-delay of the chicken pineal circadian clock is associated with the induction of cE4bp4, a potential transcriptional repressor of *cPer2* gene. *Proceedings of the National Academy of Science U S A.* **98** 8089-8094

22 Ebihara S & Kawamura H (1981) The role of the pineal organ and the suprachiasmatic nucleus in the control of circadian locomotor rhythms in the Java sparrow (*Padda oryzivova*) *Journal of Comparative Physiology A* **141** 207-214

23 Ebihara S, Uchiyama K & Oshima I (1984) Circadian organization in the pigeon Columbia *livia*: the role of the pineal organ and the eye *Journal of Comparative Physiology A* **154** 59-69

24 Florez JC, Seidenman KJ, Barrett RK, Sangoram AM & Takahashi JS (1996) Molecular cloning of chick pineal tryptophan hydroxylase and circadian oscillation of its mRNA levels. *Molecular Brain Research* **42** 25-30

25 Fuchs JL (1983) Effects of Pinealectomy and subsequent melatonin implants on activity rhythms in the house finch (*Carpodacus mexicanus*). *Journal of Comparative Physiology A* **153** 413-419

26 Gaston S & Menaker M (1968) Pineal function: the biological clock in the sparrow? *Science* **160** 1125-1127

27 Glossop NR & Hardin PE (2002) Central and peripheral circadian oscillator mechanisms in flies and mammals. *Journal of Cell Science* **115** 3369-3377

28 Green CB, Besharse JC & Zatz M (1996) Tryptophan hydroxylase mRNA levels are regulated by the circadian clock, temperature, and cAMP in chick pineal cells. *Brain Research* **738** 1-7

29 Greve P, Voisin P, Grechez-Cassiau A, Bernard M, Collin JP & Guerlotte J (1996) Circadian regulation of hydroxyindole-O-methyltransferase mRNA in the chicken pineal gland *in vivo* and *in vitro*. *Biochemical Journal* **319** 761-766

30 Gwinner E (1978) Effects of pinealectomy on circadian locomotor activity rhythms in European starlings, *Sturnus vulgaris*. *Journal of Comparative Physiology* **126** 123-129

31 Gwinner E & Benzinger I (1978) Synchronization of a circadian rhythm in pinealectomized European starlings, *Sturnus vulgaris*. *Journal of Comparative Physiology [A]* **127** 209-213

32 Gwinner E & Brandstatter R (2001) Complex bird clocks. *Philosophical Transactions of the Royal Society of London B Biological Science* **356** 1801-1810

33 Hayashi Y, Sanada K, Hirota T, Shimizu F & Fukada Y (2003) p38 mitogen-*activated* protein kinase regulates oscillation of chick pineal circadian clock. *Journal of Biological Chemistry* **278** 25166-25171

34 Heigl S & Gwinner E (1995) Synchronization of circadian rhythms of house sparrows by oral melatonin: effects of changing period. *Journal of Biological Rhythms* **10** 225-233

35 Holthues H, Engel L, Spessert R & Vollrath L (2004) Circadian gene expression patterns of melanopsin and pinopsin in the chick pineal gland. *Biochemistry and Biophysics Research Communications* **326** 160-165

36 Iuvone PM, Bernard M, Alonso-Gomez A, Greve P, Cassone VM, & Klein DC (1997) Cellular and molecular regulation of serotonin N-acetyltransferase activity in chicken retinal photoreceptors. *Biological Signals* **6** 217-224

37 Iuvone PM, Brown AD, Haque R, Weller J, Zawilska JB, Chaurasia SS, Ma M & Klein DC (2002) Retinal melatonin production: role of proteasomal proteolysis in circadian and photic control of arylalkylamine N-acetyltransferase. *Investigative Ophthalmology and Visual Science* **43** 564-572

38 Janik D, Dittami J, & Gwinner E (1992) The effect of pinealectomy on circadian plasma melatonin levels in house sparrows and European starlings. *Journal of Biological Rhythms* **7** 277-286

39 Kasal CA, Menaker M & Perez-Polo JR (1979) Circadian clock in culture: N-acetyltransferase activity of chick pineal glands oscillates *in vitro*. *Science* **203** 656-658

40 Klein DC (2004) The 2004 Aschoff/Pittendrigh lecture: Theory of the origin of the pineal gland--a tale of conflict and resolution. *Journal of Biological Rhythms* **19** 264-279

41 Klein DC, Coon SL, Roseboom PH, Weller JL, Bernard M, Gastel JA, Zatz M, Iuvone PM, Rodriguez IR, Begay V, Falcon J, Cahill GM, Cassone VM, & Baler R (1997) The melatonin rhythm-generating enzyme: molecular regulation of serotonin N-acetyltransferase in the pineal gland. *Recent Progress in Hormone Research* **52** 307-357

42 Lu J & Cassone VM (1993) Pineal regulation of circadian rhythms of 2-deoxy[^{14}C]glucose uptake and 2[^{125}I]iodomelatonin binding in the visual system of the house sparrow, *Passer domesticus. Journal of Comparative Physiology A* **173** 765-774

43 Lu J & Cassone VM (1993) Daily melatonin administration synchronizes circadian patterns of brain metabolism and behavior in pinealectomized house sparrows, *Passer domesticus. Journal of Comparative Physiology A* **173** 775-782

44 Natesan A, Geetha L & Zatz M (2002) Rhythm and soul in the avian pineal. *Cell & Tissue Research* **309** 35-45

45 Nyce J & Binkley S (1977) Extraretinal photoreception in chickens: entrainment of the circadian locomotor activity rhythm. *Photochemistry & Photobiology* **25** 529-531

46 Okano T & Fukada Y (2003) Chicktacking pineal clock. *Journal of Biochemistry (Tokyo)* **134** 791-797

47 Okano T, Yamamoto K, Okano K, Hirota, T, Kasahara T, Sasaki M, Takanaka Y & Fukada Y (2001) Chicken pineal clock genes: implication of BMAL2 as a bidirectional regulator in circadian clock oscillation. *Genes &* Cells **6** 825-836

48 Okabayashi N, Yasuo S, Watanabe M, Namikawa T, Ebihara S & Yoshimura T (2003) Ontogeny of circadian clock gene expression in the pineal and the suprachiasmatic nucleus of chick embryo. *Brain Research* **990** 231-234

49 Pittendrigh CS (1993) Temporal organization: reflections of a Darwinian clock-watcher. *Annual Review of Physiology* **55** 16-54

50 Reppert SM & Weaver DR (2002) Coordination of circadian timing in mammals. *Nature* **418** 935-941

51 Takahashi JS, Hamm H & Menaker M (1980) Circadian rhythms of melatonin release from individual superfused chicken pineal glands *in vitro*. *Proceedings of the National Academy of Science USA* **77** 2319-2322

52 Tauber E, Last KS, Olive PJ, & Kyriacou CP (2004) Clock gene evolution and functional divergence. *Journal of Biological Rhythms* **19** 445-458

53 Tu DC, Batten ML, Palczewski K & Van Gelder RN (2004) Nonvisual photoreception in the chick iris. *Science* **306** 129-131

54 Underwood H & Siopes T (1984) Circadian organization in Japanese quail. *Journal of Experimental Zoology* **232** 557-566
55 Underwood H, Siopes T & Barrett RK (1988) Does a biological clock reside in the eye of quail? *Journal of Biological Rhythms* **3** 323-331
56 Wainwright SD & Wainwright LK (1979) Chick pineal serotonin acetyltransferase: a diurnal cycle maintained *in vitro* and its regulation by light. *Canadian Journal of Biochemistry* **57** 700-709.
57 Yamamoto K, Okano T & Fukada Y (2001) Chicken pineal Cry genes: light-dependent up-regulation of *cCry1* and *cCry2* transcripts. *Neuroscience Letters* **313** 13-16
58 Yasuo S, Watanabe M, Okabayashi N, Ebihara S & Yoshimura T (2003) Circadian clock genes and photoperiodism: Comprehensive analysis of clock gene expression in the mediobasal hypothalamus, the suprachiasmatic nucleus, and the pineal gland of Japanese Quail under various light schedules. *Endocrinology* **144** 3742-3748
59 Yoshimura T, Suzuki Y, Makino E, Suzuki T, Kuroiwa A, Matsuda Y, Namikawa T & Ebihara S (2000). Molecular analysis of avian circadian clock genes *Molecular Brain Research* **78** 207-215
60 Young MW & Kay SA (2001) Time zones: a comparative genetics of circadian clocks. *Nature Reviews Genetics* **2** 702-715
61 Zatz M, Mullen DA & Moskal JR (1988) Photoendocrine transduction in cultured chick pineal cells: effects of light,dark, and potassium on the melatonin rhythm. *Brain Research* **438** 199-215
62 Zatz M, Gastel JA, Heath JR 3rd & Klein DC (2000) Chick pineal melatonin synthesis: light and cyclic AMP control abundance of serotonin N-acetyltransferase protein. *Journal of Neurochemistry* **74** 2315-2321
63 Zimmerman NH & Menaker M (1979) The pineal gland: a pacemaker within the circadian system of the house sparrow. *Proceedings of the National Academy Sciences USA* **76** 999-1003

Functional Avian Endocrinology
A Dawson & P J Sharp (Eds)

Transcriptional profiling in liver of hormonally-manipulated chickens

X Wang, W Carré, L Rejtö[1] and LA Cogburn

Department of Animal and Food Sciences, 531 South College Avenue, University of Delaware, Newark, DE 19717 USA
[1]Department of Food and Resource Economics, University of Delaware, Newark, DE 19717 USA

Introduction

For decades, we have focused on the investigation of a single or, on occasion, multiple endocrine factors that regulate avian growth and development (8, 12). Much of our current understanding of the importance of a functional thyroid axis came from studies where thyroid-active substances were fed to young broiler chickens (12-14, 17). The metabolically-active thyroid hormone, T_3, plays a major role in maintenance of metabolic rate, body temperature and body composition (12) *via* paracrine adjustments in the insulin/glucagon (I/G) molar ratio (28). Short-term treatment of pre-market broiler chickens with a low level of dietary T_3 (0.25 ppm) reduces accumulation of excessive body fat and increases accretion of protein in skeletal muscle (12). On the other hand, the incorporation of a goitrogen [propylthiouracil (PTU)] into feed of young chickens leads to hypothyroidism, hyper-secretion of insulin and GH, retarded growth rate and mild obesity (12). The obesity of hypothyroid chickens is largely due to the absence of T_3 negative feedback on GH secretion (23) which in turn contributes to elevated insulin secretion and a higher I/G molar ratio.

Early studies in young broiler chickens showed that exogenous cGH fails to increase either growth rate or plasma IGF-I, albeit the accumulation of body fat was increased (7, 12, 15, 37, 39). This is unlike the typical mammalian response to exogenous GH which includes increased growth rate, elevated plasma IGF-I and a leaner body mass (11, 26, 40). Thus, slight obesity, elevated GH, higher I/G molar ratio, and down-regulation of the hepatic GH receptor (GHR) are common features of the GH resistance observed in young chickens (12, 36). The importance of a functional somatotropic axis is clearly revealed in the GHR-deficient dwarf chicken, which has a retarded growth rate, short stature, mild obesity and hypothyroidism (17). Yet, the sex-linked dwarf chicken grows to two-thirds normal body weight, completely independent of GH action (16).

We have gained a more comprehensive view of endocrine regulation of growth and development from endocrine perturbation studies, where thyroid status was manipulated (*hypo- vs. hyper-*thyroidism) or where dietary T_3 was combined with exogenous growth hormone (cGH) (12). Hyperthyroidism (dietary T_3) reduces plasma GH levels and the I/G molar ratio, which lead to a marked depletion of body fat (12, 14). Decades ago, we discovered a very interesting synergism between two combined exogenous hormones – cGH + T_3 – which results in the dramatic depletion of body fat at market age (17).

The purpose of the present study was to use newly-developed chicken genomic tools (annotated cDNA microarrays, bioinformatic/ statistical tools, and a draft genome sequence) to examine gene expression patterns in liver in response to chronic treatment with thyroid-active substances (T_3 and PTU) alone, exogenous cGH alone, or GH combined with dietary T_3 (GH+T_3). This original microarray study has allowed the identification of several clusters of functional genes that respond to perturbations in the thyroid and GH axes, which ultimately interact to control growth and metabolism (19). With these transcriptional snapshots, we can begin to build functional maps of metabolic and regulatory pathways that regulate growth and development in the chicken – a model organism recently endowed with powerful tools of functional and structural genomics.

EST sequencing and development of liver-specific chicken cDNA microarrays

The first step toward advancing functional genomics in chickens was completion of a large international catalog of expressed sequences tags (ESTs) sequenced from numerous tissue cDNA libraries. We have constructed and normalized five single and multi-tissue chicken cDNA libraries for high throughput sequencing of ESTs and development of microarrays (19). About 37,500 ESTs were sequenced from chicken cDNA libraries which represent several important (immune, metabolic, somatic and neuroendocrine) physiological systems. The CAP3 sequence assembly program (32) was used to assemble contigs from all chicken ESTs in public collections (~409K ESTs) on March 1, 2003, including 37K EST sequences in the UD collection. Our CAP3 assembly of a chicken gene index (Version 2) revealed 31,472 contigs and 83,099 unassembled singlets ESTs. These contig (high-fidelity *in silico* cDNAs) and singlet sequences were then used in BlastX and BlastN analysis; the highest Blast hit score was used for annotation of the chicken ESTs. Non-redundant sets of chicken ESTs (unigenes) were selected from our

EST collection for PCR amplification of cDNA inserts and production of several chicken cDNA microarrays including the present 3.1K liver-specific array, an 8K metabolic systems array, a 6K neuroendocrine array, and the 14K Del-Mar Chicken Integrated Systems microarray (18). For the liver specific microarray, the non-redundant hepatic unigene set (3.1K) was selected from a total of 5,611 ESTs sequenced from the primary and normalized liver cDNA libraries (18, 19). The non-redundant liver-specific cDNAs were PCR amplified and robotically printed onto 8 x 12cm nylon membranes in a 96-well format (36 spots/grid). Our CAP3 chicken EST database of assembled contigs and unassembled singlets is searchable by nucleotide sequence or keywords and gene lists are provided for each chicken microarray developed (http://udgenome.ags.udel.edu/~cogburn/).

Liver-specific microarrays and analysis of hepatic gene expression patterns

Total RNA (25μg) was isolated from the liver of four birds per treatment group and reverse transcribed in the presence of ^{33}P-dCTP. The ^{33}P-labeled cDNA targets from each chicken (N=24) were hybridized overnight to individual liver-specific microarrays (Fig. 1). Images were captured overnight on a PhosphorImager screen and scanned by a PhosphorImager (Storm 840, Molecular Dynamics) into a tiff file. The tiff files were then imported into Pathways 4 software (Invitrogen) for determination of spot intensities and normalization (i.e., the spot intensity of each gene was normalized by dividing by the mean intensity of all spots on the microarray). The Pathways 4 software provides treatment contrast, a simple Student's *t*-test and fold-difference filters for analysis of microarrays. A subroutine was written for the Pathways 4 program to provide an output file of raw spot intensities which were used in our custom Bayesian analysis of miroarrays (BAM) program (18). Parameters of the stochastic model are first estimated by the maximum likelihood method (1). After estimating the model parameters, a Bayesian method (3) allows us to estimate the "real" gene expression level from rank normalized data, which minimizes the false discovery rate (41). The BAM model also accounts for systematic error in production of microarrays associated with the 36 cDNA source plates and the 96 pins used for printing the microarrays. The posterior probabilities are then used to identify differentially expressed genes.

Appropriate treatment contrasts were used to identify clusters of thyroid hormone (TH-) and/or GH-responsive genes. Genes were clustered according to two criteria: the significance of their spot intensities and fold-difference between treatments in the initial analysis provided by the Pathways 4 program. For more robust statistical analysis, the significance level ($P<0.05$) was determined from Wilcoxon signed-rank analysis and the mixed-model BAM program (posterior probability >0.95). The data from the quantitative RT-PCR (TaqMan® assays) were analyzed using ANOVA and Wilcoxon rank sum test.

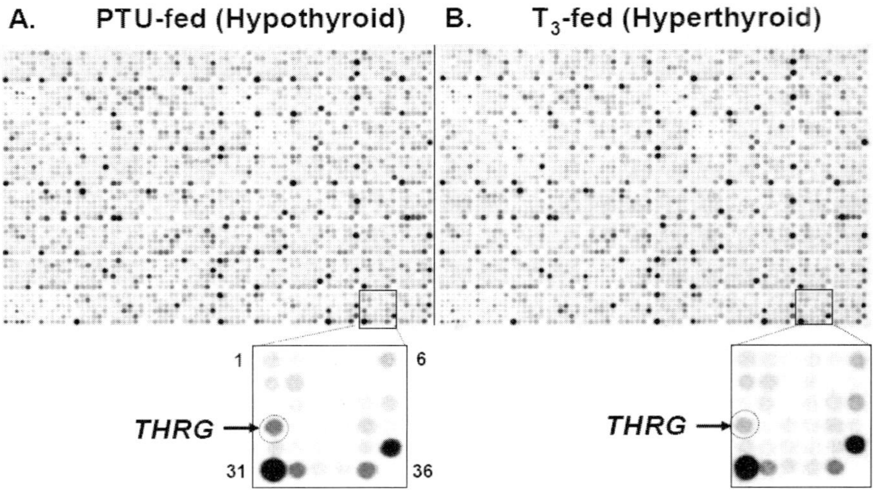

Fig. 1 Representative images of liver-specific chicken cDNA microarrays hybridized to liver RNA targets from a hypothyroid (PTU-treated) or a hyperthyroid (T_3-fed) broiler chicken. Liver total RNA (25 μg) was reverse transcribed in the presence of ^{33}P-dCTP and hybridized to a single membrane-based liver microarray. The 3,154 non-redundant cDNAs were robotically printed on a nylon membrane (8 x 12cm) in a 96-well format (36 spots/grid). The images were developed on a PhosphorImager screen which was scanned into a Tiff-file by a PhosphorImager (Storm 840, Molecular Dynamics). The gene (spot) intensities were then analyzed with Pathways 4 software (Invitrogen). Insets show that the hepatic expression of an unknown EST is higher in the hypothyroid state (**A**) and lower in the hyperthyroid state (**B**) This unknown gene has been further characterized and named thyroid hormone repressible gene (*THRG*).

Manipulation of the thyroid and/or GH axis in young chickens

Four-week-old broiler cockerels (two pens of four birds/treatment) were fed thyroid-active substances for three weeks: propylthiouracil (PTU, 0.5%), T_3 (1.25 ppm) or given a daily *i.m.* injection of cGH (250 μg/kg) alone or daily cGH injection in combination with dietary T_3 (GH+T_3). Control feed (CF) was finely-ground commercial broiler-grower ration, whereas the identical ration was fed in pellet form to the buffer control (BC) and GH-injected groups. The average body weights of injected birds (BC and GH treatments) fed the commercial ration (pellets) were slightly higher that the dietary treatment groups fed the pulverized ration (CF, PTU, T_3). The abdominal fat weight of PTU-treated birds was 53% higher ($P<0.05$) than that of the CF group, whereas dietary T_3 alone reduced abdominal fat by 52%, whereas the combined GH+T_3 treatment depleted fat by 77% (Fig. 2A). Plasma T_3 levels were reduced in hypothyroid chickens (67% reduction) and greatly elevated ($P<0.05$) in the hyperthyroid (T_3-fed) state (4-fold increase) (Fig. 2B). The further enhancement of plasma T_3 levels in GH+T_3-treated birds at 7 weeks of age represents an important synergism between the enhanced somatotropic and thyrotropic axes. The greatly elevated levels of T_3 in the GH+T_3 treatment would certainly support a higher metabolic rate and enhanced lipolysis. We suspect that the enhancement of plasma T_3 levels in the GH+T_3 treatment group was due to cGH depression of type 3 deiodinase (DI3) activity which degrades T_3 to metabolically inactive T_2 (see Van der Geyton *et al.*, Chapter 33). Earlier work has clearly shown that exogenous cGH elevates plasma T_3 levels by reducing DI3 activity in liver and kidney (21, 22). Thus, the combined treatment of GH+T_3 leads to a synergism between the thyrotropic and somatotropic axes and a remarkable depletion of body fat stores, which confirms our earlier studies (12).

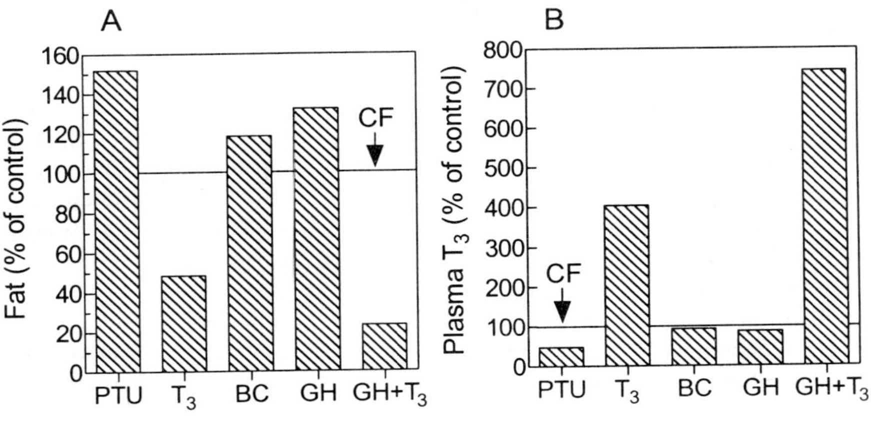

Fig. 2 Relative abdominal fat content (**A**) and plasma T_3 levels (**B**) of broiler chickens after chronic thyroid manipulation and/or cGH treatment. Birds were given either control feed (CF), 0.5% PTU, 1.25 ppm T_3 in ground feed or a single daily injection of buffer control (BC), 250 μg cGH/kg BW (GH) or GH+T_3 for three weeks (4 to 7 weeks of age). The data are presented as a percent of the CF group at 7 weeks of age (N=8 birds).

Transcriptional response to perturbation of the thyrotropic and somatotropic axes

Two representative images are shown after hybridization of liver-specific chicken cDNA microarrays with RNA isolated from liver of a hypothyroid (PTU-fed) (Fig. 1A) or a hyperthyroid (T_3-fed) (Fig. 1B) chicken. A total of 2606 clone pairs were filtered from this contrast (PTU *vs.* T_3) because their mean spot intensities were not significantly ($P<0.05$) different. The insets show the differential expression of an unknown EST, which we have named thyroid hormone repressive gene (*THRG*). *THRG* is expressed at higher levels in liver of hypothyroid (PTU-fed) chickens when compared to hyperthyroid (T_3-fed) chickens (Fig. 1B; Table 1). Gene cluster analysis, based on the differentially expression ($P<0.05$) and appropriate treatment contrast, was used to identify TH- and GH-responsive genes expressed in the liver (Fig. 3). Hypothyroidism (PTU *vs.* CF contrast) induced by dietary propylthiouracil (PTU) altered ($P<0.05$) expression of 165 liver genes (Fig. 3A). Within this cluster, a group of 144 genes was novel to PTU-induced hypothyroidism. The PTU-responsive genes include transcription factors (*Spot 14*, *C/EBPα*), several metabolic enzymes (*HMG CL, KHK, PGDS, BHMT*), transporters [*adipophilin (or ADFP), FABP*] and acute phase (*GSTα, CYP2C45, HSP90*) proteins.

Table 1 Hepatic genes with higher expression in *hypo*-thyroidism (PTU) or *hyper*-thyroidism (T_3)

Gene Name	**Microarray Fold-change**	**TaqMan® Fold-change**	**Cellular/ Molecular Function**
Genes up-regulated by *hypo*-thyroidism	(PTU/ T_3)		
Cyp2c45	+2.57		lipid metabolism
Glutathione S-transferase (GSTα)	+2.81	+8.4	detoxific-ation
Thyroid hormone repressible gene (THRG)	+2.11	+4.8	lipid metabolism
Acetyl-CoA acetyltransferase 2 (ACAT2)	+2.02		lipid metabolism
Adipophilin (ADFP)	+1.94	+9.7	lipid storage
Prostaglandin-D2 synthase (PGDS)	+1.92	+8.0	signal transduction
Longevity assurance homolog 2 (LASS2)	+1.70		cell growth
Polyamine oxidase (PAO)	+1.70		detoxific-ation

Catalase (CAT)	+1.65		peroxidase
Cysteine protease inhibitor (GHRG1)	+1.60		protein degradation
Sulfotransferase (SULT or GHRG2)	+1.62		xenobiotic metabolism
Angiopoietin-like 3 homolog (ANGPTL3)	+1.50		Lipid metabolism
Genes up-regulated by *hyper*-thyroidism	(T_3/PTU)		
Spot 14 (THRSPα)	+2.04	+2.2	transcription
Ketohexokinase β (KHK-β)	+1.80		carbohydrate metabolism
Hemopexin (HPX)	+1.73		heme transport
CCAAT/enhancer binding protein (C/EBPα)	+1.61	+1.6	transcription factor
Lactate dehydrogenase (LDH-β)	+1.57	+7.7	anaerobic glycolysis
Son of sevenless homolog 2 (SOS-2)	+1.54		apoptosis signaling
Adenylate cyclase (ADCY)	+1.50		signal transduction
Fatty acid synthase (FAS)	+1.50		lipogenesis
Glutamine synthase (GLNS)	+1.49		amino acid metabolism
Carboxypeptidase H (CPH)	+1.47		prohormone processing
Hydroxymethylglutaryl-CoA synthase (HMG CS)	+1.45		ketogenesis
Fused toes homolog (FTH)	+1.43		ubiquitin conjugation

Note: The genes in this table met both significance criteria for differential expression ($P<0.05$ and >1.4-fold). The expression of several genes was confirmed by a more sensitive independent method, quantitative RT-PCR or TaqMan® analysis.

On the other hand, hyperthyroidism (T_3 *vs.* CF contrast) induced by dietary T_3 led to the differential (up- or down-regulated) expression of 103 hepatic genes. This cluster of TH-responsive genes (51 were unique to T_3) includes *THRG, THRSP, LDHβ, HMG CS, ACAA, ApoA4, sulfotransferase* (or *GHRG2*), *IGF-I* and *GHR*. The largest cluster of thyroid-responsive genes was revealed by the extreme thyroid contrast (PTU *vs.* T_3), where 248 genes were differentially expressed (137 genes were unique to the hypo- *versus* hyper-thyroid contrast). A group of 70 common genes was found between the extreme thyroid contrast (PTU *vs.*

T_3) and the hypothyroid (PTU *vs.* CF) contrast. Only 12 genes were found in common between hypothyroid (PTU *vs.* CF) and hyperthyroid (T_3 *vs.* CF) contrasts. A group of 34 common genes was found between hyperthyroid (T_3 *vs.* CF) and the extreme thyroid (PTU *vs.* T_3) conditions.

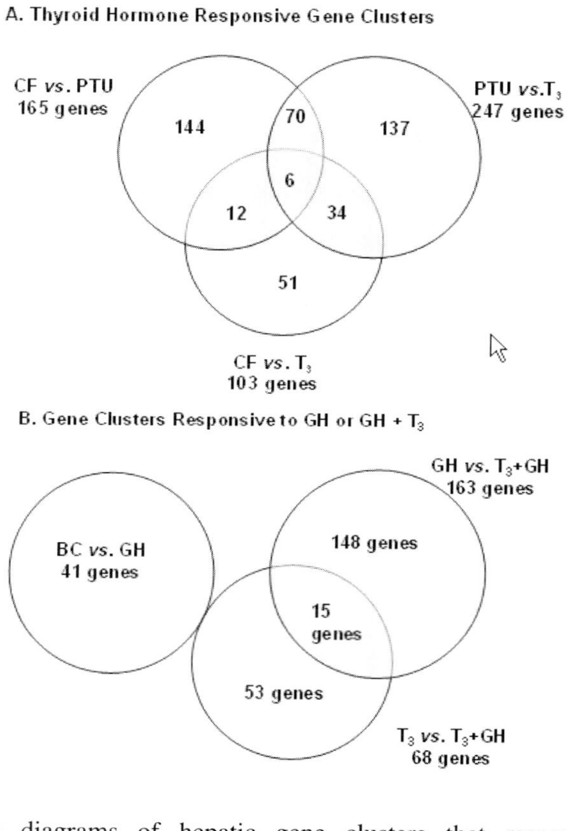

Fig. 3 Venn diagrams of hepatic gene clusters that respond to manipulation of the thyroid status (**A**) or to exogenous GH alone or GH combined with dietary T_3 (**B**). These gene clusters were derived from a Pathways 4 software analysis based on a single significance criterion (Student's *t*-test, $P<0.05$). Common genes shared between treatment contrasts are presented in the overlapping arcs.

Only 41 genes were differentially ($P<0.05$) expressed in the GH *vs.* BC contrast (Fig. 3B). A larger cluster of 166 genes was found in the GH *vs.* GH+T_3 treatment contrast. This cluster contains genes that are up-regulated by GH, where a group of 148 genes was unique to the combined GH+T_3 treatment. Similarly, only 68 differentially-expressed genes were found in the T_3 *vs.* GH+T_3 contrast where a group of 53 genes was unique to the combined treatment. Fifteen differentially expressed genes were common between the GH *vs.* GH+T_3 contrast and the T_3 *vs.* GH+T_3 contrast.

Table 2. Hepatic genes that respond to exogenous GH alone *vs.* GH combined with dietary T_3

Gene Name	Microarray Fold-change	Cellular/Molecular Function
Genes up-regulated in GH *vs.*GH+T_3 contrast	**(GH/GH+T_3)**	
B-homocysteine methyltransferase (BHMT)	+1.8	methyltransferase activity
S-cyclophilin (S-CYLP)	+1.8	protein folding
F-box only protein 34 (FBXO34)	+1.7	unknown
Zinc finger protein 198 (ZNF198)	+1.6	signal transduction via FGF-R
β-Adaptin	+1.6	intracellular trafficking
Vesicle associated protein 1 (VAP1)	+1.6	intracellular trafficking
Acetyl-Coenzyme A acyltransferase 2 (ACAA2)	+1.6	β-oxidation of fatty acids
Transmembrane superfamily (TM4SF)	+1.5	cell development / growth
Sterol carrier protein 2 (SCP2)	+1.4	fatty acid transport
Recombination protein (REC14)	+1.4	signal transduction
TcD37 homolog (Prune)	+1.4	pyrophosphatase activity
Notch homolog 2 (NOTCH2)	+1.4	intercellular signaling
Genes up-regulated in GH+T_3 *vs.* GH contrast	**(GH+T_3/GH)**	
Lactate dehydrogenase-β (LDH-β)	+1.7	gluconeogenesis
Symplekin (SYMPK)	+1.7	mRNA cleavage/polyadenylation
Prothrombin (F2)	+1.7	blood coagulation
EGF-R substrate 8-like protein 2 (EPS8L2)	+1.7	signal transduction
Cytochrome P450 (Cyp2G1)	+1.7	oxidoreductase activity
Apoptosis-regulating basic protein (ARBP)	+1.4	unknown function
Hemopexin (HPX)	+1.4	heme transport
E3 protein (LAPTM5)	+1.4	hematopoiesis
Unknown EST	+1.4	unknown
Spot 14 (THRSPα)	+1.4	signal transduction

Note: The genes in this table meet both criteria for differential expression ($P<0.05$ and >1.4-fold).

A more stringent analysis based on two criteria, a significant *t*-test (P<0.05) and a >1.4-fold cutoff filter, was used to focus on major hepatic genes that respond to endocrine perturbation (Tables 1 and 2). Hypothyroidism induced by dietary propylthiouracil (PTU) up-regulated expression of 26 liver genes, including *Cyp2C45, GSTα, THRG, adipophilin* (or adipocyte differentiation factor protein, *ADFP*), *ACAT2, PGDS, BHMT, catalase* (*CAT*), longevity assurance homolog 2 (*LASS2*), and two GH-regulated genes [cysteine protease inhibitor (*GHRG1*) and sulfotransferase (*GHRG2*)] (Table 1). A cluster of 46 genes was up-regulated [i.e., a higher T_3/PTU ratio] by hyper-thyroidism; these genes include transcription factors (*Spot 14, C/EBPα*), signaling factors *(SOS-2, ADCY*), several metabolic enzymes (*LDHβ, KHK-β, FAS, GLNS, CPH, HMG CS*), and transport proteins (*HPX, L-FABP* and *FTH*).

GH treatment alone (GH *vs.* BC) yielded only five differentially expressed genes (*HSP73, Cyp2H1, V-ATPase, U2AF35 and Gal NAc4ST-2*) that met the more stringent significance criteria (P<0.05 and >1.4-fold difference). Similarly, only 25 genes in the GH *vs.* GH+T_3 contrast met the more stringent criteria (Table 2). The genes with higher expression in the GH alone *vs.* GH+T3 contrast included several enzymes (*BHMT, ACAA2* and *Prune*), signal transduction factors [*Zinc finger protein 198* (*ZNF198*), *REC14* and *NOTCH2*] and signaling/transport proteins (*β-Adaptin, VAP1, TM4SF, SCP2*). In addition, several genes were expressed at higher levels in the GH+T_3 *vs.* GH contrast [*LDH-β, SYMPK, prothrombin* (*F2*), *Cyp2G1, ARBP, HPX,* and *THRSPα*]. The expression of some of these genes was also increased in the hyperthyroid (T_3-fed) state (Table 1).

The transcriptional response of 14 genes (Table 1) was verified by a more sensitive independent method, quantitative RT-PCR (TaqMan®) analysis (Fig. 4). The hepatic expression of *GSTα* (Fig. 4A) and *PGDS* (Fig. 4B) was 8-fold higher (P<0.05) in PTU-fed hypothyroid chickens when compared to other treatments. The hepatic expression of the novel thyroid hormone repressible gene (*THRG*) was depressed 5-fold by T_3 alone or when T_3 was combined with GH (Fig. 4C). Also, the abundance of *THRG* is depressed by synthetic fatty acids and increased by either dexamethazone (10) or prolonged fasting (18). On the other hand, dietary T_3 alone or T_3+GH greatly increased (7- to 8-fold) the hepatic expression of *LDH-β,* a key enzyme in the gluconeogenic/glycolytic pathway which reversibly converts lactate to pyruvate. Thus, these four genes and most of those listed in Table 1 appear to respond to manipulation of thyroid status. The T_3-regulated genes encompass a wide range of cellular functions: transcription (*THRSPα, C/EBP, PPARγ*), metabolism (*ACAT2, LDH-β, HMG CS, ADFP*), signal transduction

(*GHR, PDGS*), protein turnover (*GSTβ, BHMT*), intracellular trafficking, etc. Several of the T_3-responsive genes identified in this experiment with chickens (Table 1) were also induced in liver of hypothyroid rats given an acute injection of T_3 (43).

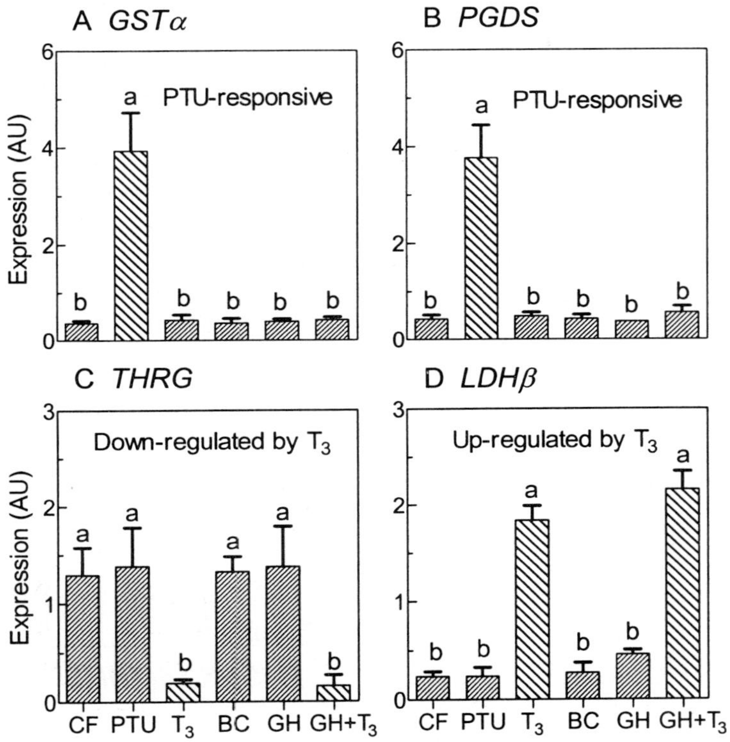

Fig. 4 Thyroid hormone-responsive gene expression verified by TaqMan analysis. The same total RNA samples used for microarray analysis were also used for quantitative RT-PCR (TaqMan) analysis. Each value [in arbitrary expression units (AU)] represents the mean±SEM of four birds/treatment group, where means with different superscripts differ significantly ($P<0.05$).

Elevation of *GSTα* and *CYP2C45* (2) could be a consequence of detoxification of PTU rather than a response to hypothyroidism. However, GST and several cytochromes involved in detoxification and xenobiotic metabolism were also induced by chronic GH treatment in hypox rats (24). Glutathione S-transferase (*GSTα*) catalyzes the conjugation of glutathione to electrophilic compounds, which protects cellular macromolecules from xenobiotics (31). In liver, GSTα exhibits glutathione peroxidase activity which protects cells from reactive oxygen species. Glutathione-independent PGDS is a thyroid-hormone responsive gene (25) that uses GSH as a cofactor to catalyze the isomerization of PGH_2 (an intermediate) into PGD_2 Prostaglandin D_2 and its derivative (PGJ2) are potent activating ligands of peroxisome proliferator-activated receptors (*PPAR*α, -β and -γ (44). The PPARs are transcription factors that belong to the nuclear receptor superfamily which controls expression of numerous metabolic enzymes (4). In our study, the expression of PPARγ was elevated by 4-fold in PTU-fed chickens (data not shown). Activated PPARs enhance the transcription of adipophilin (*ADFP*) (see Fig.5C) (6, 27), while ADFP is a lipid droplet component and a reliable indicator of lipid load in numerous cells (29, 33). The present microarray study shows that the abundance of *PGDS* and *ADFP* (Fig. 4B) is elevated in the hypothyroid (PTU-fed) condition which promotes fat deposition in the chicken (Fig. 2A).

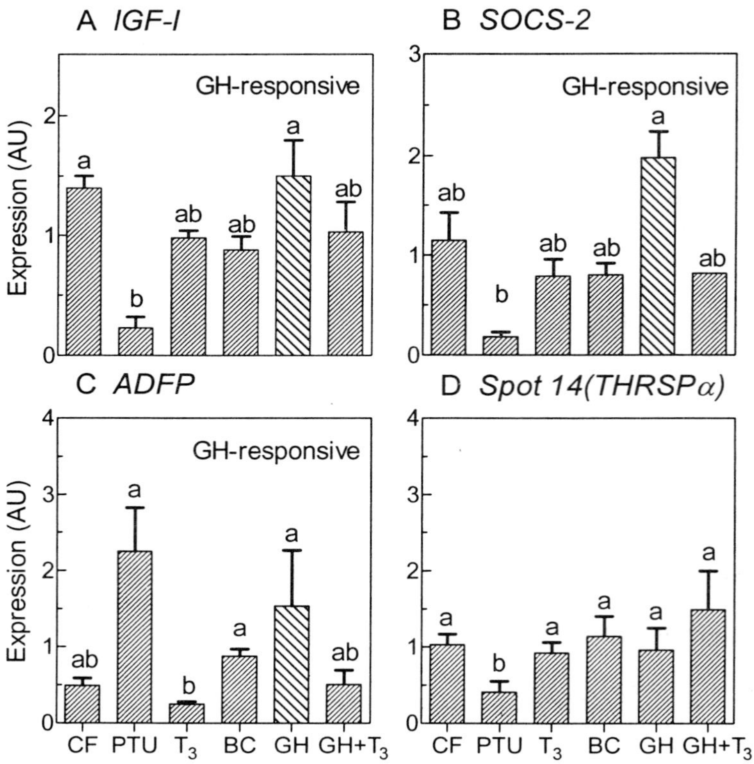

Fig. 5 Growth hormone-responsive gene expression verified by TaqMan analysis These four genes are also responsive to PTU-induced hypothyroidism. The same total RNA samples used for microarray analysis were also used for quantitative RT-PCR (TaqMan) analysis. Each value (arbitrary expression units, AU) represents the mean$\pm$SEM of four birds/treatment group, where means with different superscripts differ significantly ($P<0.05$).

The expression of some GH-responsive genes was confirmed by TaqMan analysis (Fig. 5). Hepatic levels of both *GHR* (not shown) and *IGF-I* (Fig. 5A) mRNA were depressed in hypothyroid chickens. The effect of hypothyroidism on hepatic *IGF-I* expression is also reflected by depressed plasma IGF-I levels (14). In the present study, daily GH injection enhanced expression of both *IGF-I* (Fig. 5A) and suppressor of cytokine signaling 2 (SOCS2) (Fig. 5B). Hepatic abundance of *ADFP* was increased by GH injection or dietary PTU, which also elevates plasma GH levels (Fig. 5C). The expression of *FABP* (not shown) and *THRSP*α (*Spot 14*) (Fig. 5 D) are also depressed in the (PTU-induced) hypothyroid state. The expression of *THRSP*α was highest in birds given the GH+T_3 treatment combination. The abundance of *GHR* and *IGF-I* mRNAs in liver is depressed by PTU, whereas PTU-induced hypothyroidism increases hepatic expression of catalase (*CAT*) and longevity assurance homolog 2 (*LASS2*) (Table 2). These genes are involved in GH-GHR-IGF-I signal transduction pathway; they also belong to the "longevity" pathway (www.biocarta.com) which focuses on the IGF-I receptor as a mediator of increased lifespan in calorie-restricted animals (30).

Mapping of functional metabolic pathways in chickens

We have used existing models of metabolic and regulatory pathways in other organisms to build working models of metabolic and regulatory pathways that are specific to the chicken (18, 19). The transcriptional responses to thyroid perturbation were used to build a working model of thyroid control over several metabolic processes (Fig. 6). Many of the same genes presently identified as TH- and/or GH-responsive genes also appear to respond to other metabolic perturbations (18). Ligand-activated thyroid hormone (T_3) receptors (*TR*α and *TR*β) belong to the nuclear receptor superfamily of ligand-activated transcription factors (*TR, RXR, PPAR, COUP-TF, THRSP)* that control several metabolic pathways. Orphan receptors are nuclear receptors for which the ligand is not yet known. Steroids, vitamins, thyroid hormones, retinoids, and some metabolites (glucose, sterols) serve as activating ligands for formation of homo- (i.e., *THRSP-THRSP*) or hetero-dimers (i.e., T_3-TR-RXR) of transcription factors.

Spot 14 (or *THRSP*) is an important transcription factor that controls expression of major metabolic enzymes in response to a number of perturbations (developmental, nutritional and hormonal). The transcription of *THRSP* is activated by ligand-activated receptors (T_3-*TR*, PGJ_2-PPARγ, COUPTF) which form homo- and hetero-dimers and serve as activators of transcription (Fig. 6). The *THRSP* promoter region contains three thyroid response elements (TREs) that work

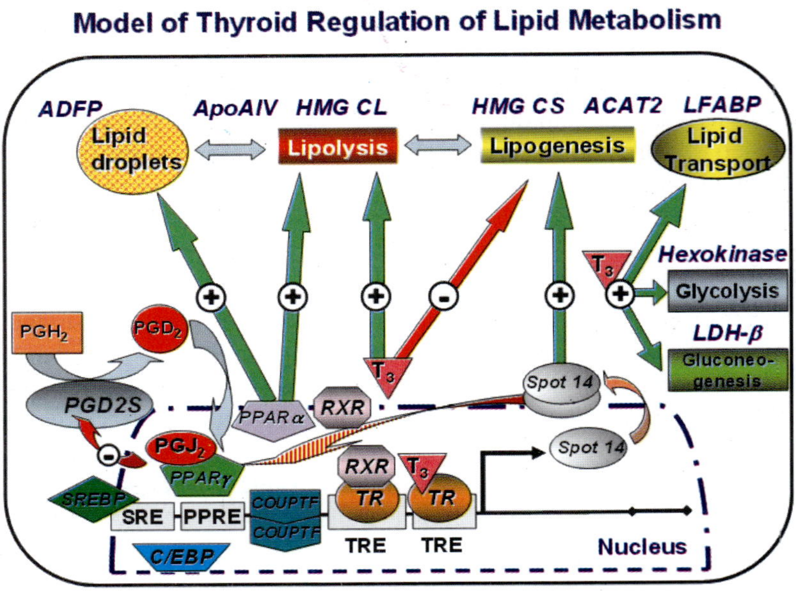

Fig. 6 Working model of thyroid regulation of energy metabolism in chicken liver. This diagram shows some regulatory factors that interact with the *Spot 14* (*THRSP*) promoter region and control its transcription. Thyroid hormone (T_3) interacts with one of its nuclear receptors (*TRα* or *TRβ*) to exert its influence on transcription of genes that maintain energy homeostasis. An example of a thyroid-responsive gene is presented above each metabolic process. The TRs belong to the ligand-inducible transcriptional activators in the nuclear receptor superfamily (*THRSPs, PPARs, COUPTF, SREBP, C/EBP*) whose ligands [steroids, thyroid hormone (T_3), metabolites, vitamins, prostaglandins (PGJ_2)] activate gene transcription. The ligand (T_3)-activated transcription factor (TR) binds to its response element (TRE) in the promoter region of target genes (*THRSP*). Transcription of duplicated chicken *Spot 14* genes (*THRSPα* and *–β*) is controlled by other members of the nuclear receptor superfamily (*TRs, COUPTF, PPARs, RXR*). Ligand-induced transcription factors form homo- and hetero-dimers which bind to their response elements (SRE, PPRE) on promoters and activate transcription of genes that regulate growth and metabolism.

synergistically and interact with far upstream region (FUR) elements to maximize T_3 responses in cultured hepatocytes (38). Apparently, the human *THRSP* promoter responds more robustly to T_3 than glucose, while the rat *THRSP* promoter region is more responsive to glucose than T_3 (9). Homodimers of *THRSP* interact with and activate chicken ovalbumin upstream promoter-transcription factor 1 (COUP-TF1) in promoting transcription of L-type pyruvate kinase (*L-PK*) through an interaction with specificity protein 1 (Sp1) (20). Furthermore, the mammalian *THRSP* promoter contains multiple response elements that respond to thyroid hormone (TRE) (39), carbohydrates (ChoRE) (35) and sterols (SRE, sterol response element), particularly SREBP1c (34). Thus, multiple response elements exert *THRSP*'s control over the expression of key lipogenic, glycolytic and gluconeogenic enzymes in a tissue-specific and fuel-dependent manner (5). We have discovered and characterized duplicated Spot 14 genes (*THRSP*α and β paralogs) in the chicken (42). A unique insertion/deletion (indel) polymorphism was found in each paralog near the DNA binding motif. These *THRSP* polymorphisms could certainly increase the complexity of transcriptional responses to homeo- and hetero-dimers of this transcription factor. Furthermore, the polymorphic *THRSP* genes are located in quantitative trait loci for fatness on the long arm of GGA1 in a fast-growing x slow-growing F_2 resource population. Therefore, *THRSP* appears to play a key role in regulating the expression of several lipogenic enzymes in the chicken [see Fig. 4 in (19)].

Summary

We have developed several tissue-specific and systems-wide chicken cDNA microarrays for analysis of multiple transcriptional snapshots under various perturbations. Our current interest lies in unraveling the gene networks controlling growth and development. Altered gene expression patterns reflect changes in metabolism, transport and storage of nutrients in response to manipulation of the thyroid and/or GH axes. It is also clear that normal growth and development is regulated by complex gene interactions between the thyrotropic and somatotropic axes which have not been clearly defined. These initial studies demonstrate the power of DNA microarrays for transcriptional profiling and discovery of key regulatory genes. With this new information, we have begun to build functional maps of the metabolic and regulatory pathways that regulate important growth traits in the chicken.

Acknowledgement

This work was supported by a grant from the USDA-IFAFS Animal Genome Program (Award Number 00-52100-9614).

References

1 Alter O, Brown PO & Botstein D (2000) Singular value decomposition for genomic-wide expression data processing and modeling. *Proceedings National Academy Sciences of the U.S.A.* **97** 10101-10106

2 Baader M, Gnerre C, Stegeman JJ & Meye UA (2002)Transcriptional activation of cytochrome P450 CYP2C45 by drugs is mediated by the

chicken xenobiotic receptor (CXR) interacting with a phenobarbital response enhancer unit. *Journal of Biological Chemistry* **277** 15647-53

3 Baldi P & Long AD (2001) A Bayesian framework for the analysis of microarray expression data: regularized t-test and statistical inferences of gene changes. *Bioinfomatics* **17** 509-516

4 Bocher V, Pineda-Torra I, Fruchart J-C & Staels B (2002) PPARs: Transcription factors controlling lipid and lipoprotein metabolism. *Annals of the New York Academy of Sciences* **967** 7-18

5 Brown SB, Maloney M & Kinlaw WB (1997) "Spot 14" protein functions at the pretranslational level in the regulation of hepatic metabolism by thyroid hormone and glucose. *Journal of Biological Chemistry* **272** 2163-2166

6 Buechler C, Ritter M, Duong CQ, Orso E, Kapinsky M & Schmitz G (2001) Adipophilin is a sensitive marker for lipid loading in human blood monocytes. *Biochimica et Biophysica Acta* **1532** 97-104

7 Burke WH, Moore JA, Ogez JR & Builder SE (1987) The properties of recombinant chicken growth hormone and its effects on growth, body composition, feed efficiency, and other factors in broiler chickens. *Endocrinology* **120** 651-658

8 Buyse J & Decuypere E (1999) The role of the somatotrophic axis in the metabolism of the chicken. *Domestic Animal Endocrinology* **17** 245-255

9 Campbell MC, Anderson GW & Mariash CN (2003) Human Spot 14 glucose and thyroid hormone response: Characterization and thyroid hormone response element identification. *Endocrinology* **144** 5242-5248

10 Carre W, Bourneuf E, Douaire M & Diot C (2002) Differential expression and genetic variation of hepatic messenger RNAs from genetically lean and fat chickens. *Gene* **299**: 235-243

11 Chung CS & Etherton TD (1986) Characterization of porcine growth hormone (pGH) binding to porcine liver microsomes: chronic administration of pGH induces pGH binding. *Endocrinology* **119** 780-786

12 Cogburn LA (1991) Endocrine manipulation of body composition in broiler chickens. *Critical Reviews of Poultry Biology* **3** 283-305

13 Cogburn LA, Burnside J & Scanes SC (2000) Physiology of growth and development. In *Sturkie's Avian Physiology*, Fifth Edn, pp.635-656. Ed GC Whittow. New York: Academic Press

14 Cogburn LA, Liou SS, Alfonso CP, McGuinness MC & McMurtry JP (1989) Dietary thyrotropin-releasing hormone stimulates growth rate and increases the insulin:glucagon molar ratio of broiler chickens. *Proceedings of the Society for Experimental Biology and Medicine* **192** 127-134

15 Cogburn LA, Liou SS, Rand AL & McMurtry JP (1989) Growth, metabolic and endocrine responses of broiler cockerels given a daily subcutaneous injection of natural or biosynthetic chicken growth hormone. *Journal of Nutrition* **119** 1213-1222

16 Cogburn LA, Mao JNC & Burnside J (1997) The growth hormone receptor in growth and development. In *Perspectives in Avian Endocrinology*, pp.101-118. Eds S Harvey & R Etches. Bristol, U.K.: Journal of Endocrinology Press

17 Cogburn LA, Mao NC, Agarwal S & Burnside J (1995) Interaction between somatotropic and thyrotropic axes in regulation of growth and development of broiler chickens. *Archives fur Geflugelkunde Sonderheft* **1** 18-21
18 Cogburn LA, Wang X, Carre W, Rejto L, Aggrey SE, Duclos MJ, Simon J & Porter TE (2004) Functional genomics in chickens: developement of integrated-systems microarrays for transcriptional profiling and discovery of regulatory pathways. *Comparative and Functional Genomics* **5** 253-261
19 Cogburn LA, Wang X, Carre W, Rejto L, Porter TE, Aggrey SE & Simon J (2003) Systems-wide chicken DNA microarrays, gene expression profiling and discovery of functional genes. *Poultry Science* **82** 939-951
20 Compe E, Sousa GE, Francois K, Roche R, Rahmani R, Torresani J, Raymondjean M & Planells R (2001) Spot 14 protein interacts and co-operates with chicken ovalbumin upstream promoter-transcription factor 1 in the transcription of the L-type pyruvate kinase gene through a specificity protein 1 (Sp1) binding site. *Biochemical Journal* **358** 175-183
21 Darras VM, Berghman LR, Vanderpooten A & Kuhn ER (1992) Growth hormone acutely decreases type III iodothyronine deiodinase in chicken liver. *FEBS Letters* **310** 5-8
22 Darras VM, Rudas P, Visser TJ, Hall TR, Huybrechts LM, Vanderpooten A, Berghman LR, Decuypere E & Kuhn ER (1993) Endogenous growth hormone controls high plasma levels of 3,3',5-triiodothyronine (T3) in growing chickens by decreasing the T3-degrading type III deiodinase activity. *Domestic Animal Endocrinology* **10** 55-65
23 Denver RJ & Harvey S (1991) Thyroidal inhibition of chicken pituitary growth hormone: alterations in secretion and accumulation of newly synthesized hormone. *Journal of Endocrinology* **131** 39-48
24 Flores-Morales A, Stahlberg N, Tollet-Egnell P, Lundeberg J, Malek RL, Quackenbush J, Lee NH & Norstedt G (2001) Microarray analysis of the in vivo effects of hypophysectomy and growth hormone treatment on gene expression in the rat. *Endocrinology* **142** 3163-3176
25 Garcia-Fernandez LF, Rausell E, Urade Y, Hayaishi O, Bernal J & Munoz A (1997) Hypothyroidism alters the expression of prostaglandin D2 synthase/beta trace in specific areas of the developing rat brain. *European Journal Neuroscience* **9**:1566-1573
26 Gluckman PD & Breier BH (1988) The regulation of the growth hormone receptor. In *Biotechnology in Growth Regulation*, pp.27-33. Eds RB Heap, CG Prosser & GE Lamming. London: Butterworths
27 Gupta RA, Polk DB, Krishna U, Israel DA, Yan F, DuBois RN & Peek RM, Jr. (2001) Activation of peroxisome proliferator-activated receptor gamma suppresses nuclear factor kappa B-mediated apoptosis induced by helicobacter pylori in gastric epithelial cells. *Journal of Biological Chemistry* **276** 31059-31066
28 Hazelwood RL (1984) Pancreatic hormones, insulin/glucagon molar ratios, and somatostatin as determinates of avian carbohydrate metabolism. *Journal of Experimental Zoology* **232** 647-652
29 Heid HW, Moll R, Schwetlick I, Rackwitz HR & Keenan TW (1998) Adipophilin is a specific marker of lipid accumulation in diverse cell types and diseases. *Cell and Tissue Research* **294** 309-321

30 Holzenberger M, Dupont J, Ducos B, Leneuve P, Geloen A, Even PC, Cervera P & le Bouc Y (2003) IGF-1 receptor regulates lifespan and resistance to oxidative stress in mice. *Nature* **421** 182-187
31 Hsieh CH, Liu LF, Tsai SP & Tam MF (1999) Characterization and cloning of avian-hepatic glutathione S-transferases. *Biochemical Journal* **343** 87-93
32 Huang X & Madan A (1999) CAP3: a DNA sequence assembly program. *Genome Research* **9** 868-877
33 Jiang HP & Serrero G (1992) Isolation and characterization of a full-length cDNA coding for an adipose differentiation-related protein. *Proceeding National Academy of Sciences of the United States of America* **89** 7856-7860
34 Jump DB, Thele AP & Mater MK (2001) Functional interaction between sterol regulatory element-binding protein-1c, nuclear factor Y, and 3,5,3'-triiodothyronine nuclear receptors. *Journal of Biological Chemestry* **276**:34419-34427
35 Koo S-H & Towle HC (2000) Glucose regulation of mouse S14 gene expression in hepatocytes. *Journal of Biological Chemistry* **275** 5200-5207
36 Leung FC, Taylor JE, Steelman SL, Bennett CD, Rodkey JA, Long RA, Serio R, Weppelman RM & Olson G (1984) Purification and properties of chicken growth hormone and the development of a homologous radioimmunoassay. *General and Comparative Endocrinology* **56** 389-400
37 Leung FC, Taylor JE, Wien S & Van Iderstine A (1986) Purified chicken growth hormone (GH) and a human pancreatic GH-releasing hormone increase weight gain in chickens. *Endocrinology* **118** 1961-1965
38 Liu H-C & Towle HC (1994) Functional synergism between multiple thyroid hormone response elements regulates hepatic expression of the rat S14 gene. *Molecular Endocrinology* **8** 1021-1037
39 Moellers RF & Cogburn LA (1995) Chronic intravenous infusion of chicken growth hormone increases body fat content of young broiler chickens. *Comparative Biochemistry and Physiology* **110A** 47-56
40 Sauerwein H, Breier BH, Bass JJ & Gluckman PD (1991) Chronic treatment with bovine growth hormone upregulates high-affinity hepatic somatotropic receptors in sheep. *Acta Endocrinologica* **124** 307-313
41 van den Oord EJCG & Sullivan PF (2003) False discoveries and models for gene discovery. *Trends in Genetics* **19** 537-542
42 Wang X, Carre W, Zhou H, Lamont SJ & Cogburn LA (2004) Duplicated Spot 14 genes in the chicken: characterization and identification of polymorphisms associated with abdominal fat traits. *Gene* **332** 79-88
43 Weitzel JM, Hamann S, Jauk M, Lacey M, Filbry A, Radtke C, Iwen KAH, Kutz S, Harneit A, Lizardi PM & Seitz HJ (2003) Hepatic gene expression patterns in thyroid hormone-treated hypothyroid rats. *Journal of Molecular Endocrinology* **31** 291-303
45 Yu K, Bayona W, Kallen CB, Harding HP, Ravera CP, McMahon G, Brown M & Lazar MA (1995) Differential activation of peroxisome proliferator-activated receptors by eicosanoids. *Journal of Biological Chemistry* **270**: 23975-23983

Functional Avian Endocrinology
A Dawson & P J Sharp (Eds)

Gene expression profiling in the developing neuroendocrine system of the chick

T E Porter and L E Ellestad

Molecular and Cell Biology Program, and Department of Animal and Avian Sciences, University of Maryland, College Park, MD 20742, USA

Introduction

Levels of messenger RNA (mRNA) in samples are commonly measured one-gene-at-a-time using gene-specific probes and Northern blotting, ribonuclease protection assays, or real time PCR. Although these techniques allow quantification of gene expression, they do not readily afford a comprehensive analysis of the many genes that may be affected among experimental groups. The advent of DNA microarray technology has allowed simultaneous analysis of differences in expression levels of thousands of genes among samples. Research in avian species has been traditionally hampered by a paucity of publicly available probes, often requiring researchers to clone a gene of interest based on sequence homology with known mammalian gene sequences. Experiments in the neuroendocrine system are further hampered by the small size of the neuroendocrine tissues and hence the amount of RNA available for analysis. Recently, a tremendous amount of genetic information has become available for one avian species – the chicken. Public databases such as GenBank now contain partial sequence for approximately 500,000 chicken cDNA clones, and most of the chicken genome has been sequenced. Moreover, several groups have assembled large numbers of chicken cDNA clones and printed these as cDNA microarrays (1, 3-6, 10, 17, 25). These advances have allowed for the first time, large-scale functional genomics analyses of an avian species. In addition, several approaches have been used to increase the amount of RNA available for each sample through amplification of a small amount of starting RNA, thereby allowing microarray analysis of minute samples. We describe here our recent advancements in functional genomics in the chicken neuroendocrine system. We have produced cDNA microarrays containing more than five thousand genes expressed in the neuroendocrine system, validated an RNA amplification protocol for use with samples from individual embryonic pituitary glands, and used these tools to study changes in gene expression profiles within the hypothalamus and anterior pituitary gland of chickens during late embryonic and early post-hatch development. Our approach and findings will be summarized in this review.

Production of neuroendocrine system cDNA microarrays

High-density cDNA microarrays for the chicken neuroendocrine system were produced in collaboration with Dr. Larry Cogburn at the University of Delaware. The cDNA libraries used to make these arrays have been described elsewhere (5, 6). Briefly, RNA was isolated from the hypothalamus, anterior pituitary gland, and pineal gland of chickens from embryonic day 12 through post-hatch day 63 and pooled. As a result, the RNA pool represented three neuroendocrine tissues and a wide range of developmental stages. This pooled RNA sample was then used to produce a cDNA library, and approximately 5,500 clones from this library were selected at random and sequenced. The library was then normalized to eliminate redundant clones, and an additional 3,300 clones from the normalized library were selected at random and sequenced. All sequences were then compared against one another to identify overlapping sequences (contigs), which likely represent the same cDNA. Clones without sequence identity with other cDNAs (singlets) were also identified. From the 8,869 clones sequenced, 5,128 unique cDNAs (Unigenes) were identified within the neuroendocrine system library. Each Unigene sequence was compared with GenBank by the BLASTX and BLASTN procedures, and the clones annotated with the name of the highest BLAST score. The Unigene clones were then amplified by PCR, and the PCR products were printed onto microscope slides as cDNA microarrays. In addition, multiple copies of selected housekeeping genes, salmon DNA, and water were spotted as quality controls. An annotated list of clones and their location on the chicken neuroendocrine system 5K microarrays is available in the Gene Expression Omnibus (GEO; GPL1744) at:
http://www.ncbi.nlm.nih.gov/projects/geo/query/acc.cgi?acc=GPL1744&targ=&view=full.

Amplification of RNA from small samples

A primary research focus in our laboratory is development of the anterior pituitary gland, with particular emphasis on the last three cell types to differentiate: the thyrotrophs, somatotrophs, and lactotrophs. However, chicken embryonic pituitary glands yield insufficient total cellular RNA for microarray analysis. Approximately 20-50μg of total RNA (8) or 1-5μg of poly(A) RNA (7) from each sample are required for microarray analysis. To overcome this limitation and allow gene expression profiling with small amounts of tissue or even individual cells, several protocols for amplifying mRNA that are variations on the Eberwine procedure (18) have been developed (2, 12, 15, 23, 26). The Eberwine protocol uses an oligo(dT) primer containing a T7 RNA polymerase promoter sequence to reverse transcribe mRNA. Synthesis of the complementary strand is then followed by *in vitro* transcription with T7 RNA polymerase, resulting in large amounts of amplified RNA (aRNA).

The protocol we have adapted in our laboratory is described here to assist other scientists working with small tissue samples, such as comparative endocrinologists, to more easily integrate RNA amplification into their own research. Starting RNA quality is important for microarray analysis, and we routinely use RNeasy® Mini and Midi Kits (Qiagen; Valencia, CA) for our experiments. To initiate first strand synthesis for amplification of RNA, 0.5μg of total RNA is mixed with 1μl of 50μM T7-oligo(dT) primer: (5'-GCCAGTGAATTGTAATACGACTCACTATAGGGAGGCGGT_{24}-3'), (Affymetrix; Santa Clara, CA) in a final volume of 11μl. The mixture is incubated at 70°C for 6min, followed by incubation at 4°C for 2min. Subsequently, 4μl 5X first-strand buffer (Invitrogen; Carlsbad, CA), 2μl 0.1M DTT (Invitrogen), 1μl 10mM dNTPs (Fermentas; Hanover, MD), 1μl RNAse inhibitor (40U/μl, Invitrogen) and 1μl SuperScript II reverse transcriptase (200U/ul, Invitrogen) are added, and the samples are incubated at 42°C for 1h. Next, for second strand synthesis, 30μl 5X second-strand buffer (Invitrogen), 3μl 10mM dNTPs (Fermentas), 4μl DNA polymerase I (10U/μl, Invitrogen), 1μl RNAse H (2U/μl, Invitrogen), 1μl *E. coli* DNA ligase (10U/μl, Invitrogen) and 92μl RNAse-free water are added, and the mixture is incubated at 16°C for 2h. T4 DNA Polymerase is then added (2μl, 5U/μl; Invitrogen), and the samples are incubated at 16°C for 10 min. The double-stranded cDNA is phenol-chloroform extracted using a Phase Lock Gel™ Light tube (Eppendorf; Hamburg, Germany) and washed 2 times with 500μl of RNAse-free water in a Microcon-30 column (Millipore; Billerica, MA). The cDNA is subsequently lyophilized to dryness, and then used as template for an *in vitro* transcription reaction using the T7 MEGAscript kit (Ambion; Austin, TX) according to the manufacturer's protocol. The samples are incubated for 5h at 37°C, and the aRNA is phenol-chloroform extracted using a Phase Lock Gel™ Heavy tube (Eppendorf) and passed through a Spin Column-30 (Sigma; Saint Louis, MO). The resulting aRNA is quantified by optical density at 260nm and with the RiboGreen RNA Quantitation Kit (Invitrogen) and is visualized by ethidium bromide staining in a formaldehyde agarose gel. The resulting aRNA appears in the gel evenly distributed in size between 0.15 and 2.5 kilobases. From 500ng of starting total RNA, the procedure typically yields 10-15μg of aRNA, which is equivalent to the same amount of mRNA.

In establishing the RNA amplification procedure in our laboratory, we verified that microarray results obtained from aRNA agree with those obtained from total RNA. Pituitary RNA was amplified by the above procedure in two independent reactions. Two samples from a pool of total RNA and two aRNA samples amplified from the same pool were each analyzed with separate microarrays, and the mean log_2-transformed raw pixel intensities from each spot were highly correlated (r^2=0.96). The high degree of correlation between microarray results using either total RNA or using aRNA validated the RNA amplification procedure for use with our microarrays.

Microarray hybridization and data analysis

This section will briefly summarize the general approach used for the studies described in more detail later in this chapter. In performing the microarray analyses, each target RNA sample was first labeled with fluorescently tagged nucleotides for detection with a laser scanner. Cy3- or Cy5-labeled cDNA was made from 1μg aRNA (e.g. pituitary samples) or 25μg total RNA (e.g. hypothalamus samples) in a two-step process using the Amino Allyl cDNA Labeling Kit (Ambion) according to the manufacturer's protocol. Total RNA was primed with an oligo(dT) primer, while aRNA was primed using random primers due to the antisense orientation of the aRNA. After hybridization and washing, the slides were scanned with an Affymetrix 418 confocal laser scanner to generate two TIFF images for each slide, one for Cy3 and one for Cy5.

As the absorption/emission spectra of Cy3 and Cy5 are sufficiently different, two samples of cDNAs labeled with Cy3 and Cy5 can be hybridized simultaneously to the same microarray. In our experiments, we used a reference design (22), so that for each study a pool of all of the samples was generated and labeled with Cy5 as an internal reference sample. Each experimental sample was then labeled with Cy3 and hybridized to an array with an aliquot of the reference sample labeled with Cy5.

Many software packages exist for analyzing microarray results. The Institute for Genomic Research (TIGR) offers TM4, a suite of freely available microarray data analysis applications, through their website (www.tigr.org/software/tm4) (21). We used two of the programs in this suite, Spotfinder, an image processing program, and MIDAS, a program for data normalization and analysis. For the developmental studies described below, the two TIFF images from the microarray scanner were loaded into Spotfinder version 2.2.4, and then the numeric values for each spot in the microarray were exported as a tab-delimited MEV file for data normalization using MIDAS version 2.18. Data from the Cy3 channel (representing each experimental sample) were Lowess normalized by block, followed by standard deviation regularization first by block then by slide, with Cy5 (the reference sample) as the reference. Finally, the normalized output from each slide was exported as a tab-delimited MEV (text) file. This normalization process reduces systematic variation introduced in microarray printing and processing.

Following normalization of the microarray results, the data were subjected to statistical analysis. Normalized data were analyzed by one-way analysis of variance (Statistical Analysis System, version 8.02; SAS Institute, Cary, NC) to identify spots that were differentially expressed on at least one of the developmental ages. Data were analyzed as the $\log_2$-ratio, or $\log_2$(normalized Cy3/raw Cy5), and differences were considered significant at $P<0.05$. As a further filter for false positives, spots were considered differentially expressed only if the highest mean among the experimental groups was greater than 175% of the lowest mean.

Significant spots were then subjected to self-organizing maps (SOMS) analysis using GeneCluster 2-version 2.0 (20), a software program developed by the Whitehead Institute/MIT Center for Genome Research and available free from their website: (www.broad.mit.edu/cancer/software/genecluster2/gc2). SOMS organizes the data so that general patterns of gene expression can be visualized, and it is particularly useful for evaluating gene expression profiles in time-course experiments.

Gene expression profiles in the hypothalamus around hatching

Hatching represents a major physiological change for the developing chick. Upon hatching, the primary source of nourishment shifts from egg yolk to ingested feed; the chick acquires the capacity to regulate body temperature and is exposed to increased light intensity and a daily photoperiod that affects its behavior. Each of these processes – regulation of feed intake, maintenance of body temperature, photoreception and entrainment to light – involves neuronal pathways located in the hypothalamus (13, 14, 16). In our first study with our neuroendocrine system 5K microarrays, we characterized changes in gene expression in the hypothalamus during late embryonic to early post-hatch development. Total RNA was extracted from the hypothalami of chicks on embryonic day (e) 17, e19, post-hatch day (d) 1 and d3. Each sample (25μg) of total RNA was compared against an internal reference sample pooled from all samples, using the neuroendocrine system 5K microarrays as described above. The microarray results were normalized and the data analyzed by analysis of variance. Among the 5,128 cDNAs represented on the microarrays, the expression of 344 spots changed significantly ($P<0.05$; n=3 samples/age) during this period of development. Of these, 105 changed substantially (greatest mean >

175% of the lowest mean). We included this filter to focus on significant changes that could be readily quantified by standard techniques and to decrease the incidence of false positives. Profiles of expression patterns among these 105 genes were identified by SOMS analysis. Expression levels for the 105 genes were placed into 20 clusters. These ranged in patterns of expression from a general increase to a general decrease in expression levels during development. Genes exhibiting an increase in expression levels after hatching included genes involved in myelination, such as myelin basic protein and myelin proteolipid protein. Other genes exhibited transient increases or decreases in levels of expression on e19 or d1. Some of our findings are presented in Fig. 1. Shown are results for myelin basic protein (MBP), dopamine and cAMP-regulated phosphoprotein (DARPP), and bone morphogenic protein 7 (BMP7). MBP and DARPP were in clusters 5 and 16, respectively, each containing 7 genes. Each of these clusters of genes exhibited the greatest increase in expression after hatching. BMP7 was part of cluster 7 that contained 4 genes which showed decreased levels of expression on d1. Levels of MBP, DARPP, and BMP7 mRNA were also measured in the same samples by quantitative real-time RT-PCR (Real-Time PCR). Results from Real-Time PCR agreed with those from the microarray analysis and with the expression profiles for their respective clusters (Fig. 1). Based on these findings, we believe that our microarray and SOMS analyses accurately describe gene expression profiles. Our findings from this study indicate that substantial structural and functional changes occur in the hypothalamus around hatching.

Expression profiles in the anterior pituitary associated with hormone ontogeny

The results from the hypothalamic study described above, demonstrate the utility of our neuroendocrine system microarrays. To determine whether differences in gene expression can be detected using amplified RNA (aRNA) and our neuroendocrine system microarrays, a preliminary study exploring differences in gene expression between e17 and d3 was conducted. Pituitary RNA from e17 and d3 chickens (500ng) was amplified, and 1μg of aRNA analyzed with the neuroendocrine system 5K microarrays. Expression of 116 cDNAs changed significantly ($P<0.05$, $n=3$, highest mean >175% of lowest mean), and these represented 113 genes due to duplication of cDNAs or genes on the array. Of these, expression levels of 90 genes increased and levels for 23 genes decreased between e17 and d3. Results for three genes were confirmed by Real-Time PCR. The microarray results and Real-Time PCR data from both aRNA and non-amplified total RNA from three genes are shown in Fig. 2. Growth hormone (GH) expression increased, and alpha-1-collagen mRNA levels decreased during development, while

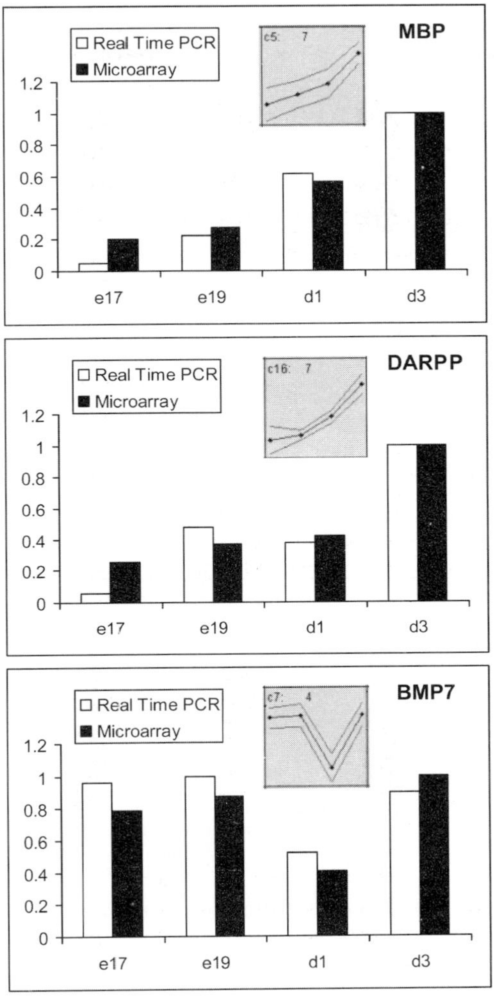

Fig. 1 Comparison of microarray analysis for hypothalamic expression levels for three genes with results from Real-Time PCR from embryonic day 17 (e17), and 19 (e19) through post-hatch day1 (d1) and 3 (d3). Results are presented for Myelin Basic Protein (MBP), Dopamine and cAMP Regulated Phosphoprotein (DARPP), and Bone Morphogenic Protein 7 (BMP7). The expression profile cluster from self organized maps (SOMS) analysis are presented in the insert for each gene.

glyceraldehyde-3-phosphate dehydrogenase (GAPDH) did not change between e17 and d3. Results from the microarrays and from Real-Time PCR performed on aRNA and non-amplified RNA were in congruence. These findings indicated that all three analyses yielded similar findings and further validated the RNA amplification procedure and the neuroendocrine system microarrays.

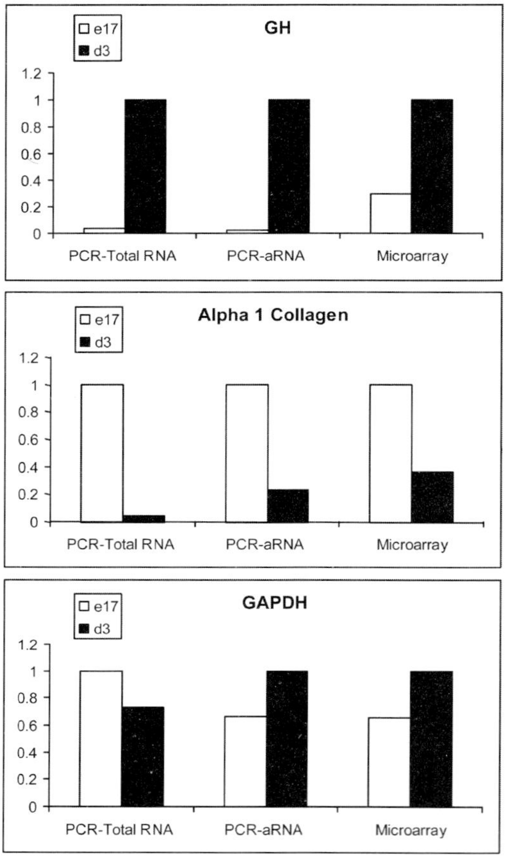

Fig. 2 A comparison of results from microarray analyses with amplified RNA (aRNA) with Real-Time PCR measurements of aRNA and non-amplified total RNA from e17 and d3 pituitary samples. Results are shown for growth hormone (GH), alpha 1 collagen, and glyceraldehyde 3 phosphodehydrogenase (GAPDH).

Next, a study was conducted to investigate changes in gene expression in the anterior pituitary gland that occur around the age of differentiation and proliferation of the last three major pituitary cell types, the thyrotrophs, somatotrophs, and lactotrophs. Thyrotrophs are first detectable on embryonic day (e) 6.5 and substantially increase on e11.5 (24), and levels of thyroid-stimulating hormone β (TSHβ) mRNA increase between e11 and e19 (9). Somatotrophs are first apparent on e14 and become a significant population by e16 (19), and GH mRNA levels increase between e14 and e16 (11). Lactotrophs differentiate by e17 (27), and prolactin (PRL) gene expression increases between e16 and e18 (11). Pituitary RNA was extracted and amplified from embryonic chicks of four ages, e10, e12, e14, and e17, and aRNA was labeled and hybridized to microarrays. Expression of 245 cDNAs, representing 233 genes, changed significantly during this period of embryonic development ($P<0.05$, $n=4$, highest mean $>175\%$ of lowest mean). The data were organized by SOMS into 18 clusters. Real-Time PCR analysis verified expression patterns for genes in 3 of these clusters. Fig. 3 shows the microarray and Real-Time PCR results from total RNA for TSHβ, GH, and PRL. In agreement with previous reports, an increase in TSHβ mRNA levels occurred between e10 and e12, GH expression increased between e12 and e14, and PRL mRNA levels did not increase until e17. These results confirmed that the neuroendocrine system cDNA microarrays can be used to detect changes in pituitary gene expression and are therefore potentially useful in identifying global gene expression patterns that have not yet been elucidated. Particularly, examination of the clusters whose expression profiles indicate potential involvement in thyrotroph, somatotroph, and lactotroph differentiation should lead to the identification of candidate genes that may play a role in these processes. For example, TSHβ, GH, and PRL were in clusters containing 27, 12, and 10 genes, respectively. Among the other genes in these clusters are cDNAs for numerous transcription factors, signaling molecules, and morphogenic factors.

Summary

We have begun applying functional genomics to studies of gene expression in the developing chick neuroendocrine system. Microarrays containing cDNAs for more than five thousand genes expressed in the chicken hypothalamus, pineal and pituitary have been produced, and a protocol for amplifying small amounts of RNA from avian neuroendocrine tissues has been established. We have used our chicken neuroendocrine system 5K microarrays to characterize profiles of gene expression in the hypothalamus from e17 to d3 and in the anterior pituitary gland from e10 through d3. Clusters of gene expression were identified indicating that structural and functional maturation continues in the hypothalamus after hatching. In the anterior pituitary, clusters of genes were identified whose expression suggests potential involvement in the initiation of TSHβ, GH and PRL gene expression. Our future studies are aimed at characterizing gene expression profiles in embryonic pituitary cells treated with corticosterone to induce GH cell differentiation, in the hypothalamus and pituitary of chickens divergently selected for high and low body weight or for high and low body fat, and in the hypothalamus of chickens subjected to fasting and refeeding.

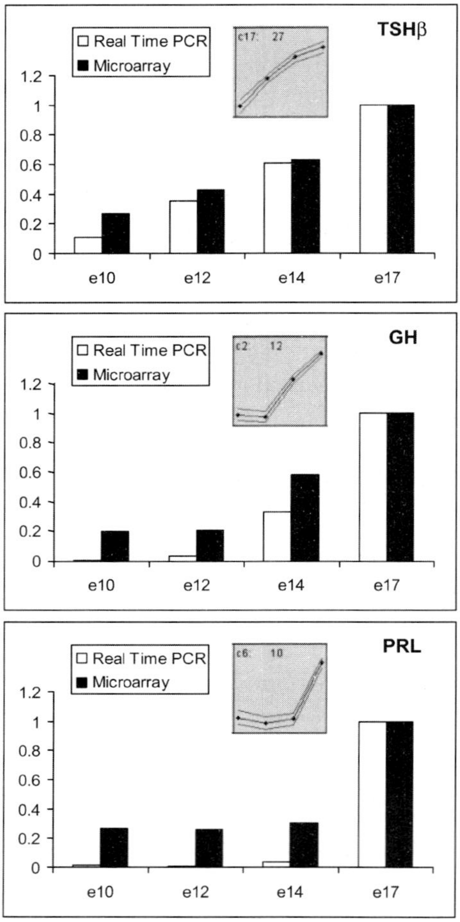

Fig. 3 A comparison of microarray and real-time PCR analyses of gene expression in the anterior pituitary from embryonic day 10 (e10) through e17. The results for thyroid-stimulating hormone β-subunit (TSHβ), growth hormone (GH), and prolactin (PRL) mRNAs are shown. Microarray analysis was performed on amplified RNA, while real-time PCR was performed on total RNA. The clusters containing each gene from SOMS analyses of gene expression profiles are shown in the inserts.

Acknowledgements

The chicken neuroendocrine system 5K microarrays were produced in collaboration with Dr. Larry Cogburn at the University of Delaware, and the authors are grateful for the assistance of Dr. Cogburn, Dr. Wilfrid Carre, Dr. Xiaofei Wang, Mr. Mike Muchow and Mr. Sultan Jenkins in the production of the microarrays. This research was supported by a competitive grant from the United States Department of Agriculture (#00-52100-9614) and by the Maryland Agricultural Experiment Station.

References

1 Abdrakhmanov I, Lodygin D, Geroth P, Arakawa H, Law A, Plachy J, Korn B & Buerstedde J (2000) A large database of chicken bursal ESTs as a resource for the analysis of vertebrate gene function. *Genome Research* **10** 2062-2069

2 Aoyagi K, Tatsuta T, Nishigaki M, Akimoto S, Tanabe C, Omoto Y, Hayashi S, Sakamoto H, Sakamoto M, Yoshida T, Terada M, Sasaki H (2003) A faithful method for PCR-mediated global mRNA amplification and its integration into microarray analysis on laser-captured cells. *Biochemical and Biophysical Research and Communications* **300** 915-920

3 Bailey MJ, Beremand PD, R RH, Bell-Pedersen D, Thomas TL & Cassone VM (2003) Transcriptional profiling of the chick pineal gland, a photoreceptive circadian oscillator and pacemaker. *Molecular Endocrinology* **17** 2084-2095

4 Boardman PE, Sanz-Ezquerro J, Overton IM, Burt DW, Bosch E, Fong WT, Tickle C, Brown WRA, Wilson SA & Hubbard SJ (2002) A comprehensive collection of chicken cDNAs. *Current Biology* **12** 1965-1969

5 Cogburn LA, Wang X, Carre W, Rejto L, Aggrey SE, Duclos MJ, Simon J & Porter TE (2004) Functional genomics in chickens: development of integrated-systems microarrays for transcriptional profiling and discovery of regulatory pathways. *Comparative and Functional Genomics* **5** 253–261

6 Cogburn LA, Wang X, Carre W, Rejto L, Porter TE, Aggrey SE & Simon J (2003) Systems-wide chicken DNA microarrays, gene expression profiling, and discovery of functional genes. *Poultry Science* **82** 939-951

7 Duggan DJ, Bittner M, Chen Y, Meltzer P & Trent JM (1999) Expression profiling using cDNA microarrays. *Nature Genetics* **21** 33-37

8 Eisen MB & Brown PO (1991) DNA arrays for analysis of gene expression. *Methods in Enzymology* **303** 179-205

9 Gregory CC, Dean CE & Porter TE (1998) Expression of Chicken Thyroid-Stimulating Hormone β-Subunit Messenger Ribonucleic Acid during Embryonic and Neonatal Development. *Endocrinology* **139** 474-478

10 Jorge EC, Monteiro-Vitorello CB, Alves HJ, Silva CS, Balan RG, Patricio M & Coutinho LL (2004) EST analysis of mRNAs expressed during embryogenesis in Gallus gallus. *International Journal of Developmental Biology* **48** 333-337

11 Kansaku N, Shimada K, Terada O & Saito N (1994) Prolactin, growth hormone, and luteinizing hormone-β subunit gene expression in the cephalic and caudal lobes of the anterior pituitary gland during embryogenesis and different reproductive stages in the chicken. *General and Comparative Endocrinology* **96** 197-205

12 Kenzelmann M, Klaren R, Hergenhahn M, Bonrouhi M, Grone H-J, Schmid W & Schutz G (2004) High-accuracy amplification of nanogram total RNA amounts for gene profiling. *Genomics* **83** 550-558

13 Kuenzel WJ (1993) The search for deep encephalic photoreceptors within the avian brain, using gonadal development as a primary indicator. *Poultry Science* **72** 959-967

14 Kuenzel WJ (1994) Central neuroanatomical systems involved in the regulation of food intake in birds and mammals. *Journal of Nutrition* **124** 1355S-1370S

15 Luo L, Salunga RC, Guo H, Bittner A, Joy KC, Galindo JE, Xiao H, Rogers KE, Wan JS, Jackson MR, Erlander MG (1999) Gene expression profiles of laser-captured adjacent neuronal subtypes. *Nature Medicine* **5** 117-122

16 McNabb FMA & Olson JM (1996) Development of thermoregulation and its hormonal control in precocial and altricial birds. *Poultry and Avian Biology Reviews* **7** 111-125

17 Neiman PE, Ruddell A, Jasoni C, Loring G, Thomas SJ, Brandvold KA, Lee R, Burnside J & Delrow J (2001) Analysis of gene expression during myc oncogene-induced lymphomagenesis in the bursa of Fabricius. *Proceedings National Academy of Science USA* **98** 6378-6383

18 Phillips J & Eberwine JH (1996) Antisense RNA amplification: a linear amplification method for analyzing the mRNA population from single living cells. *METHODS: A Companion to Methods in Enzymology* **10** 283-288

19 Porter TE, Couger GS, Dean CE & Hargis BM (1995) Ontogeny of growth hormone (GH)-secreting cells during chicken embryonic development: Initial somatotrophs are responsive to GH-releasing hormone. *Endocrinology* **136** 1850-1856

20 Reich M, Ohm K, Tamayo P, Angelo M & Mesirov JP (2004) GeneCluster 2.0: an advanced toolset for bioarray analysis. *Bioinformatics* **20** 1797-1798

21 Saeed AI, Sharov V, White J, Li J, Liang W, Bhagabati N, Braisted J, Klapa M, Currier T, Thiagarajan M *et al.* (2003) TM4: a free, open-source system for microarray data management and analysis. *Biotechniques* **34** 374-378

22 Simon R, Radmacher MD & Dobbin K (2002) Design of studies using DNA microarrays. *Genetic Epidemiology* **23** 21-36

23 Stoyanova R, Upson JJ, Patriotis C, Ross EA, Henske EP, Data K, Boman B Clapper, M Knudson AG & Bellacosa A (2004) Use of RNA amplification in the optimal characterization of global gene expression using cDNA microarrays. *Journal of Cellular Physiology* **201** 359-365

24 Thommes RC, Martens JB, Hopkins WE, Caliendo J, Sorrentino MJ & Woods JE (1983) Hypothalamo-adenohypophyseal-thyroid interrelationships in the chick embryo. IV. Immunocytochemical demonstration of TSH in the hypophyseal pars distalis. *General and Comparative Endocrinology* **51** 434-443

25 Tirunagaru VG, Sofer L, Cui J & Burnside J (2000) An expressed sequence tag database of t-cell-enriched activated chicken splenocytes: sequence analysis of 5251 clones. *Genomics* **66** 144-151

26 Wang E, Miller LD, Ohnmacht GA, Liu ET & Marincola FM (2000) High-fidelity mRNA amplification for gene profiling. *Nature Biotechnology* **18** 457-459

27 Woods KL & Porter TE (1998) Ontogeny of prolactin-secreting cells during chick embryonic development: effects of vasoactive intestinal peptide. *General and Comparative Endocrinology* **112** 240-246

Part two

Gonadotrophin releasing and inhibiting hormones

Osprey
Steve Scheich

Functional Avian Endocrinology
A Dawson & PJ Sharp (Eds)

The gonadotrophin releasing hormone neurone: key to avian reproductive function

PJ Sharp and N Ciccone

Division of Genetics and Genomics, Roslin Institute (Edinburgh), Roslin, Midlothian EH25 9PS, Scotland, U.K.

Introduction

In birds, hypothalamic chicken gonadotrophin releasing hormone-I (GnRH-I) neurones and their projections to the median eminence form the common pathway through which endocrine factors (e.g. gonadal hormones) and environmental (e.g. photoperiodic, stress-inducing) factors control follicle stimulating (FSH) and luteinizing hormone (LH) release. A comprehensive account of the biology of avian GnRH was published in 1997 (11) and is briefly updated here.

Neuroanatomy

Most GnRH-I cell bodies, numbering several hundred, are loosely associated around the midline of the preoptic-anterior hypothalamus, extending from the ventral preoptic area to the septal area, with discrete clusters in the bed nucleus of the pallial commissure and at the caudal border of the septal area (26, 45, 47). GnRH-I terminals are particularly abundant in the median eminence consistent with a role in the control of anterior pituitary function. The presence of a GnRH-I fibre network throughout the hypothalamus suggests pathways through which GnRH-I neuronal activities may be co-ordinated (26, 45, 47). Some GnRH-I cell bodies occur outside the hypothalamus, for example in the lobus parolfactorius and nucleus accumbens, and GnRH-I immunoreactive ring-like structures are seen in the forebrain, including the hippocampus and nucleus taeniae (45). GnRH-I-containing cells first appear in the chicken embryo after 3 days incubation in the olfactory placodal epithelium and then multiply and migrate along the olfactory nerve into the telencephalon (32). Migrating GnRH-I neurones first enter the telencephalon at about 6 days and multiply until they reach their final locations at around 9 days of incubation (32). The identity of neural inputs to GnRH-I neurones and of receptors and neurotransmitters associated with them is not completely known, although there is evidence of interactions between GnRH-I neurones and dopamine (8),

NPY (7), VIP (22, 46) and enkephalins (30). GnRH-I cell bodies (33) and their terminals (52) are enveloped by glial cell projections and changes in glial cell envelopment may play a role in controlling neural input.

Functional significance of GnRH-I

Chicken GnRH-I (Glu^{1}-His^{2}-Trp^{3}-Ser^{4}-Try^{5}-Gly^{6}-Leu^{7}-Gln^{8}-Pro^{9}-Gly^{10} NH_2) stimulates LH release (11) but not FSH release (13) as measured using the most recently developed assay for chicken FSH (25). The different responses of LH and FSH to GnRH-I may be facilitated by the presence in the anterior pituitary of separate cell types containing LH or FSH (35). The stimulatory effects of GnRH-I on LH release depends on circulating levels of gonadal steroids, stage of somatic maturation and stage of the breeding cycle. As illustrated by the domestic hen (49) the LH response to GnRH is less in juvenile than in adult non-breeding birds. In breeding females, the LH response to GnRH decreases (49), presumably as a result of the inhibitory feed-back action of increased circulating steroids from fully developed ovaries. In somatically mature birds gonadal regression induced by the development of broodiness is associated with an increase in responsiveness to GnRH (e.g. 38), possibly as a consequence of decreased circulating concentrations of gonadal steroids, while in contrast, gonadal regression induced by the development of photorefractoriness is not associated with an increase in LH responsiveness to GnRH (51). This may be a consequence of the withdrawal of a self-priming effect of photoinduced GnRH release (28). In addition to inducing the release of LH, GnRH-I also stimulates LH synthesis as demonstrated by its ability to stimulate LHβ (53) and gonadotrophin α subunit mRNAs (Fig. 1) in cultured avian pituitary cells. The observation that one mode of administration of GnRH-I which stimulates gonadotrophin α subunit mRNAs *in vitro* does not stimulate LHβ mRNA (Fig. 1) suggests that the two LH subunit mRNAs may be differentially controlled by GnRH-I at different doses and pulse frequencies.

Although GnRH-I may not stimulate FSH release directly it may do so indirectly by controlling FSH synthesis. In support of this view GnRH-I stimulates FSHβ subunit mRNA in cultured hen pituitary gland fragments (Fig. 1). Once the FSHβ subunit is synthesised it may heterodimerise with the α subunit to form FSH and then be released constitutively. A constitutive release of FSH was first suggested by studies in the Japanese quail where it was found that the amount of FSH released from pituitary glands cultured with hypothalamic extract was three times more than the original content (17). The factors which control FSH synthesis in birds have not been extensively investigated. In mammals, FSH synthesis is controlled by interactions between GnRH and a pituitary activin/follistatin autocrine or paracrine loop (3, 4). Follistatin plays a key role in FSH synthesis by binding activinβ_B which drives FSHβ subunit gene expression (4). Follistatin mRNA is stimulated by GnRH in a pulse frequency-dependent manner and is reciprocally related to FSHβ mRNA (3). In preliminary studies to determine whether a similar mechanism occurs in birds, it was demonstrated that the chicken anterior pituitary gland contains follistatin (Fig. 1) and activinβ_B mRNAs, although the amounts of activinβ_B mRNA were too low to be easily quantified (N. Ciccone, unpublished). However a dose and frequency of GnRH-I administration which stimulates FSHβ mRNA did not stimulate follistatin mRNA (Fig. 1).

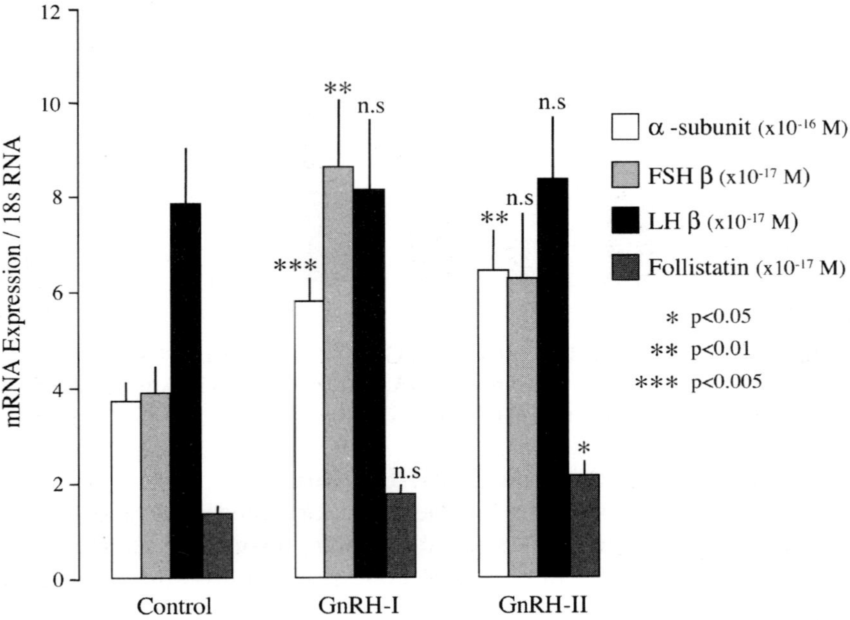

Fig. 1 The effect of sixteen, 5 minute pulses of GnRH-I or GnRH-II (10^{-7} M) at 1.5h intervals in a static incubation of laying hen anterior pituitary gland fragments on common α, LHβ, FSHβ and follistatin mRNAs. (Data from 5)

The GnRH receptor which is likely to mediate the stimulatory effects of GnRH-I on gonadotrophe function has been cloned in the chicken (42). The mRNA encoding this receptor is present in the chicken anterior pituitary gland (Fig. 2) where its expression is regulated by oestrogen (42). GnRH receptor mRNA also occurs in the chicken mediobasal and preoptic hypothalamus where it may mediate GnRH cell to cell communication and /or effects of GnRH on behaviour (27). The question of whether this receptor mediates the actions of GnRH-I or GnRH-II, or both neuropeptides, or whether there are separate receptors for GnRH-I and -II remains to be answered. It is relevant to note that GnRH receptor mRNA occurs in the cortex (Fig. 2), a brain area containing GnRH-I and -II fibres (26, 45, 47) which may mediate behavioural effects of these neuropeptides (27).

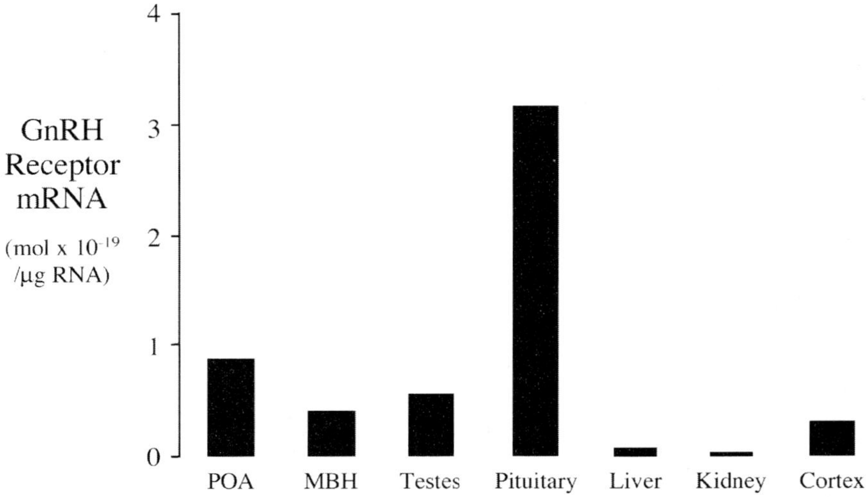

Fig. 2 Distribution of fully processed GnRH-I receptor mRNA in tissues from sexually mature cockerels. POA, preoptic hypothalamus, MBH, mediobasal hypothalamus. (Data from 42)

Gonadal steroid feedback on GnRH-I neurones

GnRH-I neurones form part of the inhibitory feed back mechanism through which circulating gonadal steroids maintain a homeostatic relationship with gonadotrophin secretion. In support of this view, in the cockerel, hypothalamic GnRH-I peptide (48) and mRNA (42), increase after castration, and these increases are reversed after treatment with oestrogen (42, 48). It is relevant to note that gonadal steroids also exert an inhibitory action on LH release at the level of the anterior pituitary gland as demonstrated by the observation that oestrogen, testosterone and progesterone inhibit GnRH-I induced LH release from cultured chicken pituitary cells (21). The stimulatory feedback of progesterone on the preovulatory LH release is likely to be mediated by an increased release of GnRH-I (15, 43) associated with a decrease in GnRH-I peptide in the basal hypothalamus (24, 50). However, GnRH-I mRNA does not increase in association with an induced or a spontaneous LH preovulatory surge in the hen (IC Dunn, unpublished data) suggesting that progesterone-induced LH release may not depend on increased GnRH-I synthesis.

Functional significance of GnRH isoforms

The best studied isoform of GnRH-I in the avian hypothalamus is GnRH-II ((His^5,Try^7,Tyr^8)GnRH-I)), which occurs in all non-agnathan vertebrates and is considered to be ancestral for most of the 15 isoforms of GnRH known to date (10). GnRH-II peptide is widely distributed in the hypothalamus and in other brain areas, notably in the telencephalon (45, 47), especially the area corticoidea dorsolateralis (PJ Sharp, unpublished) with cell bodies located in the mid-mesencephalon and caudal basal hypothalamus (e.g. 45, 47). The developmental biology of GnRH-II neurones has not been investigated, but is likely to be different from that of GnRH-I neurones because GnRH II peptide first increases in the brain of chick embryos at day 14 of incubation, 8 days later than the first observed increase for GnRH-I peptide (29). GnRH-II stimulates LH release at equal or lower doses than GnRH-I (e.g 16, 49), and the greater potency of GnRH-II is particularly pronounced in laying females (40). Further, GnRH-II stimulates gonadotrophin α subunit and follistatin mRNA *in vitro* (Fig. 1) suggesting a potential role in gonadotrophin synthesis.

It is uncertain whether GnRH-II plays a physiological role in the regulation of gonadotrophin secretion at the level of the anterior pitutary gland. Although several studies report that GnRH-II is absent or not abundant in the median eminence (e.g. 41), a combined immunocytochemical and HPLC study in the quail shows that in this bird, the median eminence contains abundant GnRH-II terminals (6). This observation supports the view that GnRH-II plays a role in the regulation of gonadotrophe function. Consistent with this view, in the turkey, hypothalamic GnRH-I and GnRH-II peptide contents are higher in laying than in non-laying photosensitive, incubating, or photorefractory hens (36) while in the chicken, during the hen ovulatory cycle, GnRH-II, but not GnRH-I decreases in the anterior hypothalamus when preovulatory concentrations of plasma LH are maximal or declining (50). On the other hand, active immunization against GnRH-I but not against GnRH-II induces ovarian regresssion in the laying hen (41) and there is no consistent evidence that hypothalamic GnRH-II peptide content is always correlated closely with changes in gonadotrophin secretion. For example, hypothalamic GnRH-I but not GnRH-II content increases at onset of puberty (41, 50) and after castration in the cockerel (48), and GnRH-I peptide, but not GnRH-II peptide content is depressed by oestrogen treatment in the castrated cockerel (48). Further, the preovulatory release of LH in the chicken is associated with a decrease in basal hypothalamic GnRH-I, but not

GnRH-II peptide (50). However, it is possible that changes in hypothalamic GnRH-II are indirectly associated with changes in reproductive function. An example is the control of sexual behavior, since intra-cerebro-ventricular (i.c.v.) injection of GnRH-II, but not GnRH-I in female White Crowned sparrow enhances courtship behaviour (27). A further possibility is that hypothalamic GnRH-II neural circuitry regulates reproductive behaviour according to energetic status (44). This is suggested by the observation, in female musk shrews, that i.c.v. administration of GnRH-II but not GnRH-I treatment reinstates mating behaviour previously inhibited by food restriction (20).

A third isoform of GnRH-I, lamprey GnRH-III, ((Tyr^3, Leu^5,Glu^6, Try^7 Lys^8) GnRH-I), which stimulates LH release has been reported in the hypothalamus and forebrain of song birds using immunocytochemistry (2). The possibility that this neuropeptide might be a FSH releasing factor has not been confirmed in the rat and hamster (31).

Changes in GnRH-I mRNA associated with puberty, broodiness and reduced food intake

Changes in GnRH-I neuronal function are involved in the onset of non-photoperiodically-induced puberty. For example, in the domestic cockerel, hypothalamic GnRH-I peptide content (Fig. 3, 23, 41), releasable GnRH-I (23), and plasma LH (23, 41) increase at the onset of puberty. This increase in GnRH-I peptide is associated with a decrease in hypothalamic GnRH-I mRNA (Fig. 3) which can be ascribed to the inhibitory feed-back effects of testicular steroids on GnRH-I mRNA, possibly after aromatization to oestrogen in the hypothalamus (12, 42). This view is supported by the observation that GnRH-I mRNA does not decrease in castrated cockerels at the age intact birds start becoming sexually mature, although there is an increase in GnRH-I peptide (Fig. 3). This latter observation demonstrates that increased GnRH-I peptide content and release at the onset of puberty are not dependent on increased GnRH-I mRNA. The possibility that GnRH-I peptide may increase at the onset of puberty as a result of an increase in the rate of further processing of proGnRH-I has been investigated by comparing the ratio of numbers of cell bodies immunolabelled specifically for GnRH-I or proGnRH-I peptides before and after the onset of puberty (A Mess & PJ Sharp, unpublished). No differences were observed, suggesting that mechanisms controlling the degradation of fully processed GnRH-I may play a key role in regulating the onset of puberty.

In breeding adult birds, it is likely that changes in food availability may regulate reproductive activity through regulating GnRH-I gene expression. This possibility is suggested by an observation in laying broiler breeder hens subjected to food restriction to prevent the development of obesity. A robust increase in GnRH-I mRNA is seen in these hens after food restriction is lifted (N.Ciccone unpublished observation).

The decrease in plasma LH, ovarian regression and decrease in hypothalamic GnRH-I peptide (36) seen in incubating hens is associated with a reduction in GnRH-I mRNA (14). This is seen when circulating ovarian steroids are low and unlikely to exert a significant inhibitory effect on GnRH mRNA (12, 42). In incubating birds it is likely that the reduction in GnRH-I mRNA is a consequence of an inhibitory effect of elevated concentrations of plasma prolactin which may counter any tendency of GnRH-I mRNA to increase as a result of a decrease in ovarian hormone inhibitory feed back (39).

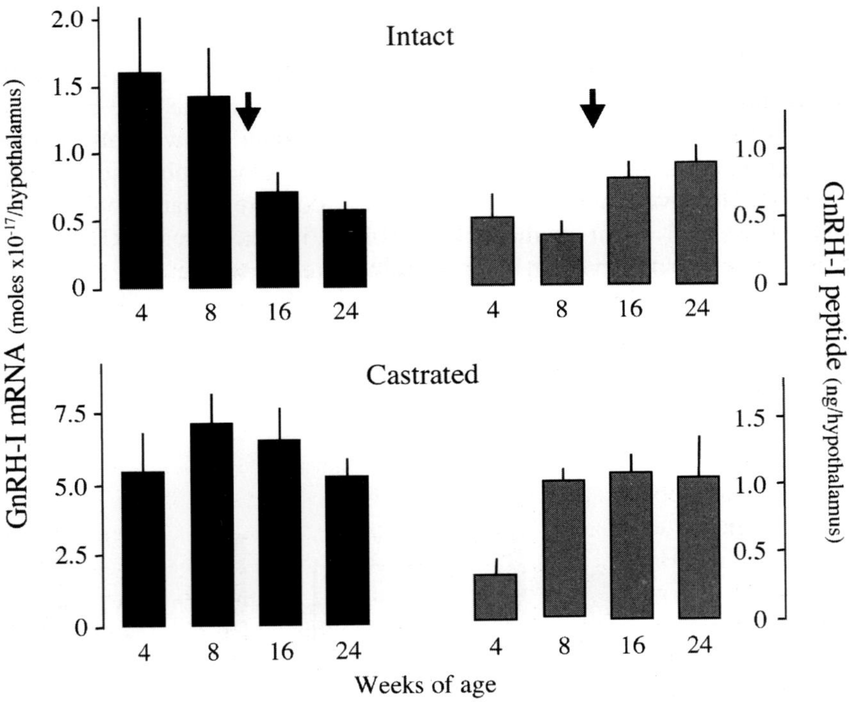

Fig. 3 Changes in hypothalamic GnRH-I mRNA and GnRH-I peptide around the onset of puberty (↓) in intact and castrated cockerels maintained from hatch on a fixed 14h photoperiod. The cockerels were castrated at 5 weeks. (Unpublished data; I C Dunn, R Gladwell, RT Talbot & PJ Sharp).

Photoinduced changes in GnRH-I mRNA

In photoperiodic birds, photoinduced LH secretion is generally associated with an increase in hypothalamic GnRH-I peptide (e.g. 9, 34). As demonstrated in the Japanese quail, photoinduced LH release begins between 20-26h after dawn of the first long day, and this correlates with an increase in GnRH-I release as demonstrated using an *in vitro* superfusion system (34). There is evidence that this acute photoinduced release of GnRH-I is mediated through a circadian-gated signalling pathway located in the mediobasal hypothalamus that stimulates GnRH-I release from terminals in the median eminence (see Chapter 11). Early evidence for this mechanism came from the demonstration that electrolytic lesions in the quail posterodorsal hypothalamus (37) that did not reduce GnRH-I terminals in the median eminence (18), blocked the photoperiodic response. This observation does not exclude the possibility that prolonged photostimulation may stimulate GnRH-I synthesis to increase the supply of GnRH-I to the median eminence. This is suggested by the observation that GnRH-I mRNA increases after seven days photostimulation in peripubertal cockerels and somatically mature out-of-lay hens (12). However a close examination of the time course of this photoinduced increase in GnRH mRNA in quail shows that it begins to increase 20 hours after dawn of the first long day of photostimulation (Fig. 4). However there is evidence in the cockerel, that photoinduced LH release can occur in the absence of an increase in GnRH mRNA (12). This was observed in prepubertal cockerels where it is suggested that the GnRH-I neuronal system may not be fully mature (12).

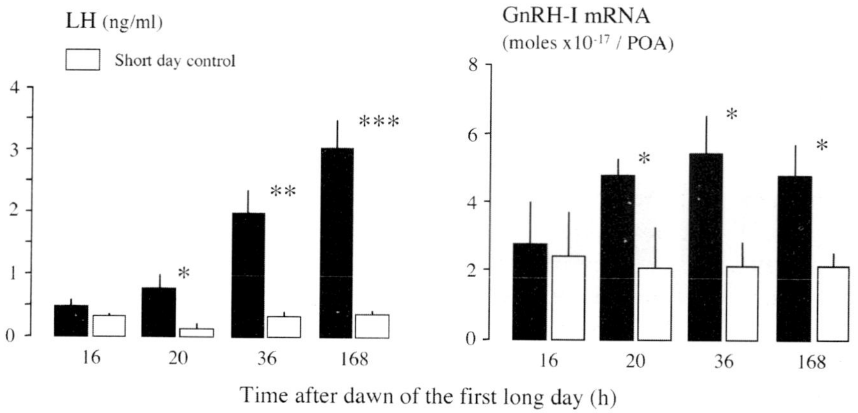

Fig. 4 Changes in plasma LH and hypothalamic GnRH-I mRNA in somatically mature male Japanese quail between 16-168h after transfer from short (8L:16D) to long (20L:4D) days. (Data from 1)

The development of reproductive photorefractoriness is marked by a steep decrease in plasma LH and gonadal regression accompanied by, rather than preceded by, a decrease in hypothalamic GnRH-I peptide (e.g. 9). It is likely that at this stage of the reproductive cycle, i.e. at the initiation of photorefractoriness, GnRH-I peptide release is actively inhibited by high plasma prolactin and stores of GnRH-I progressively decrease as result of degradation (9, 39). It is not known whether GnRH-I gene expression decreases at the initiation of photorefractoriness in passerine species such as the starling. There is no necessity for it to do so if photorefractoriness is initiated by an inhibitory effect of high plasma prolactin on GnRH-I release (9, 39). However after reproductive photorefractoriness is initiated, and plasma prolactin levels are decreasing or are low, there is a dramatic decrease in hypothalamic GnRH-I peptide (e.g. 9) that can be accounted for by a reduction in GnRH-I synthesis. This is probably caused by a reduction in GnRH-I gene expression, but this has not been demonstrated. However some relevant information is available for the Japanese quail. When breeding photostimulated Japanese quail are transferred to short days, the onset of gonadal regression precedes a detectable decrease in hypothalamic GnRH-I mRNA (19). It is concluded that short day induced gonadal regression is initiated by a decreased release of GnRH-I which is not a consequence of decreased GnRH-I mRNA synthesis. A similar relationship between GnRH-I peptide and mRNA may occur when gonadal regression is a consequence of the development of photorefractoriness.

Conclusion: control of GnRH release at the cell body and terminal?

A picture of the avian GnRH-I neurone is emerging to suggest that its activities are controlled both at the level of the cell body to regulate GnRH-I gene expression, and at its terminals in the median eminence to regulate GnRH-I release. Inhibitory control of GnRH-I gene expression is a key element in the steroid negative feedback system that maintains homeostasis between concentrations of plasma gonadotrophins and gonadal steroids. Independently of the inhibitory action of gonadal steroids, GnRH neuronal activity is also controlled by environmental factors that affect reproductive activity and of these, changes in photoperiod have been most extensively investigated. A striking feature of the GnRH-I neuronal response to photostimulation in sexually mature birds is that GnRH-I gene expression and GnRH-I release appear to be closely co-ordinated to initiate increased LH secretion. Conversely, decreased GnRH-I release caused by a reduction in photoperiod or by the development of photorefractoriness may not be so closely linked to a decrease in GnRH-I gene expression. A major challenge is to identify the regulatory neural pathways from the biological clock that measures photoperiodic time (Chapter 11) to GnRH-I neurones both at the level of GnRH-I cell body and of the GnRH-I terminals in the median eminence.

Acknowledgements

We are grateful for contributions from Beth Baines, Tim Boswell, Ian Dunn, Rose-Marie Karlsson, Yuh-Man Sun and Richard Talbot, and for support from a BBSRC Core Strategic Grant and CASE Studentship.

References

1 Baines E (2001) Photoperiodic control of hypothalamic GnRH gene expression in Japanese quail. PhD Thesis University of Edinburgh

2 Bentley GE, Moore IT, Sower SA & Wingfield JC (2004) Evidence for a novel gonadotrophin releasing hormone in hypothalamic and forebrain areas in song birds. *Brain Behavior and Evolution* **63** 34-46

3 Besecke LM, Guendner MJ, Schneyer AL, Bauer-Dantoin AC, Jameson JL & Weiss J (1996) Gonadotrophin-releasing hormone regulates follicle-stimulating hormone-β gene expression through an activin/follistatin autocrine or paracrine loop. *Endocrinology* **137** 3667-3673

4 Bilezikjian, LM, Blount AL, Leal AMO, Donaldson CJ, Fischer WH & Vale WW (2004) Autocrine/paracrine regulation of pituitary function by activin, inhibin and follistatin. *Molecular and Cellular Endocrinology* **225** 29-36

5 Ciccone N (2005) Molecular analysis of reproductive neuroendocrine function in broiler breeders. PhD Thesis University of Edinburgh

6 Clerens S, D' Hondt E, Berghman LR, Vandesande F & Arckens L (2003) Identification of GnRH-II in the median eminence of Japanese quail (*Coturnix coturnix japonica*). *General and Comparative Endocrinology* **131** 48-56

7 Contijoch AM, Malamed S, Mcdonald JK & Advis JP (1993) Neuropeptide Y regulation of LHRH release in the median eminence: immunocytochemical and physiological evidence in hens. *Neuroendocrinology* **57** 135-145

8 Contijoch AM, Gonzalez C, Singh HN, Malamed S, Troncosco S & Advis JP (1992) Dopaminergic regulation of luteinizing hormone-releasing hormone release at the median eminence: immunocytochemical and physiological evidence in hens. *Neuroendocrinology* **55** 290-300

9 Dawson A, Talbot RT, Dunn IC & Sharp PJ (2002) Changes in basal hypothalamic chicken releasing hormone–I and vasoactive intestinal polypeptide associated with a photoinduced cycle in gonadal maturation and prolactin secretion in intact and thyroidectomized starlings (*Sturnus vulgaris*). *Journal of Neuroendocrinology* **14** 533-539

10 Dubios EA, Zandbergen MA, Peute J, & Goos HJT (2002) Evolutionary development of three gonadotrophin releasing hormone systems in vertebrates. *Brain Research Bulletin* **57** 413-418

11 Dunn IC & Millam JR (1998) Gonadotrophin releasing hormone: forms and functions in birds. *Poultry and Avian Biology Reviews* **9** 61-85

12 Dunn IC & Sharp PJ (1999) Photo-induction of hypothalamic gonadotrophin releasing hormone –I mRNA in the domestic chicken: a role for oestrogen? *Journal of Neuroendocrinology* **11** 371-375

13 Dunn IC, Lewis PD, Wilson PW & Sharp PJ (2003) Acceleration of maturation of FSH and LH responses to photostimulation in prepubertal domestic hens by oestrogen. *Reproduction* **126** 217-225

14 Dunn IC, Beattie KK, Maney D, Talbot RT, Wilson PW & Sharp PJ (1996) Regulation of chicken gonadotrophin releasing hormone-I mRNA in incubating, nest –deprived and laying hens. *Neuroendocrinology* **63** 504-513

15 Fraser HM & Sharp PJ (1978) Prevention of positive feedback in the hen (*Gallus domesticus*) by antibodies to luteinizing hormone releasing hormone. *Journal of Endocrinology* **76** 181-182

16 Guemene D & Williams JB (1999) LH responses to chicken luteinizing hormone –releasing hormone I and II in laying, incubating, and out of lay turkey hens. *Domestic Animal Endocrinology* **17** 1-15

17 Hattori A, Ishii S & Wada M (1986) Different mechanisms controlling FSH and LH release in Japanese quail *(Coturnix coturnix japonica*): evidence for an inherently spontaneous release and production of FSH. *Journal of Endocrinology* **108** 239-245

18 Juss TS (1993) Neuroendocrine and neural changes associated with the photoperiodic control of reproduction.In *Avian Endocrinology* Ed PJ Sharp, Journals of Endocrinology Ltd: Bristol, UK

19 Karlsson RM (2001) Quantification of gonadotrophin-releasing hormone –I gene expression in the brain of the Japanese quail. MSc Thesis University of Uppsala

20 Kauffman AS & Rissman (2004) The evolutionarily conserved gonadotrophin releasing hormone–II modifies food intake. *Endocrinology* **145** 686-691

21 King JA, Davisdson JS, Mehl AE, Wakefield IK, Andersson PB & Millar RP (1989) Gonadal steroid modulation of signal transduction and luteinizing hormone release in cultured chicken pituitary cells. *Endocrinology* **124** 1830-1840

22 Kiyoshi K, Kondoh M, Hirunagi K & Korf H-W (1998) Confocal laser scanning and electronmicroscopic analyses of the relationship between VIP-like and GnRH-like immunoreactive neurons in the lateral septal-preoptic area of the pigeon. *Cell and Tissue Research* **293** 39-46

23 Knight PG (1983) Variations in hypothalamic luteinizing hormone releasing hormone content and release *in vitro* and plasma concentrations of luteinizing hormone and testosterone in developing cockerels. *Journal of Endocrinology* **99** 311-319

24 Knight PG, Wilson SC, Gladwell RT & Cunningham FJ (1984) Hypothalamic contents of LHRH and catecholamines during the ovulatory cycle of the hen *(Gallus domesticus). Journal of Reproduction and Fertility* **71** 289-295

25 Krishnan KA, Proudman JA, Bolt DJ & Bahr JM (1993) Development of an homologous radioimmunoassay for chicken follicle stimulating hormone

and measurement of plasma FSH during the ovulatory cycle. *Comparative Biochemistry and Physiology* **105A** 729-734
26 Kuenzel WJ & Blahser S (1991) The distribution of gonadotrophin-releasing hormone (GnRH) neurons and fibres throughout the chick brain *(Gallus domesticus). Cell and Tissue Research* **264** 481-495
27 Maney DL, Richardson RD & Wingfield JC (1997) Central administration of chicken gonadotrophin releasing hormone-I enhances courtship behavior in a female sparrow. *Hormones and Behavior* **32** 11-18
28 McNaughton FJ, Dawson A & Goldsmith AR (1995) A comparison of the responses to gonadotrophin releasing hormone of adult and juvenile, and photosensitive and photorefractory European starlings. *Sturnus vulgaris. General and Comparative Endocrinology* **97** 135-144
29 Millam JR Craig-Veit CB & Pettitte JN (1993) Brain content of GnRH I and II during embyonic development in chickens. *General and Comparative Endocrinology* **92** 311-317
30 Millam JR, Wang R, Craig-Veit CB, & Siopes TD (2002) Apposition of enkephalinergic axons with cGnRH I-containing perikarya in turkey brain. *General and Comparative Endocrinology* **127** 128-135
31 Montaner A, Mongiat L, Lux-Lantos, VAR, Park MK, Fischer WH, Craig AG, Rivier JE, Lescheid D, Lovejoy D, Libertun C, Sherwood NM & Somoza GM (2001) Structure and biological activity of gonadotrophin-releasing hormone isoforms isolated from rat and hamster brains *Neuroendocrinology* **74** 202-212
32 Mulrenin EM, Witkin JW & Silverman AJ (1999) Embryonic development of the gonadotrophin releasing hormone (GnRH) system in the chick: a spatio-temporal analysis of GnRH neuronal generation, site of origin, and migration. *Endocrinology* **140** 422-433
33 Parry DM & Goldsmith AR (1993) Ultrastructural evidence for changes in synaptic input to the hypothalamic luteinizing hormone releasing hormone neurons in photosensitve and photorefractory starlings. *Journal of Neuroendocrinology* **5** 387-395
34 Perera AD & Follett BK (1992) Photoperiodic induction *in vitro*: dynamics of gonadotrophin releasing hormone release from hypothalamic explants of Japanese quail. *Endocrinology* **131** 2898-2908
35 Proudman JA, Vandesande F & Berghman LR (1999) Immunohistochemical evidence that follicle stimulating hormone and luteinizing hormone reside in separate cells in the chicken pituitary. *Biology of Reproduction* **60** 1324-1328
36 Rozenboim I, Silsby JL, Tabibzadeh C, Pitts GR, Youngren OM & El Halawani ME (1993) Hypothalamic and posterior pituitary content of vasoactive intestinal peptide and gonadotrophin-releasing hormones I and II in the turkey hen. *Biology of Reproduction* **49** 622-626
37 Sharp PJ & Follett BK (1969) The effect of hypothalamic lesions on gonadotrophin release in Japanese quail *(Coturnix coturnix japonica). Neuroendocrinology* **5** 205-218
38 Sharp PJ & Lea RW (1981) The response of the pituitary gland to luteinizing hormone releasing hormone in broody bantams. *General and Conparative Endocrinology* **45** 131-331

39 Sharp PJ, Dawson A & Lea RW (1998) Control of luteinizing hormone and prolactin secretion in birds. *Comparative Biochemistry and Physiology C* **119** 275-282
40 Sharp PJ Dunn IC & Cerolini S (1992) Neuroendocrine control of reduced persistence of egg laying in domestic hens: evidence for the development of photorefractoriness. *Journal of Fertlity and Reproduction* **94** 221-235
41 Sharp PJ, Talbot RT, Main GM, Dunn IC, Fraser HM & Huskinson NS (1990) Physiological roles of chicken LHRH-I and -II in the control of gonadotrophin release in the domestic chicken. *Journal of Endocrinology* **124** 291-299
42 Sun YM, Dunn IC, Baines E, Talbot RT, Illing N, Millar RP & Sharp PJ (2001) Distribution and regulation by oestrogen of fully processed and variant transcripts of gonadotrophin releasing hormone I and gonadotrophin releasing hormone receptor mRNAs in the male chicken. *Journal of Neuroendocrinology* **13** 37-49
43 Tanaka K & Kamiyoshi M (1976) Induction of ovulation in hens by an analogue of LH-RH. *Journal of Reproduction and Fertility* **46** 243-244
44 Temple JL, Millar RP & Rissman EF (2003) An evolutionary conserved form of gonadotrophin releasing hormone coordinates energy and reproductive behavior. *Endocrinology* **144** 13-19
45 Teruyama R, & Beck MM (2000) Changes in immunoreactivity to anti-GnRH-I and-II are associated with photostimulated sexual status in male quail. *Cell and Tissue Research* **300** 413-426
46 Teruyama T & Beck MM (2001) Double immunocytochemistry of vasoactive intestinal peptide and GnRH-I in male quail: photoperiodic effects. *Cell and Tissue Research* **303** 403-41
47 van Gils J, Absil, P Grauwels L, Moons L, Vandesande F, & Balthazart J (1993) Disribution of luteinizing hormones I and II in quail and chicken brain as demonstrated with antibodies directed against synthetic peptides. *Journal of Comparative Neurobiology* **334** 304-323
48 Wilson SC, Gladwell RT & Cunningham FJ (1990) Differential responses of hypothalamic LHRH-I and II to castration and gonadal steroids or tamoxifen treatment in cockerels. *Journal of Endocrinology* **125** 139-146
49 Wilson SC, Cunningham FJ, Chairil RA & Gladwell RT (1989) Maturational changes in the LH respose of the domestic fowl to synthetic chicken LHRH-I and –II. *Journal of Endocrinology* **123** 311-318
50 Wilson SC, Chairil RA, Cunningham FJ & Gladwell RT (1990) Changes in the hypothalamic contents of GnRH-I and –II and pituitary responsiveness to synthetic chicken LHRH-I and II during the progesterone –induced surge of LH in the laying hen. *Journal of Endocrinology* **127** 487-496
51 Wingfield JC, Crim JW, Mattocks PW & Farner DS (1979) Responses of photosensitive and photorefractory male white-crowned sparrows (*Zonotrichia leucophrys gambelli*) to synthetic mammalian luteinizing hormone releasing hormone. *Biology of Reproduction* **19** 1046-1056
52 Yamamura TK, Hirungi S, Ebihara T & Yoshimura T (2004) Seasonal morphological changes in the neuro-glial interaction between gonadotrophin-releasing hormone nerve terminals and glial endfeet in Japanese quail. *Endocrinology* **145** 4264-4267

53 You S, Foster LK, Silsby JL, El Halawani ME & Foster DN (1995) Sequence analysis of the turkey LHβ subunit and its regulation by gonadotrophin-releasing hormone and prolactin in cultured cells. *Journal of Molecular Endocrinology* **14** 117-129

Functional Avian Endocrinology
A Dawson & PJ Sharp (Eds)

Structure, action and functional significance of GnIH

K Tsutsui, G Bentley[1] and N Ciccone[2]

Laboratory of Brain Science, Faculty of Integrated Arts and Sciences, Hiroshima University, Higashi-Hiroshima 739-8521, Japan
[1] Department of Integrative Biology, Valley Life Sciences Building, University of California, Berkeley, Berkeley CA 94720-3140, USA
[2] Division of Genetics and Genomics, Roslin Institute, Midlothian, EH25 9PS, UK

Introduction

Neuropeptide control of gonadotropin secretion at the level of the anterior pituitary gland is primarily through the stimulatory action of the hypothalamic decapeptide, gonadotropin-releasing hormone (GnRH), which was originally isolated from mammals and subsequently from nonmammals including birds. Although studies of the different forms of GnRH and its receptors have revealed a wealth of information as to their function, an inhibitory hypothalamic neuropeptide for gonadotropin release has, until recently, remained unknown in any vertebrate. In 2000, a novel neuropeptide was found specifically to inhibit gonadotropin release in quail. This peptide has now been named gonadotropin-inhibitory hormone (GnIH). A gonadotropin inhibitory system is an intriguing concept and provides us with an unprecedented opportunity to study the regulation of avian reproduction from an entirely novel standpoint. In this chapter, we summarize the advances made in our understanding of the structure, action and functional significance of GnIH, a newly discovered hypothalamic neuropeptide, in birds.

Identification of GnIH in the brain

Since the discovery of the molluscan cardioexcitatory neuropeptide Phe-Met-Arg-Phe-NH_2 (FMRFamide) in the ganglia of the venus clam (17), neuropeptides that possess the RFamide motif at their C-termini (i.e., RFamide peptides) have been characterized in various invertebrates. Immunohistochemical studies using antiserum against FMRFamide suggested that vertebrate nervous systems possess unknown neuropeptides similar to FMRFamide (18, 19). Immunohistochemical findings indicated that some of the FMRFamide-like immunoreactive neurons project to the hypothalamic region close to the pituitary gland,

and thus were predicted to play a role in the regulation of pituitary function. We therefore looked for a novel RFamide peptide in the avian brain.

An RFamide peptide was isolated from the brain of Japanese quail (*Coturnix japonica*) using high performance liquid chromatography (HPLC) and a competitive enzyme-linked immunosorbent assay (ELISA), employing the antibody against the dipeptide, Arg-Phe-NH_2 (22). Acetic acid extracts of quail brain were passed through C-18 reversed-phase cartridges, and the retained material was subjected to reversed-phase and cation-exchange HPLC. Amino acid sequence analysis of the isolated substance by automated Edman degradation with a gas-phase sequencer revealed the following sequence: Ser(62)-Ile(252)-Lys(233)-Pro(226)-Ser(38)-Ala(194)-Tyr(173)-Leu(148)-Pro(104)-Leu(108)-Arg(45)-Phe(52) with the detected amount (pmol) of each amino acid indicated in parentheses. A protonated molecule ion $(M+H)^+$ peak in the fast atom bombardment-MS (FAB-MS) of this peptide at *m/z* 1389.4 indicated that the peptide was amidated at the C-terminus. Synthetic and native peptides showed identical retention times on a C-18 reversed-phase column and a cation-exchange column. The mixture of the synthetic and native peptides eluted as a single peak from each column. Thus the isolated native peptide was confirmed as a 12 amino acid sequence (SIKPSAYLPLRFamide) with RFamide at the C-terminus (22). This neuropeptide had not been previously reported in vertebrates, although the C-terminal LPLRFamide was identical to chicken pentapeptide LPLRFamide peptide (7). The chicken peptide may be a degraded fragment of the dodecapeptide, because Dockray *et al.* (8) suggested that a longer peptide may be present as a mature peptide in the chicken brain. Subsequently, the isolated novel peptide was shown to be located in the quail hypothalamo-hypophysial system and to decrease gonadotropin release from cultured anterior pituitary in a dose-dependent manner (22). We therefore designated this novel RFamide peptide as gonadotropin-inhibitory hormone (GnIH) (22).

Localization of GnIH peptide in the brain

The localization of GnIH in the brain is essential to understand its actions. As a first step, we dissected the quail brain into several regions, and quantified the concentration of GnIH by ELISA using a rabbit polyclonal antibody raised against quail GnIH (22). The concentration of GnIH in the diencephalon was much higher than that in the mesencephalon. In contrast, GnIH concentrations in the cerebrum and cerebellum were below the level of detectability. Subsequently, we investigated the precise localization of GnIH in the quail brain by immunohistochemistry (22, 23, 25). Clusters of GnIH-immunoreactive

(-ir) neurons were found in the paraventricular nucleus (PVN) in the hypothalamus. In addition to the PVN, some scattered small cells were immunoreactive in the septal area. GnIH-ir nerve fibers were widely distributed in the diencephalic and mesencephalic regions particularly in the ventral paleostriatum, septal area, preoptic area, hypothalamus, and optic tectum. The most prominent fibers were seen in the median eminence (ME) of the hypothalamus, and in the dorsal motor nucleus of the vagus in the medulla oblongata.

We further investigated GnIH localization in the brain of seasonally-breeding songbird species (4, 16). Dense populations of GnIH-ir neurons were also found in the PVN of song sparrows (*Melospiza melodia*), house sparrows (*Passer domesticus*), white-winged crossbills (*Loxia leucoptera*), pine siskins (*Carduelis pinus*), redpolls (*Carduelis flammea*), rufous collared sparrows (*Zonotrichia capensis*) and Gambel's white-crowned sparrows (*Zonotrichia leucophrys gambelii*). The PVN was the only location where immunoreactive neurons were located, regardless of sex or species Data are only published for 3 species in this list (4, 16). Thus the presence of GnIH in the PVN appears to be a conserved property among several families and at least two orders of birds (Galliformes and Passeriformes). In addition to the dense population of GnIH-ir neurons within the hypothalamus of all the avian species studied so far, there were extensive networks of branching beaded fibers emanating from those cells, presumably transporting GnIH. Some of the fibers extended to terminals in the ME, consistent with a role for GnIH in pituitary gonadotropin regulation. In house sparrows, song sparrows and Gambel's white-crowned sparrows, other fibers extended through the brain caudally at least as far as the brainstem and possibly into the spinal cord, consistent with multiple regulatory roles for GnIH.

Neuroanatomical distribution of GnIH and GnRH

The cell bodies for each population of neurons containing chicken GnRH-I (cGnRH-I), GnIH or cGnRH-II are in discrete locations. The preoptic area (POA) contains cGnRH-I-ir neurons, the PVN contains GnIH-ir neurons and the cGnRH-II neurons are located in the midbrain (13). Thus we can confidently distinguish between two forms of GnRH based upon their location in the brain and their appearance. Closer inspection of each area indicates close proximity of GnIH fibers to the cGnRH-I neurons and fibers in the POA in songbirds. The PVN also contains cGnRH-I fibers which pass directly through and in close proximity to the population of GnIH neurons and fibers as they project to the ME. GnIH-ir fibers are also in close proximity to cGnRH-II neurons in the midbrain (4). Further, confocal microscopy indicates that the cGnRH-I and GnIH peptides are located within the same 0.2 micron optical plane and possibly in contact with one another, although electron microscopy will be necessary to determine contact conclusively. Contact appears to occur between GnIH fibers and GnRH-I and –II neurons, and between GnIH and GnRH-I fibers in the ME (4).

Taken together, there is potential for GnIH to influence the GnRH system at the neuron and fiber terminal levels. Furthermore, when song sparrows were subjected to a simulated annual cycle of changing photoperiod, GnIH-ir neuron area was significantly greater at the onset of photorefractoriness (long days) when compared to photosensitive (short days) or photostimulated (long days) birds (4). Thus there is potential for dynamic interactions of GnIH and GnRH peptides in different reproductive states and neuroanatomical locations. It is clear that further study is needed to elucidate the seasonal dynamics of GnIH synthesis and release and its relation to seasonal changes in GnRH in songbirds.

Identification of GnIH precursor cDNA and localization of its transcript

A cDNA that encoded the GnIH precursor polypeptide was identified in the quail brain using a combination of 3' and 5' rapid amplification of cDNA ends (21). The deduced GnIH precursor consisted of 173 amino acid residues that encoded one GnIH and two putative GnIH-related peptide (GnIH-RP-1 and -RP-2) sequences that included -LPXRF (X = L or Q) at their C-termini (see Fig. 1). All these peptide sequences were flanked by a glycine C-terminal amidation signal and a single basic amino acid on each end serve an endoproteolytic site.

We also cloned a cDNA that encoded GnIH in the brain of Gambel's white-crowned sparrow (16). The deduced sparrow GnIH precursor also consisted of 173 amino acid residues, encoding one sparrow GnIH and two sparrow GnIH-related peptides (sparrow GnIH-RP-1 and GnIH-RP-2) that included -LPXRFamide (X = L or Q) at their C-termini (see Fig. 1). Although the homology of sparrow and quail GnIH precursors was approximately 66%, the C-terminal structures of GnIH, GnIH-RP-1 and GnIH-RP-2 were all identical in two species (16, 21). Subsequently, a cDNA encoding GnIH and GnIH-RPs was also reported in the chicken from a gene database (see Fig. 1).

In situ hybridization further revealed the cellular localization of GnIH mRNA solely in the PVN of quail and sparrow hypothalami (16, 25). As already discussed, immunohistochemical analysis using the quail and sparrow also showed that quail and sparrow GnIH-ir cell bodies and terminals were localized in the PVN and median eminence, respectively. Thus only the PVN expresses GnIH and, in birds, the immunoreactive peptide found in fibers in multiple brain areas appears to originate from the PVN only (16, 25).

```
Quail   MEIISTQKFILLTLATVAFLTPHGACLDELMKSSLESREDDDDKYYET 48
Sparrow MKVISTKKFILFALATVVFLTSNSMCLNEAMKSRLQSREDNDDKDYEI 48
Chicken MEVISTQKFILLTLATVAFLTPHGMCLDELMKSSLESREEDDDKYYEI 48

Quail   KDSILEEKQRSLNFEEMKDWGSKNFMKVNTPTVNKVPNSVANLPLRFG 96
Sparrow KDNILEEKQRSLNFEEMEDWGSKDIIKMNPFTASKMPNSVANLPLRFG 96
Chicken KDSILEEKQRSLNFEEMKDWGSKNFLKVNTPTVNKVPNSVANLPLRFG 96

                         GnIH-RP-1

Quail   RSNPEERSIKPSAYLPLRFGRAFGESLSRRAPNLSNRSGRSPLARSSI 144
Sparrow RNYPEERSIKPFSNLPLRFGRAFGENTPNHSPRVSHRLERSPLVKGSS 144
Chicken RSNPEERSIRPSAYLPLRFGRAFGESLSRRAPNLSYRLGRSPLARSSI 144

               GnIH

Quail   QSLLNLPQRFGKSVPISLSQGVQESEPGM                    173
Sparrow QSLLNLPQRFGKSLAVNLPRDIQEFEPGI                    173
Chicken QSLLNLPQRFGKSVPINLSQGVQESEPGM                    173

   GnIH-RP-2
```

Fig. 1 Comparison of the structure of GnIH precursor polypeptide in Gambel's white-crowned sparrow (16), quail (21) and chicken. (Chicken GnIH precursor polypeptide is cited from GenBank, accession numbers AB120325, BAC87781). Putative sequences of GnIH, GnIH-RP-1 and GnIH-RP-2 are boxed. Shaded residues indicate mature GnIH (SIKPSAYLPLRF-NH_2) and GnIH-RP-2 (SSIQSLLNLPQRF-NH_2,) first identified in quail (21, 22). The glycine preceding the C-terminal cleavage site may serve as a C-terminal amidation signal.

Actions of GnIH on gonadotropin release and synthesis in domesticated birds

In view of the immunohistochemical finding indicating that GnIH-ir neurons project to the ME close to the pituitary, we analyzed the effect of SIKPSAYLPLRFamide, GnIH, on the release of luteinizing hormone (LH), follicle-stimulating hormone (FSH) and prolactin (PRL) using cultured quail anterior pituitaries (22). GnIH significantly inhibited LH release, after 100-min incubation. The inhibitory effect on LH release was dose-dependent and its threshold concentration ranged between 10^{-9} and 10^{-8}M. A possible similar tendency for GnIH to inhibit FSH release was also detected. However, there was no effect of GnIH on PRL release. Based on these results, we therefore named this novel RFamide peptide isolated from the quail brain GnIH (22). In addition to the *in vitro* effects upon pituitary LH release, we also showed that GnIH was effective in inhibiting circulating LH *in vivo*. When administered intraperitoneally to quail via osmotic pumps at the doses of 0.3 and 3.0μg per h, GnIH significantly reduced plasma LH (Ubuka *et al.*, unpublished observation).

In addition to the inhibitory effects of GnIH on gonadotropin release, there is evidence that this novel RFamide inhibits gonadotropin biosynthesis *in vitro*. Addition of a physiological dose (10^{-7}M) of GnIH to short-term (120 min) cultures of diced pituitary glands from adult cockerels suppressed common α and FSH β subunit mRNAs, with no effect on LH β subunit mRNA. The suppressive effect of GnIH on gonadotropin mRNA was associated with an inhibition of both LH and FSH release in the adult chicken. It remains to be established whether this suppressive effect of GnIH is a consequence of an inhibition of common α and FSH β subunit gene transcription, or of decreased stability of the gonadotropin subunit mRNAs. The possibility that GnIH also suppresses LH β mRNA cannot be excluded because the half life of the chicken LH β mRNA is longer than 9h (Ciccone, unpublished observation) and, therefore, considerably longer than the 120 min *in vitro* experiment. Further, mammalian LH β mRNA half life is estimated to be around 40h (5). At a higher dose (10^{-5}M) GnIH failed to suppress either LH or FSH release *in vitro*. This observation can possibly be ascribed to a desensitization of the gonadotropes. By analogy, the phenomenon of desensitization has been established for the response of the chicken (11) and mammalian (27) gonadotrope to continuous GnRH-I exposure.

Seasonal dynamics and functional significance of GnIH in songbirds

Once we had identified the white-crowned sparrow GnIH precursor, we were able to synthesize putative white-crowned sparrow GnIH ("sparrow" GnIH) with the deduced amino acid sequence SIKPFSNLPLRF-NH_2. We used this peptide for *in vivo* analysis of gonadotropin release in field-caught white-crowned sparrows during their breeding season in northern Alaska (68°37'N, 149°36'W). Based on laboratory experiments (see 16), we used sparrow GnIH at doses of 500ng (low dose) and 1000ng (high dose) per 20μl total injection volume. Birds were captured in mist nets using conspecific song playback. They were removed from the nets and saline or GnIH of either dose was injected into the jugular vein within 3 min of capture. Blood was with drawn at 2 and 10 min post-injection and assayed for LH. Birds that were injected with either dose of GnIH had lower plasma LH at 2 min than the saline-injected group (16). In contrast, saline-injected birds had high plasma LH (approx 5ng/ml) 2 min post-injection. At 10 min, plasma LH in the saline group showed no significant change from the 2 min values but was not different from the GnIH-injected birds.

Taken together, our results indicate that, despite amino acid sequence differences, sparrow GnIH and quail GnIH (16) have similar inhibitory effects on the reproductive axis in wild sparrow species. Overall, GnIH appears to be a modulator of gonadotropin release *in vivo* as well as *in vitro*. Taking the immunohistochemical data into consideration as well, there is potential for GnIH to act on the reproductive axis at the level of the GnRH neurons, the ME, and the pituitary, and thus possibly over different time-frames.

Regulation of GnIH expression in the brain

To understand the physiological role of GnIH in avian reproduction, we need data on the changes in GnIH expression during development and under different physiological conditions. Therefore we recently, characterized developmental changes in GnIH expression in the quail hypothalamo-hypophysial system. Our data indicated that, as appears to be the case for GnRH (14), GnIH begins its function around hatch and acts as a hypothalamic factor to regulate gonadotropin release in quail (23). Furthermore, it is possible that GnIH neurons have similar developmental origins and migrate during development in a similar way to GnRH neurons, as Yamamoto *et al.* (29) have suggested that FMRFamide-immunoreactive neurons also originate in the olfactory placode and migrate into the brain.

Regulatory mechanism(s) governing GnIH expression are now being investigated. Although many bird species are photoperiodic, a dogma has existed that birds do not use seasonal changes in melatonin secretion to time their reproductive effort, and a role for melatonin in birds has remained enigmatic (10, 28). Despite the accepted dogma, there is strong evidence that melatonin is involved in regulation of several seasonal processes, including gonadal activity and gonadotropin secretion (1, 2, 3, 9, 15, 20). In light of these reports and considering GnIH's inhibitory effects on gonadotropin secretion (16, 22), we investigated the action of melatonin on GnIH expression in the quail brain (24). Pinealectomy combined with orbital enucleation (Px+Ex) decreased the expression of GnIH precursor mRNA and the mature GnIH peptide in the diencephalon including the PVN and ME. Melatonin administration to Px+Ex birds caused a dose-dependent increase in expression of GnIH precursor mRNA and production of mature peptide. The expression of GnIH was photoperiodically controlled and increased under short day (SD) photoperiods (24), when the duration of melatonin secretion increases (6, 12). Finally, Mel_{1c}, a melatonin receptor subtype was found to be expressed in GnIH-ir neurons in the PVN (24). Thus melatonin appears to act directly on GnIH neurons via its receptor to induce GnIH expression.

Conclusions and future directions

Probing for undiscovered neuropeptides that play important roles in the regulation of pituitary function in birds is essential for the progress of avian endocrinology. The discovery of GnIH has opened avenues for a new research field in avian endocrinology. A gonadotropin inhibitory system such as GnIH provides us with an unprecedented opportunity to study the regulation of avian reproduction from an entirely novel standpoint. Future directions should focus on the mode of action of GnIH on the inhibition of gonadotropin release. We also need to clarify the action of GnIH-RPs on the release of pituitary hormones. Furthermore, the widespread distribution of GnIH and GnIH-RPs indicates that these peptides may participate in neuroendocrine, behavioral, autonomic and sensory functions in birds. Future studies will bring to light previously unknown physiological functions of GnIH and GnIH-RPs.

Acknowledgements

We are grateful to all our collaborators, particularly Drs. K Ukena, T Ubuka, T Osugi, S O'Brien, IT Moore, IC Dunn, T Boswell, E Saigoh, M Kikuchi, H Satake, H Minakata, S Ishii, PJ Sharp and JC Wingfield. This work was supported in part by Grants-in-Aid for Scientific Research from the Ministry of Education, Science and Culture, Japan (13210101, 15207007 and 16086206), NIH, USA (RO1 MH065974) and BBSRC/CASE (Cobb-Vantress) studentship, UK.

References

1 Bentley GE (2001) Unraveling the enigma: the role of melatonin in seasonal processes in birds. *Microscopy Research and Technique* **53** 63-71

2 Bentley GE & Ball GF (2000) Photoperiod-dependent and -independent regulation of melatonin receptors in the forebrain of songbirds. *Journal of Neuroendocrinology* **12** 745-752

3 Bentley GE, Van't Hof TJ & Ball GF (1999) Seasonal neuroplasticity in the songbird telencephalon: a role for melatonin. *Proceedings of the National Academy of Sciences of the United States of America* **13** 4674-4679

4 Bentley GE, Perfito N, Ukena K, Tsutsui K & Wingfield JC (2003) Gonadotropin-inhibitory peptide in song sparrows (*Melospiza melodia*) in

different reproductive conditions, and in house sparrows (*Passer domesticus*) relative to chicken-gonadotropin-releasing hormone. *Journal of Neuroendocrinology* **15** 794-802

5 Bouamoud N, Lerrant Y, Ribot G & Counis R (1992) Differential stability of mRNAs coding for α and gonadotrophin β subunits in cultured rat pituitary cells. *Molecular and Cellular Endocrinology* **88** 143-151

6 Cockrem JF & Follett BK (1985) Circadian rhythm of melatonin in the pineal gland of the Japanese quail (*Coturnix coturnix japonica*). *Journal of Endocrinology* **107** 317-324

7 Dockray GJ, Reeve JR Jr, Shively J, Gayton RJ & Barnard CS (1983) A novel active pentapeptide from chicken brain identified by antibodies to FMRFamide. *Nature* **305** 38-330

8 Dockray GJ & Dimaline R (1985) FMRFamide- and gastrin/CCK-like peptides in birds. *Peptides* **3** 333-337

9 Guyomarc'h C, Lumineau S, Vivien-Roels B, Richard J & Deregnaucourt S (2001) Effect of melatonin supplementation on the sexual development in European quail (*Coturnix coturnix*). *Behavioural Processes* **53** 121-130

10 Juss TS, Meddle SL, Servant RS & King VM (1993) Melatonin and photoperiodic time measurement in Japanese quail (*Coturnix coturnix japonica*). *Proceedings of the Royal Society of London Series B, Biological Sciences* **254** 21-28

11 King JA, Davidson JS & Millar RP (1986) Desensitisation to gonadotrophin releasing hormone in perifused chicken anterior pituitary cells. *Endocrinology* **119** 1510-1518

12 Kumar V & Follett BK (1993) The circadian nature of melatonin secretion in Japanese quail (*Coturnix coturnix japonica*). *Journal of Pineal Research* **14** 192-200

13 Millam JR, Faris PL, Youngren OM, el Halawani ME & Hartman BK (1993) Immunohistochemical localization of chicken gonadotropin-releasing hormones I and II (cGnRH-I and -II) in turkey hen brain. *Journal of Comparative Neurology* **333** 68-82

14 Millam JR, Craig-Veit CB & Petitte JN (1993) Brain content of cGnRH I and II during embryonic development in chickens. *General and Comparative Endocrinology* 92:311-317

15 Ohta M, Kadota C & Konishi H (1989) A role of melatonin in the initial stage of photoperiodism in the Japanese quail. *Biology of Reproduction* **40** 935-941

16 Osugi T, Ukena K, Bentley GE, O'Brien S, Moore IT, Wingfield JC & Tsutsui K (2004) Gonadotropin-inhibitory hormone in Gambel's white-crowned sparrows: cDNA identification, transcript localization and functional effects in laboratory and field experiments. *Journal of Endocrinology* **182** 33-42

17 Price DA & Greenberg MJ (1977) Structure of a molluscan cardioexcitatory neuropeptide. *Science* **197** 670-671

18 Raffa RB (1988) The action of FMRFamide (Phe-Met-Arg-Phe-NH_2) and related peptides on mammals. *Peptides* **9** 915-922

19 Rastogi RK, D'Aniello B, Pinelli C, Fiorentino M, Di Fiore MM, Di Meglio M & Iela L (2001) FMRFamide in the amphibian brain: a comprehensive

survey. *Microscopy Research and Technique* **54** 158-172

20 Rozenboim I, Aharony T & Yahav S (2002) The effect of melatonin administration on circulating plasma luteinizing hormone concentration in castrated White Leghorn roosters. *Poultry Science* **81** 1354-1359

21 Satake H, Hisada M, Kawada T, Minakata H, Ukena K & Tsutsui K (2001) Characterization of a cDNA encoding a novel avian hypothalamic neuropeptide exerting an inhibitory effect on gonadotropin release. *Biochemical Journal* **354** 379-385

22 Tsutsui K, Saigoh E, Ukena K, Teranishi H, Fujisawa Y, Kikuchi M, Ishii S & Sharp PJ (2000) A novel avian hypothalamic peptide inhibiting gonadotropin release. *Biochemical and Biophysical Research Communications* **275** 661-667

23 Ubuka T, Ueno M, Ukena K & Tsutsui K (2003) Developmental changes in gonadotropin-inhibitory hormone in the Japanese quail (*Coturnix japonica*) hypothalamo-hypophysial system. *Journal of Endocrinology* **178** 311-318

24 Ubuka T, Bentley GE, Ukena K, Wingfield JC & Tsutsui K (2005) Melatonin induces the expression of gonadotropin-inhibitory hormone in the avian brain. *Proceedings of the National Academy of Sciences of the United States of America* In press

25 Ukena K, Ubuka T & Tsutsui K (2003) Distribution of a novel avian gonadotropin-inhibitory hormone in the quail brain. *Cell and Tissue Research* **312** 73-79

26 Underwood H, Binkley S, Siopes T & Mosher K (1984) Melatonin rhythms in the eyes, pineal bodies, and blood of Japanese quail (*Coturnix coturnix japonica*). *General and Comparative Endocrinology* **56** 70-81

27 Weiss J, Jameson L, Burrin JM & Crowley WF (1990) Divergent responses of gonadotropin subunit messenger RNAs to continuous versus pulsatile gonadotrophin-releasing hormone in vitro. *Molecular Endocrinology* **4** 557-564

28 Wilson FE (1991) Neither retinal nor pineal photoreceptors mediate photoperiodic control of seasonal reproduction in American tree sparrows (*Spizella arborea*). *Journal of Experimental Zoology* **259** 117-127

29 Yamamoto N, Uchiyama H, Ohki-Hamazaki H, Tanaka H & Ito H (1996) Migration of GnRH-immunoreactive neurons from the olfactory placode to the brain: a study using avian embryonic chimeras. *Developmental Brain Research* **95** 234-244

Functional Avian Endocrinology
A Dawson & P J Sharp (Eds)

Regulation of FSH secretion: GnRH independent?

J D Kirby, J A Vizcarra[1], L R Berghman[2], J A Proudman[3], J Yang[4] and C G Scanes[5]

Department of Poultry Science, University of Arkansas, Fayetteville, AR 72701
[1]Animal and Food Sciences, Texas Tech University, Lubbock, Texas 79409
[2]Department of Poultry Science, Texas A&M University, College Station, Texas 77843
[3]Biotechnology and Germplasm Laboratory, USDA/ARS, Beltsville, MD20705
[4]Department of Obstetrics and Gynecology, University of Texas Southwestern Medical Center, Dallas, Texas 75390
[5]Office of Research, Mississippi State University, Mississippi State, MS 39762

Introduction

Selection for increased growth rate and yield over the past 50 + years has dramatically altered the "physiology" of the modern broiler breeder chicken. One consequence of the dramatic improvements in growth rate has been a tremendous reduction in the broiler breeder's reproductive capacity-numbers of eggs produced and fertility (10). This period of genetic selection has also resulted in a reduction in the quality and numbers of viable sperm produced and in the overall duration of fertility in broiler breeders (25). Other consequences of this selection include the requirement for quantitative feed restriction of breeders (both male and female) and a significant reduction in the reproductive "life-expectancy" of breeders. In commercial flocks, it is very common for fertility to begin to decline immediately following "peak" egg production (2). As a result, producers are frequently bringing new, younger, males into their flock, a process known as spiking, to rescue fertility. Based on our observations, we propose that a primary cause for the decline in fertility observed in post-peak production flocks is the loss of the robust pattern of episodic release of FSH and LH by the pituitary (44).

Specific radioimmunoassays for chicken and/or other avian follicle stimulating hormone (FSH) have been developed (11, 26). Despite this, and the sequencing of chicken FSHβ sub-unit cDNA (42) and gene (Yang *et al.* [GenBank Accession number AF467082]), there is a general paucity of information regarding FSH synthesis and release in birds, including chickens. While GnRH has a clear and demonstrable role in the regulation of FSH synthesis and secretion in mammals, the role of GnRH, if any, in regulating FSH in the fowl is less well defined.

Role of FSH

As might be expected, there are marked effects of FSH on gonadal functioning in poultry. For instance, FSH can stimulate proliferation of granulosa cells (7) and is required for the long term culture of granulosa cells (41). Moreover, FSH plays an important role in the differentiation of granulosa cells; stimulating progesterone production and expression of steroidogenic acute regulatory protein and P450side chain cleavage in granulosa cells from small yellow follicles (21). Moreover, FSH enhances expression of inhibin alpha and beta subunits *in vitro* (7, 28, 40). In mammals, FSH signaling is essential for initiation and maintenance of spermatogenesis and high production of spermatozoa due to intimate contacts between Sertoli cells and differentiating spermatogonia (8). Sertoli cells exclusively express FSH-R. There is evidence that a similar situation exists in poultry (18).

Information from a longitudinal study revealed that males were failing to maintain testis size and semen production, as shown in Fig. 1.

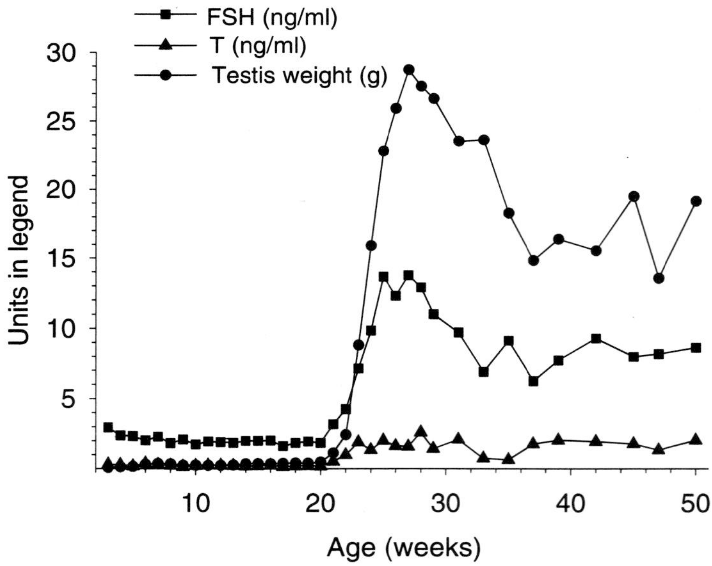

Fig. 1 Effects of photostimulation (14L:10D) at 20 weeks on circulating FSH and testosterone levels and testis weight in broiler breeder males. Ten birds were killed weekly from one week of age until 50 weeks of age. (Data from Vizcarra *et al.*, submitted)

Note that testis size closely follows changes in FSH levels, increasing in mass with increased FSH and decreasing in mass with reduced FSH, with testosterone showing little change over time. The cause for concern associated with declining testis weights is shown in Fig. 2. Testis weights in the fowl are highly correlated to daily sperm production (DSP), with an R^2 of 0.89 (Vizcarra *et al.*, submitted). Thus, there is a direct and easily

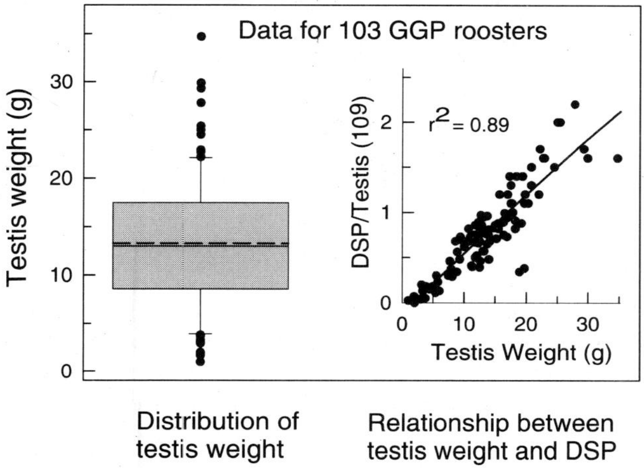

Fig. 2 Distribution of testis weights (box and whisker plot, left panel) and the relationship between daily sperm production (DSP) and testis weight (right panel) in a population of 103 great grandparent stock of broiler breeders. The highly predictive nature of mass for sperm production (r^2=0.89) is shown. (Data from Vizcarra et al, submitted, figure provided by R.P. Ammann)

measurable relationship between testis size and reproductive status. Further, we have shown that the efficiency of spermatogenesis (DSP/gram of testis) reaches a peak at about 11 grams, and that those testes less than 11 grams producing fewer sperm per gram of testis per day are more likely to produce abnormal sperm. Circulating FSH levels are a very good indicator of testis function in mature (28 weeks of age and older) broiler breeder males (Fig. 3). Based on this experiment, with a regression R^2 of 0.81, and several other replicates with regression R^2 values of 0.75 to 0.9, we have concluded that FSH is critical, not just in the initiation of spermatogenesis at puberty, but also for the long term maintenance of quantitatively normal spermatogenesis in mature males, a critical component for the maintenance of adequate fertility levels.

The effects of reduced FSH levels on testis function and histology in mature males are shown in Fig. 4. In Panel A, a seminiferous tubule from a normal male (testis ~20g, FSH ~12ng/ml) is shown. Notice the maturing elongated spermatids in the lumen of the tubule. In Panel B, a seminiferous tubule of a hypogonadotropic male (testis~6g, FSH <6ng/ml) is shown. In

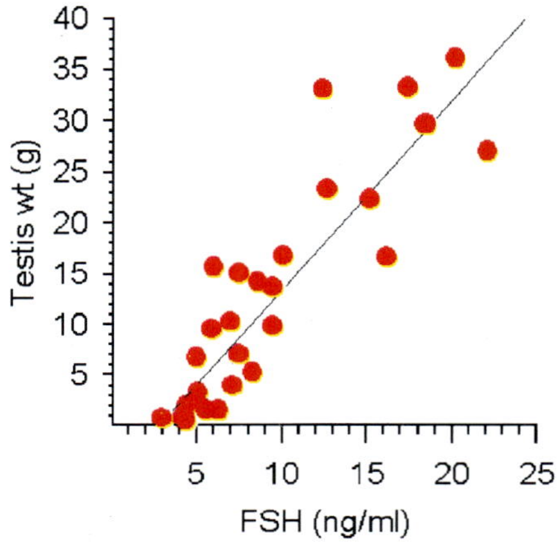

Fig. 3 The relationship between testis weight and plasma FSH levels in 28 week old broiler breeder males. This relationship is supported by an r^2 of 0.81 for the regression line shown. Circulating FSH levels are highly predictive of testis weight in mature broiler breeders. (Data from Vizcarra *et al.*, submitted).

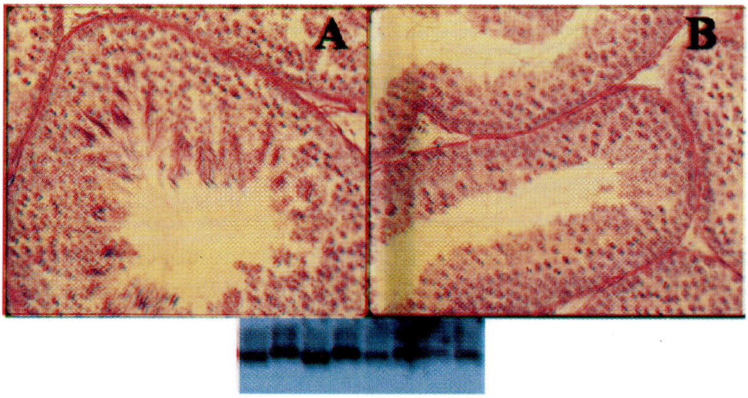

Fig. 4 Cross sections of seminiferous tubules from the testes of (A) a normal FSH male (FSH >12 ng/ml plasma, testis weight >20 g) and (B) of a low FSH male (FSH <6ng/ml plasma, testis weight <6g). The lower panel shows a northern blot of testis RNAs from 4 normal (left 4 lanes) and 4 low FSH (right 4 lanes) probed with a chicken FSH receptor cDNA probe. Following normalization for 28s rRNA binding, normal FSH males had 2.5 X the level of FSHr mRNA as seen in low FSH males. (Data from 18).

the low FSH male, very few elongated spermatids are observed. Thus, even though spermatids from these males would be counted using standard estimators of DSP, due to the presence of spermatids with condensed nuclei, very few mature, functional sperm are evolved into the lumen of the seminiferous tubules. One additional point is that FSH receptor mRNA expression is 2.5 fold higher in the testes of normal males than in low FSH males (18).

FSH and LH are secreted in asynchronous pulses, with pulse frequencies of 0.33 and 0.54 per hour, respectively (44). The relationship between reproductive status and the pulsatility of FSH and LH has received only limited attention at this time. However, the patterns of LH and testosterone, sampled at 10 min intervals for 12 h, were evaluated in a low FSH male (baseline FSH <6.7ng/ml with no observed pulses and left testis weight of 6.3g) and from a normal male (baseline FSH >10.5ng/ml with 4 FSH pulses and a left testis weight of 18.5g). This showed firstly, a loss of pulsatile testosterone secretion in response to LH pulses in the low FSH male; and secondly, a significant decline in both the baseline and amplitude of LH pulses, with no apparent reduction in LH pulse frequency in the absence of FSH pulses (44).

Sulfamethazine, which alters hypothalamic function, has been shown to accelerate puberty in male fowl by rapidly increasinggonadotropin levels in male fowl (27). The effects of 6 weeks of SMZ treatment on low FSH males are shown in Fig. 5. The SMZ-treated males had significantly ($p<0.05$) elevated FSH levels and daily sperm production, with small increases in LH and testosterone (not shown). These data further suggestthat the causes of reduced testis function in broiler breeder males are centrally mediated, and can be reversed by increasing circulating FSH levels.

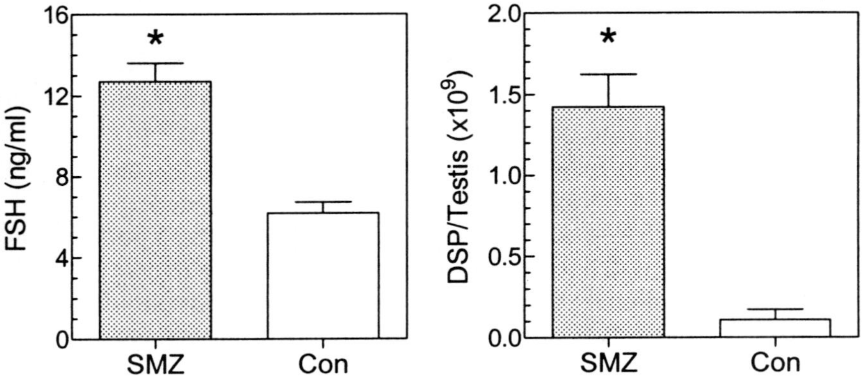

Fig. 5 Effects of dietary supplementation with 0.2% sulfamethazine (SMZ) on plasma FSH (a) and daily sperm production (DSP) (b) in caged, low FSH, broiler breeder males reared on full feed from hatch until seven weeks of age. Males were provided with SMZ in their diets for eight weeks starting at photostimulation at 20 weeks of age. (Data from Kirby and Vizcarra, unpublished).

Hypothalamic control of FSH synthesis and release

It is not well established how the hypothalamus controls FSH synthesis or release. The profile of changes in plasma FSH during growth and sexual maturation has been reported in both male and female chickens (20; and Fig. 1) as have effects of feed restriction (3). Moreover, despite the failure of chicken gonadotropin releasing hormone-I (cGnRH-I) to elevate plasma concentrations of FSH, photostimulation acting *via* the CNS increases plasma FSH and pituitary FSH in young chickens (9). To date, three members of this decapeptide family have been localized in birds; they include cLHRH-I (cGnRH-I, 43, 32), cLHRH-II (cGnRH-II, 43, 32) and lamprey GnRH-III (lGnRH-III) (1). What is not well characterized are their effects on FSH release and synthesis. Independent control of chicken LH and FSH is supported by the dis-synchronous pulsatile release of the two gonadotropins (44). Moreover, unlike the situation in mammals where LH and FSH are synthesized in the same cell but packaged into different secretory granules (31), it has been shown that LH and FSH are produced in different cells in the chicken (23, 24). To date, the hypothalamic mechanisms mediating changes in the circulating concentrations of FSH following interventions such as photoperiodic stimulation or castration are not clear. While in mammals, there is limited or no cell division of gonadotropes in the adult pituitary (reviewed in 31), this does not preclude there being changes in FSH cell number or the rate of FSH synthesis per cell in poultry.

Chicken GnRH-I and -II

It is well established that gonadotropin releasing hormone (GnRH) can stimulate the release of both LH and FSH in mammals. There are at least two decapeptide releasing hormones for LH found in the chicken hypothalamus; namely cGnRH-I and cGnRH-II (23, 24, 33). There is controversy as both to whether GnRH-I and/or GnRH-II can stimulate release of FSH in birds and whether cGnRH-II has a physiological role in the control of either gonadotropin. Mammalian GnRH and both chicken GnRH-I and GnRH-II have been reported to stimulate the release of LH and FSH from quail pituitary tissue *in vitro* and *in vivo* (15, 16). Similarly, both cGnRH-I and cGnRH-II stimulate FSH secretion by chicken pituitary cells *in vitro* (33). However, no effect of mammalian GnRH on FSH release was seen with turkey pituitary tissue *in vitro* (13). The reports on FSH responses to cGnRH-I or cGnRH-II challenges *in vivo* are equivocal. Challenging with cGnRH-I failed to evoke an increase in circulating concentrations of FSH in photostimulated young chickens in one study (9) and stimulated FSHrelease only in ovariectomized (not intact) chickens in another (3). No studies have explored the mechanisms by which photoperiodic stimulation or castration increase circulating FSH and/or pituitary FSH mRNA.

Control of LH and FSH secretion

In mammals, both LH and FSH are produced in the same gonadotroph. The consensus view (for reviews see 31, 36) is that LH is secreted both constitutively and in response to GnRH. Pulsatile or surge secretion (by GnRH stimulation) occurs *via* release of LH secretory granules. However, FSH is released largely constitutively, with the rate of release tightly coupled to synthesis. Synthesis, in turn, is regulated by negative feedback by inhibin and estradiol. Thus release of FSH is predominantly independent of pulsatile GnRH input to the pituitary gland. There is clear evidence that FSH secretory pulses in mammals are less regular than LH pulses (e.g. 39). The FSH pulses are not necessarily associated with GnRH, as for example LH pulses are abolished by blocking GnRH action while FSH pulses persist (35). Endogenous GnRH release is synchronized with LH but not necessarily FSH secretory pulses (35, 37). This position is also indicated in the male chicken due to the lack of coordinated LH and FSH pulses, with FSH pulses occurring at approximately ½ the frequency of LH pulses (44). Further, immunization of roosters against cGnRH-I or cGnRH-II led to declines in LH, but had no effect on either LH or FSH pulse frequency or on circulating FSH levels (Vizcarra and Kirby, unpublished).

Regulation of FSH secretion may be dependent on other factors, such as inhibin and activin. There is evidence for linkage between follistatin/activin and FSH pulses (34). However, it is not possible to preclude the existence of a hypothalamic releasing hormone for FSH distinct from GnRH (reviewed in 38). For instance, lesioning studies support the notion that there are separate areas within the hypothalamus controlling LH and FSH secretion (29). Indeed, McCann and Yu have proposed that lamprey GnRH-III is the endogenous FSH releasing hormone in mammals (50, 30).

Is cGnRH-II a physiological releasing hormone for LH and/or FSH?

Chicken GnRH-II is more potent than cGnRH-I in stimulating LH release *in vitro* (chickens: 33, 4; turkeys: 14). Similarly cGnRH-II is reported to be more than ten fold more potent than cGnRH-I in stimulating FSH release *in vitro* (33). *In vivo*, cGnRH-I and cGnRH-II have either similar activities in stimulating LH release (4) or cGnRH-II is more potent (46). There is strong evidence that cGnRH-I is (a) present at substantial levels in the avian median eminence (32), (b) found in concentrations associated with reproductive state (41-43) and (c) is released from the median eminence in response to the stimulation with potassium (32, 22). It is, however, not clear whether cGnRH-II has a physiological role in the control of LH and/or FSH release. There is cGnRH-II in the median eminence (1, 22) albeit at very low concentrations in some studies (32) but higher levels in others (5). No release detectable by radioimmunoassay has been seen for cGnRH-II from avian median eminence (32). There are, however, changes in hypothalamic nuclei contents of GnRH-II with reproductive state in some studies (45, 48) but not others (47). Based on studies with antagonists to mammalian GnRH, it was concluded that there is only a single class of receptor for cGnRH-I and cGnRH-II (33). However, it is not known whether this receptor is present on both LH and FSH producing cells in the chicken or if a second class of GnRH receptor may be present. In mammals, it is unlikely that the GnRH-II (acting *via* the GnRH-II receptor) is an endogenous regulator of LH release from the gonadotroph (12).

Is there a separate releasing hormone for FSH in birds?

It was recently proposed that lGnRH-III may be the distinct releasing factor for FSH (30, 17, 49). This peptide is present in the chicken hypothalamus (see Fig. 6; 1). What is not known is whether lGnRH-III can be an endogenous antagonist for cGnRH-I or cGnRH-II stimulation of the release of LH and/or FSH. Moreover, it is not known whether the median eminence releases activin and/or its binding protein, follistatin.

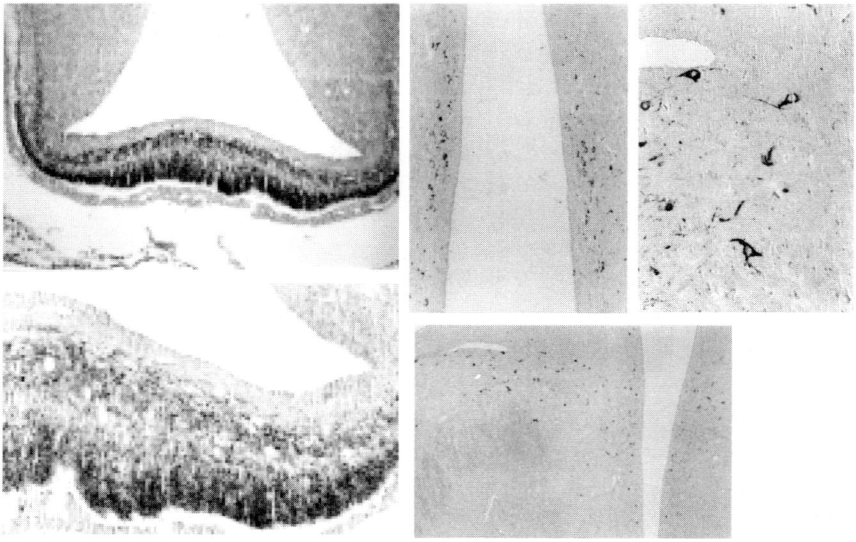

Fig. 6 Lamprey GnRH immunopositive fibers in the median eminence of the chicken (left panel) and in the cell bodies in the paraventricular nucleus (right panel upper left) and more dorsolaterally (right panel upper right) in the hypothalamus of a colchicin-treated chicken.. (From reference 1)

Characterization of the FSHβ gene and promoter

The chicken FSHβ gene and approximately 5kb of sequence proximal to the transcription initiation site was recently cloned and sequenced, and the gene has been mapped (Yang *et al*, submitted). The chicken FSHβ gene consists of three exons and two introns and is a relatively compact gene of approximately 4.2 kb. The proximal promoter region contains a CCAAT box at -100 to -95 relative to the transcription initiation site, and has a number of potential regulatory elements (AP-1, SF-1 and others) dispersed over the ~5kb proximal region. However, the distribution of these potential regulatory elements is distinct from that observed in the well characterized promoters of sheep and mice (6). In mammals, the tissue specific regulatory elements of FSH secretion appear to be unique from those that are regulated by either GnRH or activin/inhibin/follistatin. Further, while GnRH appears to be a primary regulator of FSH synthesis in mammals, very little is known of FSH regulation in birds.

As discussed earlier, regulation of FSH secretion appears somewhat complex in mammals and has been greatly facilitated through the detailed study of the proximal promoter region. To this end we have cloned and mapped the *FSHβ* gene from the Red Jungle Fowl (GenBank Accession number AF467082). The *FSHβ* gene is approximately 4.2kb with 3 exons and two introns. The first Exon is 47nt, the second 159nt and the third is 2234nt. The mRNA has a 5' untranslated region (UTR) of 48nt, a coding region of 396nt and a 3' UTR of 1997nt and is found in the pituitary. The introns in the chicken *FSHβ* gene are smaller than those in mammals (443nt and 491nt, respectively) and the gene has an overall structure to that seen in mammals and is mapped to portion of Chicken Chromosome 3 with the linked genes of *FSHβ,INS,TH,CAPN1*. This same arrangement of genes was confirmed in the draft chicken genome sequence, and is found in what appears to be a syntenic region to Human Chromosome 11 (HTTP://poultry.mph.msu.edu/resources/rimage/compare.jpg.)

Although the overall structure of the gene appears similar to the mammalian *FSHβ* genes cloned to date, the proximal promoter appears to lack significant similarity to any of the mammalian *FSHβ* promoters which have been studied. The area of greatest interest, from bases -50 to -1,000, due to the recent developments with the mouse (6) and sheep (19) is devoid of the expected fused CAAT/AP1 binding site and thus merits further study. Thus, at least at the level of GnRH regulation of the *FSHβ* promoter, it appears that the regulation of FSH synthesis may be quite distinct in the fowl.

Conclusions

FSH is a heterodimeric glycoprotein hormone produced in the pituitary that has profound effects on male reproductive function. The episodic release of

FSH is readily apparent in the circulation of cannulated and unrestrained males (44). The dynamic pattern of hormone release appears to be necessary for the initiation and maintenance of spermatogenesis in the fowl. However, the actual regulation of FSH secretion remains unclear, as the factors responsible for the physiological control of synthesis and release of this hormone have not been revealed.

The fowl provides a unique model for studying the differential regulation of gonadotropin secretion in that, (a) FSH and LH appear to be produced by separate cells in the pituitary, (b) FSH and LH are both released in a robustly pulsatile fashion, and (c) the promoter region of the FSHβ gene is quite different to that studied in mammals, to date. Based on the fact that FSH and LH pulses are asynchronous, and that *in vivo* GnRH injections have little effect on FSH levels, the role of the GnRH family in regulating FSH remains unresolved. That FSH is a critical component of the reproductive control scheme in males is of little doubt, the control of its synthesis and secretion, while elusive to date, deserves further attention.

References

1 Berghman LR, D'Hondt E, Puebla N, Dees L, Hiney J, Sower S, Vandesande F (2000) Immunocytochemical evidence for the existence of a lamprey luteinizing hormone releasing hormone type III-like peptide in the chicken hypothalamus. *British Poultry Science* **41** S56-57

2 Bramwell RK (2000) Assessing fertility in Broiler Breeder Flocks. Proceedings of the National Poultry Breeders Roundtable, St Louis, Missouri

3 Bruggeman V, Onagbesan O, Vanmontfort D, Berghman L, Verhoeven G, Decuypere E. 1998 Effect of long-term food restriction on pituitary sensitivity to cLHRH-I in broiler breeder females. *Journal of Reproduction and Fertility* **114** 267-276

4 Chou HF, Johnson AL, Williams JB (1985) Luteinizing hormone releasing activity of [Gln8]-LHRH and [His5, Trp7, Tyr8]-LHRH in the cockerel, *in vivo* and *in vitro*. *Life Sciences* **37** 2459-2465

5 Clerens S, D'Hondt E, Berghman LR, Vandesande F, Arckens L. (2003) Identification of cGnRH-II in the median eminence of Japanese quail (*Coturnix coturnix japonica*). *General and Comparative Endocrinology* **131** 48-56

6 Coss D, Jacobs SBR, Bender CE, Mellon PL. (2004) A novel AP1 site is critical for maximal induction of the FSHβ gene by GnRH. *Journal of Biological Chemistry* **279** 152-162

7 Davis AJ, Brooks CF, Johnson PA (2001) Follicle-stimulating hormone regulation of inhibin alpha- and beta(B)-subunit and follistatin messenger ribonucleic acid in cultured avian granulosa cells. *Biology of Reproduction* **64** 100-106

8 Dierich A, Sairam MR, Monaco L, Fimia GM, Gansmuller A, LeMeur M, Sassone-Corsi P (1998) Impairing follicle-stimulating hormone (FSH)

signaling *in vivo*: targeted disruption of the FSH receptor leads to aberrant gametogenesis and hormonal imbalance. *Proceedings of the National Academy of Sciences USA* **95**13612-13617

9 Dunn IC, Lewis PD, Wilson PW, Sharp PJ (2003) Acceleration of maturation of FSH and LH responses to photostimulation in prepubertal domestic hens by oestrogen. *Reproduction* **126** 217-225

10 Etches RJ (1996) Reproduction in Poultry. CAB International: Wallingford Oxon UK

11 Follett BK (1976) Plasma follicle-stimulating hormone during photoperiodically induced sexual maturation in male Japanese quail. *Journal of Endocrinology* **69** 117-126

12 Gault PM, Maudsley S, Lincoln GA (2003) Evidence that gonadotropin-releasing hormone II is not a physiological regulator of gonadotropin secretion in mammals. *Journal of Neuroendocrinology* **15** 831-839

13 Godden, PMM, Luck, MR & Scanes, CG (1977) The effects of luteinizing hormone-releasing hormone and steroids on the release of LH and FSH from isolated turkey pituitary cells. *Acta Endocrinologica* **85** 713-717

14 Guemene D, Williams JB (1992) In-vitro and in-vivo responses to chicken LHRH-I and chicken LHRH-II in male turkeys (*Meleagris gallopavo*) *Endocrinology* **32** 387-393

15 Hattori A, Ishii S, Wada M (1986) Effects of two kinds of chicken luteinizing hormone-releasing hormone (LH-RH), mammalian LH-RH and its analogs on the release of LH and FSH in Japanese quail and chicken. *General and Comparative Endocrinology* 64 446-455

16 Hattori A, Ishii S, Wada M, Miyamoto K, Hasegawa Y, Igarashi M, Sakakibara S (1985) Effects of chicken (Gln8)- and mammalian (Arg8)-luteinizing hormone-releasing hormones on the release of gonadotrophins in vitro and in vivo from the adenohypophysis of Japanese quail. *General and Comparative Endocrinology* **59** 155-161

17 Hiney JK, Sower SA, Yu WH, McCann SM, Dees WL (2002) Gonadotropin-releasing hormone neurons in the preoptic-hypothalamic region of the rat contain lamprey gonadotropin-releasing hormone III, mammalian luteinizing hormone-releasing hormone, or both peptides. *Proceedings of the National Academy of Sciences USA* **99** 2386-2391

18 Hsu CC (2001) Expression of the Follicle Stimulating Hormone receptor (FSHR) in the testis of the domestic fowl. PhD Dissertation, University of Arkansas, Fayetteville, Arkansas. 117pp.

19 Huang HJ, Sebastian J, Strahl B, Wu JC, Miller WL (2001) Transcriptional regulation of the ovine follicle-stimulating hormone-beta gene by activin and gonadotropin-releasing hormone (GnRH): involvement of twoproximal activator protein-1 sites for GnRH stimulation. *Endocrinology* **142** 2267-2274

20 Johnson PA, Brooks C (1996) Developmental profile of plasma inhibin and gonadotropins from hatch to sexual maturity in male and female chickens. *General and Comparative Endocrinology* **102** 56-60

21 Johnson AL, Solovieva EV, Bridgham JT (2002) Relationship between steroidogenic acute regulatory protein expression and progesterone

production in hen granulosa cells during follicle development. *Biology of Reproduction* **267** 1313-1320

22 Katz IA, Millar RP, King JA (1990) Differential regional distribution and release of two forms of gonadotropin-releasing hormone in the chicken brain. *Peptides* **11** 443-450

23 King JA, Millar RP (1982a) Structure of chicken hypothalamic luteinizing hormone-releasing hormone. I. Structural determination on partially purified material. *Journal of Biological Chemistry* **257** 10722-10728

24 King JA, Millar RP (1982b) Structure of chicken hypothalamic luteinizing hormone-releasing hormone. II. Isolation and characterization. *Journal of Biological Chemistry* **257** 10729-10732

25 Kirby, JD, Tressler CJ, Kirby YK (1998) Evaluation of the duration of sperm fertilizing ability in five lines of commercial broiler breeder and Delaware cross males. *Poultry Science* **77** 1688-1694

26 Krishnan KA, Proudman JA, Bolt DJ, Bahr JM. (1993) Development of an homologous radioimmunoassay for chicken follicle-stimulating hormone and measurement of plasma FSH during the ovulatory cycle. *Comparative Biochemistry and Physiology - Comparative Physiology* **105** 729-734

27 Kuenzel WK, Abdul-Maksoud MM, Elasser T, Proudman JA (2004) Sulfamethazine advances puberty in male chicks by effecting a rapid increase in gonadotropins. *Comparative Biochemistry and Physiology A Molecular Integrative Physiology* **137** 349-355

28 Lovell TM, Gladwell RT, Groome NP, Knight PG (2002) Modulatory effects of gonadotrophins and insulin-like growth factor on the secretion of inhibin A and progesterone by granulosa cells from chicken preovulatory (F1-F3) follicles. *Reproduction* **123** 291-300

29 Marubayashi U, Yu WH, McCann SM (1999) Median eminence lesions reveal separate hypothalamic control of pulsatile follicle-stimulating hormone and luteinizing hormone release. *Proceedings of the Society for Experimental Biology and Medicine* **220** 139-146

30 McCann SM, Karanth S, Mastronardi CA, Dees WL, Childs G, Miller B, Sower S, Yu WH (2002) Hypothalamic control of gonadotropin secretion. *Progress in Brain Research* **141** 151-164

31 McNeilly AS, Crawford JL, Taragnat C, Nicol L, McNeilly JR (2003) The differential secretion of FSH and LH: regulation through genes, feedback and packaging. *Reproduction Supplement* **61** 463-76

32 Millam JR, Ottinger MA, Craig-Veit CB, Fan Y, Chaiseha Y, El Halawani ME (1998) Multiple forms of GnRH are released from perifused medial basal hypothalamic/preoptic area (MBH/POA) explants in birds. *General and Comparative Endocrinology* **111** 95-101

33 Millar RP, Milton RC, Follett BK, King JA (1986) Receptor binding and gonadotropin-releasing activity of a novel chicken gonadotropin-releasing hormone ([His5, Trp7, Tyr8]GnRH) and a D-Arg6 analog. *Endocrinology* **119** 224-31

34 Padmanabhan V, Battaglia D, Brown MB, Karsch FJ, Lee JS, Pan W, Phillips DJ, Van Cleeff J (2002) Neuroendocrine control of follicle-stimulating hormone (FSH) secretion: II. Is follistatin-induced suppression

of FSH secretion mediated via changes in activin availability and does it involve changes in gonadotropin-releasing hormone secretion? *Biology of Reproduction* **66** 1395-402

35 Padmanabhan V, Brown MB, Dahl GE, Evans NP, Karsch FJ, Mauger DT, Neill JD, Van Cleeff J (2003) Neuroendocrine control of follicle-stimulating hormone (FSH) secretion: III. Is there a gonadotropin-releasing hormone-independent component of episodic FSH secretion in ovariectomized and luteal phase ewes? *Endocrinology* **144** 1380-1392

36 Padmanabhan V, Karsch FJ, Lee JS (2002) Hypothalamic, pituitary and gonadal regulation of FSH. *Reproduction Supplement* **59** 67-82

37 Padmanabhan V, McFadden K, Mauger DT, Karsch FJ, Midgley AR Jr (1997) Neuroendocrine control of follicle-stimulating hormone (FSH) secretion. I. Direct evidence for separate episodic and basal components of FSH secretion. *Endocrinology* **138** 424-432

38 Padmanabhan V, McNeilly AS (2001) Is there an FSH-releasing factor? *Reproduction* **121** 21-30

39 Pincus SM, Padmanabhan V, Lemon W, Randolph J, Rees Midgley A (1998) Follicle-stimulating hormone is secreted more irregularly than luteinizing hormone in both humans and sheep. *Journal of Clinical Investigation* **101** 1318-1324

40 Safi M, Onagbesan OM, Bruggeman V, Vleugels B, Volckaert G, Decuypere E (2003) Regulation of Inhibin alpha- and beta(A)-subunit messenger ribonucleic acid levels by gonadotropins and IGF-I in cultured chicken granulosa cells. *General and Comparative Endocrinology* **131** 159-167

41 Schmierer B, Schuster MK, Shkumatava A, Kuchler K (2003) Activin and follicle-stimulating hormone signaling are required for long-term culture of functionally differentiated primary granulosa cells from the chicken ovary. *Biology of Reproduction* **68** 620-627

42 Shen ST, Yu JY (2002) Cloning and gene expression of a cDNA for the chicken follicle-stimulating hormone (FSH)-beta-subunit. *General and Comparative Endocrinology* **125** 375-386

43 van Gils J, Absil P, Grauwels L, Moons L, Vandesande F, Balthazart J (1993) Distribution of luteinizing hormone-releasing hormones I and II (LHRH-I and -II) in the quail and chicken brain as demonstrated with antibodies directed against synthetic peptides. *Journal of Comparative Neurology* **334** 304-323

44 Vizcarra JA, Kreider DL, Kirby JD (2004) Episodic gonadotropin secretion in the mature fowl: serial blood sampling from unrestrained male broiler breeders (*Gallus domesticus*). *Biology of Reproduction* **70** 1798-1805

45 Wilson SC, Chairil RA, Cunningham FJ, Gladwell RT (1990) Changes in the hypothalamic contents of LHRH-I and -II and in pituitary responsiveness to synthetic chicken LHRH-I and -II during the progesterone-induced surge of LH in the laying hen *Journal of Endocrinology* **127** 487-496

46 Wilson SC, Cunningham FJ, Chairil RA, Gladwell RT (1989) Maturational changes in the LH response of domestic fowl to synthetic chicken LHRH-I

and -II *Journal of Endocrinology* **123** 311-318

47 Wilson SC, Gladwell RT, Cunningham FJ (1990) Differential responses of hypothalamic LHRH-I and -II to castration and gonadal steroid or tamoxifen treatment in cockerels. *Journal of Endocrinology* **125** 139-146

48 Wilson SC, Gladwell RT, Cunningham FJ (1991) Diurnal changes in the plasma concentrations of LH and hypothalamic contents of LHRH-I and LHRH-II in the domestic hen. *Endocrinology* **130** 457-462

49 Yu WH, Karanth S, Sower SA, Parlow AF, McCann SM (2000) The similarity of FSH-releasing factor to lamprey gonadotropin-releasing hormone III (l-GnRH-III). *Proceedings of the Society for Experimental Biology and Medicine* **224** 87-92

50 Yu WH, Karanth S, Mastronardi CA, Sealfon S, Dean C, Dees WL, McCann SM (2002) Lamprey GnRH-III acts on its putative receptor *via* nitric oxide to release follicle-stimulating hormone specifically. *Experimental Biological Medicine (Maywood)* **227** 786-793

Functional Avian Endocrinology
A Dawson & P J Sharp (Eds)

GnRH, photorefractoriness and breeding schedules of cardueline finches

S A MacDougall-Shackleton, M E Pereyra[1] and T P Hahn[1]

Departments of Psychology and Biology, University of Western Ontario, London, Ontario N6A 5C2 Canada
[1]Section of Neurobiology, Physiology and Behavior; Animal Communication Lab; University of California Davis, Davis, CA 95616-8761 USA

Introduction

Natural selection ensures that animals time reproduction such that offspring are produced when resources they require are most abundant. However, the proximate environmental cues that animals rely on to time breeding are typically surrogates for those environmental factors that actually influence offspring survival (2). Seasonally breeding birds time reproduction through coordinated physiological responses to the annual change in photoperiod (8). Generally photoperiod acts as an initial predictive cue (26) for the timing of reproduction, and other cues such as temperature or food availability modulate the primary photoperiod-driven response.

Seasonally breeding songbirds usually breed in the spring when the insects they feed their young are most abundant. This seasonal spring breeding comes about through combined stimulatory and inhibitory effects of exposure to long days. In early spring, birds are said to be photosensitive, and exposure to a photophase greater than a critical length (a long day) induces two processes. First, long day exposure results in photostimulation in which the hypothalamo-pituitary-gonad (HPG) axis is stimulated, the gonads recrudesce and breeding commences. Second, long day exposure eventually results in the development of unresponsiveness to long days, or photorefractoriness. Thus, although days are still long, the HPG axis is no longer stimulated, the gonads regress, breeding ceases and molt typically ensues (8). This cycle of photostimulation and photorefractoriness results in breeding seasons that are asymmetrical with respect to the annual photocycle; indeed gonadal regression may occur even prior to the summer solstice (21).

Seasonally breeding songbirds typically demonstrate absolute photorefractoriness (13) characterized by i) eventual spontaneous gonadal regression when birds are housed under constant long days, and ii) failure to exhibit gonadal recrudescence in response to longer days or even constant light once the gonads have regressed. Exposure to short days facilitates dissipation of photorefractoriness to reinstate photosensitivity (8, 21, 25).

Although photorefractoriness is defined with respect to declining responsiveness to photoperiod, it appears to also involve unresponsiveness to all potential reproductively stimulating environmental cues (3). This suggests that the HPG axis may serve as the final common neuroendocrine pathway for the transduction of environmental stimuli into reproductive physiological changes.

GnRH and photorefractoriness

Reproduction in vertebrates is controlled by the neuropeptide gonadotropin-releasing hormone. In birds the hypophysiotropic form is called chicken gonadotropin-releasing hormone-I or cGnRH-I (4, see Chapter 5); hereafter we refer to this form simply as GnRH. In a variety of songbird species absolute photorefractoriness has been shown to coincide with an extreme down-regulation of hypothalamic GnRH content. This decrease in GnRH follows gonadal regression (8) and releasable stores of GnRH persist into the early stages of photorefractoriness (20). However, later in photorefractoriness hypothalamic content of GnRH may be two orders of magnitude lower than in photostimulated birds (8). This suggests that unresponsiveness to environmental cues during photorefractoriness may be mediated in part by the HPG axis being "switched off" at the level of the hypothalamus.

In contrast to absolutely photorefractory songbirds, Japanese quail (*Coturnix japonica*) never become absolutely photorefractory. Quail maintain the capacity to respond to long days throughout the year, and birds held on constant long days maintain mature gonads indefinitely (23). Quail also exhibit no decline of hypothalamic GnRH when in non-breeding condition (9). The differences in GnRH plasticity – up- and down-regulation of GnRH immunoreactivity on a seasonal basis – between Japanese quail and songbirds that become absolutely photorefractory led to the hypothesis that interspecific variation in GnRH plasticity may be linked to interspecific variation in photorefractoriness and hence breeding schedules (4, 11).

Evolution of photorefractoriness

Absolute photorefractoriness terminates breeding in many seasonally breeding songbirds. However, because absolute photorefractoriness precludes a reproductive response it also reduces reproductive flexibility. That is, species that become absolutely photorefractory would be unable to exploit for breeding seasonal resources that become available during their photorefractory phase. For absolute photorefractoriness to remain adaptive for any given species, the benefits of eliminating reproductive competence must outweigh the potential costs of doing so.

Because the annual change in food resources required for young varies among species, the costs and benefits of photorefractoriness are also likely to vary. The presumed benefit of photorefractoriness in seasonal environments is to ensure termination of breeding, despite continued favorable conditions for meeting the immediate demands of nesting, in time for young to fledge and achieve independence, and for adults to molt and, if necessary, prepare for migration (15, 21). For species that feed their young on a food resource that is unpredictable or constantly available, photorefractoriness would be potentially costly and less likely to evolve or be maintained. For species that feed their young a food resource with a predictable annual cycle of availability, photorefractoriness would be less costly and more likely to evolve. Because most mid to high latitude songbirds feed their young insects that are in peak abundance during spring and summer, seasonal breeding terminated by photorefractoriness should be widespread in this taxon.

Despite these predictions, there are relatively few comparative data on photoperiodic responses among different species of songbirds. Most of what we know regarding photoperiodism in songbirds derives from studies of a few taxa, most notably European starlings and white-crowned sparrows, but also dark-eyed juncos, American tree sparrows, house finches, house sparrows, and a few others (8, 21, 25). All of these focal taxa are strictly seasonal breeders that become absolutely photorefractory. More recently we, and others, have begun a comparative examination of photoperiodism among cardueline finch species (summarized in Table 1). This group of songbirds contains species that range from strictly seasonal breeders to flexible breeders to opportunistic breeders. For each species we are accruing information regarding breeding schedules, photorefractoriness and seasonal changes in GnRH. Because the phylogenetic relationships among these species are relatively well understood we should be able to determine the co-evolution of these traits.

A review of Table 1 indicates greater diversity among species in photoperiodic mechanisms than may have been previously surmised. In particular, the two criteria used to define absolute photorefractoriness (spontaneous regression of the gonads on constant long days, and subsequent unresponsiveness to even constant light) appear to vary independently among species. The degree of GnRH plasticity also varies among species that differ in breeding schedules. Below, we examine how these three traits vary among cardueline finch species and propose how they may have evolved within this taxon.

Table 1 Summary of breeding schedules, indicators of photorefractoriness and seasonal changes in GnRH in some species of cardueline finch, as well as a non-cardueline songbird (European starling) and a non-songbird (Japanese quail). "Spontaneous regression" indicates whether or not a species exhibits spontaneous regression of the gonads when held on constant long days. "Unresponsive to 24L" indicates whether or not a species responds to a very long photoperiod of 24 h light once the gonads have regressed. "Δ GnRH" indicates whether or not a species exhibits large down-regulation of hypothalamic GnRH when photorefractory. Blank cells indicate no information available. Superscripts indicate references.

Species	Breeding Schedule	Spontaneous Regression?	Unresponsive to 24L?	Δ GnRH
Red Crossbill *Loxia curvirostra*	Opportunistic	NO[10]	NO[18]	
White-winged Crossbill *Loxia leucoptera*	Opportunistic	NO[12]		NO[22]
Common Redpoll *Carduelis flammea*	Seasonal	YES[12]		YES[22]
Pine Siskin *Carduelis pinus*	Flexible	YES[12]	NO[18]	YES[22]
American Goldfinch *Carduelis tristis*	Seasonal	YES[19]		YES[19]
Canary *Serinus canaria*	Flexible	YES[5, 24]		NO[5]
House Finch *Carpodacus mexicanus*	Seasonal	YES[13]	YES[13]	YES[6]
Cassin's Finch *Carpodacus cassinii*	Seasonal	YES*	YES[18]	

Gray-crowned Rosy-finch *Leucosticte tephrocotis*	Seasonal		NO[18]	
European Starling *Sturnus vulgaris*	Seasonal	YES[21]	YES[21]	YES[7]
Japanese Quail *Coturnix japonica*	Seasonal	NO[23]	NO[23]	NO[9]

*T.P. Hahn, unpublished. Based on molt data only.

Response to constant long days

One of the criteria for absolute photorefractoriness is that birds will eventually spontaneously regress their gonads when held on constant long days. This phenomenon is widespread among seasonally breeding songbirds in general as well as among cardueline finch species (Fig. 1). A notable exception to this is the response of crossbill species. White-winged and red crossbills (*Loxia leucoptera* and *L. curvirostra*) held on constant long days for up to 148 and 225 days, respectively, failed to completely regress their gonads (10, 12). In contrast, all other species of cardueline regress their gonads following exposure to 120 or fewer long days (12, 13, 19). Thus, by this criterion crossbills are not absolutely photorefractory and are more similar to Japanese quail than to seasonally breeding songbirds.

The distribution among species of spontaneous gonadal regression in response to long days suggests that it is the ancestral condition for cardueline finches (Fig. 1). Further observations support this. Although the photoperiodic responses of canaries may vary among strains and potentially the sexes (5) it is clear that, for males at least continued exposure to long days results in regressed gonads (5, 24). Further, although no studies have tracked gonad changes in gray-crowned rosy-finches or Cassin's finches on constant long days, both species do exhibit some asymmetry in gonad size relative to the annual photocycle (18). That is, in late summer gonads are smaller than they were on days of the same duration earlier in the summer. This suggests the development of at least some unresponsiveness to a long photophase. Further, Cassin's finches held on constant long days spontaneously initiate and complete prebasic molt (TP Hahn, unpublished data), as is typical of all taxa that become absolutely photorefractory. Finally, non-cardueline songbirds such as European starlings and house sparrows regress their gonads when held on constant long days (21). Thus, the most parsimonious hypothesis at this point is that spontaneous regression of gonads on long days is an ancestral trait of the cardueliness, if not all songbirds.

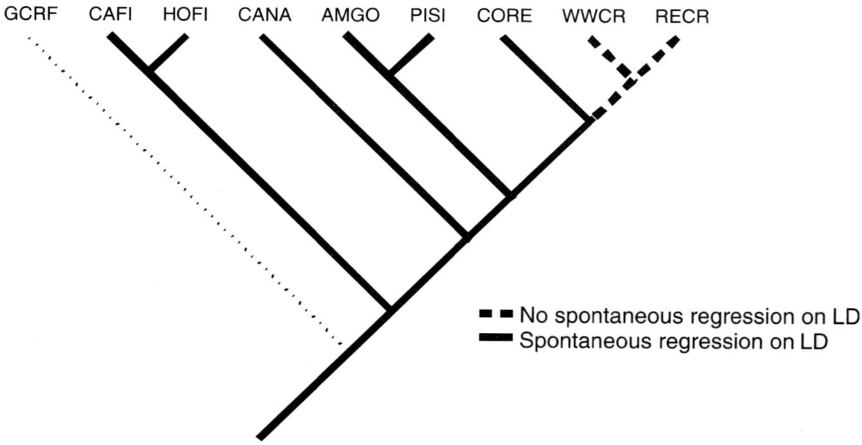

Fig. 1 Variation among cardueline finch species in one criterion for absolute photorefractoriness. The solid lines indicate species that regress their gonads when held on constant long days; the bold dashed lines indicate those that do not. The thin dotted lines indicate species for which no data are available. GCRF = gray-crowned rosy-finch, CAFI = Cassin's finch, HOFI = house finch, CANA = canary, AMGO = American goldfinch, PISI = pine siskin, CORE = common redpoll, WWCR = white-winged crossbill, RECR = red crossbill. Phylogeny based on (1).

Responses to 24L challenge

Regression of the gonads on constant long days indicates the development of photorefractoriness. However *absolute* photorefractoriness has been argued to require a lack of response to even longer photophases (24L : 0D in the extreme) once the gonads have regressed (13, 14). European starlings and other species exhibit both criteria of absolute photorefractoriness, and in some cases these criteria may be treated as two manifestations of the same underlying mechanism. Recent studies of cardueline finches, however, reveal that some species that exhibit one of these criteria may not exhibit the other (Table 1).

Although crossbills exhibit neither criterion for absolute photorefractoriness and house finches exhibit both, species such as pine siskins suggest that these two criteria are dissociable. Pine siskins regress their gonads when held on constant long days (12). However, when pine siskins that have regressed their gonads on declining photoperiod and are molting are exposed to 24 h light they recrudesce their gonads rapidly (18). Red crossbills and gray-crowned rosy-finches have similarly been shown to respond to 24 h light when putatively photorefractory (18). American goldfinches treated in this way did not detectably recrudesce their gonads, but did exhibit an increase in circulating luteinizing hormone (SA MacDougall-Shackleton and TP Hahn, unpublished). This responsiveness to very long photoperiods even when gonads are regressing and feathers are molting appears widespread among the cardueline finches (Fig. 2).

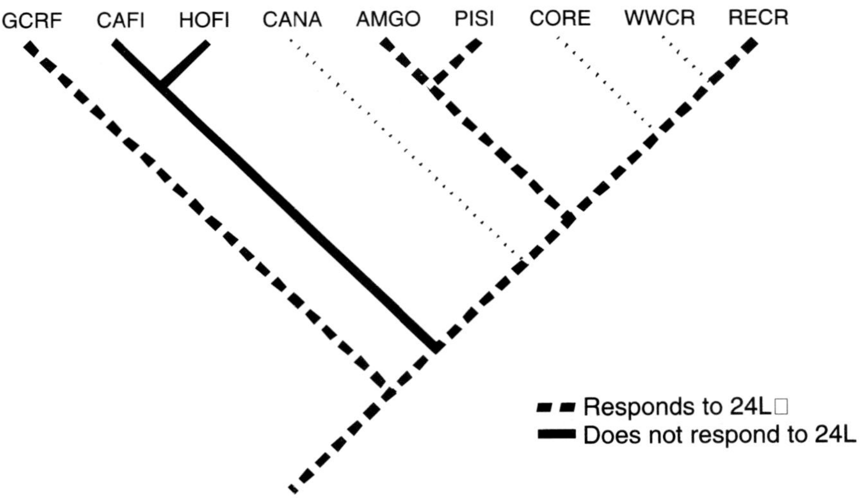

Fig. 2 Variation among cardueline finch species in a second criterion for absolute photorefractoriness. The solid lines indicate species that do not respond even to 24 h light when their gonads are regressed; the bold dashed lines indicate those that do. The thin dotted lines indicate species for which no data are available. See Fig. 1 for abbreviations.

The distribution of this second criterion for absolute photorefractoriness suggests that it was not the ancestral condition among cardueline finches. Rather, it has only been robustly demonstrated in the finches of the genus *Carpodacus* (house and Cassin's finches). Thus complete unresponsiveness to a long photophase may be a derived trait evident in only some lineages of carduelines. For such a trait to evolve the costs of mounting or maintaining a reproductive response at the wrong time of year would have to be high compared to the potential benefits of exploiting "out of season" resources for breeding.

Further work in a greater number of species is required to determine the evolution of complete unresponsiveness to long days in songbirds. Further work will also need to address whether unresponsiveness to 24 h light coincides with unresponsiveness to other environmental cues. Although species such as pine siskins maintain the ability to respond to very long (24 h) days even when their gonads have recently regressed this particular kind of stimulus would not be common in their natural habitat. Whether or not pine siskins can respond to other stimulatory cues – such as an abundance of a particular food type – when their gonads are regressing remains to be determined.

A further question regarding prolonged responsiveness to long days is why this trait might exist in strictly seasonal breeders such as gray-crowned rosy-finches. One explanation is that the prolonged responsiveness was inherited as a neutral trait. If ancestral cardueline finches did possess persistent responsiveness to environmental cues this would have predisposed this taxon towards the evolution of flexible and opportunistic breeding species. Alternatively, if persistent responsiveness to long days also coincides with persistent responsiveness to other environmental cues, this trait may be an adaptation. Rosy-finches are extreme environment specialists breeding at high altitude and/or latitudes (17). Although gray-crowned rosy-finches are strictly seasonal breeders, the precise time of year that is optimal for breeding varies substantially from year to year in their alpine breeding habitat. Thus persistent responsiveness to long days and other cues may permit this species to exploit transient resources regardless of when they become available. Similarly, late-summer breeding American goldfinches, flexibly breeding pine siskins and opportunistically breeding crossbills would all benefit from prolonged responsiveness to environmental cues. Further work is required to determine whether persistent responsiveness to 24 h light does correspond to responsiveness to other cues, and how this responsiveness varies among species.

Although many questions remain, data from cardueline finches suggest that the two criteria for absolute photorefractoriness are not merely different manifestations of the same phenomenon. Rather, exposing birds to constant long days and testing their responsiveness to very long days when gonads are regressed may assess different mechanisms within a single integrated photoperiodic response system. The data presented above suggest that pine siskins would be an excellent species with which to test this hypothesis.

GnRH plasticity

Species that become absolutely photorefractory typically exhibit an extreme down-regulation of hypothalamic GnRH during photorefractoriness. Some species of cardueline finch are no exception. House finches exhibit both criteria of absolute photorefractoriness, and birds captured in the fall during molt exhibit minimal GnRH immunoreactivity in the preoptic area and septum compared to birds captured in the spring (6). In contrast, crossbills likely never become absolutely photorefractory and maintain relatively tonic levels of GnRH year round (16). Data from a variety of species (Table 1) indicate that maintenance of tonic levels of GnRH may be relatively rare, and is unlikely to be the ancestral condition in cardueline finches (Fig. 3).

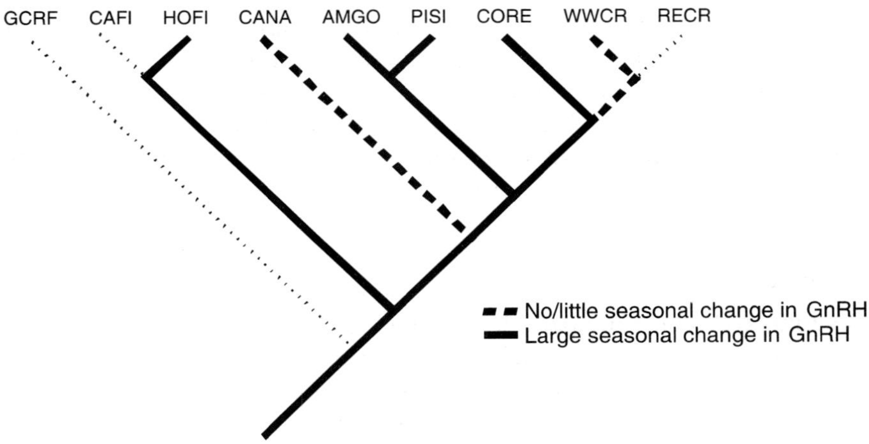

Fig. 3 Variation among cardueline finch species in seasonal plasticity of GnRH. The solid lines indicate species that exhibit large seasonal or photoperiod driven changes in GnRH immunoreactivity; the bold dashed lines indicate those that do not. The thin dotted lines indicate species for which no data are available. See Fig. 1 for abbreviations.

Large seasonal changes in GnRH have now been documented in many species of seasonally breeding songbirds, including several cardueline finch species. In fact only two exceptions have been reported among the cardeulines. In two independent studies, opportunistically breeding white-winged crossbills have exhibited very little seasonal or photoperiod-driven changes in GnRH immunoreactivity despite large changes in gonad size and song-control regions in the brain (16, 22). The second species for which no photoperiodic response of GnRH immunoreactivity was observed is the canary. Bentley *et al.* found that testis size of males varied as a function of photoperiod, but neither follicle size in females nor GnRH immunoreactivity did. GnRH immunoreactivity was similar for photosensitive, photostimulated and photorefractory canaries. Thus, at least some strains of canaries may possess reproductive neuroendocrine mechanisms more similar to opportunistic breeders than to seasonal breeders.

By and large, the pattern of interspecific variation in GnRH plasticity is more similar to the pattern of criterion 1 for absolute photorefractoriness – regression of gonads on long days – than to the pattern for criterion 2 – failure to respond to 24L after regression (see Figs. 1-3). That is, both a failure to regress the gonads on constant long days and relatively stable levels of hypothalamic GnRH appear primarily constrained to opportunistically breeding species of crossbills. Studies of more species are required, but a tentative hypothesis is that down-regulation of hypothalamic GnRH co-evolved with spontaneous gonadal regression rather than a complete insensitivity to long days. Termination of GnRH release, not down-regulation of GnRH content, leads to gonadal regression (8). Thus, the co-evolution of GnRH plasticity and spontaneous gonadal regression suggests that if these two traits are mechanistically linked it is through a third unascertained physiological process.

Synthesis

Determination of photoperiodic mechanisms in even a single species requires numerous studies and years of experiments. That said, enough data have now accumulated for a variety of cardueline finch species to allow tentative hypotheses regarding the co-evolution of photorefractoriness and breeding schedules. The data reviewed above suggest the following conclusions:

The two criteria for absolute photorefractoriness are dissociable

Although gonadal regression following constant long days and subsequent unresponsiveness to even constant light comprised the initial definition of absolute photorefractoriness (13, 14), not all species exhibit both of these criteria. Species such as pine siskins regress their gonads on constant long days, but maintain responsiveness to the longest of days even when they are regressing their gonads and molting (12, 18). Further work is required to determine if pine siskins are photorefractory in any absolute sense. First, some yet untested very long days may be capable of sustaining the reproductive system indefinitely. Second, it is unclear whether very long days short of constant light are capable of interrupting gonadal regression and instating recrudescence. These and other data will be required to determine the similarities and differences between the photoperiodic responses of pine siskins and that of relatively photorefractory Japanese quail. That said, the current data from pine siskins allow us to reject the hypothesis that the two criteria for absolute photorefractoriness are always one and the same.

Responsiveness to very long photoperiods occurs in species with a variety of breeding schedules

A surprising finding is that species with a variety of breeding schedules maintain responsiveness to very long photoperiods of 24 h light even during gonadal regression and molt. Such species include those with strictly seasonal (gray-crowned rosy-finch), flexible (pine siskin) and opportunistic (red crossbill) breeding schedules (18). One explanation is that the last common ancestor of the cardueline finches maintained persistent responsiveness to photoperiod. If so, this could explain the propensity of species with flexible and opportunistic breeding schedules in this taxon. Moreover, this scenario would result in some species having adaptively lost persistent responsiveness (e.g. *Carpodacus* finches), others adaptively exploiting this trait (e.g. opportunistic and flexible breeders), and others maintaining this a neutral trait though phylogenetic inertia (e.g. rosy-finches). Further studies of how and when various species are able to respond to environmental cues and how this affects reproductive success would be required to resolve questions as to the adaptive nature of this trait.

Down-regulation of GnRH during photorefractoriness appears to have co-evolved with the trait of spontaneous gonadal regression, not unresponsiveness to very long days

Extreme photoperiod-dependent changes in GnRH immunoreactivity are widespread among songbirds. This GnRH plasticity occurs even in several species that retain the ability to respond to 24 h light during gonadal regression (18, 22). Thus it seems unlikely that down-regulation of hypothalamic GnRH results in complete unresponsiveness to environmental cues. Rather, both the extreme decrease in hypothalamic GnRH and spontaneous gonadal regression likely result from a third mechanism associated with the central development of photorefractoriness.

Further work is also required to determine how species such as pine siskins can rapidly respond to long days despite the down-regulation of hypothalamic GnRH during gonadal regression. At this point, studies demonstrating the ability to respond to 24L (18) and those demonstrating gonadal regression on constant long days have used birds with differing photoperiodic histories, so further work is needed to clarify under what conditions gonadal regression and responsiveness to 24L occur. Also, the degree to which hypothalamic GnRH is down-regulated appears continuously variable, despite the binary presentation here (Table 1, Fig. 3). It appears that down-regulation of GnRH occurs in species with a variety of breeding schedules, but that have all demonstrated gonadal regression on constant long days. Determining the relationship between variation in responsiveness to 24 h light and variation in GnRH plasticity will require further study.

Acknowledgments

Our research programs have been funded by the National Science Foundation USA (IBN-019-6093 to TPH, postdoctoral fellowship to MEP), the National Institutes of Health USA (NRSA to SAM-S) and the Natural Sciences and Engineering Research Council Canada (Discovery Grant to SAM-S).

References

1 Arnaiz-Villena A, Guillén J, Ruiz-del-Valle, Lowy E, Zamora J, Varela P, Stefani D & Allende LM (2001) Phylogeography of crossbills, bullfinches, grosbeaks and rosefinches. *Cellular and Molecular Life Sciences* **58** 1159-1166

2 Baker JR (1938) The evolution of breeding seasons. In *Evolution: Essays on Aspects of Evolutionary Biology*, pp 161-177. Ed GB DeBeer. Clarendon Press: Oxford

3 Ball GF (1993) The neural integration of environmental information by seasonally breeding birds. *American Zoologist* **33** 185-199

4 Ball GF & Hahn TP (1997) GnRH neuronal systems in birds and their relation to the control of seasonal reproduction. In *GnRH neurons: gene to behavior*, pp 325-342. Eds IS Parhar & Y Sakuma. Brain Shuppan: Tokyo

5 Bentley GE, Audage NC, Hanspal EK, Ball GF & Hahn TP (2003) Photoperiodic response of the hypothalamo-pituitary-gonad axis in male and female canaries, *Serinus canaria*. *Journal of Experimental Zoology* **296A** 143-151

6 Cho RN, Hahn TP, MacDougall-Shackleton SA & Ball GF (1998) Seasonal variation in brain GnRH in free-living breeding and photorefractory house finches *(Carpodacus mexicanus)*. *General and Comparative Enocrinology* **109** 244-250

7 Dawson A, Follett BK, Goldsmith AR & Nicholls TJ (1985) Hypothalamic gonadotrophin-releasing hormone and pituitary and plasma FSH and prolactin during photostimulation and photorefractoriness in intact and thyroidectomized starlings *(Sturnus vulgaris)*. *Journal of Endocrinology* **105** 71-77

8 Dawson A, King VM, Bentley GE & Ball GF (2001) Photoperiodic control of seasonality in birds. *Journal of Biological Rhythms* **16** 365-380

9 Foster RG, Panzica GC, Parry DM & Viglietti-Panzica C (1988) Immunocytochemical studies on the LHRH system of the Japanese quail: influence by photoperiod and aspects of sexual differentiation. *Cell and Tissue Research* **253** 327-335

10 Hahn TP (1995) Integration of photoperiodic and food cues to time changes in reproductive physiology by an opportunistic breeder, the red crossbill, *Loxia curvirostra* (Aves: carduelinae). *Journal of Experimental Zoology* **272** 213-226

11 Hahn TP, Boswell T, Wingfield JC & Ball GF (1997) Temporal flexibility in avian reproduction: patterns and mechanisms. In *Current Ornithology* vol 14, pp 39-80. Eds V Nolan Jr, ED Ketterson & CF Thompson. Plenum Press: New York

12 Hahn TP, Pereyra ME, Sharbaugh SM & Bentley GE (2004) Physiological responses to photoperiod in three cardueline finch species. *General and Comparative Endocrinology* **137** 99-108

13 Hamner WM (1968) The photorefractory period of the house finch. *Ecology* **49** 211-227

14 Hamner WM & Stocking J (1970) Why don't bobolinks breed in Brazil? *Ecology* **51** 743-751

15 Lofts B & Murton RK (1968) Photoperiodic and physiological adaptations regulating avian breeding cycles and their ecological significance. *Journal of Zoology (London)* **155** 327-394

16 MacDougall-Shackleton SA, Deviche PJ, Crain RD, Ball GF & Hahn TP (2001) Seasonal changes in brain GnRH immunoreactivity and song-control nuclei volumes in an opportunistically breeding songbird. *Brain Behavior and Evolution* **58** 38-48

17 MacDougall-Shackleton SA, Johnson RE & Hahn TP (2000) Gray-crowned rosy-Finch (*Leucosticte tephrocotis*). In *The Birds of North America*, No. 559. Eds A Poole & F Gill. The Birds of North America Inc: Philadelphia

18 MacDougall-Shackleton SA, Katti M & Hahn TP (In review) Tests of absolute photorefractoriness in four species of cardueline finch that differ in reproductive schedule.

19 Marsh RH, MacDougall-Shackleton SA & Hahn TP (2002) Photorefractoriness and neural response to day length in American goldfinches, *Carduelis tristis*. *Canadian Journal of Zoology* **80** 2100-2107

20 Meddle SL, Maney DL & Wingfield JC (1999) Effects of N-methyl-D-aspartate on luteinizing hormone release and Fos-like immunoreactivity in the male white-crowned sparrow *(Zonotrichia leucophrys gambelii)*. *Endocrinology* **140** 5922-5928

21 Nicholls TJ, Goldsmith AR & Dawson A (1988) Photorefractoriness in birds and comparison with mammals. Physiological Review **68** 133-176.

22 Pereyra ME, Sharbaugh SM & Hahn TP (2005) Interspecific variation in photo-induced GnRH plasticity among nomadic Cardueline finches. *Brain, Behavior and Evolution* In press

23 Robinson JE & Follett BK (1982) Photoperiodism in Japanese quail: the termination of seasonal breeding by photorefractoriness. *Proceedings of the Royal Society of London B* **215** 95-116

24 Storey CR & Nicholls TJ (1976) Some effects of manipulation of daily photoperiod on the rate of onset of a photorefractory state in canaries (*Serinus canarius*). *General and Comparative Endocrinology* **30** 204-208

25 Wilson FE & Donham RS (1988) Daylength and control of seasonal reproduction in male birds. In *Processing of Environmental Information in Vertebrates*, pp 101-120. Ed MH Stetson, Springer-Verlag: Berlin
26 Wingfield JC (1983) Environmental and endocrine control of reproduction: an ecological approach. In *Avian Endocrinology: Environmental and Ecological Perspectives*, pp. 265-288. Eds SI Mikami, K Homma & M Wada. Japan Scientific Society Press, Springer-Verlag: Berlin

Part three

Reproductive neuroendocrine circuitary

Cactus Wren
Eyal Shochat

Functional Avian Endocrinology
A Dawson & P J Sharp (Eds)

Audio-vocal pathways controlling GnRH release

M-F Cheng

Department of Psychology, Rutgers University, 101 Warren Street, Newark, New Jersey 07102 U.S.A.

Introduction

Photoperiodicity, food availability, and mating are associated with GnRH release (35, 38) and courtship interplay leading to egg-laying provides a compelling example of extra-hypothalamic control of the hypothalamus-pituitary-gonadal axis. This is particularly apparent in the efficacy of the stimulation of GnRH release in the female in response to species-specific auditory and visual mating signals. The causal relationship between courtship display and GnRH release is well established, however, the underlying neuroanatomical and physiological mechanisms have not yet been well elucidated. Little is known about (a) the characteristics of sensory units that detect and process courtship signals, (b) the pathways transmitting neural signals to GnRH neurons or GnRH-related interneurons, and (c) where and how the conversion of neural signals to GnRH release take place. In this chapter, I will report the progress we have made in the ring dove (*Streptopelia risoria*) regarding these questions.

Nest-cooing and GnRH release

Dove courtship consists of bouts of vocal and visual displays. Typically, the male initiates courtship with bowing and cooing (bow coo), which is often accompanied by aggressive pecking should the female remain in close proximity. Later, the male abruptly stops chasing the female and adopts an oblique posture with the head tipping forward while rapidly flipping his wings and emitting a coo song (nest coo). The female is attracted to the male and joins him in performing the nest coo herself (7). Spectral analysis of the female nest coo shows that most of the energy used to produce it is expended on the harmonics rather than on the fundamental frequency (17, 19). This might be how female doves distinguish female nest coos from male nest coos (Fig. 1).

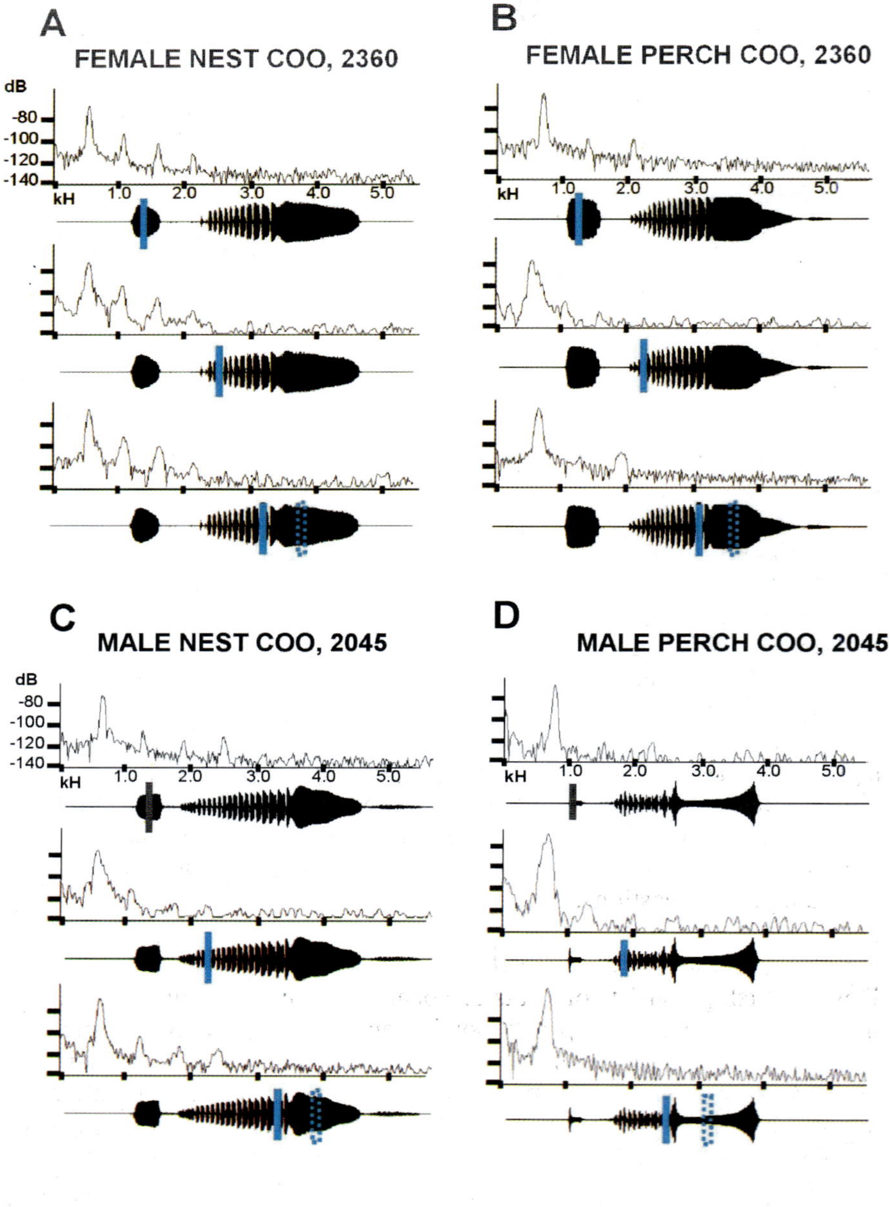

Fig. 1 Coo songs of male and female ring doves. A-D: Coo spectra (Fourier transforms) of a nest coo and a perch coo (undirected song) from, respectively, (A, B) a female and (C, D) male dove recorded during courtship within a sound-isolation chamber. Three spectra are illustrated from each coo song and were generated from three 30 ms time bins. The temporal location of each bin is identified by a bar within the accompanying coo oscillogram (data from 19).

Sustained performance of the nest coo by a female ring dove is a harbinger of egg-laying. Indeed, we have established that the female's own nest coo and not the male's nest coo triggers GnRH release, as indicated by increased ovarian follicular size (8, 10, 11). Female doves were isolated, and the nerves controlling their syrinxes were severed. When the female's own coo songs were played back to her, it stimulated egg-laying more effectively than male coo songs (8). However, when females were not devocalized, the male nest coo was the most potent stimulus (10). These seemingly paradoxical findings can be explained by the observation that females performed more nest coos when hearing male nest coo playback than when hearing female nest coo playback. When female doves were deafened, some were cued into performing nest coo by seeing male nest cooing. Although it took considerably longer for such females to initiate the nest coo, once the nest coo was set in motion, egg-laying ensued normally (10). This observation suggests that proprioceptive stimuli associated with nest coo are effective stimuli for GnRH release. Thus we conclude that the female's performance of her own coo song, in response to acoustic and proprioceptive feedback, is the most effective stimulus for GnRH release.

The importance of these playback studies is that we now know of a quantifiable sensory stimulus that controls GnRH release. This finding has enabled us to identify the neural pathway that mediates auditory stimulation of GnRH release. An early attempt to identify this pathway involved lesioning the ventromedial nucleus (VMN) of the hypothalamus (also known as the posterior medial hypothalamus [PMH]) or the midbrain vocal control nuclei, the nucleus intercollicularis (ICo). Lesions of either the VMN (27) or the ICo (21) abolished the vocal aspect of the nest coo, associated gestures (wing-flip and oblique posture) and follicular growth, suggesting that they are parts of a neural circuit underlying nest coo control of GnRH release.

An audio-vocal circuit through the hypothalamus

A tract tracing study of the midbrain ICo identified an auditory nucleus that surrounds the thalamic auditory relay nucleus (nucleus ovoidalis [Ov]). Injection of the anterograde tracer, PHAL, into the medial ICo, including the caudomedial shell nucleus of the torus semicircularis (CM), revealed massive en passant terminal fields encircling the Ov (13, 23). This robust projection immunolabelled prominently against methionine enkephalin (13, 24, 33). Neurons in the area surrounding the Ov are distinct from those of the Ov in morphology, chemical properties, connectivit, and function. We therefore designated this region as the Ov shell. The Ov shell neurons are aligned with their long axons and dendrites parallel to the Ov border. Single- and multiple-unit recordings through the Ov core and shell regions revealed that coo-responsive units are distributed mostly in the shell (23). This finding, together with the observations that the Ov shell, but not the Ov core, contains estrogen-binding sites (29) and that estrogen is obligatory for nest-coo performance (6, 7), suggest that Ov shell, rather than the Ov core, processes the female auditory nest coo signals.

The Ov shell gives rise to two axonal projections to the hypothalamus: one to the preoptic anterior/ medial hypothalamus (POA/AMH) and the other to the VMN (23). Since the POA/AMH region contains GnRH-I cell bodies (13, 35), the POA/AMH pathway may mediate nest coo control of GnRH release. To test this hypothesis, we first confirmed the projections by injecting a retrograde tracer, Fluorogold, into the hypothalamus. Next, we determined that the pathways were indeed auditory by using antidromic experiments (14). We searched for units in the Ov shell that responded to sounds and then stimulated the hypothalamus with electrical pulses. Collisions of action potentials traveling down from the Ov or the Ov shell neurons with those generating from the hypothalamus suggested that both were traveling along the same pathways. Thus, Ov shell receives and processes signals induced by nest cooing and relays these (23) acoustic inputs down to the hypothalamus (14). These convergent findings lead us to hypothesize that there are acoustic units in the hypothalamus dedicated to processing the coo song and that the outcome of nest coo processing is GnRH release.

Female nest coo-responsive units in the POA/AMH

We designed an experiment to demonstrate that nest coo stimulation (sensory stimulus), simultaneously evokes discharge (neural processing), and GnRH release (luteinizing hormone [LH] output). To this end, we recorded electrical activities from the POA/AMH or from the VMN of anesthetized female doves either kept in silence to measure spontaneous discharge, or exposed to auditory stimulation to measure evoked discharge (15). Concurrently blood was drawn from the venous outflow of the pituitary gland at 10-20-min intervals for 120 min. In total, 935 neurons from 36 birds were recorded. Interestingly, most excitatory units (75.2%) were found in the POA/AMH regions where GnRH-I cell bodies are located (13, 36), whereas most inhibitory units were found in the posterior portion of VMN, where mostly GnRH-I fibers are found.

As shown in Fig. 2, some of the excitatory units responded exclusively to female nest coo stimulation. The male nest coo or reversed female nest coo did not elicit bursts in previously responsive units. Clearly these excitatory units can detect the temporal pattern of acoustic signals. Interestingly, female nest coo-specific units recorded from POA exhibited two bursts of four to seven spikes each

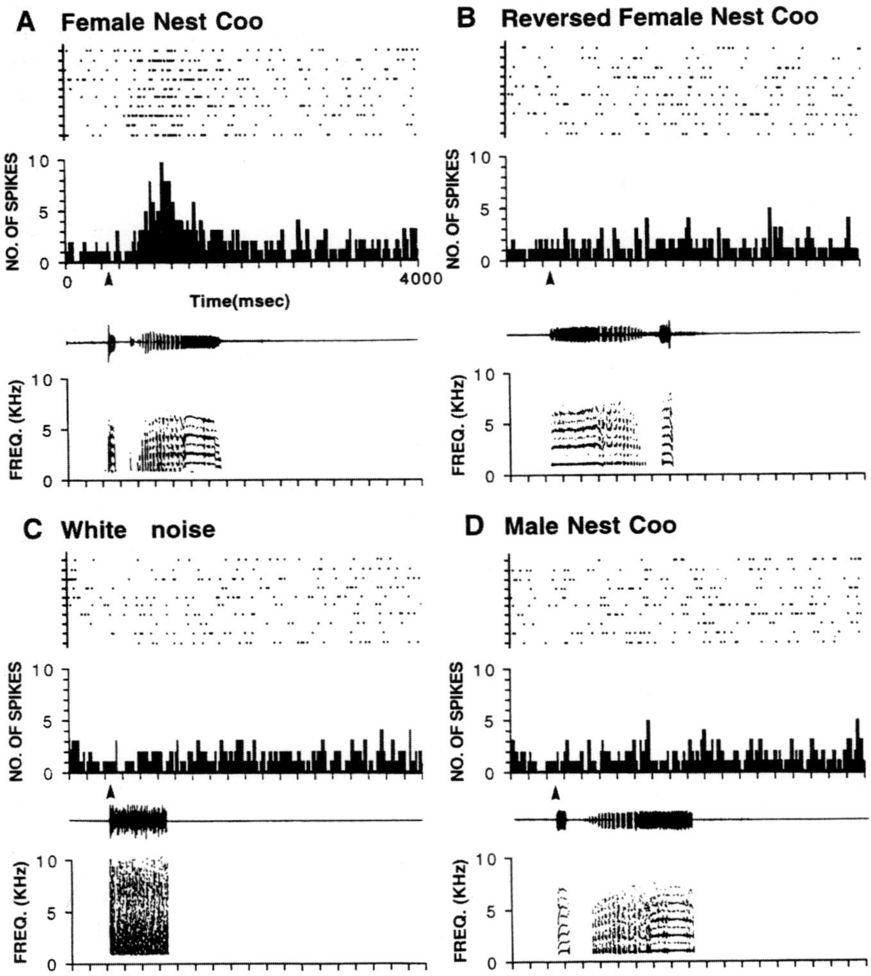

Fig. 2 Response of one unit in the preoptic area of a female ring dove to different acoustic stimuli. A, The female nest coo as the stimulus; the activity of the unit increases significantly. *Top,* Dot raster plot showing 10 sweeps of the unit's response. Each dot represents one spike. Bin size is 20msec. *Middle*, Histogram of the unit's response. *Bottom*, Computer amplitude display and spectrogram of the female nest coo: amplitude over time. B, Reversed female nest coo as the stimulus; no change in neuronal activity can be detected in the number of spikes over time or in the histogram representation. C, White noise as the stimulus: no change in neuronal activity. D, Male nest coo as the stimulus: no change in neuronal activity. All histograms represent the sum of 10 repetitions of the stimulus. A *closed triangle* denotes the onset of the stimulus. (data from 15)

separated by a silent period of 200-600 msec, a pattern corresponding temporally to a two-note coo (Fig. 3A). Similarly patterned selective responses have been observed in the song-specific neurons of zebra finches (28). Other researchers have detected the temporal features of the stimulus waveform in electrosensory cells (26). Radioimmunoassay of LH showed that the LH level was dramatically elevated after activation of female nest coo-specific units in the POA/AMH. This LH level was threefold higher than the level recorded after male nest coo stimulation (Fig. 3B). These female nest coo-specific units therefore are likely to control GnRH release (15).

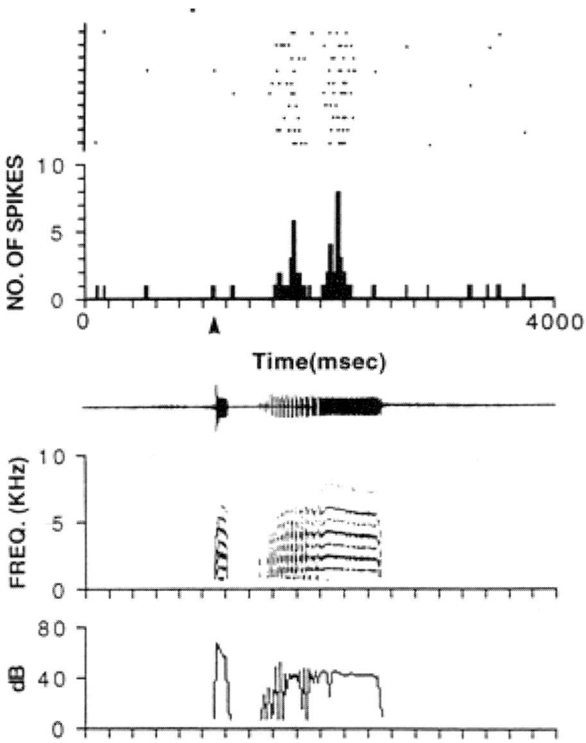

Fig. 3A Effects of nest coo stimulation on neuronal response of anterior medial hypothalamus units. A typical female nest coo-specific unit found in preoptic area. Note the two burst spiking discharge. (Data from 15).

Transduction of neuronal signals to endocrine output

Although we do not know the molecular basis of the conversion of neuronal activation to neuroendocrine release, we found that GnRH release is not a graded response to neuronal discharge. To determine whether GnRH release responds to the intensity of neural signals, we

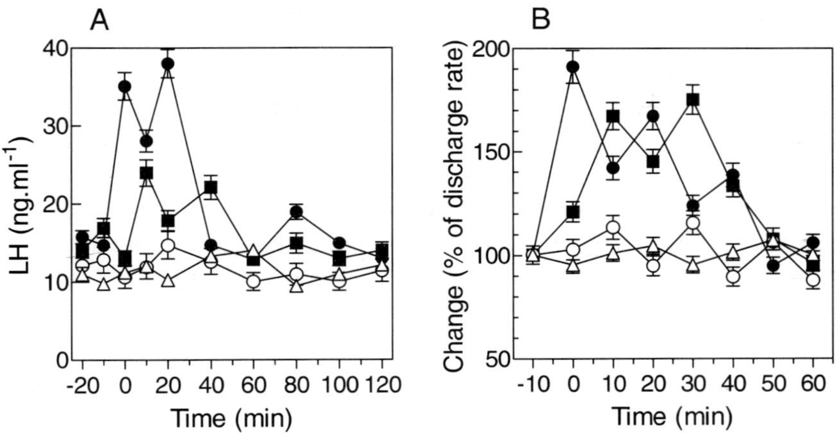

Fig. 3B Effects of nest coo stimulation on LH levels in female ring doves. There is a significant effect of different acoustic stimuli on LH levels ($F_{3, 36}$ =6.91; p<0.01) and on discharge rate ($F_{3, 31}$ =4.67; p<0.01). A, LH levels were significantly higher for the female birds (n=3) hearing female nest coo play backs (*closed circles*) than for those receiving no playback (*closed triangles*) at time points 0, 10, and 20 min, in which 0=40 min after the onset of female nest coo stimulation (t=3.08; P<0.01). B, Percentage changes of firing rate of units recorded during each sound stimulation. *Closed circles*, Female nest coo, responses of nine units; *closed squares*, male nest coo, responses of seven units; *open triangles*, no playback, responses of five units; *open circles*, reversed male nest coo and reversed female nest coo continued, responses of seven units. The percentage changes of neuronal response to the female nest coo and to the male nest coo are strikingly similar (t=1.39; p>0.05).

examined the percentage change in discharge in the POA-AMH region generated by the male or female nest coo. The sum totals of excitatory and inhibitory discharges arriving at GnRH networks were not different between male and female nest coos (Fig. 3B). This raises the possibility that female nest coo-specific neurons do not regulate GnRH release in a linear response manner, but instead serve as a signal detection device which functions as an on/off switch for GnRH release (15). Thus, when GnRH is induced by female nest coo, it may function as a signal to move from no, or low response to a state of optimal response (44). This can be achieved with very few neurons, which may explain why there are so few female nest coo-specific units. Also relevant here is the observation that the nest coo level that results in egg-laying shows striking individual differences between birds (11). Thus, the threshold of GnRH release may vary from one female to another. We still need to investigate whether these individual differences reflect differences in the number of female nest coo-specific units in each bird.

Inhibitory nature of neuroendocrine control of GnRH neurons

Most neuroendocrine control of GnRH neurons appears to be inhibitory, as illustrated by the fact that the ICo-Ov shell – hypothalamic projections are enkephalinergic (13, 24). An inhibitory regulator of gonadotropin release, gondotropin-inhibitory hormone (GnIH), was identified recently (2, 20, 31, 39, 40) and it is suggested that it may act at the level of the hypothalamus (5), exerting inhibitory effects on the reproductive axis (31). GnRH release therefore may be stimulated by lifting the inhibitory action of GnRH in addition to a direct stimulatory action. Judging from the patterns of exponential follicular growth in the doves (12) and opportunistic breeders, which reflect a sudden surge of GnRH release, nonlinear GnRH release maybe common in birds. This threshold-sensitive release is compatible with a system that is normally under inhibitory control that can be disinhibited by prescribed signals. We suggest that the prescribed signals for GnRH release in the case of the ring dove are specific discharge patterns of female nest coo-specific neurons which may act through GnIH to lift an inhibitory action on GnRH release.

Audio-vocal pathways controlling GnRH release

Auditory stimulation of GnRH release is only part of the story. As shown in the playback study, the proprioceptive feedback associated with nest coo also contributes to GnRH release. The Ov shell receives main ascending inputs from the CM, the medial border of the midbrain auditory relay. The CM was found to differ from the midbrain auditory relay with respect to its connections and immunohistochemistry (32). It receives auditory input from brainstem cochlear nuclei as well as somatosensory (proprioceptive) input from the ascending dorsal column system (41, 42, 43). Thus, the Ov shell input to the POA/AMH may include somatosensory input associated with nest coo, which, along with the direct innervation from medial ICo (13), contributes to GnRH release.

The DM (dorsomedial nucleus of ICo) - nXIIts pathway has been identified as the conserved motor pathway controlling vocalization in both vocal (e.g., song birds; 30) and nonvocal learners (e.g., ring doves; 22, 43). We confirmed the involvement of the DM in the performance of nest coo by using c-fos immunohistochemistry (5). The nest coo song was blocked by VMN lesions, which suggests that the VMN has synaptic connections at the midbrain vocal control nucleus, i.e. the DM. After Fluorogold (retrograde tracer) was infused into the anterior ICo inclusive of the DM, or the medial ICo, the backfilled cells in the VMN suggested a monosynaptic connection from the VMN to DM, confirming an earlier finding using horseradish peroxidase retrograde tracer (9). However, a careful study using anterograde tracing from the VMN suggests polysynaptic connections. Robust VMN projections were found throughout the anterior and medial ICo. The VMN projection to the DM was, however, relatively weak (5). Since the DM receives significant efferent projections from other parts of the anterior ICo and the medial ICo (37), the VMN may exert its influence on the vocal motor pathway via mostly polysynaptic projections.

In summary, during a breeding cycle, the female dove performs nest coo initially in response to male nest coo then engages in repeated bouts of nest coo resulting in GnRH release and ovarian follicular growth. We propose that this process is mediated by the CM-Ov shell-POA/AMH auditory (and collateral somatosensory) pathways. The parallel vocal pathways of Ov shell-VMN-ICo (DM) mediate vocal output (nest cooing), the feedback of which initiates yet another round of sensory processing by the CM-Ov shell-POA/AMH pathways.

Are coo-responsive neurons renewable?

In adult songbirds, experimentally induced cell death of specific high vocal center (HVC) neurons in zebra finches resulted in compensatory replacement of the same type of cells (34). Does electrolytically induced death of female nest coo-specific neurons in the adult ring dove also result in compensatory replacement? Unlike the HVC, the hypothalamus is not normally neurogenic. Bilateral electrolytic lesions in the ventromedial nucleus (VMN) of the hypothalamus, however, resulted in elevated cell proliferation along the ventricular zone of the lateral ventricle; two months later some of these cells migrated and differentiated into mature neurons at the lesion site (4). Are these newly generated neurons functional? Specifically, do they exhibit the female nest coo-specific discharge pattern when stimulated by auditory nest coo? Systematic single-unit recordings were made from the lesioned VMN, before and after bilateral lesion, for varying periods up to 3 months (18). Concurrently, the lesioned area was monitored for changes in cell types by using bromodeoxyuridine (BrdU) to label newly divided cells and NeuN to identify mature neurons. As shown in Fig. 4, during the first month, a necrotic cavity created by the lesion was prominent, and there was no sign of electrical activity. Only BrdU-labeled cells were detected at the peripheral region of the lesion site. When the first electrical activity occurred, it had an abnormal discharge pattern and no neurons responded to female nest coo stimulation. Shortly after detection of newborn neurons (BrdU+/NeuN+ double-labeled cells), 2-3 months after lesion, the necrosis was no longer visible, and the first mature discharge pattern and then female nest coo-specific discharge were recorded. The discharge at the lesion site was no different from that of sham lesion (Fig. 5). The double-labeled cells were confirmed using

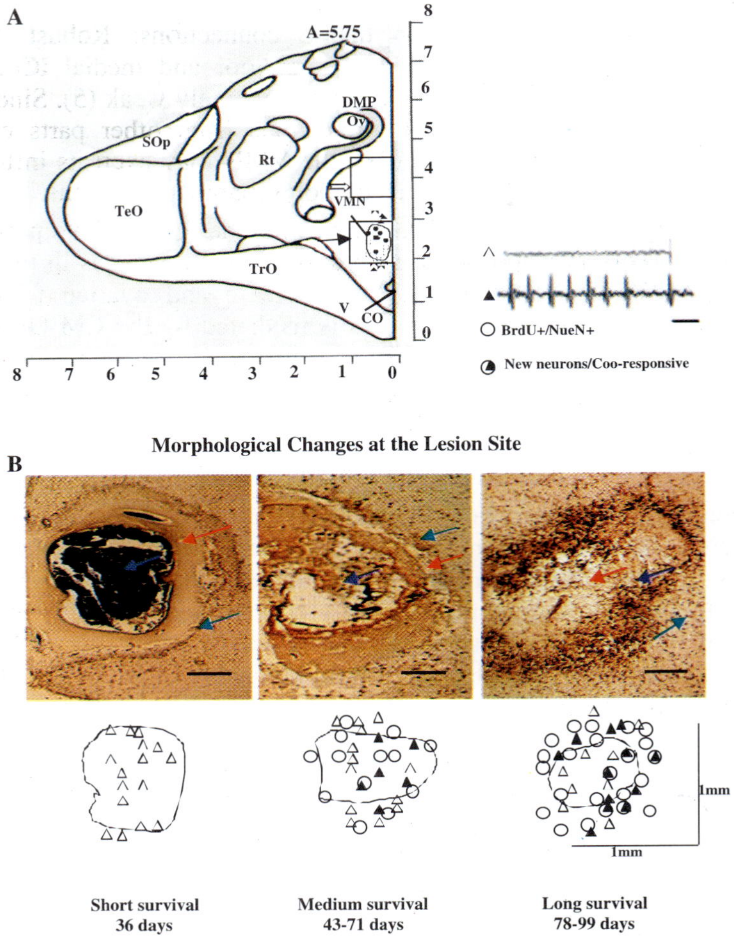

Fig. 4 Lesion-induced neurogenesis participates in recovery of VMN neurons. Birds received BrdU injections (40mg/kg) at 4 day intervals after surgery. Concurrently single unit recordings were made the day before each injection. Birds were killed at different survival times for perfusion and immunohistochemical analysis. **A**. Location of lesion, recording and area of immunoreactive cell counts. The *solid arrow* points to the square area in which immunoreactive cell counts were made. The *open arrow* points to the square area in which control cell counts were made. **B**: Morphological changes at the lesion site over time, and parallel changes in electrical activity and double-labeled cells. *Open triangles* denote absence of any electrical activity. *Solid triangles* denote normal neural discharge in responding to nest coo stimulation. *Open circles* denote new cells double-labeled with NeuN a neuron-specific marker, and BrdU, i.e. new neurons. A *Solid triangle* inside a *circle* denotes a coo-responsive unit co-localized with a new neuron confirming that new neurons exhibit nest coo specific units characteristic of VMN adult ring doves (data from 18).

laser confocal scanning microscopy (Fig. 6A). The newly generated neurons were further validated by negative labelling for GFAP, a marker for glial cells (Fig. 6B). The number of coo-responsive units was statistically significantly correlated with the number of newborn neurons ($r=0.9876$, $p<0.01$). Indeed, markers for coo-responsive units (dye deposited at the conclusion of each successful recording) that were colocalized with BrdU+/NeuN+ double-labeled cells (Fig. 6C) confirmed that lesion-induced new neurons exhibited a normal female nest coo-specific discharge pattern, which had been lost after lesion (18).

The hypothalamus is a complex system, and the functions attributable to the VMN of the ring dove are no less complex. However, the VMN of the adult ring dove exhibits two biological functions that are amenable to quantitative measurement, namely control of endocrine output and response to vocalization. The VMN also contains sensory units, i.e., acoustic units that respond preferentially to species-specific vocal stimulation. Given this background in assessing recovery of VMN functions, it is important to distinguish types of cells involved in the recovery of nest coo behavior. Nest coo behavior may involve sensory neurons, motor projection neurons, or interneurons that activate projection neurons (or secretory neurons, such as those that stimulate gonadotrophin-releasing hormones). Destruction of any of these neurons can block nest coo behavior. Similarly, recovery of the behavior may require new neurons to replace one or more of these neurons. We have established that coo-responsive units can be replaced and have preliminary data suggesting that some of the newly generated cells are GnRH immunoreactive cells. We have not determined, however, whether GnRH immunoreactive cells and coo-responsive cells are the same or in different groups of cells. The ultimate test for the functionality of these newly generated neurons rests with a demonstrable increase of LH output when these units are stimulated.

Summary and future directions

This chapter focused on audio-vocal regulation of GnRH with the understanding that audio-vocal pathways are part of a larger circuitry underlying courtship behavior. Mating behavior is social behavior and as such is intimately linked to the limbic brain (19). The hypothalamus in the ring dove as well as vocal learner receives input from the avian amygdala complex (16). It should also be noted that methionine enkephalin immunoreactivity has been found throughout the vocal system and within the Ov shell, in the hypothalamus and limbic pallium of oscine songbirds and budgerigars (1, 3, 25, 33). This neurochemical link suggests that regulation of GnRH by audio-vocal pathways is not limited to ring doves. A future challenge will be to elucidate how complex songs mediated by the frontal cortex regulate GnRH release.

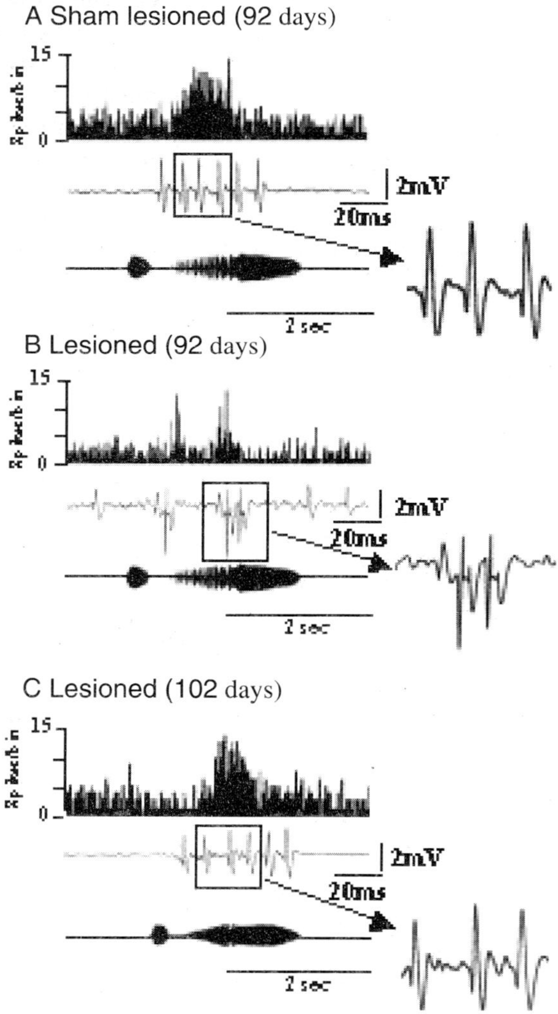

Fig. 5 Responses of female nest coo-specific units in VMN of female ring doves surviving 92-102 days. A, Normal spontaneous discharge and evoked discharge in response to female nest coo stimulation in control group (sham lesion). *Top*, histogram of the unit's respons: the discharge is highlighted. *Middle*, evoked specific response by female nest coo stimulation. *Bottom*, computer amplitude display of female nest coo: amplitude over time. B, A unit on the way to a normal function: normal spontaneous discharge but abnormal discharge in response to female nest coo stimulation. C. A unit exhibiting an excitatory response to female nest coo stimulation not distinguishable from the sham control bird shown in A. All histograms represent the sum of 10 repetitions of the stimulus. Bin size, 20ms (Data from 18).

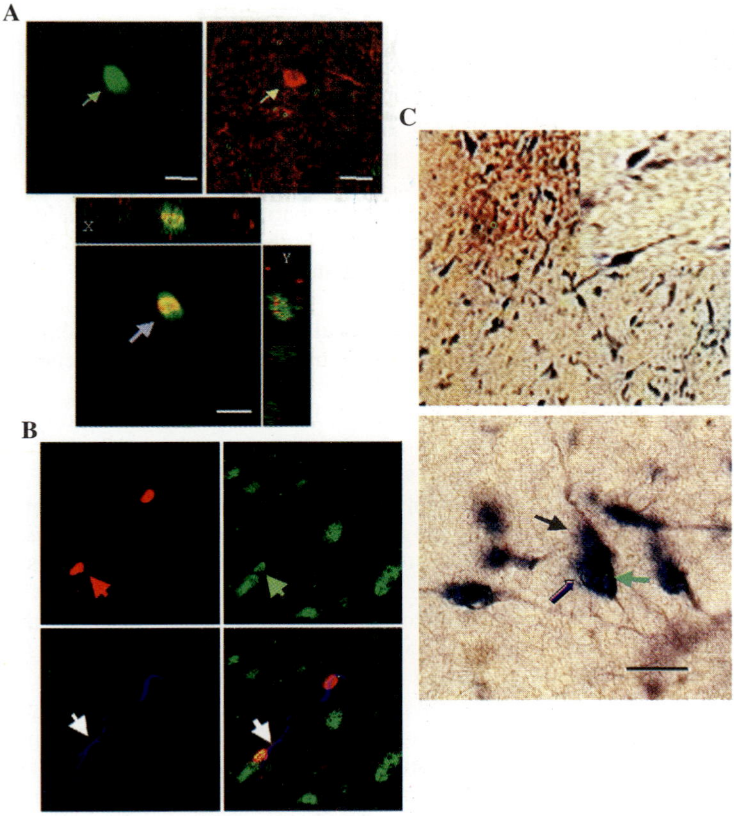

Fig. 6 Samples of double and triple labeled cells in the VMN. *Left panel* (A, B) Confocal images of double labeled cells. A. A cell at the lesion site in the VMN double-labeled with BrdU (*red*) and NeuN (*green*) rotated in orthogonal planes (*x,y,z*). Scale bar, 10μm. B. A cell double-labeled with BrdU and NeuN (*yellow arrow*) which is not triple labeled for GFAP (*white arrow*). Scale bar 10μm. Thus $BrdU^+/NeuN^+$ double-labeled cells are new neurons not new glia cells. *Right panel* (C) Photomicrographs showing the histologic location of a single recording electrode in relation to a double-labeled cell and the lesion site. *Top*, at low magnification, a $BrdU^+/NeuN^+$ cell (*green arrow*) at the recording site (blue stained area), near the center of the lesion (*thick blue arrow*). A $BrdU^+/NeuN^+$ cell at the recoding site (black arrow) is highlighted to show the dendritic spine (*red arrow*) of the cell. *Bottom* At high magnification, a merged image (black arrow) of the recoding site (*blue arrow*) with a BrdU+/ NeuN+ cell (*green arrow*) displaying distinct morphologic features of a dendritic spine (*red arrow*). Scale bars: 80μm (A); 10μm (B) (Data from 18).

Acknowledgements

MFC was supported by National Institute of Mental Health and a Rutgers Busch Biomedical Research Grant.

References

1 Ball GF, Faris PL, Hartmen BK & Wingfield, JC (1988) Immunohistochemical localization of neuropeptides in the vocal control regions of two songbird species. *Journal of Comparative and Neurology* **268** 171-180

2 Bentley GE, Perfito N, Udena K, Tsutsui K & Wingfield JC (2003) Gonadotropin-inhibitiry peptide in song sparrows (*Melospina meldodia*) in different reproductive conditions, and in house sparrow (*Passer domesticus*) relative to chicken-gonadotropin-releasing hormone. *Journal of Neuroendocrinology* **15** 794-802

3 Bottjer SW & Alexander A (1995) Localization of met-enkephalin and vasoactive intestinal polypeptide in the brains of male zebra finches. *Brain Behavior and Evolution.* **45** 153-177

4 Cao J, Wenberg K & Cheng MF (2002) Lesion induced new neurons incorporation in the adult hypothalamus of the avian brain. *Brain Research* **943** 80-92

5 Chen G (2004) The contribution of neurogenesis to behavioral recovery after hypothalamic lesion in the adult ring dove (*Streptopelia risoria*). PhD thesis, The Graduate School, Rutgers University, Newark, New Jersey (submitted).

6 Cheng MF (1973) Effect of estrogen on ovariectomy on the reproductive behavior of female ring doves (*Streptopelia risoria*). *Journal of Comparative and Physiological Psychology* **83** 234-239

7 Cheng MF (1979) Progress and prospects in ring dove research: A personal view. In *Advances in the Study of Behavio*r **9** 97-129. Eds JS. Rosenblatt, RA Hind, C Beer & M-C Busnel, Academic Press, New York

8 Cheng MF (1986) Female cooing promotes ovarian development in ring doves. *Physiological Behavior* **37** 371-374

9 Cheng MF, Akesson TR & DeLanerolle N (1987) Retrograde HRP demonstration of afferent projections to the midbrain and nest calls in the ring dove. *Brain Research Bulletin* **18** 45-48

10 Cheng MF, Desiderio C, Havens M & Johnson A (1988) Behavioural stimulation of ovarian growth. *Hormones and Behavior* **22** 388-402

11 Cheng MF (1992) For whom does the female dove coo? A case for the role of vocal self-stimulation. *Animal Behavior* **43** 1035-1044

12 Cheng MF (1993) Vocal, auditory and endocrine systems: three-way connectivities and implications. *Poultry Science Reviews* **5** 37-47.

13 Cheng MF & Zuo M (1994) Proposed pathways for vocal self-stimulation: met-enkephalinergic projections linking the midbrain vocal nucleus, auditory-responsive thalamic regions and neurosecretory hypothalamus. *Journal of Neurobiology* **25** 361-379

14. Cheng MF & Peng J (1997) Reciprocal talk between the auditory thalamus and hypothalamus: an antidromic study. *NeuroReport* **8** 653-658

15 Cheng MF, Peng JP & Johnson P (1998) GnRH neurons preferentially respond to female nest coo stimulation: Demonstration of direct acoustic stimulation of luteinizing hormone release. *Journal of Neuroscience* **18** 5477-5489

16 Cheng MF, Chaiken M, Zuo M, & Miller H (1999) Nucleus taenia of the avian amygdala: anatomical and functional studies in ring doves (*Streptopelia risoria*) and starlings (*Sturnus vulgaris*). *Brain, Behaviour and Evolution* **53** 243-270

17 Cheng MF (2003) Vocal self-stimulation: From the ring dove story to *emotion*-based vocal communication. In *Advances in the Study of Behavior*, **33** 309-353. Eds PJB Slater, JS Rosenblatt, CT Snowdon, TJ Roper & M Naguib. Elsevier Academic Press

18 Cheng MF, Peng JP, Chen G, Gardner JP & Bonder EM (2004) Functional restoration of acoustic units and adult-generated neurons after hypothalamic lesion. *Journal of Neurobiology* **60** 197-213.

19 Cheng MF & Durand SE (2004) Song and limbic brain: a new function for the bird's own song. In *Behavioral Neurobiology of Birdsong*, pp 1016:611-627. Eds HP Zeigler & P Marler. Annals of the New York Academy of Sciences

20 Ciccone NA, Tsutsui K, Sharp PJ & Dunn IC (2004) Does gonadotrophin inhibiting hormone (GnIH) play a functional role in avian reproduction? *British Poutlry Science Supplement* **1** S28-29

21 Cohen J (1981) Hormones and midbrain mediation of courtship behavior in the male ring dove (*Streptopelia risoria*). *Journal of Comparative Physiological Psychology*. **95** 512-528

22 Cohen J & Cheng MF (1981) The role of the midbrain in courtship behavior of the female ring doves (*Streptopelia risoria*): evidence from radiofrequency lesion and hormone implant studies. *Brain Research* **207** 279-30

23 Durand S, Tepper J & Cheng MF (1992) The shell region of the nucleus ovoidalis: A subdivision of the avian auditory thalamus. *Journal of Comparative Neurology* **323** 495-518

24 Durand SE, Zuo M, Zhou S & Cheng MF (1993) Avian auditory pathways show met-enkephalin-like immunoreactivity. *NeuroReports* **4** 727-730.

25 Durand SE, Liang W & Brauth SE (1998) Methionine enkephalin immunoreactivity in the brain of the budgerigar: Similarities and differences with respect to oscine songbirds *Journal of Comparative Neurology* **393** 145-168

26 Gabbiani F, Metzner W, Wessel R & Koch C (1996) From stimulus encoding to feature extraction weakly electric fish. *Nature* **384** 564-567

27 Gibson M & Cheng MF (1977) The neural mediation of estrogen dependent courtship behavior in the female ring dove. *Journal of Comparative and PhysiologicalPsychology* **93** 855-867

28 Lewicki MS (1966) Intracellular characterization of song-specific neurons in the zebra finch auditory forebrain. *Journal of Neuroscience* **16** 5854-5863

29 Martinez-Vargas MC, Stumpf WE &. Sar M (1976) Anatomical distribution of estrogen target cells in the avian CNS: A comparison with the mammalian CNS. *Journal of Comparative Neurology* **167** 83-104

30 Nottebohm F, Stokes TN & Leonard CM (1976) Central control of song in the canary *Serinus canarius*. *Journal of Comparative Neurology* **165** 457-486

31 Osugi T, Ukena K, Bentley TE, O'Brien S, Moore IT, Wingfield JC & Tsutsui K (2004) Gonadotropin-inhibitory hormone in Gambel's white-crowned sparrow (*Zonotrichia leucophyrs gambelii*): cDNA identification, transcript localization and functional effects in laboratory and field experiments. *Journal of Endocrinology* **182** 33-42

32 Puelles L, Robles T, Martinez de la Torre M & Martinez S (1994) New subdivision schema for the avian torus semicircular: neurochemical maps in the chick. *Journal of Comparative Neurology* **340** 98-152

33 Ryan SM, Arnold AP &. Elde RP (1981) Enkephalin-like immunoreactivity in vocal control regions of the zebra finch brain. *Brain Research* **229** 236-240

34 Scharff C, Kirn JR, Grossman M, Macklis JD & Nottebohm F (2000) Neuronal death affects neuronal replacement and vocal behavior in adult songbirds. *Neuron* **25** 481-492

35 Silver R, Ramos C, Machca H, Silverin B (1992) Immunocytochemical distribution of GnRH in the brain of adult and posthatching great tit (*Parus major*) and ring dove (*Streptopelia risoria*). *Ornis Scandinavica* **23** 222-232

36 Sharp PJ & Sreekumar KP (2001) Photoperiodic control of prolactin secretion. In *Avian Endocrinology* , pp 245-255. Eds A Dawson & CM Chaturvedi. New Dehli, India: Narosa Publishing House.

37 Striedter GF & Vu ET (1998) Bilateral feedback projections to the forebrain in the premotor network for singing in zebra finches. *Journal of Neurobiology* **34** 27-40

38 Sullivan SD, DeFazio A & Moenter SM (2003) Metabolic regulation of fertility through presynaptic and postsynaptic signaling to gonadotropin-releasing hormone neurons. *Journal of Neuroscience* **23** 8578-8585

39 Tsutsui K, Saigoh E, Ukena K, Teranishi H, Fujosawa U, Kicuchi M, Ishii S & Sharp PJ (2000) A novel avian hypothalamic peptide inhibiting gonadotropin release. *Biochemical and Biophysical Research Communications* **275** 661-667

40 Ukena K, Ubuka T & Tsutsui K (2003) Distribution of a novel avian gonadotropin-inhibitory hormone in the quail brain. *Cell and Tissue Research* **312** 73-79

41 Wild JM. (1989) Avian somatosensory system: II. Ascending projections of the dorsal column and external cuneate nuclei in the pigeon. *Journal of* Comparative *Neurology* **287** 1-18

42 Wild, JM (1995) Convergence of somatosensory and auditory projections in the avian torus semicirculatis, including the central auditory nucleus. *Journal of Comparative Neurology* **358** 465-486

43 Wild JM, Li D & Eagleton C (1997) Projections of the dorsomedial nucleus of the intercollicular complex (DM) in relation to respiratory-vocal nuclei in the brainstem of pigeon (*Columba livia*) and zebra finch (*Taeniopygia guttata*). *Journal of Comparative Neurology* **377** 392-413

44 William RJ & Zipser D (1989) A learning algorithm for continually running fully recurrent neural networks. *Neural Computation* **1** 270-280

Functional Avian Endocrinology
A Dawson & P J Sharp (Eds)

The possible role of the lateral septal organ in housing encephalic photoreceptors responsible for regulating gonadal development

W J Kuenzel, M B Ferrari[1], T Rathinam and H Li

Poultry Science Center, University of Arkansas, Fayetteville, AR 72701, USA
[1]School of Biological Sciences, University of Missouri, Kansas City, MO 64110, USA

Introduction

Many species of birds, particularly those that breed in north temperate regions, respond to increased photoperiods by showing rapid gonadal development (3, 28, 38). Seasonal cycles of reproduction have therefore been studied by exposing birds to artificial day lengths for defined periods of time. Results clearly suggested the presence of a photoneuroendocrine system (PNES) responsible for activating reproductive development (4, 32, 40). Sensory neural elements located in three structures were proposed as likely candidates for sensing and responding to photostimulation: eyes, pineal gland and encephalic photoreceptors (EPRs, receptors in the brain). Early findings addressing avian reproductive cycles showed that in contrast to mammals, birds do not require their eyes to display seasonally developed gonads (2) nor do they need a pineal gland (13). In addition to ducks and quail reported initially, other species showing little or no need for eyes for reproductive development included the House Sparrow, *Passer domesticus* (26, 29); Japanese Quail, *Coturnix coturnix japonica* (5, 30, 31); White-crowned Sparrow, *Zonotrichia leucophrys gambelii*, Golden-crowned Sparrow, *Z. atricapilla* (9); Tree Sparrow, *Spizella arborea* (45, 46) and chicken, *Gallus gallus* (10, 34). Species showing no pineal requirement for normal development of the testes included Japanese Quail (12, 13), chicken (11), House Sparrow (29), and Tree Sparrow (46).

Encephalic photoreceptors have been major candidates to examine for possessing the required sensory elements for detecting seasonal or artificial changes in day length (8, 40). Several techniques were utilized to locate EPRs including electrolytic lesioning, isolating brain areas *via* knife cuts and photostimulating brain regions and specific brain loci. The result has been the identification of two brain regions shown to affect significantly photoperiod-induced testes development or progesterone-induced ovulation (reviewed in 15). One site is the medio-basal hypothalamus (MBH, Fig. 1). Large lesions of this region have been shown to block photoperiod-induced testicular development (4). A detailed spectroradiometric investigation determined which wavelength of light was most effective for releasing LH from the pituitary in birds. The light source was attached to the top of the head of a Japanese Quail and had to penetrate through the skull, brain and illuminate the MBH (6). The authors concluded that the action spectrum for the photoperiodic response in the bird matched rhodopsin visual pigments (6). A monoclonal antibody (RET-P1) against a membrane fraction of the rat retina was successful in identifying potential photoreceptors in the MBH of the dove, quail and duck (41). In addition, it was determined that the opsin-like immunoreactive (ir) neurons were a sub-population of vasoactive intestinal polypeptide (VIP) ir cerebrospinal fluid-contacting neurons (CSFcn) in the tuberal region of the hypothalamus (41). The CSFcn have been proposed to be EPRs for a number of years (42).

The second brain site containing EPRs was reported to be in the lobus parolfactorius (LPO) using radioluminous paint pellets (33). Recently the LPO has been renamed the medial striatum (MSt) by an international forum (37). The use of immunocytochemistry and the RET-P1 antibody further refined the location and showed the neurons to be in the septal region, just medial to the MSt (41). We have more specifically identified the site as the lateral septal organ (LSO, Fig. 1), a circumventricular organ (22) containing the unusual CSFcn that immunostain with antibodies to VIP (17) as well as show gene expression for that peptide (21).

Breakthrough results have been published in the past decade demonstrating that in lizards (*Anolis carolinensis*) only CSFcn in the septal area of the deep brain, homologous to the avian lateral septum, immunostained with antisera against cone opsin (7). Rhodopsin ir and gene expression were detected in the lateral septal organ (LSO) of the pigeon (43). A subsequent study showed that an antibody to the α-subunit of rod-type transducin (Gt1α) was found in CSFcn in the LSO of the pigeon (44). In addition, using electron microscopy direct synaptic connections have been demonstrated between VIP immunoreactive

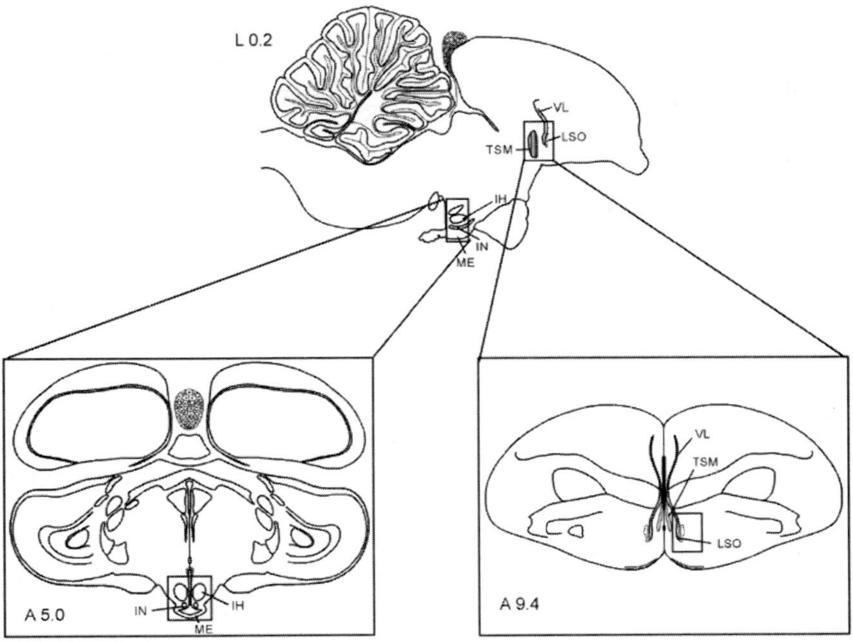

Fig. 1 Sections of a chick brain in planes of the Kuenzel and Masson chicken stereotaxic atlas (20) showing the two proposed sites that contain extra-retinal photoreceptors (EPRs). The site shown on cross-sectional plane A 5.0 is located in the mediobasal hypothalamus (MBH) while the second located on plane A 9.4 is found in the septal region, dorsal to the hypothalamus. Abbreviations: IH = Inferior hypothalamic nucleus, IN = infundibular nucleus, LSO = lateral septal organ, ME = median eminence, TSM = septopallio-mesencephalic tract, VL = lateral ventricle.

axons and gonadotropin releasing hormone (GnRH) perikarya located in the septal brain region (14) and opsin positive terminals form axo-dendritic synapses onto GnRH dendrites (39) in the lateral septal and preoptic area of the pigeon and Ring Dove, respectively. Evidence suggests that the proposed EPRs (VIP or opsin ir neurons) innervate GnRH neurons in the septal, preoptic brain area which in turn project to the median eminence. The anatomical information suggests a direct neural pathway in birds between nerve cells that respond to photoperiodic information and the classical hypothalamo-pituitary gonadal axis responsible for causing rapid development of the gonads in birds subjected to long days.

Up to the present time, only anatomical studies have suggested the importance of the LSO in containing CSFcn essential for responding to photostimulation in birds. We therefore have designed functional studies to test the hypothesis that the LSO contains specialized neurons that are part of the PNES responsible for responding to changes in day length and activating gonadal development.

Electrolytic lesion (knockout) study

If the LSO is a principal component of the PNES, then removal of the organ by applying bilateral electrolytic lesions should significantly attenuate the response of the testes in birds subjected to a significant increase in photoperiod. To conduct the study, we utilized a compound, sulfamethazine (SMZ), known to stimulate gonadal development in broiler chicks (19, 24). Our data to date suggest that SMZ operates via the PNES as a compound which when given to broiler chicks, exposed to a long photoperiod induces maximum growth and development of the gonads, and significantly increases plasma luteinizing hormone (LH) and follicle stimulating hormone (16). In the absence of SMZ treatment, gonadal growth in chicks exposed to long days over a two-to-four week period is negligible. Sulfamethazine treatment also results in an increase in number of visible neuropeptide Y neurons in the hypothalamus (24) which produce a peptide known to augment the release of LH from the anterior pituitary (reviewed in 18). Adding SMZ to a standard broiler ration at 0.2% enables us to determine in an efficient manner whether the PNES is operating at a near maximum rate in treated birds as testes weight is consistently greater within three weeks compared to controls fed the traditional diet and subjected to a long photoperiod. Without the compound, we would have to continue our studies for several weeks to assess whether the surgical treatment had a significant impact on gonadal growth and development following photostimulation.

Using a stereotaxic instrument, electrodes were targeted for the LSO and bilateral lesions made. Groups of chicks sustaining lesions and sham-operated controls (birds receiving the identical surgical treatment without having the electrolytic current passing into brain tissue) were then returned to their cages and fed either a standard ration or one with SMZ added to the diet. Following surgery, the photoperiod was increased from 8 hours of light per day to 16h/day. Birds were subjected to the light treatments and diets for three weeks. Thereafter birds were sacrificed, testes weighed and brains perfused and sectioned for histochemical analysis of the extent of damage to the medial portion of the LSO. Results of the study are shown in Fig. 2. Birds with bilateral damage of 30% or greater of the medial LSO (group L-Xb+S, Fig. 2) showed significant attenuation of gonadal development compared to sham-operated (Sh) controls fed the SMZ (S) diet (35). Of interest was that lesions (X) that missed (m) the target and were placed caudal and ventral to the LSO (group L-Xm(c+v)+S, Fig. 2) likewise resulted in a significant reduction in gonadal size, suggesting a destruction of axons from the VIP CSFcn in the LSO that project ventromedially toward their neuronal targets. In contrast, lesions that destroyed the lateral LSO or were positioned rostral to the LSO (group L-Xm(r+l)+S, Fig. 2), were not shown to reduce testes weight compared to controls fed the SMZ diet (group L-Sh+S, Fig. 2). Note that gonadal response in the LSO experimental (group L-Xb+S) was not reduced to the level of sham-operated (Sh) controls fed a standard (C) chick ration (group L-Sh+C, Fig. 2). Therefore either the bilateral lesions did not remove all CSFcn from the LSO in experimental birds or another brain location, such as the MBH (Fig. 1) may house EPRs that likewise function within the PNES to stimulate gonadal development. A problem with surgically eliminating proposed EPRs in the MBH is that many axons from GnRH neurons pass through the MBH to innervate the median eminence. Hence it is technically very difficult to disrupt the proposed EPRs in the MBH that occur in and about the infundibular (IN) and inferior hypothalamic (IH) nuclei (Fig. 1) without damaging fibers of passage comprising GnRH neurons terminating in the median eminence (Fig. 1). Nonetheless the MBH will need to be targeted to eliminate neurons containing rhodopsin-like ir without impacting GnRH neurons if an accurate assessment of their contribution to the PNES is to be done.

Direct physiological evidence that CSFcn in the LSO respond to light

The second set of experiments that address the question whether or not CSFcn in the LSO function as EPRs was completed using chick embryos. Research performed in rods and cones has shown that a decrease in intracellular free calcium level ([Ca^2+]i) occurs during photostimulation due to light-induced activation of a retinal phototransduction cascade (25, 27, 36, 48). We hypothesized that if CSFcn in the LSO serve as EPRs, then they should utilize a similar phototransduction pathway as retinal photoreceptors. We therefore monitored changes in [Ca^2+]i following light stimulation to putative brain photoreceptors in embryonic chick brain slices. Detailed

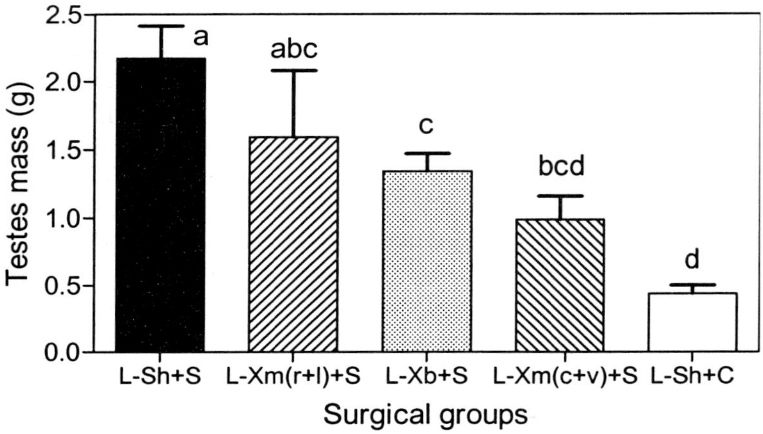

Fig. 2 Mean testes mass (± SEM) of photo-stimulated broiler chicks with surgical lesions targeted at the lateral septal organ (LSO), and fed sulfamethazine (SMZ) to stimulate testicular development, and sham-operated controls. The experimental groups were: birds fed SMZ (S) with a bilaterally lesioned LSO (L-Xb+S, n=10); birds fed SMZ with lesions placed rostral and lateral to the intact LSO (L-Xm(r+l)+S, n=7), and birds fed SMZ (S) with lesions placed caudal and ventral to the intact LSO (L-Xm(c+v)+S, n=7). Sham-operated (Sh) control birds were either fed SMZ (L-Sh+S, n=24) or a control (C) diet (L-Sh+C, n=23). Histograms with different letters are significantly different ($p<0.05$).

procedures for maintaining viable brain slices for imaging over time can be found in Li et al. (23). Following rapid dissection of the brain from the calvarium, the brain was placed in a cold bath with fresh, oxygenated artificial cerebrospinal fluid and rapidly sectioned into 300μm slices using a vibratome. Slices containing the LSO and a second circum-ventricular organ, the subseptal organ (SSO), were identified and placed in a darkened glass tube and incubated with a fluorescent calcium indicator, fluo-4 acetoxymethyl (Molecular Probes, Eugene, OR) for 1 hr. Slices were transferred to an imaging chamber made from glass bottom cell culture dishes. Chambers were positioned on the stage of an upright epifluorescent microscope (BH2, Olympus America) and imaged with water immersion lenses. A xenon arc light source and appropriate filters provided the correct excitation, emission and imaging during each experiment. Each slice and neuron visualized were photostimulated for 1sec every minute for 30min. Digital images were taken each time during photostimulation. Control of the experiment, image acquisition, and data analysis were done with the software package MetaFluor /MetaMorph (Universal Imaging, Downingtown, PA).

Results of the intracellular calcium imaging experiment are shown in Fig. 3. More details can be found in Li et al. (23). Note that CSFcn in the LSO showed a more rapid decline in the intensity of the calcium indicator dye than did control neurons found at a distance from the lateral ventricle and hence not CSFcn (Fig. 3A). The average of regression coefficients, that reflects the rate of change in fluorescence intensity of intracellular calcium, was compared between the two groups of neurons over the 30min period of photostimulation. Results are shown in Fig. 3C. Cerebrospinal fluid-contacting neurons in the LSO displayed a significantly greater change in calcium intensity (t-test, $p \leq 0.01$). A second group of control neurons analyzed were CSFcn, however, located in another circumventricular organ, the SSO. The SSO has not been shown in previous studies to contain neurons having a photoperiodic function associated with seasonal events. Similar to the first experiment, CSFcn in the LSO showed a more rapid rate of decrease in fluorescence intensity than did CSFcn located in the SSO (Fig. 3B). The mean change in calcium fluorescence intensity was then compared between CSFcn in the two circumventricular organs. The mean of regression coefficients for CSFcn in the LSO was significantly greater than that determined for the SSO (Fig. 3D, $p \leq 0.01$). In both sets of experiments there was a greater than 50% average rate of change in fluorescence intensity of VIP CSFcn in the LSO compared to the respective control groups. We conclude that the CSFcn in the LSO respond to phototimulation by showing a decrease in intracellular calcium, similar to rods and cones in the vertebrate eye and is physiological evidence that the neurons may serve as EPRs in birds.

Conclusions

The pathway defining the entire PNES remains elusive in the avian brain. There has been considerable controversy as to the location/s of receptors responsible for responding to photoperiodic information impacting an individual species. Considerable anatomical data have suggested that EPRs, particularly occurring in the MBH, are a major component of the PNES. The recent elegant work in the Japanese Quail (47, 49) has shown circadian clock genes as well as the gene encoding type two iodothyronine deiodinase (Dio2) located in the MBH. The latter catalyses the conversion of thyroxine to the active metabolic hormone triiodothyronine (T3) and is induced by light (49). The intriguing results

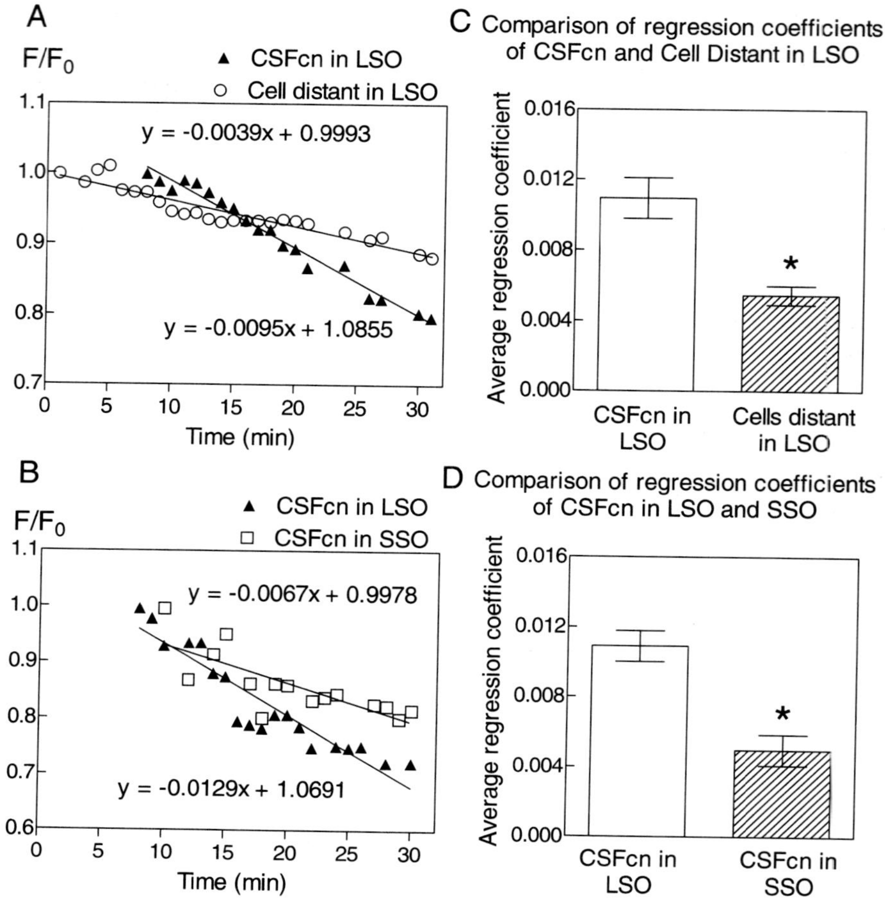

Fig. 3 Representative changes in percent fluorescence intensity of an indicator of free intracellular calcium (F/F_0) in cerebrospinal fluid-contacting neurons (CSFcn) in the LSO and control neurons during 30 min photostimulation and a comparison of average regression coefficients. A decrease in intracellular free calcium is diagnostic for a photoresponsive cell. (A) Percent fluorescence intensity, starting at 1, was plotted over time for CSFcn in the LSO and cell distant from lateral ventricle as well as CSFcn in the SSO (B). Linear regression lines were generated and the equations are shown. (C) Histograms showing mean values for regression coefficients (± SEM) of CSFcn in the LSO and cells distant from the lateral ventricle (n=6 per treatment), as well as CSFcn in the LSO and the SSO (D, n=9 per treatment) $^*P \leq 0.01$

in quail have lead to one possible conclusion that the major neural components for sensing and recording photoperiodic time measurement in birds occur in the MBH (1). Our data suggest that CSFcn in the LSO are another source of EPRs serving as sensory detectors and a component of the PNES that affects traits regulated by photoperiod such as gonadal development. Further studies are needed to substantiate the proposed role of the LSO in housing EPRs and, more importantly, to determine how that group of specialized neurons interacts with the clock and Dio2 genes in the MBH to track seasonal photoperiodic time measurement. It will take a concerted effort to decipher this neural pathway critical for the fitness and survival of many avian species.

Acknowledgments

Authors are grateful to Christopher Golden for technical help and Dr. Ronald McNew for statistical expertise. The work was supported by N.S.F. Grant # IBN-0315793 to WJK and MBF.

References

1 Ball G & Balthazart J (2003) Birds return every spring like clockwork, but where is the clock. *Endocrinology* **144** 3739-3741

2 Benoit J (1964) The role of the eye and the hypothalamus in photostimulation of gonads in the duck. *Annals of the New York Academy of Sciences* **117** 204-217

3 Farner DS (1964) The photoperiodic control of reproductive cycles in birds. *American Scientist* **52** 137-156

4 Follett BK & Davies DT (1975) Photoperiodicity and the neuroendocrine control of reproduction in birds. *Symposium Zoological Society of London* **No. 35** 199-224

5 Follett BK, Davies DT & Magee V (1975) The rate of testicular development in Japanese quail (*Coturnix coturnix-japonica*) following stimulation of the extra retinal photoreceptor. *Experentia* **31** 48-50

6 Foster RG & Follett BK (1985) The involvement of a rhodopsin-like photopigment in the photoperiodic response of the Japanese quail. *Journal of Comparative Physiology* **A 157** 519-528

7 Grace MS, Alones V, Menaker M & Foster RG (1996) Light perception in the vertebrate brain: an ultrastructural analysis of opsin- and vasoactive intestinal polypeptide-immunoreactive neurons in iguranid lizards. *Journal of Comparative Neurology* **367** 575-594

8 Groos G (1982) The comparative physiology of extraocular photoreception. *Experentia Generalia* **38** 989-1128

9 Gwinner EG, Turek FW & Smith SD (1971) Extraocular light perception in photoperiodic responses of the white-crowned sparrow (*Zonotrichia*

leucophrys) and of the golden-crowned sparrow (*Z. atricapilla*) *Zeitschrift für Vergleichende Physiologie* **75** 323-331

10 Harrison PC (1972) Extraretinal photocontrol of reproductive responses of leghorn hens to photoperiods of different length and spectrum. *Poultry Science* **51** 2060-2063

11 Harrison PC & Becker WC (1969) Extraretinal photocontrol in pinealectomized domestic fowl. *Proceedings of the Society for Experimental Biology and Medicine* **132** 161-164

12 Homma K & Sakakibara Y (1971) Encephalic photoreceptors and their significance in photoperiodic control of sexual activity in Japanese quail. In *Biochronometry* pp 331-341. Ed M Menaker. Washington DC: National Academy of Science

13 Homma K, McFarland LZ & Wilson WO (1967) Response of the *reproductive* organs of the Japanese quail to pinealectomy and melatonin injections. *Poultry Science* **46** 314-319

14 Kiyoshi K, Kondoh M, Hirunagi K & Korf HW (1998) Confocal laser scanning and *electron*-microscopic analyses of the relationship between VIP-like and GnRH-like immunoreactive neurons in the lateral septal-preoptic area of the pigeon. *Cell and Tissue Research* **293** 39-46

15 Kuenzel WJ (1993) The search for deep encephalic photoreceptors within the avian brain, using gonadal development as a primary indicator. *Poultry Science* **72** 959-967

16 Kuenzel WJ, Abdel-Maksaud MM, Elsasser T & Proudman JA (2004) Sulfamethazine advances puberty in male chicks by effecting a rapid increase in gonadotropins. *Comparative Biochemistry and Physiology Part A* **137** 349-355

17 Kuenzel WJ & Blähser S (1994) Vasoactive intestinal polypeptide (VIP)-containing neurons: distribution throughout the brain of the chick (*Gallus domesticus*), with focus upon the lateral septal organ. *Cell and Tissue Research* **275** 91-107

18 Kuenzel WJ & Fraley GS (1995) Neuropeptide Y: its role in food intake and reproductive function in birds. *Poultry and Avian Biology Review* **6** 185-209

19 Kuenzel WJ, Macko Walsh K & Proudman JA (1997) Neuroendocrine control of early sexual development. In *Perspectives in Avian Endocrinology* pp 81-90. Eds S Harvey & R Etches. Bristol: Journal of Endocrinology Ltd

20 Kuenzel WJ & Masson M (1988) *A stereotaxic atlas of the brain of the chick* (*Gallus domesticus*). pp 166. Baltimore: The Johns Hopkins University Press

21 Kuenzel WJ, McCune SK, Talbot RT, Sharp PJ & Hill JM (1997) Sites of gene expression for vasoactive intestinal polypeptide throughout the brain of the chick (*Gallus domesticus*). *Journal of Comparative Neurology* **381** 101-118

22 Kuenzel WJ & van Tienhoven A (1982) Nomenclature and location of avian hypothalamic nuclei and associated circumventricular organs. *Journal of Comparative Neurology* **206** 293-313
23 Li H, Ferrari MB & Kuenzel WJ (2004) Light-induced reduction of cytoplasmic free calcium in neurons proposed to be encephalic photoreceptors in chick brain. *Developmental Brain Research* **153** 153-161
24 Macko Walsh K & Kuenzel WJ (1997) Effect of sulfamethazine on sexual precocity and neuropeptide Y neurons within the tuberoinfundibular region of the chick brain. *Brain Research Bulletin* **44** 707-713
25 Matthews HR, Torre V & Lamb TD (1985) Effects on the photoresponse of calcium buffers and cyclic GMP incorporated into the cytoplasm of retinal rods. *Nature* **313** 582-585
26 McMillan JP, Underwood HA, Elliott JA, Stetson MH & Menaker M (1975) Extraretinal light perception in the sparrow. IV. Further evidence that the eyes do not participate in photoperiodic photoreception. *Journal of Comparative Physiology* **97** 205-213
27 McNaughton PA, Cervetto L & Nunn BJ (1986) Measurement of the intracellular free calcium concentration in salamander rods. *Nature* **322** 261-263
28 Menaker M & Keatts H (1968) Extraretinal light perception in the sparrow II. Photoperiodic stimulation of testis growth. *Proceedings of the National Academy of Sciences, USA* **60** 146-151
29 Menaker M, Roberts R, Elliott J & Underwood H (1970) Extraretinal light perception in the sparrow, III: The eyes do not participate in photoperiodic photoreception. *Proceedings of the National Academy of Sciences, USA* **67** 320-325
30 Oishi T & Lauber JK (1973) Photoreception in the photosexual response of the quail. I. Site of photoreceptor. *American Journal of Physiology* **225** 155-158
31 Oishi T, Konishi T & Kato M (1966) Investigation of photoreception mechanism to control the gonadal development in Japanese quail. Environmental. *Control and Biology* **3** 87-90
32 Oksche A & Farner DS (1974) Neurohistological studies of the hypothalamo-hypophysial system of Zonotrichia leucophrys gambelii. *Advances in Anatomy and Embryological Cell Biology* **48** 1-136
33 Oliver J, Jallegeas M, Sicard B & Baylé JD (1980) Gonadotropic activation induced by local photic stimulation of the lobus parolfactorius in male quail. *Journal of Physiology* (Paris) ***6*** 611-616
34 Ookawa T (1970) Effects of bilateral optic enucleation on body growth and gonad in young male chicks. *Poultry Science* **49** 1531-1535
35 Rathinam T & Kuenzel WJ (2005) Attenuation of gonadal response to photostimulation following ablation of neurons in the lateral septal organ of chicks. *Brain Research Bulletin* **64** 455-461
36 Ratto GM, Payne R, Owen WG & Tsien RY (1988) The concentration of

cytosolic free calcium in vertebrate rod outer segments measured with fura-2. *Journal of Neuroscience* **8** 3240-3246

37 Reiner A, Perkel DJ, Bruce LL, Bulter AB, Csillag A, Kuenzel W, Medina L, Paxinos G, Shimizu T, Striedter G, Wild M, Ball GF, Durand S, Güntürkün O, Lee DW, Mello CV, Powers A, White SA, Hough G, Kubikova L, Smulders TV, Wada K, Dugas-Ford J, Husband S, Yamamoto K, Yu J, Siang C & Jarvis ED (2004) Revised nomenclature for avian telencephalon and some related brainstem nuclei. *Journal of Comparative Neurology* **473** 377-414

38 Rowan W (1926) On photoperiodism, reproductive periodicity, and the annual migrations of birds and certain fishes. *Proceedings of the Boston Society of Natural History* **38** 147-189

39 Saldanha CJ, Silverman AJ & Silver R (2001) Direct innervation of GnRH neurons by encephalic photoreceptors in birds. *Journal of Biological Rhythms* **16** 39-49

40 Scharrer E (1964) Photo-neuro-endocrine systems: general concepts. *Annals of the New York Academy of Sciences* **117** 13-22

41 Silver R, Witovsky P, Horvath P, Alones V, Barnstable CJ & Lehman MN (1988) Coexpression of opsin-and VIP-like-immunoreactivity in CSF-contacting neurons of the avian brain. *Cell Tissue Research* **253** 189-198

42 Vigh B & Vigh-Teichmann I (1998) Actual problems of the cerebrospinal fluid-contacting neurons. *Microscopy Research and Technique 41* 57-83

43 Wada Y, Okano T, Adachi A, Ebihara S & Fukada Y (1998) Identification of rhodopsin in the pigeon deep brain. *FEBS Letters* **424** 53-56

44 Wada Y, Okano T & Fukada Y (2000) Phototransduction molecules in the pigeon deep brain. *Journal of Comparative Neurology* **428** 138-144

45 Wilson FE (1990) Extraocular control of seasonal reproduction in female tree sparrows (*Spizella arborea*). *General and Comparative Endocrinology* **77** 397-402

46 Wilson FE (1991) Neither retinal nor pineal photoreceptors mediate photoperiodic control of seasonal reproduction in American Tree sparrows. (*Spizella arborea*). *Journal of Experimental Zoology* **259** 117-127

47 Yasuo S, Watanabe M, Okabayashi N, Ebihara S & Yoshimura T (2003) Circadian clock genes and photoperiodism: comprehensive analysis of clock gene expression in the mediobasal hypothalamus, the suprachiasmatic nucleus, and the pineal gland of Japanese Quail under various light schedules. *Endocrinology* **144** 3742-3748

48 Yau KW & Nakatani K (1985) Light-induced reduction of cytoplasmic free calcium in retinal rod outer segment. *Nature* **313** 579-582

49 Yoshimura T, Yasuo S, Watanabe M, Iiogo M, Yamamura T, Hirunagi K & Ebihara S (2003) Light-induced hormone conversion of T_3 to T_4 regulates photoperiodic response of gonads in birds. *Nature* **426** 178-181

Functional Avian Endocrinology
A Dawson & PJ Sharp (Eds)

Photoperiodic time measurement

T Yoshimura, S Yasuo, T Yamamura, N Nakao and S Ebihara

Division of Biomodeling, Graduate School of Bioagricultural Sciences,
Nagoya University, Nagoya 464-8601, Japan

Introduction

Most of animals living in temperate zones are seasonal breeders and offspring take advantage of the abundance of food and optimal climate for survival. Day length is the primary synchronizing environmental cue for most animals. However, the mechanism for measuring day length (photoperiodic time measurement) remains a profound mystery.

There are two major hypothetical models to describe the mechanism of photoperiodic time measurement, the external and the internal coincidence models (20, 21). In the external coinicidence model, light has two distinct roles. First, there is a light-entrained circadian clock that keeps time. Second, light induces a photoperiodic response only at a specific time of day referred to as the "photoinducible phase". This photoinducible phase is determined by the light entrained circadian clock. In the internal coincidence model, two circadian oscillators exist, and the phase relationship between the two circadian oscillators, which varies depending on the photoperiod, is important for the photoperiodic response. In either model, the circadian clock is involved and it has already been established that the circadian clock is involved in photoperiodic time measurement (21). Recent progress in understanding the molecular mechanism of the circadian clock (14, 24) has provided a way to examine the relationship between the circadian clock and photoperiodism (28-30).

Where is the clock?

In birds, circadian pacemakers are localized in the eye, the pineal gland, and the suprachiasmatic nucleus (SCN) (5, 31). However, lesions in these pacemakers do not block the gonadal photoperiodic response (2, 10, 26), which indicates the possibility that the circadian clock regulating photoperiodism may exist in another region of the brain. A most likely candidate area is the mediobasal hypothalamus (MBH), including the infundibular nucleus (IN), the inferior hypothalamus (IH), and the median eminence (ME). This is because lesions in the IN or ME block photoinduced luteinizing hormone (LH) release (2, 23), even though the GnRH neuronal system is left intact (9). In addition, the MBH contains deep brain photoreceptors, which are thought to be involved in photoperiodic light perception (22, 25). Furthermore, Meddle and Follett have shown the expression of Fos-like immunoreactivity in the IN and ME after photostimulation (15, 16). Therefore, although almost all the components required for the photoperiodic gonadal regulation seem to be localized in the MBH, evidence for the presence of a circadian clock has been lacking.

Recently, Yasuo *et al.* (28) reported that all known avian clock genes are expressed in the MBH of Japanese quail, as well as in the circadian pacemakers, the SCN and the pineal gland (Fig. 1). The temporal patterns of expresssion of clock genes did not differ between long day and short day in the MBH, whereas various changes were observed in the SCN and the pineal gland. The rhythmic expression of clock genes was also characterised under a night-time interrupted light schedule. Expression profiles of clock genes in the SCN and the pineal gland under a night-time interruption schedule, which caused testicular growth, were not consistent with those under long day conditions. These results indicate that different expression profiles of circadian clock genes in the SCN and the pineal gland are not an absolute requirement for encoding and decoding of photoperiodic information. In contrast, patterns of expression of circadian clock genes in the MBH were stable under night-time interrupted schedules. These results indicate that the circadian clock involved in photoperiodic time measurement is located in the MBH, and this clock seems to have the role in keeping a steady-state photoinducible phase for the photoperiodic time measurement.

Identification of the genes which regulates photoperiodic time measurement in birds

All the molecular machinery required for photoperiodic time measurement seems to be localized in the MBH. If this region is the center for the control of the photoperiodic response, it is reasonable to assume that some molecular event must take place in the MBH when birds respond to photoperiodic change. Light pulses at the photoinducible phase are known to mimic long day stimulus (7). Yoshimura *et al.* (32) used differential subtractive hybridization analysis to analyze differences in gene expression in the MBH in Japanese quail given a 1h light pulse during the photoinducible phase and birds held on short days given no light pulse. They found one gene (*type 2 iodothyronine deiodinase* (*Dio2*)) among 150 clones, whose expression was induced by the light pulse. Induction of this gene was observed in the dorsal hypothalamus (DH) and was specific to the photoinducible

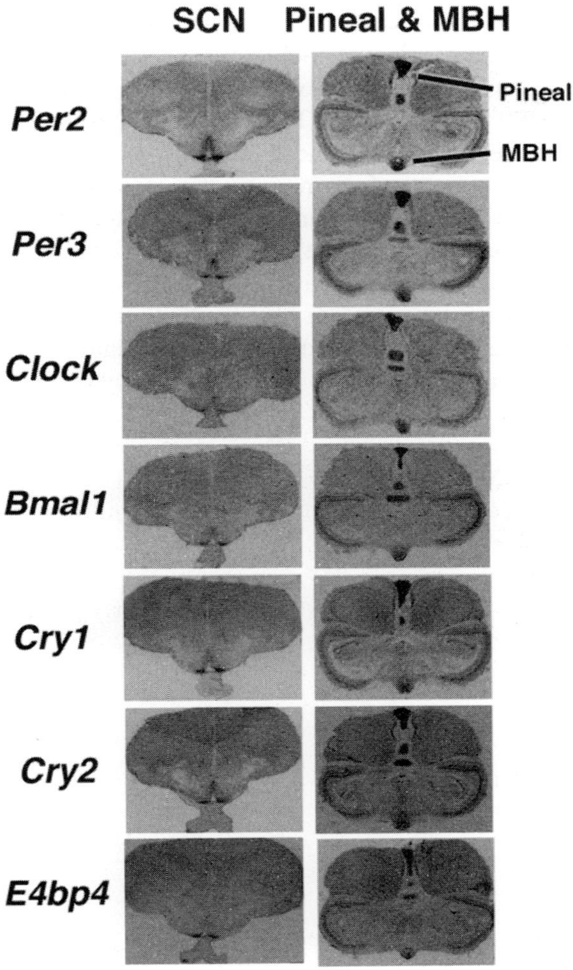

Fig. 1 Representative autoradiographs of clock gene (*Per 2, Per3, Clock, Bmal1, Cry1, Cry2* and *E4bp4*) expression in the suprachiasmatic nucleus (SCN), the pineal gland, and the mediobasal hypothalamus (MBH) in Japanese quail. (Data from 28).

phase (Fig. 2a, c). *Dio2* expression was also observed in the basal tuberal hypothalamus (BTH) that consists of the infundibular nucleus (IN) and median eminence (ME) (Fig.2b). In the BTH expression was high under long day conditions and low under short day conditions (Fig. 2d) and no acute induction of *Dio 2* by a 1h light pulse was observed in the BTH. The findings for these two brain regions, (DH and BTH) accorded with the results of lesion study conducted by Sharp and Follett (23), which suggested an important role for these two regions in mediating the reproductive photoperiodic response.

The gene *Dio2* encodes the enzyme, type 2 deiodinase. This enzyme catalyzes outer ring deiodination of thyroid hormones and converts the prohormone thyroxine (T_4) to a bioactive form, triiodothyronine (T_3) (11). It is known that thyroid hormones are involved in the regulation of the photoperiodic response, and the discovery of a photoinduced change in the expression of this gene confirms the role of thyroid hormones in the response (3, 4, 19). Indeed, T_3 content was approximately 10 times higher in the MBH under the long days than under the short days as a result of the action of this enzyme. Such a difference was not observed in other parts of the brain such as the optic tectum and cerebellum, and light-induced Dio2 seemed to control local T_3 concentrations in the MBH.

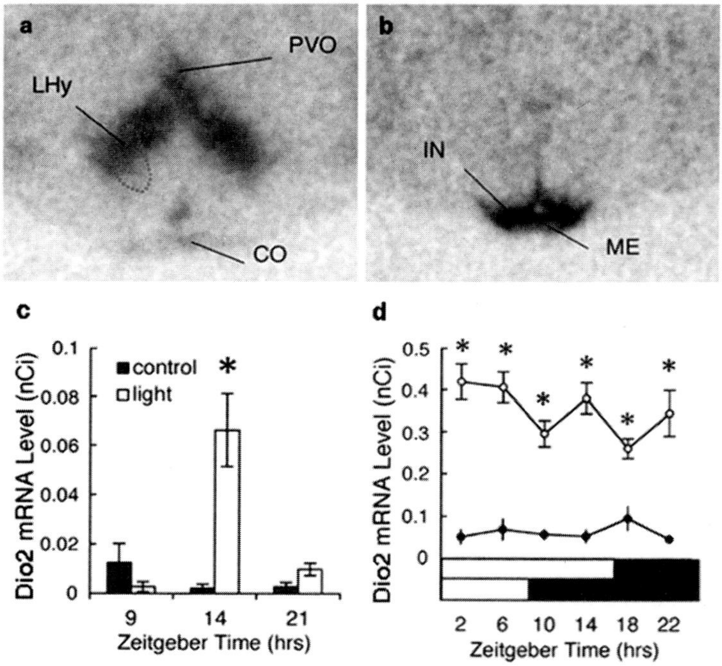

Fig. 2 Representative autoradiographs of type 2 iodothyronine deiodinase (*Dio 2)* gene expression in the (a), the dorsal hypothalamus ((DH) and (b), the basal tuberal hypothalamus (BTH) in Japanese quail after photostimulation and quantification of *Dio2* expression at different diel times in the (c) dorsal hypothalamus and (d) in the basal tuberal hypothalamus. Note that *Dio2* is upregulated acutely in the presence of light in the DH only during the photoinducuble phase (ie *Zeitgeber* time 14) whereas the gene is chronically upregulated at all times after photostimulation in the BTH. CO: optic chiasma; PVO: paraventricular organ; LHy: lateral hypothalamic area; IN: infundibular nucleus; ME: median eminence. Open circle: long day condition, closed circle: short day condition. (Data from 32).

Target site and role of long day-induced triiodothyronine

The genomic action of thyroid hormone is mediated by the thyroid hormone receptors (TRs) and retinoid X receptors (RXRs) (33). Expression of TRα, TRβ and RXRα was observed in the BTH, which suggested that locally light-generated T_3 may act on the BTH. Thyroid hormones are known to be involved in the development and plasticity of the central nervous system (1). Therefore, it is possible that thyroid hormone dependent morphological changes occur in the BTH when short-day birds are transferred to long-days. Immunoelectron microscopy using GnRH antibody revealed that the distance between GnRH nerve terminals and basal lamina of the median eminence was longer under short days than under long days. In addition, conventional transmission electron microscopy revealed the ensheathment of nerve terminals by the glial endfeet under short-day conditions (27, Fig.3). It is therefore suggested that photoperiodicically-induced GnRH release is controlled at the GnRH terminals by glia, rather than at the point of GnRH synthesis in the cell body (3, 16). Thus, morphological changes in the neuro-glial interaction between GnRH nerve terminals and glial endfeet may regulate seasonal GnRH secretion.

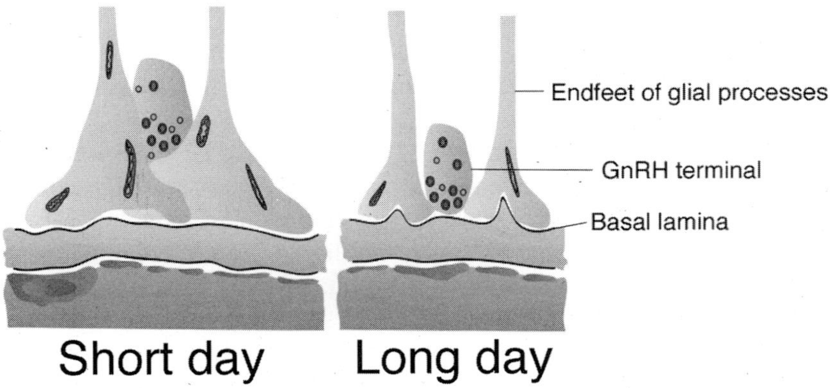

Fig. 3 Schematic to show suggested morphological changes in the relationship between GnRH nerve terminals and endfeet of glial processes in the median eminence under short and long days. GnRH nerve terminals are in close proximity to the basal lamina under long days but not under short days when they are fully ensheathed by glial processes.

Effect of thyroid hormone and Dio2 inhibitor infusion

To assess the functional significance of Dio2, the effects of intracerebroventricular infusion of thyroid hormones and Dio2 inhibitor were examined on testicular growth in short day quail (32).

Physiological doses of T_3 mimicked photoperiodically induced testicular growth in a dose dependent manner, while T_4 had little effect. Interestingly, however, no testicular growth was observed at the largest dose of T_3 infusion. Since thyroid hormone is also known to be involved in the regulation of photorefractoriness, this result suggests the possibility that down regulation mechanism of thyroid hormone action may regulate photorefractoriness. This hypothesis will need to be confirmed by future study. In addition, intracerebro-ventricular injection of the Dio2 inhibitor iopanoic acid, reduced testicular growth under long day conditions. Taken together, these observations suggest that light-induced *Dio2* expression in the MBH is critical for the regulation of the photoperiodic response of the gonads in Japanese quail.

Photoperiodic time measurement in the pars tuberalis

Prolactin (PRL) is associated with incubation behavior and parental behavior (8). Prolactin secretion is also regulated by photoperiod in many species and in mammals, the pars tuberalis (PT) of the pituitary stalk is known to be the center for the photoperiodic control of prolactin secretion (18). In birds, vasoactive intestinal peptide (VIP) is known to be the PRL-releasing hormone (6), and a role for the avian PT in the PRL regulation has not been established .

In recent years, many studies on clock gene expression in the mammalian PT has been reported and these are thought to be involved in the photoperiodic control of PRL secretion (12, 13, 17). Therefore, Yasuo *et al.* examined clock gene expression in the PT of Japanese quail (29). Among all the clock genes, *Per2* and *Cry1* were strongly expressed, and the expression of both genes was rhythmic under long day, short day, and nighttime interrupted light conditions. Detailed analysis of expression patterns revealed that the peak of *Per2* expression occurred early in the day both under long day and short day conditions, while the peak of *Cry1* expression was observed around light offset and phase delayed under a lengthened photoperiod. Thus, the phase relationship of these two genes changed under different photoperiods. Light induction of the *Cry1* gene during the photoinducible phase seemed to be the reason for the different phase relationship between *Per2* and *Cry1*. The different phase relationships between *Per2* and *Cry1* are consistent with the internal coincidence model of photoinduction (20, 21). Since the protein products of *Per* and *Cry* genes are known to form complexes that regulate the transcription of target genes (14, 24), different phase relationships may regulate the expression of other clock genes and down stream genes which might be related to PRL secretion (Fig. 4).

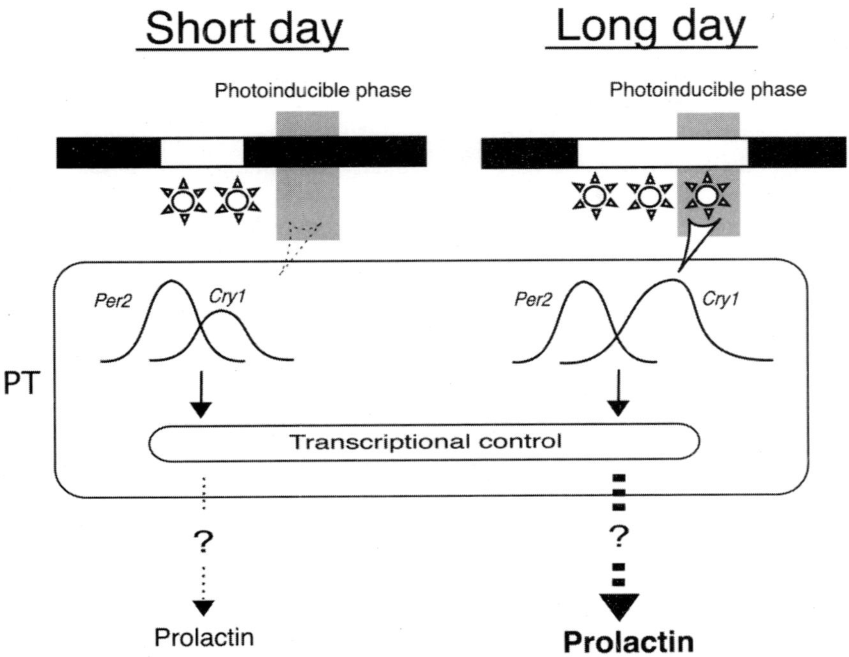

Fig. 4 A model for decoding photoperiodic information in the pars tuberalis (PT) of Japanese quail. Peak *Per2* expression occurs in the early day in both long and short days, whereas the *Cry1* peak is delayed by light given during the photoinducible phase. Thus, the phase relationship between *Per2* and *Cry1* varies depending on the photoperiod. The different *Per2-Cry1* association may affect the transcription of other circadian clock genes and downstream genes that regulate photoperiodic PRL secretion. (Redrawn from 29).

Conclusions

The mechanism underlying photoperiodic time measurement has remained a profound mystery for a long time. Recent progress has provided a new insight into the molecular mechanism underlying the for photoperiodic response. The role of thyroid hormones in the regulation of photoperiodism had been considered to be permissive because of the conflicting results derived from different experimental designs and species (3). Recent progress may provide a way of explaining these conflicting results in the future. These recent findings are just the end of the beginning for understanding photoperiodism and the core mechanism underlying photoperiodic time measurement awaits further clarification.

Acknowledgements

TYo is supported by the Program for Promotion of Basic Research Activities for Innovative Biosciences (PROBRAIN).

References

1 Bernal J (2002) Action of thyroid hormone in brain. *Journal of Endocrinology Investigation* **25** 268-288

2 Davies DT & Follett BK (1975) The neuroendocrine control of gonadotrophin release in Japanese quail. I. The role of the tuberal hypothalamus. *Proceedings of Royal Society of London Series B* **191** 285-301

3 Dawson A, King VM, Bentley GE & Ball GF (2001) Photoperiodic control of seasonality in birds. *Journal of Biological Rhythms* **16** 365-380

4 Dawson A & Thapliyal JP (2001) The thyroid and photoperiodism. In *Avian Endocrinology* Eds. A Dawson & CM Chaturvedi. pp.141-151 New Delhi, Narosa Publishing House

5 Ebihara S, Oshima I, Yamada H, Goto M & Sato K (1987) Circadian organization in the pigeon. In *Comparative aspects of circadian clocks*. pp. 88-94. Eds T Hiroshige & K Honma. Sapporo: Hokkaido University Press

6 El Halawani ME, Youngren OM & Pitts GR (1997) Vasoactive intestinal peptide as the avian prolactin-releasing factor. In *Perspectives in avian endocrinology*. pp.403-416. Eds S Harvey & RJ Etches. Bristol: Journal of Endocrinology Limited

7 Follett BK & Sharp P J (1969) Circadian rhythmicity in photoperiodically induced gonadotrophin release and gonadal growth in the quail. *Nature* **223** 968-971

8 Goldsmith AR (1983) Prolactin in avian reproductive cycles. In *Hormones and behaviour in higher vertebrates*. pp.375-387. Eds J Balthazart, E Prove & R Gilles. Berlin: Springer-Verlag

9 Juss TS (1993) Neuroendocrine and neural changes associated with the photoperiodic control of reproduction. In *Avian endocrinology*. pp.47-60. Ed PJ Sharp. Bristol: Society for Endocrinology

10 Konishi H, Iida K, Ohta M & Takahashi M. (1988) A possible role for the eyes in the photoperiodic response of quail. *General and Comparative Endocrinology* **72** 461-466

11 Leonard JL & Visser TJ (1986) Biochemistry of deiodination. In *Thyroid Hormone Metabolism* pp. 189-229 Ed G Hennemann. Marcel Dekker: New York

12 Lincoln G, Messager S, Andersson H & Hazlerigg D (2002) Temporal expression of seven clock genes in the suprachiasmatic nucleus and the pars tuberalis of the sheep: evidence for an internal coincidence timer. *Proceedings of National Academy of Sciences USA* **99** 13890-13895

13 Lincoln GA, Andersson H & Hazlerigg D (2003) Clock genes and the long-term regulation of prolactin secretion: evidence for a photoperiod/circannual timer in the pars tuberalis. *Journal of Neuroendocrinology* **15** 390-397

14 Lowrey PL, Shimomura K, Antoch MP, Yamazaki S, Zemenides PD, Ralph MR, Menaker M & Takahashi JS (2000) Positional syntenic cloning and functional characterization of the mammalian circadian mutation *tau*. *Science* **288** 483-492

15 Meddle SL & Follett BK (1995) Photoperiodic activation of Fos-like immunoreactive protein in neurons within the tuberal hypothalamus of Japanese quail. *Journal of Comparative Physiology A* **176** 79-89
16 Meddle SL & Follett BK (1997) Photoperiodically driven changes in Fos expression within the basal tuberal hypothalamus and median eminence of Japanese quail. *Journal of Neuroscience* **17** 8909-8918
17 Messager S, Ross AW, Barrett P, Morgan PJ (1999) Decoding photoperiodic time through *Per1* and ICER gene amplitude. *Proceedings of National Academy of Sciences USA* **96** 9938-9943
18 Morgan PJ & Williams LM (1996) The pars tuberalis of the pituitary: a gateway for neuroendocrine output. *Reviews in Reproduction* **1** 153-161
19 Nicholls TJ, Follett BK, Goldsmith AR & Pearson H (1988) Possible homologies between photorefractoriness in sheep and birds: the effect of thyroidectomy on the length of the ewe's breeding season. *Reproduction Nutrition Development* **28** 375-385
20 Pittendrigh CS & Minis DH (1964) The entrainment of circadian oscillations by light and their role as photoperiodic clocks. *American Naturalist* **98** 261-294
21 Pittendrigh CS (1972) Circadian surfaces and the diversity of possible roles of circadian organization in photoperiodic induction. *Proceedings of National Academy of Sciences USA* **69** 2734-2737
22 Saldanha CJ, Leak RK & Silver R (1994) Detection and transduction of daylength in birds. *Psychoneuroendocrinology* **19** 641-656
23 Sharp PJ & Follett BK (1969) The effect of hypothalamic lesions on gonadotrophin release in Japanese quail (*Coturnix coturnix japonica*). *Neuroendocrinology* **5** 205-218
24 Shearman LP, Sriram S, Weaver DR, Maywood ES, Chaves I, Zheng B, Kume K, Lee CC, van der Horst GT, Hastings MH & Reppert SM (2000) Interacting molecular loops in the mammalian circadian clock. *Science* **288** 1013-1019
25 Silver R, Witkovsky P, Horvath P, Alones V, Barnstable CJ & Lehman MN (1988) Coexpression of opsin- and VIP-like-immunoreactivity in CSF-contacting neurons of the avian brain. *Cell and Tissue Research* **253** 189-198
26 Siopes TD & Wilson WO (1974) Extraocular modification of photoreception in intact and pinealectomized *Coturnix. Poultry Science* **53** 2035-2041
27 Yamamura T, Hirunagi K, Ebihara S & Yoshimura T (2004) Seasonal morphological changes in the neuro-glial interaction between gonadotropin-releasing hormone nerve terminals and glial endfeet in Japanese quail. *Endocrinology* **145** 4264-4267
28 Yasuo S, Watanabe M, Okabayashi N, Ebihara S & Yoshimura T (2003) Circadian clock genes and photoperiodism: Comprehensive analysis of clock gene expression in the mediobasal hypothalamus, the suprachiasmatic nucleus, and the pineal gland of Japanese quail under various light schedules. *Endocrinology* **144** 3742-3748
29 Yasuo S, Watanabe M, Tsukada A, Takagi T, Iigo M, Shimada K, Ebihara S & Yoshimura T (2004) Photoinducible phase-specific light induction of

Cry1 gene in the pars tuberalis of Japanese quail. *Endocrinology* **145** 1612-1616

30 Yoshimura T, Suzuki Y, Makino E, Suzuki T, Kuroiwa A, Matsuda Y, Namikawa T & Ebihara S (2000) Molecular analysis of avian circadian clock genes. *Molecular Brain Research* **78** 207-215

31 Yoshimura T, Yasuo S, Suzuki Y, Makino E, Yokota Y, Ebihara S (2001) Identification of the suprachiasmatic nuleus in birds. *American Journal of Physiology* **280** R1185-R1189

32 Yoshimura T, Yasuo S, Watanabe M, Iigo M, Yamamura T, Hirunagi K & Ebihara S (2003) Light-induced hormone conversion of T_4 to T_3 regulates photoperiodic response of gonads in birds. *Nature* **426** 189-181

33 Zhang J & Lazar MA (2000) The mechanism for action of thyroid hormones. *Annual Review of Physiology* **62** 439-466

Part four

Non-photoperiodic seasonality

Rufous-winged Sparrow
Pierre Deviche

Functional Avian Endocrinology
A Dawson & P J Sharp (Eds)

Environmental control of reproduction in Sonoran Desert *Aimophila* sparrows

P Deviche and T Small

School of Life Sciences, Arizona State University, Tempe, Arizona 85287-4501, USA

Introduction

Many studies on the control of avian reproductive cycles have used temperate zone passerines as experimental models (28). With some exceptions such as crossbills (*Loxia* spp.; 9), most of these species live in environments where resources necessary to raise young vary predictably during the course of the year and they use annual changes in day length as a main proximate factor to control the activity of their reproductive system (28). In other environments such as deserts, resources necessary for successful breeding are available less predictably than in temperate regions (18). Many species breeding in these environments exhibit correspondingly high reproductive plasticity (18), but the extent to which they depend primarily on photoperiod as compared to other environmental factors to time their reproductive cycles is poorly known. Little information is available also on the nature and mode of action of nonphotic factors on these cycles.

Environmental information and the timing of reproduction

In many species of birds, favorable conditions for breeding occur only during a brief period of the year (6). To maximize reproductive success these species initiate reproductive development prior to the onset of the optimal conditions for raising offspring and such that offspring arrival and development coincide with maximal availability of food resources (5). Photoperiod is commonly used as a main proximate environmental cue to stimulate the reproductive system in anticipation of favorable breeding conditions (29). In most seasonal breeders studied to date, exposure of photosensitive birds to long days (LD, more than approximately 12 hours of light per day) stimulates gonadotropin-releasing hormone (GnRH) release from the hypothalamus, which induces the secretion of gonadotropins and ensuing gonadal development (5). Continued exposure to LD, however, eventually causes desensitization to this stimulus. This phenomenon, called *absolute* photorefractoriness, causes regression of the reproductive system followed by molt (6, 7). Return to a photosensitive condition is gradual and requires exposure to short days (SD, less than approximately 12 hours of light per day; 8, 27). In contrast to absolutely photorefractory species, the domestic Japanese Quail, *Coturnix coturnix japonica,* never loses photosensitivity completely. When exposed to sufficiently long and constant photoperiod, quail maintain developed gonads (11). However, a decrease in day length to a duration longer than that which induced gonadal development can cause testicular regression, and exposure of photoregressed birds to LD stimulates gonadal recrudescence without an intervening period of non-responsiveness. Japanese Quail are said to be *relatively* photorefractory (11, 23). Relative photorefractoriness as defined in quail has been suggested to exist in non-domesticated passerines (14), but in these birds has not been unequivocally demonstrated through captive experiments.

The effects of photoperiod on the reproductive system can be accelerated or inhibited by supplementary cues such as social factors, temperature, and food availability that provide information about the physical and social environments (16, 24, 28, 30). The neuroendocrine pathways that mediate the influence of these cues and the mechanisms by which their effects are translated into appropriate neuroendocrine responses are poorly understood (1, 4, 12).

Flexibility and opportunism/flexibility

Deserts represent approximately 25-30% of the land on Earth and are generally characterized by low and irregular precipitation and low primary productivity. Water and food availability is often erratic and unpredictable, and favorable conditions for breeding occur less predictably than in temperate zones (18). In response to these conditions many desert birds exhibit flexible breeding patterns or breed opportunistically in association with periods of precipitation (18). Gonadal cycles in these species may be controlled primarily by non-photoperiodic factors. For example, gonads of the Australian desert Zebra Finch, *Taeniopygia guttata*, are partially developed throughout the year, but may regress completely during periods of severe drought (17, 31). In this species, irregular precipitation stimulates reproduction (17), possibly as a result of increased food availability (32). Although it is generally thought that Zebra Finches are non-photoperiodic, exposure of males to day length exceeding that experienced in the wild is gonadostimulatory (3). Thus, Zebra Finches and possibly other allegedly non-photoperiodic species can respond to changes in day length (13, 15, 26). Whether the neuroendocrine mechanisms that control photoperiodic responses in these species are identical to those known in typically photoperiodic species is unclear.

Characteristics of the Sonoran Desert

The Sonoran Desert is located in Southwestern North America and Northern Mexico (26). Average high temperatures in the Arizona part of this desert range from 15 °C in winter to well over 35 °C during several months before as well as during the summer monsoon season (Fig. 1). Precipitation follows a bimodal pattern with summer monsoons (July-September) producing approximately 50% of the total annual rainfall (10). The amount and timing of monsoon precipitation vary greatly between years. For example, annual precipitation in the Tucson, Arizona area has ranged from 21 to 67 cm (1922-2002; 20), and the beginning of the monsoon season in this area has varied between 17 June and 25 July (1945-2004). Winter rains can also vary greatly in timing and quantity (26).

The bimodal pattern of rainfall results in two periods of the year when breeding is common in the Sonoran Desert: Spring, after winter rains and when ambient temperature is rising and day length is increasing; and during and after the summer monsoons, when day length is decreasing. Vleck (1993; 26) analyzed the breeding patterns of 49 Sonoran Desert bird species. Twenty-five of these species were found to breed only in the spring as is the case of many temperate and high latitude area species. Five species were identified that breed only in the summer, and 19 species were found to be capable of breeding in spring and summer. This diversity of breeding patterns provides unique opportunities to compare species with different breeding strategies but living in a common environment in order to better understand the environmental factors and the mechanisms used to regulate the timing of reproduction.

Rufous-winged Sparrow breeding biology

We have used the male Rufous-winged Sparrow, *Aimophila carpalis*, a year-round resident of the Sonoran desert, as an experimental model. Rufous-winged Sparrows develop their gonads in the spring, concurrent with increasing day length (Fig. 2; 22). Breeding occurs, but only infrequently, as early as March during years of unusually high winter or early spring precipitation (22). In most years, however, *A. carpalis* breeds well after the summer solstice, when photoperiod is already declining (22). Thus, Rufous-winged Sparrows have characteristics of seasonal breeders (predictable period of vernal gonadal recrudescence) and of flexible breeders (spring or summer reproduction starting at variable dates coincident with seasonally irregular rainfall). A breeding pattern that is rather similar to that of the Rufous-winged Sparrow exists in the non-desert-associated American Goldfinch, *Carduelis tristis*, which develops gonads in spring and reproduces in summer (19, 21).

Gonadal development in American Goldfinches is photoperiod-driven and as in most other species examined to date, is terminated by absolute photorefractoriness (19).

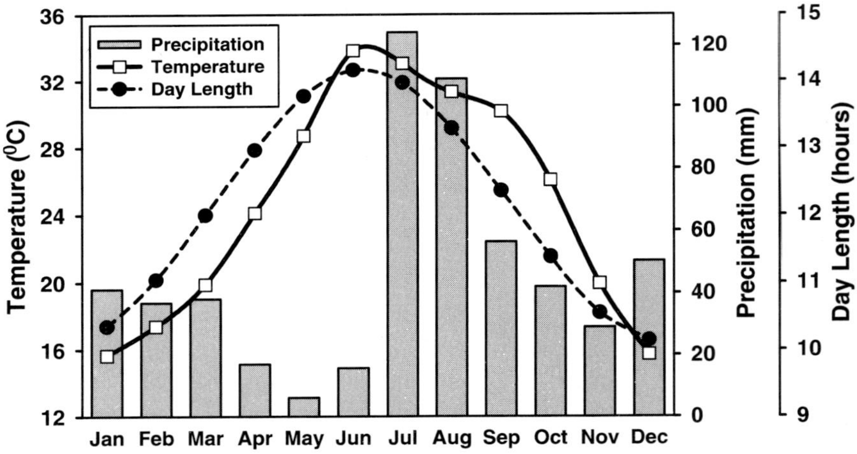

Fig. 1 Average daily maximum temperature, precipitation (mm water), and day length in the Green Valley, Southeast Arizona area. Temperature and precipitation data represent averages over at least 30 years. Source: The Weather Channel.

Annual reproductive cycle of the Rufous-winged Sparrow

Testicular recrudescence in Rufous-winged Sparrows begins in March and testes are fully developed by May - early June (Fig. 2A). Testicular regression occurs in September, when birds also undergo prebasic molt (Fig. 2B). Plasma luteinizing hormone (LH) and testosterone (T) do not follow the same seasonal profile as gonads. Circulating concentrations of these hormones remain relatively low throughout spring and until mid-summer and then increase two- to three-fold within two weeks of the beginning of the monsoon rains, when birds begin to breed. Plasma LH remains elevated throughout the summer reproductive period (mid-July to end of August) and then declines at the same time that gonads begin to regress, when plasma T has already decreased markedly relative to its seasonal peak. Thus, in contrast to the situation in many other species, plasma LH and T in *A. carpalis* are seasonally elevated as photoperiod is declining. These observations are consistent with the hypotheses that (a) gonadal growth and plasma LH and T secretion are controlled by different mechanisms; (b) increasing day length is not the proximate factor that stimulates the secretion of these hormones, and (c) autumnal testicular involution and decline of reproductive hormone plasma concentrations, which all take place when day length declines, are the end results of one or several photoperiod-mediated process(es).

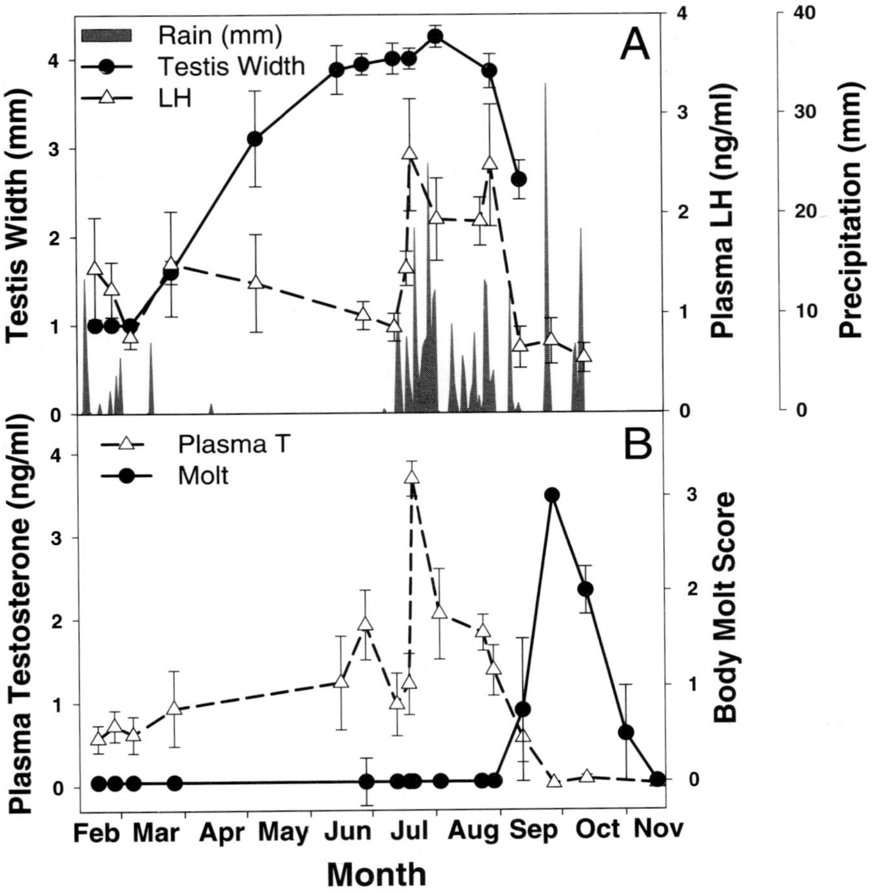

Fig. 2. (A): Testis length and plasma luteinizing hormone (LH) concentrations (means ± standard errors) of free-ranging adult male Rufous-winged Sparrows from a population in Southeast Arizona between March and November, 2003. The figure also shows local daily precipitation (mm water; data courtesy H. Hanson, National Weather Service during the same period. (B): Seasonal changes in plasma testosterone concentrations (means ± standard errors) and body molt score (medians ± 0.5 interquartile intervals; ranging from 0, no body molt to 3, heavy, generalized body molt) in the same birds as in panel A.

Photoperiodic control of reproductive activity in the Rufous-winged sparrow

We performed photoperiodic manipulations on captive adult male Rufous-winged Sparrows to characterize their responses to changes in day length. Males that we chronically exposed to SD (8 hours of light, 16 hours of darkness, 8L:16D) did not develop their testes (Fig. 3A), but transfer from SD to LD for two months induced partial (transfer to 13L:11D) or full (transfer to 14L:10D or 16L:8D) gonadal development. Males that we transferred from 16L:8D to shorter day length for two months partially (14L:10D) or completely (13L:11D and 8L:16D) regressed their testes (Fig. 3B) and started to molt (personal observations). Furthermore, males with regressed testes as a result of exposure to SD for three weeks redeveloped their gonads when transferred to LD (personal observations), indicating that they were photosensitive at the time of this transfer. Thus, male *A. carpalis* are photoperiodic and natural changes in day length may suffice to account for the seasonal cycle of gonadal development and regression seen in free-ranging birds.

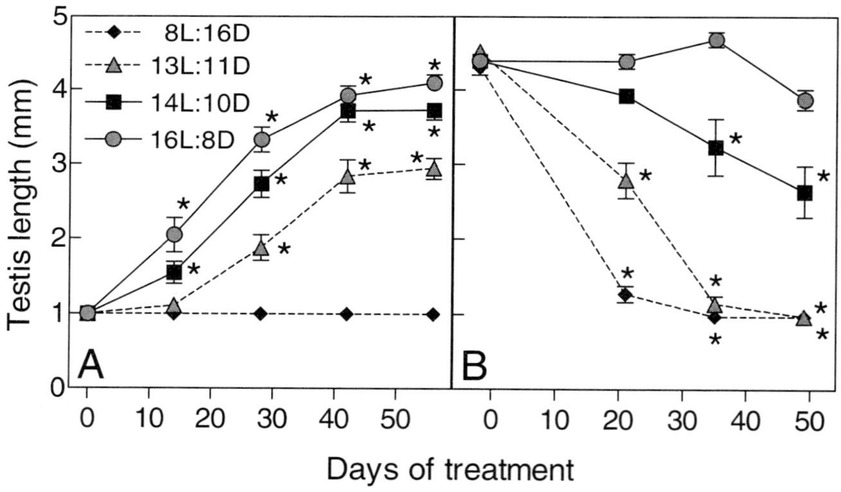

Fig. 3 Testis length (means ± standard errors) of captive adult male Rufous-winged Sparrows (A) transferred from 10 weeks of 8L:16D to 8L:16D, 13L:11D, 14L:10D, or 16L:8D on day 0; (B) birds transferred from 11 weeks of 16L:8D to 16L:8D, 14L:10D, 13L:11D, or 8L:16D on day 0. * indicates a significant difference ($p \leq 0.05$) compared to the corresponding data point for males exposed to 8L:16D (A) or 16L:8D (B; Student-Newman-Keuls pairwise comparison test).

To further investigate the photic regulation of autumnal gonadal regression, we brought adult males into captivity in July when they had been naturally exposed to LD for four months, and kept them under LD (16L:8D) until December, by which time they had been exposed to LD for nine consecutive months. This duration of LD exposure is considerably longer than that experienced by free-ranging birds. These captive males maintained fully developed testes throughout and did not molt. Further, males that we brought into captivity at the end of summer (September), *while* they were regressing their testes, and exposed to artificial LD, developed testes back to full size within four weeks (personal observations). When captured these males were, therefore, photosensitive and *absolute* photorefractoriness in this species is not responsible for autumnal gonadal regression.

As Fig. 3 shows, transfer from SD to 13L:11D stimulated partial gonadal development, but exposure to the latter photoperiod after longer day length caused testicular regression. This situation parallels that in the domesticated Japanese Quail, a *relatively* photorefractory species (see above). The strikingly similar photoperiodic responses of *A. carpalis* and Japanese Quail support the hypothesis in sparrows that *relative* photorefractoriness mediates autumnal testicular regression.

One potential advantage of relative *vs.* absolute photorefractoriness is to relax dependency on a single, fixed environmental factor (photoperiod) to terminate a reproductive cycle, effectively widening the temporal window during which this cycle can be completed. The timing of testicular regression and prebasic molt in Rufous-winged Sparrows varies between years such that earlier testicular regression is associated with earlier onset or faster progression of prebasic molt (Fig. 4: 2000 and 2003 *vs.* 2001), demonstrating modulation of these parameters by supplementary (nonphotic) information. This modulation of the effects of day length on late summer reproductive system activity by non photoperiodic factors may be particularly important for desert species such as *A. carpalis* that synchronize their breeding season with the arrival of summer monsoon rains. By influencing the timing of late summer gonadal involution, monsoon season-related supplemental cues may prolong the period during which birds can take advantage of local resources to breed and, therefore, ultimately increase reproductive success. These cues may also have the opposite effect, i.e., inactivate the reproductive system and as a result curtail reproduction when environmental conditions are unfavorable.

Photoperiodic responses of other desert *Aimophila* species

The above-described photoperiodic responses are not unique to Rufous-winged Sparrows and are generally similar in two congeneric and sympatric species, the Cassin's and the Rufous-crowned Sparrows (*A.*

cassinii and *A. ruficeps*). Similar to *A. carpalis*, males of these species have fully developed gonads until early September (personal observations). Adult males that we brought into captivity and held under 14L:10D until December, when they had been exposed to day length exceeding 12L for nine consecutive months, had developed gonads (Fig. 5, left panels). Transfer from 14L:13D to 13L:11D, 12L:13D, or 8L:16D for three weeks induced prebasic molt and, in Rufous-crowned Sparrows, gonadal regression. As in *A. carpalis*, transfer of photoregressed male Cassin's and Rufous-crowned Sparrows from 8L:16D to 13L:11D stimulated gonadal growth (Fig. 5, right panels). Thus, Cassin's and Rufous-crowned Sparrows are photoperiodic. The observation in Rufous-crowned Sparrows that transfer from 14L:10D to 13L:11D induced regression but transfer from 8L:16D to 13L:11D was gonadostimulatory, supports the hypothesis that this species is *relatively* photorefractory. The data thus suggest that not only Rufous-winged, but also Rufous-crowned and possibly Cassin's Sparrows regress gonads seasonally due to relative rather than absolute photorefractoriness.

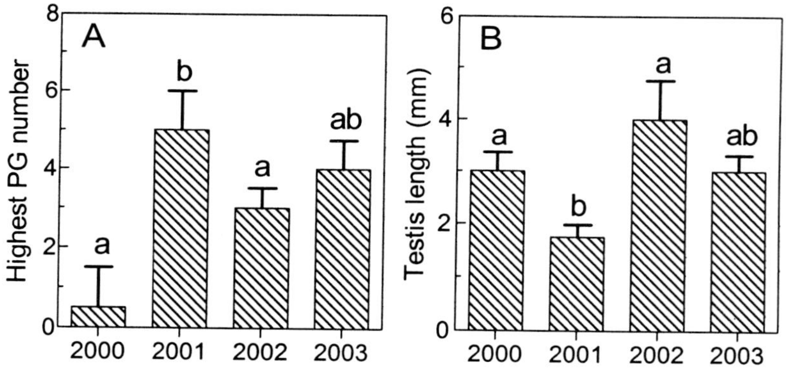

Fig. 4 Interannual variation in the time course of prebasic molt (A: highest growing primary number, ranging from 0 to 9) and gonadal regression (B: testis length) in free-ranging adult male Rufous-winged Sparrows. All data (medians + 0.5 interquartile intervals) were collected within the same five days (Sept. 6-11) of all calendar years. For each variable, medians with a same superscript letter do not differ significantly ($p > 0.05$; Student-Newman-Keuls multiple pairwise comparison test).

Control of LH secretion in Rufous-winged sparrow

We determined the photoperiod dependency of LH secretion in Rufous-winged Sparrows by measuring plasma LH in the birds used for the experiment described in Fig. 3A. Plasma LH increased after transfer from SD to 16L:8D but not 13L:11D or 14L:10D (Fig. 6). This increase indicates the potential of LD exposure after SD to stimulate LH secretion, but the physiological relevance of this finding is unknown because free-ranging Rufous-winged Sparrows never experience day length exceeding 15 hours (Fig. 1). It is possible that LD exposure stimulated LH secretion transiently before the first blood sample collection (day 25 of treatment, Fig. 6). Nevertheless, captive data do not support the hypothesis in *A. carpalis* that LH secretion is normally photoperiod-dependent. Consistent with this conclusion, plasma LH in free-ranging sparrows increases not in the spring, but in mid-summer as day length is decreasing (Fig. 2).

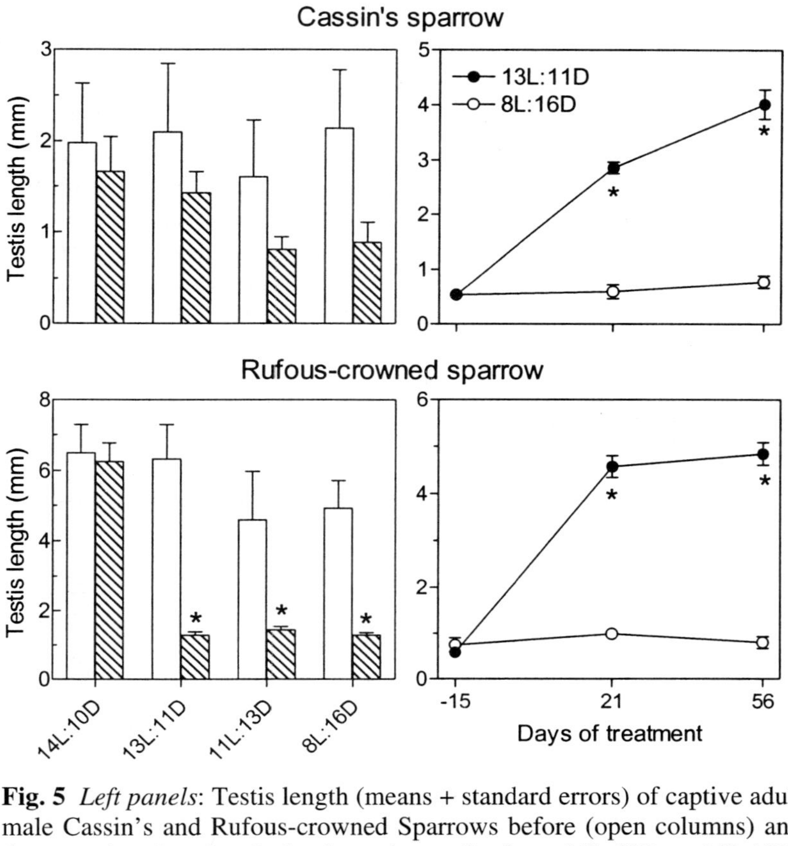

Fig. 5 *Left panels*: Testis length (means + standard errors) of captive adult male Cassin's and Rufous-crowned Sparrows before (open columns) and three weeks after (hatched columns) transfer from 14L:10D to 14L:10D, 13L:11D, 12L:12D, or 8L:16D. * indicates a significant difference ($p \leq 0.05$) with the corresponding group of males exposed to 14L:10D (Student-Newman-Keuls multiple pairwise comparison test). *Right panels*: Testis length (means ± standard errors) of captive adult male Cassin's and Rufous-crowned Sparrows transferred on Day 0 from 8L:16D to 8L:16D or 13L:11D. * indicates a significant difference ($p \leq 0.05$) with the corresponding 8L:16D group (Student-Newman-Keuls multiple pairwise comparison test).

The neuroendocrine mechanism(s) that mediate(s) the rapid increase in plasma LH and T observed at the beginning of the monsoon rain season in *A. carpalis* is(are) unknown. We recently began to investigate the hypothesis that gonadotropin-inhibiting hormone (GnIH) plays a role in this process. In Rufous-winged Sparrows as in other avian species investigated so far, GnIH-containing perikarya are located in the hypothalamic paraventricular nucleus of the brain (PVN; Fig. 7; 2, 25). Males sacrificed seven days before the beginning of the 2002 monsoon season had lower plasma LH than males sacrificed 20 days after the beginning of this season. The two groups of males had similar numbers of gonadotropin-releasing hormone-like immunoreactive (GnRH-I-li ir) perikarya in the preoptic area of the brain (personal observations), but fewer PVN GnIH-li ir perikarya during than before the monsoon season. One interpretation of this observation is that environmental factors associated with this season downregulate the brain GnIH system and the release of this neuropeptide, resulting in enhanced LH secretion. Accordingly, seasonal changes in the activity of the reproductive system in *A. carpalis* may result from a balance between stimulatory effects of GnRH and inhibitory effects of GnIH. We hypothesize that increasing day length in the spring stimulates the release of GnRH, in turn inducing the anterior pituitary gland to release of LH and FSH. However, at this time a GnIH-mediated mechanism limits LH release. Supplemental cues associated with the monsoon season remove this inhibition, thus facilitating plasma LH secretion.

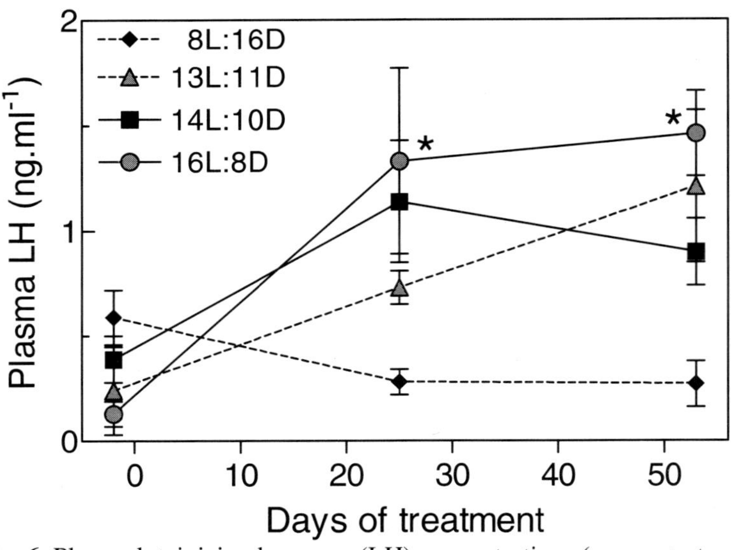

Fig. 6 Plasma luteinizing hormone (LH) concentrations (means ± standard error) of the birds used in the experiment shown on Fig. 3A. See Fig. 3 for additional comments.

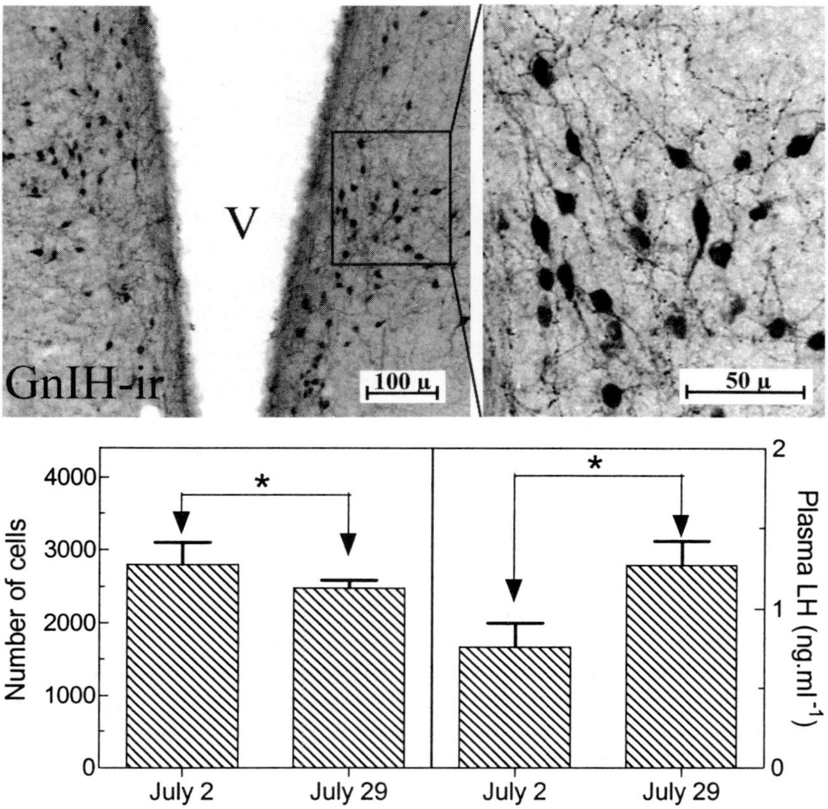

Fig. 7 *Upper panels*: Low and high magnification microphotographs showing the distribution of gonadotropin-inhibiting hormone (GnIH)-like immunoreactive (GnIH-li ir) perikarya and fibers in the paraventricular nucleus (PVN) of an adult male Rufous-winged Sparrow. *Lower panels*: numbers of PVN GnIH-li ir perikarya (left) and plasma LH concentrations in males sampled before (July 2) and during (July 29) monsoon season. Data are shown as means + standard errors. *: $p \leq 0.05$, comparison of July 2 and 29 data (Student's t test).

Conclusions

Seasonally breeding birds have long been used to investigate the role of environmental factors in the control of breeding cycles. Considerable progress has been made toward elucidating the nature of these factors and also their mechanisms of action and effects on the neuroendocrine system. While a large body of information on this subject has been collected for photoperiod, we know much less on the role of nonphotic factors on avian reproductive systems. We have addressed this issue using desert *Aimophila* sparrows as experimental models. Some of these sparrows exhibit flexible reproductive patterns such that the beginning and end of their annual breeding season vary between years to coincide with temporally and spatially irregular summer monsoon precipitations. The breeding season of these species is limited in time by the fact that they are photoperiodic. However, seasonal gonadal regression results not from absolute, but apparently from relative photorefractoriness. This results in birds being able to delay reproduction until mid-summer and to continue breeding long after photoperiod has begun to decline. Future studies on these and other desert species should lead to a better understanding of and appreciation for the plasticity of avian reproductive cycles and to new insights into the complex interactions between environmental factors in the control of these cycles.

Acknowledgments

The authors thank George Bentley and Peter Sharp for assistance with plasma LH assays, Kazuoshi Tsutsui for the generous gift the GnIH antiserum, the managers of the Santa Rita Experimental Range for providing precipitation data, and Russell Martin, Joseph Georges, and Matthew Ross for valuable field assistance.

References

1 Ball GF (1993) The neural integration of environmental information by seasonally breeding birds. *American Zoologist* **33** 185-199

2 Bentley G E, Perfito N, Ukena K, Tsutsui K & Wingfield JC (2003) Gonadotropin-inhibitory peptide in song sparrows (*Melospiza melodia*) in different reproductive conditions, and in house sparrows (*Passer domesticus*) relative to chicken-gonadotropin-releasing hormone. *Journal of Neuroendocrinology* **15** 794-802

3 Bentley GE, Spar BD, MacDougall-Shackleton SA, Hahn TP & Ball GF (2000) Photoperiodic regulation of the reproductive axis in male zebra finches, *Taeniopygia guttata. General and Comparative Endocrinology* **117** 449-455

4 Cheng MF, Peng JP & Johnson P (1998) Hypothalamic neurons preferentially respond to female nest coo stimulation: Demonstration of direct acoustic stimulation of luteinizing hormone release. *Journal of Neuroscience* **18** 5477-5489

5 Dawson A (1999) Photoperiodic control of gonadotropin-releasing hormone secretion in seasonally breeding birds. In *Neural Regulation in the Vertebrate Endocrine System*, pp. 141-159. Eds P Rao & P Kluwer. New York: Academic/Plenum Pub

6 Dawson A (2002) Photoperiodic control of the annual cycle in birds and comparison with mammals. *Ardea* **90** 355-367

7 Dawson A (1998) Photoperiodic control of the termination of breeding and the induction of moult in house sparrows *Passer domesticus. Ibis* **140** 35-40

8 Dawson A & Goldsmith AR (1997) Changes in gonadotrophin-releasing hormone (GnRH-I) in the pre-optic area and median eminence of starlings

(*Sturnus vulgaris*) during the recovery of photosensitivity and during photostimulation. *Journal of Reproduction and Fertility* **111** 1-6

9 Deviche P & Sharp PJ (2001) Reproductive endocrinology of a free-living, opportunistically breeding passerine (White-winged Crossbill, *Loxia leucoptera*). *General and Comparative Endocrinology* **123** 268-279

10 Douglas MW, Maddox RA, Howard K & Reyes S (1993) The Mexican Monsoon. *Journal of Climate* **6** 1665-1677

11 Follett BK & Pearce-Kelly A (1990) Photoperiodic control of the termination of reproduction in Japanese-Quail (*Coturnix coturnix japonica*). *Proceedings of the Royal Society of London Series B-Biological Sciences* **242** 225-230

12 Hahn TP, Boswell T, Wingfield JC & Ball GF (1997) Temporal flexibility in avian reproduction. *Current Ornithology* **114** 39-80

13 Hahn TP (1995) Integration of photoperiodic and food cues to time changes in reproductive physiology by an opportunistic breeder, the Red Crossbill, *Loxia curvirostra* (Aves, Carduelinae). *Journal of Experimental Zoology* **272** 213-226

14 Hahn TP, Pereyra ME, Sharbaugh SM & Bentley GE (2004) Physiological responses to photoperiod in three cardueline finch species. *General and Comparative Endocrinology* **137** 99-108

15 Hau M, Wikelski M & Wingfield JC (1998) A neotropical forest bird can measure the slight changes in tropical photoperiod. *Proceedings of the Royal Society of London Series B-Biological Sciences* **265** 89-95

16 Hau M, Wikelski M & Wingfield JC (2000) Visual and nutritional food cues fine-tune timing of reproduction in a neotropical rainforest bird. *Journal of Experimental Zoology* **286** 494-504

17 Keast A & Marshall AJ (1954) The influence of drought and rainfall on reproduction in Australian birds. *Proceedings of the Zoological Society of London* **124** 493-499

18 Maclean GL (1996) Breeding (1): timing of breeding. In *The Ecophysiology of Desert Birds*, pp.113-131 Berlin: Springer-Verlag

19 Marsh RH, Dougall-Shackleton SA & Hahn TP (2002) Photorefractoriness and seasonal changes in the brain in response to changes in day length in American goldfinches (*Carduelis tristis*). *Canadian Journal of Zoology* **80** 2100-2107

20 McClaran MP, Deborah AL, Wissler A & Wissler C (2002) Santa Rita Experimental Range digital database. Ogden, UT: U.S. Department of Agriculture, Forest Service, Rocky Mountain Research Station

21 Middleton ALA (1978) Annual cycle of the American Goldfinch. *Condor* **80** 401-406

22 Ohmart RD (1969) Physiological and ethological adaptations of the Rufous-winged Sparrow (*Aimophila carpalis*) to a desert environment. PhD Dissertation, University of Arizona, Tucson

23 Robinson JE & Follett BK (1982) Photoperiodism in Japanese quail: the termination of seasonal breeding by photorefractoriness. *Proceedings of the Royal Society of London Series B-Biological Science* **215** 95-116

24 Scheuerlein A & Gwinner E (2002) Is food availability a circannual zeitgeber in tropical birds? A field experiment on stonechats in tropical Africa. *Journal of Biological Rhythms* **17** 171-180

25 Ukena K, Ubuka T & Tsutsui K (2003) Distribution of a novel avian gonadotropin-inhibitory hormone in the quail brain. *Cell and Tissue Research* **312** 73-79

26 Vleck CM (1993) Hormones, reproduction, and behaviour in birds of the Sonoran desert. In *Avian Endocrinology*, pp.73-86 Ed PJ Sharp. Bristol: Journal of Endocrinology Ltd

27 Wilson FE (1990) On the recovery of photosensitivity in two passerine species, American Tree Sparrows (*Spizella arborea*) and Harris's Sparrows (*Zonotrichia querula*). *General and Comparative Endocrinology* **79** 283-290

28 Wingfield JC, Jacobs JD, Tramontin AD, Perfito N, Meddle S, Maney DL & Soma K (2000) Towards an Ecological Basis of Hormone-Behavior Interactions in Reproduction of Birds. In *Reproduction in Context*, pp. 85-128 Eds K Wallen & JE Schneider.Massachusetts: MIT Press Cambridge

29 Wingfield JC, Hahn TP, Levin R & Honey P (1992) Environmental predictability and control of gonadal cycles in birds. *Journal of Experimental Zoology* **261** 214-231

30 Wingfield JC, Hahn TP, Maney DL, Schoech SJ, Wada M. & Morton ML (2003) Effects of temperature on photoperiodically induced reproductive development, circulating plasma luteinizing hormone and thyroid hormones, body mass, fat deposition and molt in mountain white-crowned sparrows, *Zonotrichia leucophrys oriantha. General and Comparative Endocrinology* **131** 143-158

31 Zann RA (1996) Breeding periodicity. In *The Zebra Finch : a synthesis of field and laboratory studies*, pp. 114-138. New York: Oxford University Press

32 Zann RA, Morton SR, Jones KR & Burley NT (1995) The timing of breeding by zebra finches in relation to rainfall in central Australia. *Emu* **95** 208-222

Functional Avian Endocrinology
A Dawson & P J Sharp (Eds)

Effects of food availability on the reproductive system

T P Hahn, M E Pereyra, M Katti[1], G M Ward[2] and
S A MacDougall-Shackleton[3]

Section of Neurobiology, Physiology and Behavior, University of California Davis, Davis, CA, 95616, USA
[1] Dept. of Biology, California State University Fresno, Fresno, CA, 93740, USA
[2] Dept. of Ecology and Evolutionary Biology, Princeton University, Princeton, NJ, 08544, USA,[3] Departments of Biology and Psychology, University of Western Ontario, London, N6A 5C2, Ontario, Canada

Introduction

Food supply is widely accepted to be the primary ultimate determinant of the timing of breeding seasons of birds in seasonal environments (3, 21, 33, 40, 41). Food is unusual among ultimate factors in that it can also act as an important proximate regulator of reproductive physiology and behavior. Though this has been known for years, the quantity of information regarding proximate effects of food availability on avian reproductive systems lags far behind that for some other proximate factors, particularly photoperiod. In this paper, we discuss general issues regarding the effects of food on avian reproduction, with particular focus on wild taxa for which tracking changes in food availability is critical to reproductive success. We distinguish among several features of food availability: food amount, food type, food quality, and food predictability. We also discuss how proximate effects of food may be direct, through nutritional and energetic physiology (i.e., availability of fuel and building block molecules), or indirect, via a perceptual pathway (e.g., by seeing or physically handling the food). We also present a selective review of recent field and laboratory studies of wild taxa, including new data on several taxa of cardueline finches. Finally, we highlight avenues for future research in this important area of avian evolutionary physiology.

Ultimate and proximate importance of food availability in avian reproduction

It has long been known that in seasonal environments, avian reproduction is timed so that particularly demanding processes, such as egg production and feeding young, occur when food availability is high

(21). The idea that food supply determines, in an ultimate sense, timing of breeding is supported by the fact that reproductive success is highest for pairs that match rearing of young with peak food availability (e.g., 34, 28, 38). More generally, timing of the onset of nesting may be constrained in an ultimate sense by the availability of energy or nutrients required for egg production, or to feed nestlings or fledglings (33). Investment of stored fat or protein reserves ("capital" investment) can also permit advancement of nesting onset if individuals are in good physical condition (12, 26, 41). In any case, timing the onset of nesting to take maximal advantage of the season of favorable food availability requires that the reproductive axis be prepared before food availability actually permits egg laying or feeding of young; if onset of gonadal development did not occur until the food supply was sufficient to support egg production then the onset of nesting would be delayed well beyond the optimum. Birds have of course evolved to take advantage of predictive cues, most notably the seasonal change in photoperiod, to anticipate times of favorable food availability. These predictive cues activate the hypothalamo-pituitary-gonad axis and stimulate the early stages of reproductive development well before conditions actually could permit successful breeding. Consequently, ovaries will be ready to begin yolk deposition, and testes capable of releasing active spermatozoa, as soon as food supply permits breeding to commence.

Although the early stages of reproductive development are largely dependent on non-food proximate cues, food availability itself, acting either directly or indirectly on the HPG axis, can play an important proximate role in reproductive development. Proximate effects of food availability could be direct, through energy and specific nutrients available for investment in reproduction. Processes such as thermoregulation may have priority in garnering energy from the available pool. Consequently, the ease with which a bird can meet its maintenance needs and also generate an excess pool of energy and nutrient resources for investment in "optional" (i.e., not required for survival) processes can affect, in a proximate sense, when it becomes possible to initiate reproductive development. If resources are monopolized by processes essential to survival, then only when the bird achieves a surplus could resources be directed into reproduction. An alternative (not exclusive) to this direct processing would be that the nervous system could process features of food availability and transduce these stimuli into neuroendocrine signals. Some features of food availability, such as perception of the appearance or disappearance of specific food types, or having the opportunity to physically handle certain kinds of foods, would not require physiological processing of the food itself to have an effect mediated through the nervous system.

Effects of food amount

The fact that reproductive success is highest when pairs match peak breeding demands to peak food availability suggests that individual birds fine tune timing of nesting using food as a proximate cue. One way to investigate this is to estimate intake rates relative to reproductive activity or physiology. Although it can be difficult to estimate food intake rates accurately in the field, some specialist taxa are particularly amenable to this approach. In crossbills (*Loxia* sp.), intake rates can be estimated by directly measuring seed consumption rates and multiplying by average seed kernel mass (4). With the exception of an autumn nadir in breeding activity that may be independent of intake rates, and is probably related to presence of some form of refractory period (see 16, 17), crossbill breeding activity across much of the year corresponds to periods of elevated intake rates (5), suggesting that food amount itself is acting as a proximate reproductive stimulus.

Long-terms field studies provide even more solid evidence for a proximate role for food amount in avian reproduction. For instance, food supplementation provided for several months prior to the normal onset of nesting led to advancement of laying date in free-living song sparrows (2); this is just one of a number of similar types of studies with similar results (see 25 for review). One of the key issues identified by these types of long-term field studies is the important role that food amount plays in determining the interplay between onset of egg laying and clutch size (e.g., 25). For example, hunting yield of male kestrels correlates with the onset of egg laying by their mates, and females that begin nesting earlier lay larger clutches. Though coincident with an increase in hunting flight time (i.e., effort expended), the increase in hunting yield (a measure of efficiency) is likely a consequence of the availability of voles, the primary food of these birds. Experimental supplementation of kestrel pairs demonstrates the causal relationship between increased food intake and advanced laying date. Larger clutches laid by supplemented females appear to be the result of the effect of food on laying date, not a direct effect on clutch size per se (25).

In captive studies, food amount can be manipulated more precisely than in field studies. Many manipulations using domestic fowl have focused on the effects of release from rather extreme food restriction on the reproductive axis (e.g., 7). Experiments that use manipulations more similar to situations in the wild have also shown effects of food amount on the reproductive axis. Male red crossbills provided unlimited food show a more rapid LH surge when transferred to 16L:8D from 8L:16D than do those restricted to their *ad libitum* short day intake levels (16); significant LH elevation was detectable after just one long day in the unrestricted birds, but took until day 10 in the restricted birds. Further, ANCOVA revealed that the amount eaten on the first long day was significantly positively related to the magnitude of the long day induced LH surge. Despite these effects, there was no detectable difference in testis size after 30 days of photostimulation under these food treatments (16).

Effects of food amount on captive crossbills are more subtle if the manipulations are performed without photostimulation. Circulating LH levels were not affected by a release from mild food restriction in either birds held on a naturally declining photoperiod in autumn (when the birds are putatively relatively refractory, see 16, 17) or birds held on short days in winter (when rapid acquisition of full reproductive condition occurs regularly) (Ward, Katti and Hahn, unpublished). Neither females nor males released from mild food restriction in autumn showed a gonadal response to the increase in food. However, in winter, females fed *ad libitum* showed significant enhancement of ovarian development compared with restricted controls (Fig. 1). Males showed a similar trend that was not significant (Fig. 1).

Effects of food quality and food type

Food quality and food type are closely related, but distinct, features of food supply. By food quality, we refer specifically to the nutritional makeup in terms of overall energy content as well the relative contributions of fat, protein, carbohydrate, and other essential nutrients such as calcium. In many cases, a change in food type will also involve a change in food quality. For example, many songbirds switch from eating primarily seeds to insects as the spring progresses. This is a change in primary food type, and the relatively high protein content of insects represents a substantial change in food quality as well. However it is also possible for food quality to change without a change in food type. For example, as seeds mature their nutritional makeup changes (e.g. 8, 9). So even if birds keep eating the same food, nutritional quality could change.

Few studies with wild birds have specifically explored proximate effects of food quality as distinct from food type. One excellent example comes from recent work on Florida scrub jays. These birds begin nesting earlier in suburban habitats, where various high energy and high protein foods (e.g. peanuts, dogfood) are in unlimited supply, than they do in wildland habitats, where only natural foods are available (13). Although the specific differences between the diets of suburban jays and wildland jays are not known, the suburban resources likely involve increases in amount, quality, type, and predictability of food. Food supplementation experiments providing great food excess in the form of dog food, peanuts and mealworms led to advancement of clutch initiation and larger clutches in wildland breeding females (36).

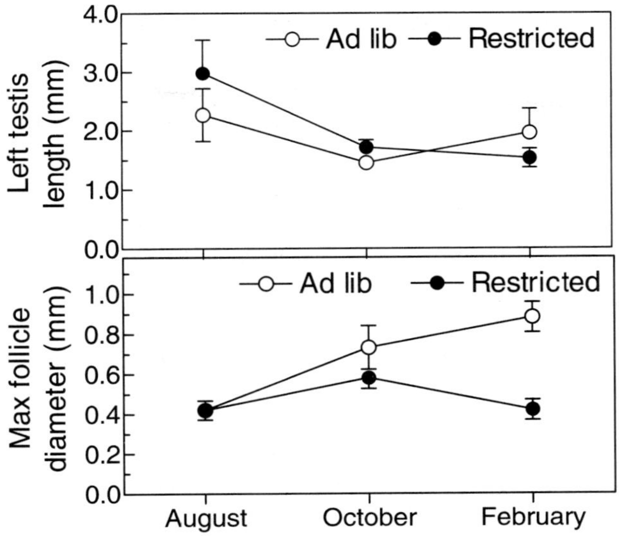

Fig. 1 Effects of release from mild food restriction on male (top) and female (bottom) red crossbill gonadal state. Birds had *ad libitum* food and were mostly in molt when laparotomized in August. All birds were restricted to their previously measured average daily intake for two weeks, at which point the *ad libitum* birds were returned to unlimited food (released from restriction) and the restricted birds remained on restricted food amount for an additional month. Laparotomies were performed at the end of these treatments in October and again following identical treatments in February. For males, treatment effect: $F_{1,9} = 0.211$, $P = 0.657$; date effect: $F_{2,18} = 7.563$, $P = 0.004$; treatment by date interaction: $F_{2,18} = 2.011$, $P = 0.163$. For females, treatment effect: $F_{1,8} = 24.025$, $P = 0.001$; date effect: $F_{2,16} = 6.623$, $P = 0.008$; treatment by date interaction: $F_{2,16} = 4.898$, $P = 0.022$.

More finely-resolved experiments compared the effects of isocaloric "high fat, high protein" supplements to those of "high fat, low protein" supplements, using identical-looking pellets for both diets (35). The high fat, high protein birds showed a lesser decline in egg mass with laying order than did controls, but laying date was advanced and clutch size increased to similar degrees over un-supplemented controls in both supplemented groups. Thus, these experiments show that food availability in a general sense has substantial effects on the reproductive axis of free-living Florida scrub jays, but the effect of diet quality *per se* is subtle. The fact that there were any differences at all between the supplemented groups also supports the interpretation that the effects were being mediated directly through metabolism, not indirectly through a perceptual process. A similar study with great tits (27) likewise demonstrated similar laying date advancement and clutch size enhancement relative to controls independent of the protein content of the food supplement provided.

Aspects of quality other than the energy or macronutrient content may be important, such as availability of certain nutrients essential for breeding. Calcium deficiency, which can be particularly prevalent in areas with acid rain, can lead to malformed eggshells and other reproductive problems (e.g., 32, 14). However calcium supplementation can also lead to advancement of laying date, suggesting that it can provide proximate information timing reproduction as well (24).

Changes in the availability of specific food types may be of particular proximate importance in the wild, and the effects may often be either independent of, or at least in addition to, concomitant changes in food quality. Many kinds of birds use particular types of foods during breeding, often because the nestlings have very specific dietary requirements. Evidence that availability of specific food types can have proximate effects on the reproductive axis comes from a variety of field and laboratory studies. One of the best recent studies (20) is particularly illuminating because it separates direct metabolic effects from indirect perceptual effects. Spotted antbirds are neotropical forest birds that regulate changes in reproductive physiology at least partly using the very modest seasonal changes in photoperiod that they experience at 9°N latitude (19). However they are also likely to make substantial adjustments to reproductive physiology and behavior using non-photic cues, such as food availability. Captive male spotted antbirds provided with live crickets in addition to a nutritionally complete standard diet, enhanced gonadal development and singing activity within three weeks compared to birds provided the standard diet alone (20). Some of these effects are at least partly attributable to non-nutritional aspects of the crickets. Male antbirds provided dead crickets (which they ate at the same rate as they ate live ones) did not increase singing activity. Furthermore, males allowed to view live crickets in dishes with transparent covers increased singing activity robustly, although relatively transiently, despite not being able to actually consume the crickets. It seems likely that seeing, pursuing, capturing and handling the live crickets all contributed to the effect on the gonads and singing behavior.

Other studies have likewise shown proximate effects of specific food types on the reproductive axis. Ligon's classic studies of the effects of green pinyon pine cones on pinyon jays are one superb example. Free-living pinyon jays often re-initiate breeding activity late in summer if they encounter a bumper crop of developing pinyon pine cones (23). Captive male pinyon jays provided with free access to green pinyon cones in late summer are significantly more likely to reverse gonadal regression than are control individuals provided a cone free diet (22).

Furthermore, males provided *ad libitum* access to mature pinyon seeds but not permitted to handle the cones were no more likely to show testicular development than were the controls, indicating that the presence and possibly the physical handling of the cones is an important aspect of this proximate effect, though differences in nutritional content of mature versus green seeds may also be involved. This is very similar to the situation just described for spotted antbirds: the effects of the cue appear to include non-nutritional effects.

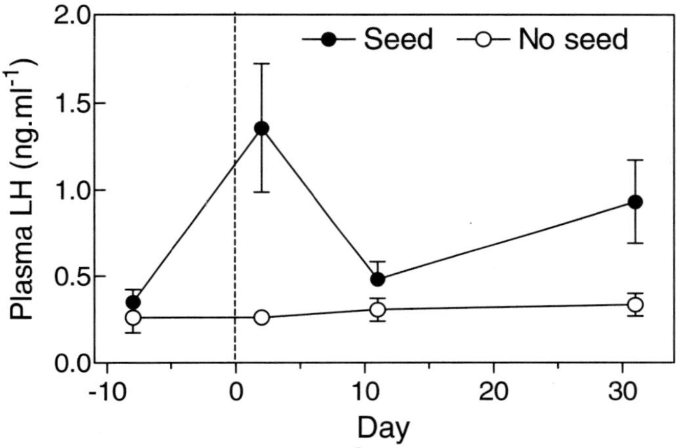

Fig. 2 Effects of access to sunflower seeds on LH secretion in male red crossbills held on short days. Birds of both treatments had unlimited access to Mazuri pellets but no sunflower seeds prior to day 0. On day 0, the "seeds" group birds were given unlimited access to whole black oil sunflower seeds in addition to the Mazuri pellets. Treatment effect: $F_{1,6} = 54.738$, $P < 0.001$; date effect: $F_{3,18} = 11.813$, $P < 0.001$; treatment by date interaction: $F_{3,18} = 11.562$, $P < 0.001$.

Crossbills are known for their specialization on conifer seeds throughout the year, but particularly when breeding (6, 1). Captive crossbills show substantial physiological responses to changes in food type. When males held on short days are provided with unlimited access to black oil sunflower seeds in addition to a standard pellet diet, they show rapid, significant enhancement of luteinizing hormone secretion compared with controls held on the *ad libitum* pellet-only diet (Fig. 2). Following 30 days of exposure to these treatments, testis length was also significantly greater in the seed group than in the no seed group (Hahn, Ward and Katti, unpublished). It is unclear whether this effect is metabolic/nutritional or psychological/perceptual, but subcutaneous fat depots did not differ in the two groups, consistent with the latter (39).

Crossbills specialize on conifer seeds year round if available. Consequently it is plausible that crossbills would be at least as responsive to cones as are pinyon jays. Despite some experiments yielding null results, in one study, cones did have a significant effect on reproductive development of captive female red crossbills held on short days. Female red crossbills provided unlimited sunflower seed chips in winter, then restricted to 70% of individual *ad libitum* intake for 1 week, and then returned to *ad libitum* sunflower seed chips plus a generous daily allotment of green (frozen since the previous late summer) western hemlock (*Tsuga heterophylla*) cones, increased their circulating LH levels 7 days later (within treatment repeated measures effect, $F_{1,14}$ = 12.074, P = 0.004; Fig. 3). The females returned from restriction to *ad libitum* sunflower chips alone did not increase LH secretion ($F_{1,14}$ = 1.857, P = 0.195). It appears likely that crossbills may often encounter abrupt changes in food type during their nomadic wanderings, and consequently that they may have evolved to be particularly sensitive to food type (cones specifically) as a proximate cue.

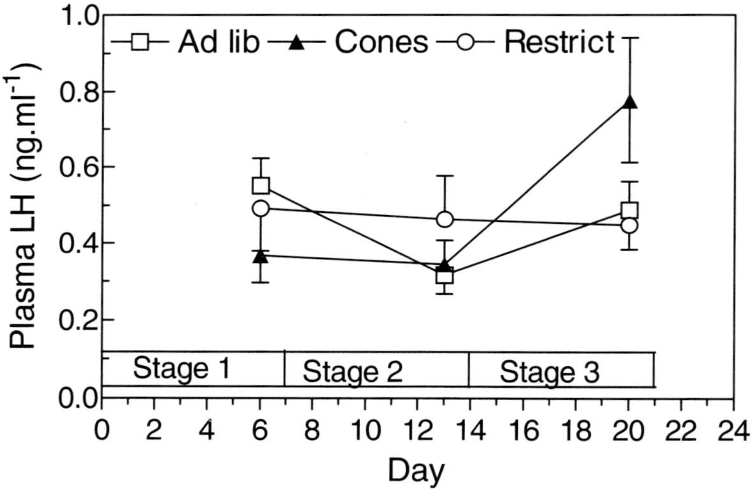

Fig. 3 Effects of access to seed-filled western hemlock cones on plasma LH levels in female red crossbills. All birds were fed unlimited sunflower seed chips during Stage 1, and were restricted to 70% of *ad libitum* sunflower chips during Stage 2. During Stage 3, the Restricted birds (open circles) remained on 70% *ad libitum* sunflower chips, the *Ad lib* birds (open squares) were returned to unlimited sunflower chips, and the Cones birds (opaque triangles) were returned to ad libitum sunflower chips and provided with a generous supply of green seed-filled hemlock cones daily.

Crossbills are not the only cardueline finches that are likely to encounter, and respond to, changes in food type. A number of carduelines breed seasonally, but can initiate reproductive activity relatively early in spring if conditions permit. Availability of emergent green vegetable matter early in spring is one factor that could play a proximate role even in high latitude breeders, such as common redpolls (*Carduelis flammea*). In captive male redpolls housed in outdoor aviaries in Fairbanks, Alaska, on natural photoperiod and temperature, the early stages of gonadal development were significantly accelerated if green alfalfa sprouts were provided daily in addition to the control diet of mixed seeds (Pereyra and Hahn, unpublished; Fig. 4). The disappearance of the treatment difference by mid April indicates that once photoperiod gets very long its stimulatory effects outweigh the relatively modest supplementary effects of the food type.

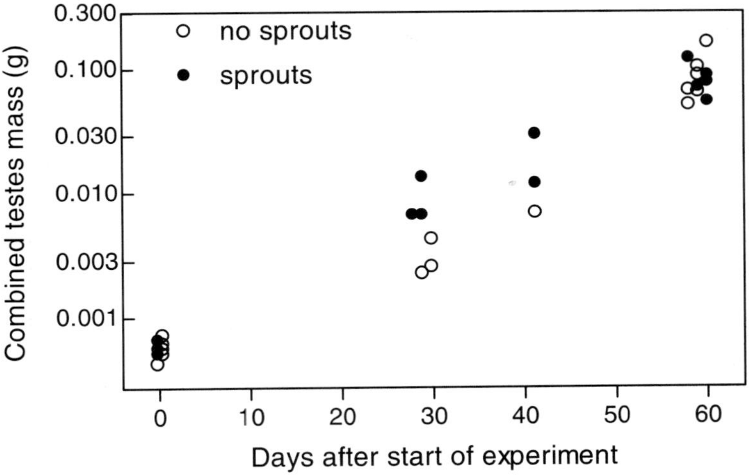

Fig. 4 Effects of provisioning with fresh alfalfa sprouts to male common redpolls from mid February through mid April. Experimental and control groups received equal allotments of mixed seeds (canary seed, thistle and sunflower seed) into which a small quantity of Mazuri was added. The control group received, in addition, 17 g of fresh alfalfa sprouts each day (data from Pereyra and Hahn, unpublished). 2-way ANOVA (treatment and date): $F_{7,23} = 15.081$, $P < 0.001$.

Food predictability

Intuitively, effects of food predictability are likely to be most apparent in the context of opportunistic, nomadic taxa that exploit food resources which vary unpredictably in space and/or time. For these taxa, the transition from an unreliable/unpredictable food resource to a reliable/predictable one (e.g., when a new bumper cone crop is discovered) could be stimulatory. Conversely, the transition from a steady, predictable food supply to one that declines rapidly without warning could be inhibitory. Proximate effects of food predictability remain relatively unexplored, but we have performed one preliminary study on house finches (*Carpodacus mexicanus*). There was a non-significant trend for male house finches photostimulated under tonic food restriction (90% of *ad libitum* intake) to show reduced gonadal development compared with finches receiving either *ad libitum* food or unpredictably varying food; the expected inhibitory effect of unpredictably varying food was not observed (Katti and Hahn, unpublished, Fig. 5). This is a puzzling result. It is possible that unlimited food only one day in four is sufficiently stimulatory under long days to maintain enhanced gonadal growth, similar to the prolonged effect of exposure to just a single long day (31, 11). We would expect that crossbills would be strongly inhibited by a food supply that varies erratically on a day-to-day basis, forgoing reproductive development until such time as food resources become more reliable. It will be interesting to perform similar experiments on crossbills and other nomadic, opportunistic carduelines to ascertain whether they respond similarly to house finches or show a qualitatively different response.

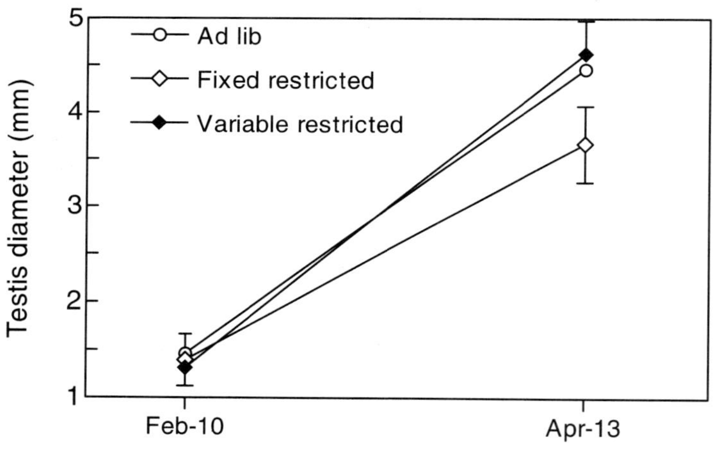

Fig. 5 Effects of variation in food amount and food predictability on photo-induced testis development of male house finches. Gonad growth was greater in the ad libitum and unpredictably varying food supply than in the fixed restricted group. The variable restricted group received daily allotments varying at random from 60% to 120% (4 day average of 90%, the same as the fixed restricted group) of individual ad libitum short day intake measured prior to photostimulation (16L:8D). Diet was Mazuri pellets only for all three groups. Repeated measures ANOVA, food effect, $F_{2,15}=1.93$, $P=0.18$; date effect, $F_{1,15}=146.57$, $P<0.0001$; food*date effect, $F_{2,15}=1.69$, $P=0.22$.

Integration of food availability with other environmental cues

A number of the studies cited above provide insight into the integration of food cues with other cues, such as photoperiod and social factors. The kestrel studies, crossbill studies, and antbird studies are three good examples. In many cases, particularly in highly seasonal environments, food cues likely function primarily as supplementary information (cf. 40, 41) once reproductive development has already been activated by seasonal changes in day length. Even fairly dramatic food restriction can have relatively modest effects on photoinduced reproductive development. European starlings exposed to 16L:8D but restricted to only 8 hours of access to food showed only very slight attenuation of gonadal growth compared with long day birds on ad libitum food, and this difference had disappeared by 8 weeks (10). The main (and somewhat surprising) difference between ad libitum and restricted starlings was a delay in gonadal regression in *restricted* birds. Perhaps the slight reduction in the reproductive "drive" imposed by the aggregate proximate cues under food restriction has an effect similar to simply using a shorter photoperiod for stimulation, leading to slightly less rapid reproductive development and a longer interval before gonadal regression begins (see 31).

Opportunists such as crossbills may be substantially more responsive to food cues in the absence of photostimulation than are more regularly seasonal taxa. They have often been presumed to regulate reproductive timing exclusively using food cues (e.g., 15, 29). The more recent evidence, including the new data reported above, demonstrates that the effects of food on crossbills depend on photoperiodic experience. Tropical birds such as spotted antbirds may also be expected to respond more dramatically to food cues in the absence of stimulation by other cues than would typical temperate zone seasonal taxa. However even the antbirds were already in a state of partial reproductive development, presumably owing to changes in photoperiod, when the food type effects described above (20) were detected.

In general, it appears likely that all manner of effects of food are generally going to be part of an integrated response to a suite of relevant proximate cues. Wingfield's (40, 41) classification of food supply as a supplementary cue that is integrated with initial predictive cues appears to us still to be the most appropriate way to think about these issues, even for opportunistic and tropical taxa.

Conclusions

Food clearly plays a dual role in the reproductive biology of birds, functioning both as ultimate determinant of optimal reproductive timing and as proximate cue signaling changes in reproductive physiology and behavior. The combination of descriptive and experimental field studies, and experimental laboratory studies, that have been accumulating reveal the importance of considering different aspects of food availability – amount, type, quality, and predictability – when attempting to unravel precisely how food affects birds' reproductive systems. This fact, as well as the important distinction between direct metabolic/nutritional effects and indirect perceptual effects, illustrates the potential for very different proximate mechanisms that may be involved in the processing of different kinds of food cues. There is still relatively little known about the specific mechanisms by which food cues are detected and transduced into endocrine and neuroendocrine signals regulating changes in reproductive physiology and behavior of wild birds. The studies reported above largely describe the reproductive physiological outcomes of exposure to food cues, downstream of whatever detection and transduction mechanisms are engaged. Future work should attempt to identify and characterize the nature and location of the primary responses to different kinds of food cues in wild birds.

Acknowledgements

We thank J Cornelius for comments on a draft of the manuscript. Some of the work was supported by NSF grant IBN-0196093 to TPH, NSERC and individual NIMH-NRSA fellowships to SAM-S, Princeton University Dean's Fund to GMW, and a NSF postdoctoral fellowship to MEP.

References

1 Adkisson CS (1996) Red crossbill (*Loxia curvirostra*). In *The Birds of North America* **No. 256** Eds A Poole & F Gill. Academy of Natural Sciences: Philadelphia & American Ornithologists' Union: Washington DC

2 Arcese P & Smith JNM (1988) Effects of population density and food on reproduction in song sparrows. *Journal of Animal Ecology* **57** 119-136

3 Baker JR (1938) The evolution of breeding seasons. In *Evolution: Essays on Aspects of Evolutionary Biology* 161-177. Ed GB DeBeer. Clarendon Press: Oxford

4 Benkman CW (1987) Food profitability and the foraging ecology of crossbills. *Ecological Monographs* **57** 251-267

5 Benkman CW (1990) Foraging rates and the timing of crossbill reproduction. *Auk* **107** 376-386

6 Benkman CW (1992) White-winged crossbill (*Loxia leucoptera*) In *The Birds of North America* **No. 27** Eds A Poole, P Stettenheim & F Gill. Academy of Natural Sciences: Philadelphia & American Ornithologists' Union: Washington DC

7 Brake J (1992) Mechanisms of and metabolic requirements for complete and rapid reproductive rejuvenation during an induced molt – a brief review. *Ornis Scandinavica* **23** 335-339

8 Chavan UD, Shahidi F, Bal AK & McKenzie DB (1999) Changes in chemical composition of beach pea (*Lathyrus maritimus*) seeds and pod shells at different stages of maturity. *Journal of Food Biochemistry* **23** 323-340

9 Chinnasamy G. & Bal AK (2003) The pattern of seed development and maturation in beach pea (*Lathyrus maritimus*). *Canadian Journal of Botany* **81** 531-540

10 Dawson A (1986) The effect of restricting the daily period of food availability on testicular growth of starlings *Sturnus vulgaris*. *Ibis* **128** 572-575

11 Dawson A (2001) The effects of a single long photoperiod on induction and dissipation of reproductive photorefractoriness in European starlings. *General and Comparative Endocrinology* **121** 316-324

12 Drent RH & Daan S (1980) The prudent parent: Energetic adjustments in avian breeding. *Ardea* **68** 225-331

13 Fleischer AL, Bowman R & Woolfenden GE (2003) Variation in foraging behavior, diet, and time of breeding of Florida scrub-jays in suburban and wildland habitats. *Condor* **105** 515-527

14 Graveland J & Van der Wal R (1996) Decline in snail abundance due to soil acidification causes eggshell defects in forest passerines. *Oecologia* **105** 351-360

15 Griscom L (1937) A monographic study of the red crossbill. *Proceedings of the Boston Society of Natural History* **41** 77-210

16 Hahn TP (1995) Integration of photoperiodic and food cues to time changes in reproductive physiology by an opportunistic breeder, the red crossbill, *Loxia curvirostra* (Aves: Carduelinae). *Journal of Experimental Zoology* **272** 213-226

17 Hahn TP (1998) Reproductive seasonality in an opportunistic breeder, the red crossbill, *Loxia curvirostra*. *Ecology* **79** 2365-2375

18 Hahn, T. P., T. Boswell, J. C. Wingfield, and G. F. Ball. 1997. Temporal flexibility in avian reproduction: Patterns and mechanisms. Current Ornithology 14: 39-80. V. Nolan, Jr., E. D. Ketterson, and C. F. Thompson (eds.), Plenum, New York and London.

19 Hau M, Wikelski M & Wingfield JC (1998) A neotropical forest bird can measure the slight changes in tropical photoperiod. *Proceedings of the Royal Society of London B* **265** 89-95

20 Hau M, Wikelski M & Wingfield JC (2000) Visual and nutritional food cues fine-tune timing of reproduction in a neotropical rainforest bird. *Journal of Experimental Zoology* **286** 494-504

21 Lack D (1968) Ecological Adaptations for Breeding in Birds. London: Methuen

22 Ligon JD (1974) Green cones of the pinon pine stimulate late summer breeding in the pinon jay. *Nature* **250** 80-82

23 Ligon JD (1978) Reproductive interdependence of pinon jays and pinon pines. *Ecological Monographs* **48** 111-126

24 Mand R, Tilgar V & Leivits A (2000) Reproductive response of great tits, *Parus major*, in a naturally base-poor forest habitat to calcium supplementation. *Canadian Journal of Zoology* **78** 689-695

25 Meijer T, Daan S & Hall M (1990) Family planning in the kestrel (*Falco tinnunculus*): The proximate control of covariation of laying date and clutch size. *Behaviour* **114** 117-136

26 Meijer T & Drent R (1999) Re-examination of the capital and income dichotomy in breeding birds. *Ibis* **141** 399-414

27 Nager RG, Ruegger C & van Noorwijk AJ (1997) Nutrient or energy limitation on egg formation: a feeding experiment in great tits. *Journal of Animal Ecology* **66** 495-507

28 Nager RG & van Noordwijk AJ (1995) Proximate and ultimate aspects of phenotypic plasticity in timing of great tit breeding in a heterogeneous environment. American Naturalist **146** 454-474

29 Newton, I. 1973. Finches. Taplinger, Englewood Cliffs, NJ, USA.

30 Nicholls TJ, Follett B & Robinson J E (1983) A photoperiodic response in gonadectomized Japanese quail exposed to a single long day. *Journal of Endocrinology* **97** 121-126

31 Nicholls T J, Goldsmith AR & Dawson A (1988) Photorefractoriness in birds and comparison with mammals. *Physiological Reviews* **68** 133-176

32 Nyholm NEI & Myhrberg H E (1977) Severe eggshell defects and impaired reproductive capacity in small passerines in Swedish Lapland. *Oikos* **29** 340-341

33 Perrins CM (1970) The timing of birds' breeding seasons. *Ibis* **112** 242-255

34 Perrins CM (1991) Tits and their caterpillar food supply. *Ibis* **133** (Suppl) 49-54

35 Reynolds SJ, Schoech SJ & Bowman R (2003) Nutritional quality of prebreeding diet influences breeding performance of the Florida scrub-jay. *Oecologia* **13**4 308-316

36 Schoech SJ (1996) The effect of supplemental food on body condition and the timing of reproduction in a cooperative breeder, the Florida scrub-jay. *Condor* **98** 234-244

37 Schoech SJ & Bowman R (2003) Does differential access to protein influence differences in timing of breeding of Florida scrub-jays (*Aphelocoma coerulescens*) in suburban and wildland habitats? *Auk* **120** 1114-1127

38 van Noordwijk AJ, McCleery RH & Perrins CM (1995) Selection for the synchronisation of great tit (*Parus major*) breeding with caterpillar growth, using temperature as a predictor. *Journal of Animal Ecology* **64** 451-458

39 Ward GM (2000) Seasonal effects of food availability on gonadal development, body mass, and fat deposition in red crossbills. Unpublished senior thesis: Princeton University

40 Wingfield JC (1980) Fine temporal adjustments of reproductive function. In *Avian Endocrinology* 367-389. Eds A Epple & MH Stetson. Academic Press: New York

41 Wingfield JC (1983) Environmental and endocrine control of reproduction: An ecological approach. In *Avian Endocrinology. Environmental and Ecological Perspectives* 265-288. Eds SI Mikami, K Homma & M Wada. Japan Scientific Society Press: Tokyo & Springer-Verlag: Berlin

Functional Avian Endocrinology
A Dawson & P J Sharp (Eds)

Non-photoperiodic seasonality in a tropical songbird

I T Moore

Department of Biology, 2119 Derring Hall, Virginia Tech, Blacksburg, Virginia 24061 USA

Introduction

Seasonal reproduction synchronizes breeding with food abundance to maximize reproductive success (17). The mechanisms mediating seasonal reproduction have been extensively studied in mid- to high latitude species (34). Despite the fact that the vast majority of avian species are tropical (up to 80% of passerine bird species: 31), this group has been much less well studied by biologists interested in mechanisms mediating behavior. In this paper I describe some of my work investigating neuroendocrine control of reproduction that occurs independent of photoperiod in equatorial sparrows.

Photoperiod has long been known to be crucial to the timing of seasonal reproduction in temperate zone bird species (9). In these mid- to high latitude species, the ultimately predictable change in photoperiod is the initial predictive cue used to time reproduction. Subsequently, supplementary cues such as food availability, temperature, and rainfall are used to fine tune the timing of breeding to the local environment (34, 35). Both initial predictive and supplementary cues are crucial as gonads take time to recrudesce and available periods to reproduce can be brief (12, 15). Extensive studies on avian species breeding at temperate latitudes have shown that plasma levels of testosterone in males, as well as estradiol and progesterone levels in females, are basal during the non-breeding period and elevated during breeding. The increase in sex steroid levels, immediately preceding and during breeding, results from the activation of the hypothalamic-pituitary-gonadal (HPG) axis (1). Reproduction is usually followed by feather molt during which sex steroid levels return to basal values.

Expression of seasonally appropriate reproductive behaviors, such as breeding and associated territoriality, are typically mediated by the seasonal changes in the neuroendocrine system described above. Singing by male song birds is one of the most obvious seasonal behaviors and plays an important role in reproduction, both for territorial defense and mate attraction (5). Song is often produced seasonally, with birds singing primarily or only during breeding. Song learning and production are regulated by a series of interconnected brain nuclei collectively known as the song control system (3. 37). Associated with seasonal changes in singing behavior are seasonal changes of the song control nuclei in the brain, which are larger during the breeding season than during the non-breeding season (32). It is thought that seasonal changes in the song control system of temperate zone species are driven by changes in photoperiod (32). The increase in day length in the spring stimulates neurons in the hypothalamus to release gonadotropin-releasing hormone (GnRH), which triggers the release of luteinizing hormone that, in turn, triggers an increase in plasma testosterone concentration. Testosterone is the primary physiological cue regulating seasonal changes in the song control system (32, 33). Seasonal plasticity of the brain, both in the song control and the GnRH systems, has only been observed in species that breed at high latitudes, with photoperiod as the relevant environmental cue (32).

Seasonal reproduction in tropical birds has not been nearly as well studied, from a mechanistic perspective, as in temperate zone species (13). Tropical species often experience little seasonal change in photoperiod, yet observational studies, both in the Old and New World tropics, have described seasonal reproduction (reviewed in 31). It is worth noting that even small changes in photoperiod can be used to time reproduction for some species such as the spotted antbird (*Hylophylax naevioides*) in Panama (14). Other seasonally breeding tropical bird populations are thought to rely on other cues that vary locally, such as food availability, rainfall, and/or temperature (31). For example, transplant experiments in the Seychelles warbler (*Acrocephalus sechellensis*) have demonstrated that populations with better food availability prolong breeding (16). Thus, as local environmental cues vary so can reproductive phenology, particularly among populations. Alternatively, tropical birds may rely on endogenous rhythms to time seasonal reproduction (10). Equatorial stonechats (*Saxicola torquata*) exhibit an endogenous annual rhythm of gonadal activity for their entire lives, when housed in constant conditions (10). In free-living individuals, however, food availability can adjust the timing of reproduction (18). Thus, tropical species may not use environmental cues in the same manner (ie. initial predictive and supplementary) as do temperate zone species. Therefore, it is unclear whether many tropical species are truly seasonal breeders. Many species might be better described as opportunistic breeders, especially in cases where reproduction can occur over an extended period of the year (4). In the case of seasonal breeding, as with the stonechats, reproduction should be predictable year-to-year with the gonads regressing after the breeding season and recrudescing in anticipation of the coming breeding season. With opportunistic breeding, timing of reproduction should vary with the presence of some crucial resource (35). In this case the gonads may be maintained in an active state for much of the year in preparation for the appearance of the resource. Tropical birds exhibit sex steroid cycles with environmental and social factors affecting peak plasma testosterone levels in males (8).

As some tropical birds can be exquisitely sensitive to even the small changes in photoperiod that may be present at tropical latitudes (14), care must be taken before reproduction can be described as occurring independent of photoperiodic cues. There are at least three possible manners to investigate if seasonal reproduction occurs independent of photoperiod: 1) maintain populations of birds in captivity in constant conditions, 2) observe populations of birds that are resident year-round on the equator where photoperiod does not change seasonally, 3) observe populations of birds that breed asynchronously but at the same latitude and therefore experience similar photoperiods. All three manners have benefits and potential drawbacks. Maintaining birds in captivity can be difficult and expensive but can also provide definitive, experimental evidence of non-photoperiodic seasonality (10). The second manner has the drawback that even close to the equator, seasonal changes in photoperiod can exist which makes it more difficult to conclusively state that a given species is not making use of the cues. Finally, asynchronous reproduction does not necessarily occur independent of photoperiod and can be representative of population differences in sensitivity to photoperiodic cues (18, 19). I have used a combination of the latter two techniques, investigating two bird populations near the equator, to describe seasonal reproduction and associated neuroendocrine mechanisms, that occur independent of photoperiod.

Equatorial seasonal reproduction

The rufous-collared sparrow (previously known as the Andean sparrow, *Zonotrichia capensis*) is a small emberizine sparrow with a species range from Southern Mexico to Tierra del Fuego. Populations in Ecuador are ideal for investigations of non-photoperiodic reproduction because they are numerous and can occur near the equator where seasonal changes in photoperiod are minimal. In addition, this species is a congener of a temperate zone species (the white-crowned sparrow, *Z. leucophrys*) that has been studied in great detail and thus permits the use of established techniques and limits the phylogenetic confounds often associated with cross-species comparisons.

Seasonal reproduction

Previous studies have described broadly seasonal reproduction in populations of *Z. capensis* from tropical latitudes (6, 20, 21, 23, 36). In addition, tropical *Z. capensis* have been experimentally demonstrated to be responsive to photoperiodic cues when transplanted to higher latitudes (22). I extended these studies of *Z. capensis* to populations closer to the equator where photoperiodic cues are almost non-existent. Over a period of four years, I have documented seasonal reproduction in a non-migratory population of *Z. capensis* residing year round on the equator (Fig. 1). This population in Papallacta, Ecuador (0°21.7'S, 78°9.0'W, 3300 m elevation) breeds (August to December) and molts (February) during distinct, consistent, and repeatable times of the year. In addition, the gonads regress completely for much of the year, as expected for seasonally breeding species but not necessarily for opportunistically breeding species who might maintain their gonads in an operational state waiting for the appropriate environmental cue to reproduce (11). Seasonal changes in photoperiod are minimal at the latitude of the Papallacta population (~3min., United States Naval Observatory, Astronomical Applications Department) while other climatic factors vary, particularly rainfall (2). During the breeding season, males in Papallacta have fully recrudesced testes and elevated plasma levels of testosterone that are as high as those observed in closely related northern species during the breeding season (8, 26, 27, 28). Their testes regress and testosterone levels are basal during non-breeding and molt (Fig. 1). This is similar to the pattern described for high latitude species (34).

Reproductive asynchrony

Populations of *Z. capensis* in the high Andes of Ecuador show seasonal reproductive cycles that are closely associated with the local environment. As such, I have documented asynchronous seasonal reproduction, over a period of two and a half years, between two populations that are in close proximity: the Papallacta population described above and a second population that is 25 km away in Pintag (0°22.6'S, 78°22.5'W, 2900 m elevation). It is worth noting that *Z. capensis*, in Ecuador, has an altitudinal range of 1500-3500m and the Andean Divide between the sites is around 4000m, resulting in the two populations being discontinuous. Individuals in the Papallacta population breed from August-December and undergo feather molt in February, while the Pintag population breeds from November-May and molt in June. The Papallacta population is on the eastern slope of the Andes while the Pintag population is in an inter-Andean valley.

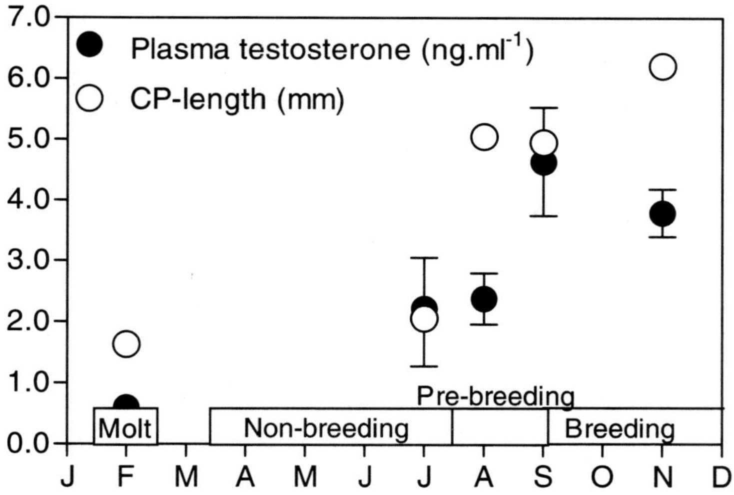

Fig. 1 Seasonal cycle of reproduction in male *Z. capensis* in Papallacta, Ecuador. The cloacal protuberance (CP) is an androgen sensitive copulatory organ that enlarges during breeding and closely parallels testis volume. Breeding and molting periods were determined through observation. Each point represents the average and SE of >10 individuals sampled during that period.

As the two populations are at almost the same latitude, the asynchronous reproduction between them provides convincing evidence that breeding occurs independent of photoperiod cues. This asynchrony appears to be associated with very different climates, particularly rainfall patterns (eg. July is the wettest month of the year in Papallacta and the driest in Pintag), due to the rain shadow cast by the Andes mountains (2). Also, the rainfall appears to occur in a different manner between the sites; Papallacta can be described as a cloud forest while Pintag is much more arid but gets heavier rains when they do occur. An important difference between the populations is that the Papallacta population begins breeding after its rainy season when the environment is drying out, while the Pintag population begins breeding at the beginning of its rainy season. However, both populations are consistent in molting during the driest periods of the year. This similarity in phenology suggests that the timing of feather molt could be crucial to organizing the annual cycle.

It is unclear if the breeding asynchrony between the two populations is genetic or environmental in nature. The Papallacta and Pintag populations of *Z. capensis* sing different song dialects and are genetically distinct from one another (25). An analysis at four microsatellite loci documented minimal rates of migration between the populations and the existence of private alleles at each of the loci (25). The relative roles of genetic versus environmental cues in timing reproduction in these populations is an obvious future line of investigation.

Seasonal neuroplasticity

To further the mechanistic investigation of seasonal reproduction of equatorial birds, I investigated whether the seasonal reproduction in the above described populations of *Z. capensis* is associated with plasticity of brain structure. Males in these populations only sing during the breeding season, and not during molt or the non-breeding period (29). This seasonal expression of song behavior suggests seasonal changes in the brain. To test this, male birds were captured from the Papallacta and Pintag populations during each one's respective breeding and non-breeding periods. The reproductive activity (size of the testes and plasma testosterone levels) and state of the neural song control system were quantified for each individual. Breeding birds had active reproductive systems and enlarged song control nuclei (nuclei: HVC, RA, and X) relative to non-breeding birds within each population (29). These reproductive and neural traits did not differ between populations when the birds were in the same reproductive state. However, during a single period, when the song control nuclei were fully grown and gonads active in the breeding Papallacta population, both the song control nuclei and gonads were regressed in the non-breeding Pintag population. This result suggests that the birds are using local environmental cues to time seasonal brain plasticity. These results demonstrate the brain's exquisite sensitivity and ability to make use of a variety of environmental cues that vary seasonally. These results also suggest that environmental changes such as global warming could have dramatic effects at the individual level.

Seasonal changes in gonadal state are mediated by seasonal changes in the hypothalamic content of GnRH with breeding birds having larger GnRH immunoreactive cells in the preoptic area of the hypothalamus (24). This difference in GnRH expression is similar to that documented in photostimulated and photorefractory birds from high latitudes and support the idea that these equatorial birds are indeed seasonally and not opportunistic breeding (7). In contrast, there are no differences in cell size or number in the recently described system of neurons expressing gonadotropin-inhibiting hormone (unpublished). Future studies will investigate whether seasonal reproduction in this system is associated with other seasonal neuroendocrine changes.

Conclusions

The description of seasonal reproduction and brain plasticity in equatorial *Z. capensis* indicates that cues other than photoperiod can drive seasonal processes. Whether or not the brain plasticity is driven by the same hormonal signals as in high latitude species (testosterone) is unknown at this point (Fig. 2). The fact that the gonads regress completely and predictably supports the idea that reproduction is seasonal and not opportunistic in these populations. Two lines of evidence demonstrate that photoperiod is not the critical seasonal cue used in these populations of *Z. capensis* to time the seasonal processes. First, both populations live year-round on the equator, where seasonal changes in photoperiod are minimal. Second, the two populations have reproductive cycles that are asynchronous, despite residing at the same latitude and thus experiencing the same photoperiod. While it is unclear exactly what environmental cues or endogenous rhythms are being used to time reproduction within each population, rainfall, food availability, and temperature are likely candidates for these and other tropical birds (31).

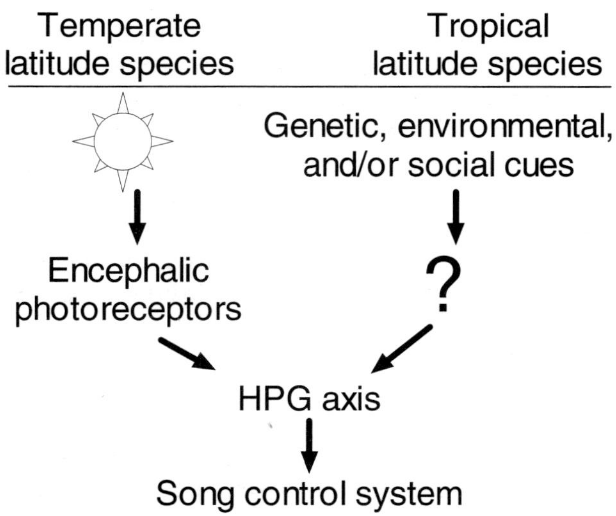

Fig. 2 Contrasting cues used to mediate seasonal brain plasticity in temperate and tropical birds. Temperate zone species use photoperiodic cues that are translated into activity of the HPG axis and then growth of the neural song control system. Tropical species may use genetic, environmental, or social cues to stimulate the HPG axis. Stimulation of the HPG axis may then mediate seasonal growth of the neural song control system in a similar manner to that described in temperate zone species, but this is as yet unknown.

Why not use photoperiod changes to time reproductive activity? At many tropical latitudes, seasonal changes in photoperiod cues may be large enough to be detected. However, it is not clear that the photoperiodic cues at these latitudes are good predictors of future food availability that is necessary for the increased energetic demands of reproduction. Even if photoperiod cues are available, they might not be as useful as in higher latitudes because available breeding periods can be variable in timing. In these cases, animals would be predicted to use local cues to time seasonal processes, or rely on endogenous rhythms, or in many cases use a combination of all possible cues as the need to successfully reproduce is too important to ignore any cue that could prove useful.

Acknowledgements

I would like to thank the following people who helped conduct the field work: Lisa Belden, Shallin Busch, Holly Goyert, Darren Lerner, Particio and Carmen Muñoz, Nicole Perfito, Particio Pillajo, Haruka Wada, and Brian Walker. I would also like to thank George Bentley and Cheryl Wotus who helped conduct the brain analyses, Fran Bonier who conducted the genetic analysis, as well as Karin Lent and Lynn Erchmann for assistance in the laboratory. Holly Goyert, Amanda Malueg, and Eric Weigel provided useful comments on earlier versions of the manuscript. Finally, I would like to thank my postdoctoral advisors, John Wingfield and Eliot Brenowitz, who provided encouraging and stimulating environments to conduct curiosity based research. This research was supported by NSF grant IBN-9905679 to John Wingfield, NIMH grant MH53032 to Eliot Brenowitz, and a Franklin Research Grant from the American Philosophical Society to Ignacio Moore. I.T.M. was supported by a NSF Postdoctoral fellowship (DBI-9904144) and a Society for Neuroscience Postdoctoral fellowship (T32 MH20069).

References

1 Ball GF & Bentley GE (2000) Neuroendocrine mechanisms mediating the photoperiodic and social regulation of seasonal reproduction in birds. In *Reproduction in Context. Social and Environmental Influences on Reproduction* 129-158. Eds K Wallen & JE Schneider. A Bradford Book, The MIT Press: Cambridge, MA, London

2 Bendix J & Rafiqpoor MD (2001) Studies on the thermal conditions of soils at the upper tree line in the paramo of Papallacta (eastern cordillera of Ecuador). *Erdkunde* **55** 257-276

3 Brenowitz EA, Margoliash D & Nordeem KW (1997) An introduction to birdsong and the avian song system. *Journal of Neurobiology* **33** 495-500

4 Brown LH & Britton PL (1980) The Breeding Seasons of East African Birds. Nairobi: The East Africa Natural History Society

5 Catchpole CK & Slater PJB (1995) Bird Song: Biological Themes and Variations. Cambridge: Cambridge University Press

6 Davis J (1971) Breeding and molt schedules of the rufous-collared sparrow in coastal Peru. *Condor* **73** 127-146

7 Dawson A, King VM, Bentley GE & Ball GF (2001) Photoperiodic control of seasonality in birds. *Journal of Biological Rhythms* **16** 365-380

8 Goymann W, Moore IT, Scheuerlein A, Hirschenhauser K, Grafen A & Wingfield JC (2004) Testosterone in tropical birds: effects of environmental and social factors. *American Naturalist* **164** 327-334

9 Gwinner E (2003) Circannual rhythms in birds. *Current Opinion in Neurobiology* **13** 770-778

10 Gwinner E & Dittami JP (1990) Endogenous reproductive rhythms in a tropical bird. *Science* **249** 906-908

11 Hahn TP (1998) Reproductive seasonality in an opportunistic breeder, the red crossbill, *Loxia curvirostra. Ecology* **79** 2365-2375

12 Hahn TP, Wingfield JC & Mullen R (1995) Endocrine bases of spatial and temporal opportunism in arctic-breeding birds. *American Zoologist* **35** 259-273

13 Hau M (2001) Timing of breeding in variable environments: tropical birds as model systems. *Hormones and Behavior* **40** 281-290

14 Hau M, Wikelski M & Wingfield JC (1998) A neotropical forest bird can measure the slight changes in tropical photoperiod. *Proceedings of the Royal Society of London, B.* **265** 89-95

15 Hunt K, Wingfield JC, Astheimer LB, Buttemer WA & Hahn TP (1995) Temporal patterns of territorial behavior and circulating testosterone in the lapland langspur and other arctic passerines. *American Zoology* **35** 274-284

16 Komdeur J (1996) Seasonal timing of reproduction in a tropical bird, the Seychelles warbler: a field experiment using translocation. *Journal of Biological Rhythms* **11** 333-346

17 Lack D (1968) Ecological adaptations for breeding in birds. London: Methuen

18 Lambrechts MM, Blondel J, Maistre M & Perret P (1997) A single response mechanism is responsible for evolutionary adaptive variation in a bird's laying date. *Proceedings of the National Academy of Sciences, USA* **94** 5153-5155

19 Lambrechts MM, Perret P, Maistre M & Blondel J (1999) Do experiments with captive non-domesticated animals make sense without population field studies? A case study with blue tits' breeding time. *Proceedings of the Royal Society of London, B* **266** 1311-131520.

20 Miller AH (1959) Reproductive cycles in an equitorial sparrow. *Proceedings of the National Academy of Sciences, USA* **45** 1095-1100

21 Miller AH (1962) Bimodal occurence of breeding in an equitorial sparrow. *Proceedings of the National Academy of Sciences, USA* **48** 396-400

22 Miller AH (1965) Capacity for photoperiodic response and endogenous factors in the reproductive cycles of an equitorial sparrow. *Proceedings of the National Academy of Sciences, USA* **54** 97-101

23 Miller AH & Miller VD (1968) The behavioral ecology and breeding biology of the Andean sparrow, *Zonotrichia capensis. Caldasia* **10** 83-154

24 Moore IT, Bentley GE, Wingfield JC & Brenowitz, EA (2003) Seasonal plasticity in the brain of a tropical songbird. *Abstract Viewer/Itinerary Planner. Washingion, DC: Society for Neuroscience. Online.* **Program No. 200.9.**

25 Moore IT, Bonier F & Wingfield JC (in review) Reproductive phenology, breeding synchrony, and genetic isolation in two tropical bird populations in close proximity. *Behavioral Ecology*

26 Moore IT, Perfito N, Wada H, Sperry TS & Wingfield JC (2002) Latitudinal variation in plasma testosterone levels in birds of the genus *Zonotrichia. General and Comparative Endocrinology* **129** 13-19

27 Moore IT, Wada H, Perfito N, Busch DS & Wingfield JC (*in press*) Territoriality and testosterone in an equatorial population of Rufous-collared sparrows, *Zonotrichia capensis. Animal Behaviour*

28 Moore IT, Walker BG & Wingfield JC (2004) The effects of combined aromatase inhibitor and anti-androgen on male territorial aggression in a tropical population of rufous-collared sparrows, *Zonotrichia capensis. General and Comparative Endocrinology* **135** 223-229

29 Moore IT, Wingfield JC & Brenowitz EA (2004) Plasticity of the avian song control system in response to localized environmental cues in an equatorial songbird. *Journal of Neuroscience* **24** 10182-10185

30 Scheuerlein A & Gwinner E (2002) Is food availability a circannual zeitgeber in tropical birds? A field experiment on stonechats in tropical Africa. *Journal of Biological Rhythms* **17** 171-180

31 Stutchbury BJM & Morton ES (2001) Behavioral Ecology of Tropical Birds. San Diego: Academic Press

32 Tramontin AD & Brenowitz EA (2000) Seasonal plasticity in the adult brain. *Trends in Neuroscience* **23** 251-258

33 Tramontin AD, Wingfield JC & Brenowitz, EA (1998) Contributions of social cues and photoperiod to seasonal plasticity in the adult avian song control system. *Journal of Neuroscience* **19** 476-483

34 Wingfield JC & Farner DS (1993) Endocrinology of reproduction in wild species. In *Avian Biology* 163-327 Eds DS Farner, JR King & KC Parkes. Academic Press: San Diego, CA.

35 Wingfield JC & Kenagy GJ (1991) Natural regulation of reproductive cycles. In *Vertebrate Endocrinology: Fundamentals and Biomedical Implications* 181-241 Eds PKT Pawg & MP Schreibruan. Academic Press, Inc: San Diego

36 Wolf LL (1969) Breeding and molting periods in a Costa Rican population of the Andean sparrow. *The Condor* **71** 212-219

37 Zeigler HP & Marler P (2004) In *Annals of the New York Academy of Sciences* Vol. 1016. New York Academy of Science: New York

Part five

Functions of melatonin

Blue-winged Teal
Eyal Shochat

Functional Avian Endocrinology
A Dawson & P J Sharp (Eds)

Melatonin regulation of migratory behaviour and physiology

L Fusani and E Gwinner[1]

Università di Siena, Dipartimento di Fisiologia, via Aldo Moro, 53100 Siena, Italy
[1]Max-Planck-Institute for Ornithology, 82346 Andechs, Germany

Introduction

Migration is a critical life history stage for a large number of bird species. This event involves major physiological changes in nearly every system including hormonal systems. There is an extensive (and sometimes contrasting) literature on the role played by pituitary, gonadal, thyroid, and adrenal hormones in a range of migration-related phenomena such as fat deposition, metabolic adaptations, orientation, and migratory disposition (5). Little attention has been given so far to the involvement of the pineal hormone melatonin in migratory physiology and behaviour. In most bird species the pineal gland contains an autonomous circadian oscillator that is responsible for the rhythmic release of melatonin (reviewed in 18, 23), and melatonin is a major component of the avian circadian system (reviewed in 7, 16). In birds the duration of the night melatonin peak reflects the photoperiod, however, in contrast to what is seen in mammals, melatonin does not appear to be important for the synchronization of avian annual cycles (8, 17). The crucial link between melatonin and migration becomes evident if we consider that a large proportion of birds are nocturnal migrants, that is they are normally active during the day but migrate mainly or exclusively at night. Hence, in these species, migration is associated with a major switch in the diel pattern of activity, with a supposed involvement of the circadian system. In addition, a modulatory effect of melatonin on the circadian pacemaker might help in adjusting to changing photoperiod in species that migrate fast and/or along the east-west axis. Preliminary experiments with bramblings (*Fringilla montifringilla*) and garden warblers indicate that resynchronization times are shorter during the migratory season than at other times (21) (Gwinner, unpublished data). In this paper we will review studies on melatonin and nocturnal migration focusing on our work on *Sylvia* warblers, and discuss how melatonin might mediate the different requirements of autumnal and spring migration.

Nocturnal activity and melatonin

Bird keepers have known for a long time that during migratory periods some bird species show a "Zugunruhe" or nocturnal migratory restlessness. In captivity, the intensity of Zugunruhe can be measured as the amount of locomotory activity recorded by a perch switch (hopping) or by an infrared sensor. There is good agreement in the literature that in birds and other organisms melatonin levels are elevated only during night time, irrespective of whether the species is diurnal or nocturnal (2, 20). Thus, it is interesting to ask whether a switch in the activity pattern from diurnal to nocturnal in a normally diurnal species is correlated with as inversion in the pattern of melatonin secretion. This hypothesis was tested with garden warblers (*Sylvia borin*) by Gwinner and collaborators in 1993 (14). Under constant photoperiodic conditions garden warblers show a circannual rhythmicity of seasonal processes, including migratory disposition (12). Garden warblers were held in a constant 12 hr photoperiod from September to June. The amount of locomotory activity was recorded continuously and blood samples were taken every 1-2 months to measure melatonin at 4 hr interval. During this 9-month experimental period, the birds went through the autumn migratory season, winter quiescence, and spring migratory season. During migratory periods the pattern of melatonin secretion was not different from that recorded during the other times (Fig.1 A). Thus, the onset of nocturnal activity does not alter the pattern of melatonin secretion, which remains high at night and low during the day. However, melatonin levels at 1:00 and 5:00 were lower during migration times than during the winter quiescent phase (Fig.1 A) (14). This suggested that the presence of nocturnal activity is correlated with a decrease in the amplitude of melatonin release. Three alternative hypotheses can be proposed to explain this correlation. First, it could be a seasonal correlation, i.e. melatonin and nocturnal activity are not functionally related but they change seasonally in parallel with each other. Secondly, the lower amplitude of melatonin during migratory periods could be a consequence of the fact that the birds are awake at night. Locomotory activity or increased light input through the open eyes may induce a decrease in the circulating concentration of melatonin (see discussion in 15). Thirdly, lowering melatonin release may facilitate nocturnal activity or serve some other migration-related purposes. In the following sections, we will discuss the results of a series of experiments that support the latter hypothesis.

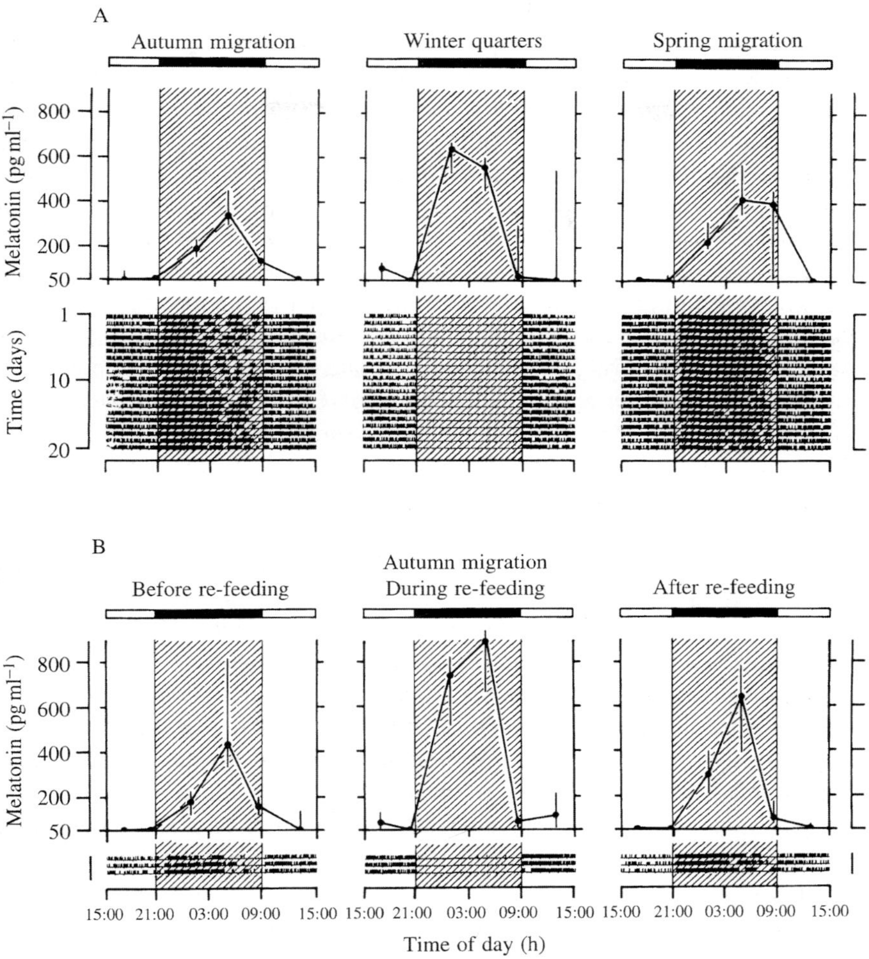

Fig. 1 Plasma melatonin (medians and quartiles, upper panels in A and B) and locomotory activity (representative birds, lower panels in A and B) of garden warblers kept under constant 12L:12D. (A) During the winter resting phase, melatonin amplitude is higher than during the autumn and spring migratory periods (14). (B) Similarly, during autumn migration, melatonin is higher when Zugunruhe is suppressed using the fasting-and -refeeding protocol than in the preceding and following periods (Gwinner, Schwabl-Benzinger, & Schwabl, unpublished results). Reproduced with permission from (15), © The Company of Biologists Ltd.

Migration-specific decrease in the amplitude of melatonin at night

In principle, there are two ways to assess whether seasonal changes in melatonin amplitude are specifically related to migration: to separate migration from its seasonal context or to remove migration from the seasonal cycle. The obvious difficulties of performing experiments of this kind were overcome by the availability of an "experiment of nature". The blackcap (*Sylvia atricapilla*), a close relative of the garden warbler, is a common Eurasian bird. Within the species there are populations with very different migratory behaviours, ranging from the non-migratory populations of some Atlantic islands to the long-distance migrants of Northeast Europe (4). We compared migratory behaviour and melatonin profiles between resident blackcaps from the Cape Verde islands and migratory blackcaps from Sweden and Kenya (9). Swedish birds were caught in late September, prior to the onset of southward autumnal migration. Kenyan birds were caught in late March, prior to the onset of northward spring migration. The birds of Cape Verde were caught in November. All birds were housed in the same room and were kept in a 12-h photoperiod (light on: 06:00) and food and water were given *ad libitum*. At one-month intervals, we took a series of 3 blood samples from the birds to measure the nocturnal profile of melatonin. In birds from migrating populations we found seasonal changes in the nocturnal peak of melatonin (9). In these birds, the pattern of melatonin was similar to that described for the garden warblers (14); melatonin amplitude was lower during the migratory period, when bird showed nocturnal activity, than at other times of the year, when birds did not show nocturnal activity (9) (Fig.2). In the birds from Cape Verde, on the contrary, we did not find significant changes in the amplitude of melatonin between seasons. In these birds the pattern of locomotory activity was less clear cut than in those from the migrant populations. Some birds showed nocturnal activity with no clear night-day transition, whereas others showed bursts of nocturnal activity alternated with periods of rest throughout the year. Non-migratory nocturnal restlessness has been reported for several species (3) and its causes are still unclear. To further investigate this point, we compared the melatonin peak in a group of birds from Cape Verde that were assigned to two groups depending on whether they exhibited nocturnal activity within the previous two weeks. All the birds were sampled within 5 days to minimize any season-related effect. We found no differences in melatonin levels between night active and inactive Cape Verde blackcaps (Fig.2). Thus, the comparison between birds of the same species but originating from populations with different migratory behaviour allowed us to conclude that in blackcaps seasonal changes in melatonin amplitude are directly related to migratory disposition (9). In addition, the lack of differences in the melatonin peak between night active and night inactive birds of a nonmigratory population indicates that the presence or absence of nocturnal activity does not predict melatonin amplitude. These results are contrary to the first two hypotheses, i.e. that the decrease in melatonin during Zugunruhe reflects parallel although independent seasonal processes, or results from behavioural changes such as increased locomotory activity or higher light input through the open eyes.

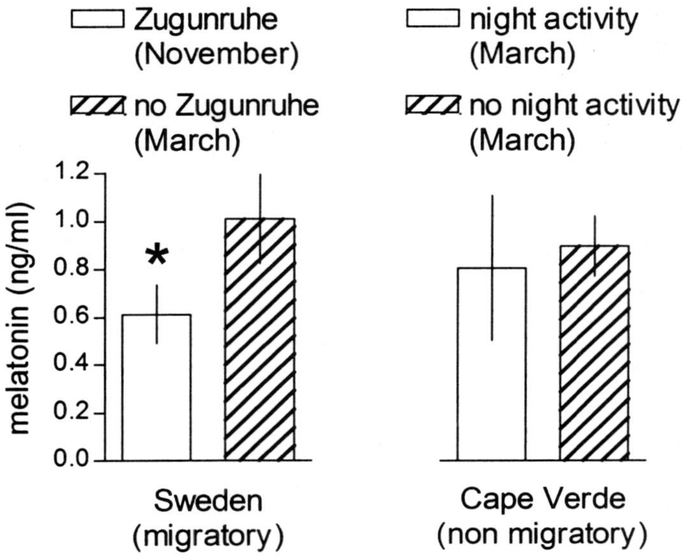

Fig. 2 Peak night melatonin (mean ± SEM) in blackcaps kept in constant 12L:12D. In migratory blackcaps from Sweden, melatonin is lower when the birds show migratory restlessness. In resident Cape Verde blackcaps, in contrast, there are no differences in melatonin levels between night active and inactive birds sampled at the same time of the year. * = $P < 0.05$, t-test. Adapted from (9)

Testing the functional relationships between reduced melatonin amplitude and Zugunruhe

Obviously, it would be very interesting to test if the manipulation of melatonin action does affect the expression or the amount of Zugunruhe. However, this approach is made difficult by the rhythmic nature of melatonin release. Pinealectomy or melatonin implants that results in constant low or high levels of melatonin disrupt or abolish circadian locomotory activity in a number of species, which masks specific effects on Zugunruhe (16, Gwinner, unpublished observations). We chose an approach that may appear even more challenging: to manipulate Zugunruhe and look at changes in melatonin using the "fasting-and-refeeding" protocol. It is known from previous studies on spotted flycatchers (*Muscicapa striata*) (6) and garden warblers (11, 13) that food reintroduction after a short period of food deprivation temporarily suppresses Zugunruhe. This experimental paradigm presumably mimics the situation of a migrating bird that has been fasting during a long flight and subsequently interrupts migration upon reaching a suitable refuelling site. This approach does not allow us to tell whether any observed change in melatonin causes changes in night activity or vice versa. However, it allows us to test if the behavioural and the hormonal effects are functionally associated. In a preliminary study with garden warblers, it was found that melatonin levels increase after inducing suppression of autumnal Zugunruhe by means of the fasting-and-refeeding protocol (Gwinner, I. Schwabl-Benzinger, & H. Schwabl, unpublished results, reported in 15) (Fig.1 B). Recently, we performed a series of experiments applying the same protocol to alter Zugunruhe in blackcaps (10). The experiments were done with birds caught in Sweden before autumn migration and in Kenya before spring migration. Blackcaps from long-distance migratory populations show robust, long-lasting Zugunruhe in captivity (4, 9). The comparison between autumn and spring, in addition, addressed the hypothesis of an involvement of melatonin in differences reported between these two migratory periods such as migration speed and duration (5). We will describe the fasting-and-refeeding protocol in brief (details in 10). Birds were held in a 12-h photoperiod (light on: 06:00) with food and water ad libitum. Experiments started at the end of October for autumnal migration and in mid May for spring migration. On Day 1, just before lights off, the food dish was removed from the cages and the animals received no food on Day 2 and 3, but fresh water was given as usual. On Day 4, at lights on, fresh food was given to the birds. In the following night, and specifically between 23:30 of Day 4 and 00:30 of Day 5, the birds were bled for melatonin measurement. In autumn, the protocol induced a dramatic reduction of Zugunruhe in the first night following food reintroduction, which was accompanied by an increase in the peak of melatonin (Fig.3). In spring the effects were similar but of a lesser extent, with a large variability in the individual response (Fig.3). Interestingly, the levels of melatonin were lower in the spring birds independently of the treatment (Fig.3), an aspect of the results that will be discussed below. These experiments demonstrated that the correlation between Zugunruhe and melatonin is a functional one: when Zugunruhe is suppressed by simulating a stopover after a long, non stop flight, melatonin increases to levels similar to those reported for non-migratory periods (9).

Conclusions

Although the data discussed in the previous sections strongly suggest that melatonin modulates nocturnal activity in migrants, a formal demonstration is still missing. Nevertheless, it seems likely that a reduction in melatonin release may be functional to several migration-related physiological adaptations. For example, reduced melatonin amplitude may serve to adapt faster to a changing photoperiod as encountered during migration. Treatment of house sparrows with exogenous melatonin reduces resynchronization times by dampening the endogenous melatonin rhythm and thus making the internal clock more "passive" (1, 19). A similar effect due to a reduction of melatonin amplitude during migration could facilitate a faster adjustment of the activity pattern to latitudinal changes in photoperiod (15). This hypothesis is in line with the facts that melatonin levels are lower in spring (10), when birds migrate faster and encounter more rapid changes in photoperiod than in autumn (5).

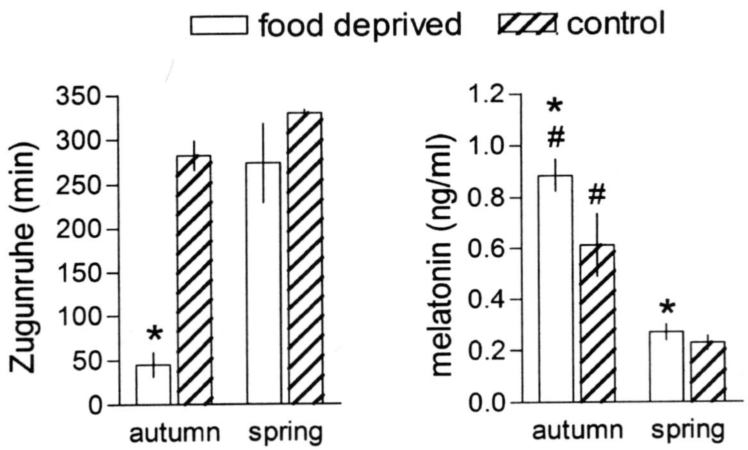

Fig. 3 Zugunruhe (left) and peak night melatonin (right) in food deprived and control birds in the night following food reintroduction. The amount of Zugunruhe was dramatically reduced in food deprived birds in autumn, but not in spring. Melatonin levels were higher in autumn than in spring and in food deprived than in control birds. Left, * = $p < 0.0001$, Two-way ANOVA followed by direct comparison (t-test); right, # = $p < 0.001$, autumn vs. spring, and * = $p < 0.05$, food deprived vs. controls, Two-way ANOVA. Adapted from (10).

At present, we do not know how the fasting-and-refeeding protocol induces an increase in melatonin and suppresses Zugunruhe. There is evidence in mammals that the diet can affect the amplitude of melatonin release (22). It has been shown in two bird species that the effects of the fasting-and-refeeding protocol are stronger when the birds are brought down to a lean, non migratory body mass (6, 11). In spring, the response of blackcaps to the fasting-and-refeeding protocol is highly variable and depends on the body mass and the amount of fat reserves reached at the end of the food-deprivation period (10) (Fig.4). Interestingly, in these conditions melatonin levels are negatively correlated with the decrease in body mass (10)(Fig.4), suggesting that melatonin might be regulated by nutrition-related factors.

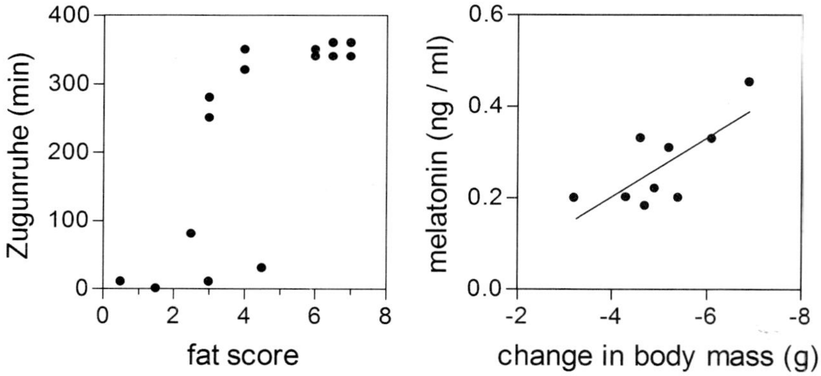

Fig. 4 Left: Zugunruhe vs. fat score of Kenya blackcaps that had been subjected to the fasting-and-refeeding protocol (see text for details). With one exception, all birds that had reached a fat score of 3 or less showed a substantial decrease in Zugunruhe (r_S = 0.74, n = 16, p< 0.001). Right: Peak night melatonin vs. the variation in body mass in the same birds. Melatonin levels were correlated with the amount of body mass lost during the fasting period (r = 0.743, n = 9, p < 0.03). Drawn from data reported in (10).

Acknowledgements

We thank Cristiano Bertolucci and Gregory F. Ball for their helpful comments on a previous version of the manuscript.

References

1 Abraham U, Gwinner E & Van't Hof TJ (2000) Exogenous melatonin reduces the resynchronization time after phase shifts of a nonphotic zeitgeber in the house sparrow (*Passer domesticus*). *Journal of Biological Rhythms* **15** 48-56

2 Arendt J (1998) Melatonin and the pineal gland: influence on mammalian seasonal and circadian physiology. *Reviews of Reproduction* **3** 13-22

3 Berthold P (1988) Unruhe-Aktivität bei Vögeln: eine Übersicht. *Vogelwarte* **34** 249-259

4 Berthold P, Querner U & Schlenker R (1990) Die Moenchsgrasmuecke. Wittenberg-Lutherstadt: Ziemsen Verlag

5 Berthold P (2001) Bird Migration: A General Survey. Oxford: Oxford University Press

6 Biebach H (1985) Sahara Stopover in Migratory Flycatchers (*Muscicapa striata*): Fat and Food Affect the Time Program. *Experientia* **41** 695-697

7 Cassone VM & Menaker M (1984) Is the avian circadian system a neuroendocrine loop? *Journal of Experimental Zoology* **232** 539-549

8 Dawson A, King VM, Bentley GE & Ball GF (2001) Photoperiodic control of seasonality in birds. *Journal of Biological Rhythms* **16** 365-380

9 Fusani L & Gwinner E (2001) Reduced amplitude of melatonin secretion during migration in the blackcap (*Sylvia atricapilla*). In *Persperctive in Comparative Endocrinology: Unity and Diversity. Proceed. 14th Intl. Congr. Comp. Endocrinol. Sorrento, Italy, May 2001*, pp. 295-300. Eds HJT Goos, RK Rastogi, H Vaudry & R Pierantoni. Bologna: Monduzzi

10 Fusani L & Gwinner E (2004) Simulation of migratory flight and stopover affects night levels of melatonin in a nocturnal migrant. *Proceedings of the Royal Society of London Series B-Biological Sciences* **271** 205-211

11 Gwinner E, Biebach H & von Kries I (1985) Food Availability Affects Migratory Restlessness in Caged Garden Warblers (*Sylvia borin*). *Naturwissenschaften* **72** 51-52

12 Gwinner E (1986) Circannual Ryhthms. Heidelberg: Springer-Verlag

13 Gwinner E, Schwabl H & Schwabl-Benzinger I (1988) Effects of food-deprivation on migratory restlessness and diurnal activity in the garden warbler *Sylvia borin. Oecologia* **77** 321-326

14 Gwinner E, Schwabl-Benzinger I, Schwabl H & Dittami J (1993) 24-Hour Melatonin Profiles in a Nocturnally Migrating Bird During and Between Migratory Seasons. *General and Comparative Endocrinology* **90** 119-124

15 Gwinner E (1996) Circadian and circannual programmes in avian migration. *The Journal of Experimental Biology* **199** 39-48

16 Gwinner E, Hau M & Heigl S (1997) Melatonin: Generation and modulation of avian circadian rhythms. *Brain Research Bullettin* **44** 439-444

17 Gwinner E & Hau M (2000) The Pineal Gland, Circadian Rhythms, and Photoperiodism. In *Sturkie's Avian Physiology*, pp. 557-568. Eds CG Whittow. San Diego: Academic Press

18 Gwinner E & Brandstaetter R (2001) Complex bird clocks. *Philosophical Transactions of the Royal Society of London. Series B: Biological Sciences* **356** 1801-1810

19 Hau M & Gwinner E (1995) Continuous Melatonin Administration Accelerates Resynchronization Following Phase-Shifts of a Light-Dark Cycle. *Physiology & Behavior* **58** 89-95

20 Kumar V, Gwinner E & Van't Hof TJ (2000) Circadian rhythms of melatonin in European starlings exposed to different lighting conditions: relationship with locomotor and feeding rhythms. *Journal of Comparative Physiology A - Sensory Neural and Behavioral Physiology* **186** 205-215

21 Pohl H (2000) Circadian control of migratory restlessness and the effects of exogenous melatonin in the brambling (*Fringilla montifringilla*). *Cronobiology International* **17** 471-488

22 Selmaoui B, Oguine A & Thibault L (2001) Food access schedule and diet composition alter rhythmicity of serum melatonin and pineal NAT activity. *Physiology & Behavior* **74** 449-455
23 Takahashi JS, Hamm H & Menaker M (1980) Circadian rhythms of melatonin release from individual superfused chicken pineal glands *in vitro*. *Proceedings of the National Academy of Sciences of the United States of America* **77** 2319-2322

Functional Avian Endocrinology
A Dawson & P J Sharp (Eds)

Melatonin-dependent song pattern of songbirds: Bypassing the gonadal control of sexual behavior

M Gahr, L Fusani[1], R Jansen, A ter Maat

Department of Developmental and Behavioral Neurobiology, Institute of Neuroscience, Vrije Universiteit Amsterdam, 1087 De Boelelaan, 1081 HV Amsterdam, The Netherlands
[1]Dipartimento di Fisiologia, Sezione di Neuroscienze e Fisiologia Applicata, Universita' di Siena, Via Aldo Moro, 53100 Siena, Italy

Introduction

Sexual behaviors of vertebrates including birds are generally thought to be controlled by the gonadal steroid hormones, androgens, estrogens and progestins. In combination with brain-autonomous genetic mechanisms, the gonadal hormones direct sexual development of brain areas involved in the control of sexual behaviors in birds and mammals. In adulthood, these hormones are crucial for the reversible modification of neural circuits, which correlates with an increased probability for sexual behaviors to occur and/or with an increased "sexyness" of the behavior (for review: 3, 15, 16). Neurobiologically well investigated examples of such gonadal steroid sensitive behaviors in birds are the copulatory behavior of male Japanese quail, courtship sequences of male ring doves, the copulation-solicitation display of female songbirds, and the singing of songbirds (for review: 3, 15, 16). These behavioral actions of sex hormones are likely mediated by binding to their cognate receptors, the androgen, the estrogen and the progestin receptors, which are ligand activated transcription factors.

Breeding periods occur either seasonally, in most temperate species, or coincide with unpredictable events such as food availability, in tropical species. Both in seasonal and opportunistic species, proximate cues activate the GnRH-system that up-regulates the production of gonadal hormones either transiently or permanently. In seasonal breeders (most non-tropical bird species), photoperiod is the dominant proximate factor (10). Increasing photoperiod stimulates gonadal maturation and secretion of gonadal hormones. This process involves the activation of hypothalamic photoreceptors but does not involve the photoperiod-dependent production of the pineal hormone melatonin, as it does in mammals (10). Similarly, gonadal regression is photoperiod-induced but does not involve melatonin. In opportunistic breeders, endogenous circannual rhythmicity in combination with non-photic environmental cues is likely to trigger gonadal maturation and gonadal hormone secretion (10).

Among songbirds, we find both seasonal and opportunistic breeders. Although the relation between singing and testosterone secretion is less clear in some tropical species, it is generally assumed that singing activity or song pattern or both of the reproductive periods depend on elevated levels of testosterone (for review: 3, 15, 46). E.g., wild canaries sing at most times of the year but their song repertoire and song temporal pattern, as well as the circumstances of singing episodes, change between the breeding season and the non-breeding season in relation with plasma testosterone levels (25). The notion that songs, which function as sexual signals, depend on testosterone and its estrogenic metabolite 17ß-estradiol is supported by numerous castration and hormone replacement studies of male and female songbirds (for review: 3, 15, 16, 46). Cerebrally, these behavioral findings correlate with the expression of androgen receptors or both androgen and estrogen receptors in areas of the neural song control circuit (15). Further, the expression of these receptors is dynamic, which might be due to androgen and estrogen-dependent autoregulation (15). Thus, gonadal hormones appear to affect sexual behavior such as singing by local action on androgen and estrogen receptor expressing vocal neurons.

Intrinsic and environmental cues, including food availability, stress or photoperiod, alter the production and cerebral availability of gonadal hormones (10) and/or the neural expression of their receptors (11, 15, 39) and in consequence modify the abundance and/or structure of sexual behaviors. However, in addition to gonadal steroid receptors, song control neurons express receptors for the pineal hormone melatonin (22), for stress hormones such mineralocorticoids and glucocorticoids (Gahr and Metzdorf, unpublished), and for cytokines (26). These neurochemical phenotypes suggest that song behavior is likely under the control of multiple endocrine signals that directly act on vocal neurons, bypassing the gonadal system. In the following, we first point out the distribution of melatonin receptors in various avian neural circuits including the song control system. Secondly we discuss the modulation of the song pattern by the action of melatonin on vocal neurons. Last, we discuss the possibility of non-gonadal endocrine control of brain function and sexual behavior in a more general context.

Distribution of melatonin receptor types in the avian brain

Melatonin acts on target areas *via* its three high affinity receptors, the Mel_{1A}, Mel_{1B} and Mel_{1C} receptors, G-protein coupled membrane receptors (36, 37). The binding of the receptor ligand iodo-melatonin and expression of the three receptors was studied in adjacent sections of the brains of the zebra finch and of the blackcap using *in situ* hybridization localization of the receptor mRNA (22, Fusani, Metzdorf, Gwinner, and Gahr, subm). Common areas of expression and melatonin binding are retinorecipient and integrative areas involved in vision. Melatonin receptors are also found in non-visual sensory areas, in the cerebellum, the hippocampus, in many motor areas, and in most pallial areas; the latter have cortex-analogous functions in birds. Similar to these brain areas, the Mel_{1A}, Mel_{1B} and Mel_{1C} are co-expressed in an area-specific manner in nuclei of the vocal control system. Further, the combined expression of the melatonin receptor types (22; Fusani, Metzdorf, Gwinner and Gahr, subm) in general is congruent with the iodo-melatonin binding brain areas of songbirds (4, 17, 47). Since the distribution of iodo-melatonin binding sites among species of different avian orders is rather similar (2, 4, 17, 37, 47), it is likely that the distribution of melatonin receptor in the avian brain is conserved with exception of taxa-specific neural adaptations such as the song control circuit (see below). Thus, in birds, melatonin might directly affect a large number of brain areas that are not directly involved in circadian pacemaking. The increasing number of brain areas of various mammals that are reported to contain melatonin receptors (localized by *in vitro* ligand binding, immunocytochemistry or mRNA expression) suggests a similar picture for mammals (for review 22). The comparison of the data from birds and mammals shows melatonin targets both in areas of the circadian system, such as the retina and the suprachiasmatic nucleus, and in the neuroendocrine system (for review 22). Further, melatonin target sites are present in areas not involved in circadian pacemaking or neuroendocrine homeostasis, including the visual system of birds, the cerebellum, and the striatum of birds and mammals, the mammalian hippocampus and cortex (for review 22).

The cellular pathways *via* which the melatonin receptor types modulate neural circuits in birds have not been studied in detail but are likely to differ between brain areas and neuron types as in mammals (18, 23, 44). In mammals, melatonin receptors are known to affect numerous signal transduction pathways (inhibition of cyclic AMP and cyclic GMP accumulation, suppression of diacylglycerol synthesis and arachidonic release, decrease of intracellular Ca^{2+}-concentration, and increase of membrane potential) with redundant and non-redundant roles of the receptor subtypes (18, 23, 44).

Melatonin receptors of the song control system

The descending motor pathway of this circuit consists of the cortex-analogous sensory-motor integration area HVC (nucleus hyperstriatalis ventrale, pars caudale), the premotor area RA (nucleus robustus archistriatalis), the medullary motornucleus nXIIts (nucleus hypoglossus pars tracheosyringealis) that innervates the sound producing syrinx, and medullary expiratory and inspiratory premotor areas that project to spinal cord respiratory motorneurons. In the descending motor pathway of the song system of zebra finches and blackcaps, HVC and RA expressed the Mel_{1B} type but neither Mel_{1A} nor Mel_{1C} receptor mRNA (22; Fusani, Metzdorf, Gwinner, and Gahr, subm). Neurons of nXIIts expressed Mel_{1C} but neither Mel_{1A} nor Mel_{1B} mRNA (22; Fusani, Metzdorf, Gwinner, and Gahr, subm). Further, the RAM (nucleus retroambigualis), an expiratory premotor area expressed Mel_{1C} receptor mRNA but neither Mel_{1A} nor Mel_{1B} (22; Fusani, Metzdor, Gwinner, and Gahr, subm). Since the expression pattern of the receptor types in the song system of the blackcaps, a seasonal breeding songbird species, and of the zebra finch, an opportunistic breeder, is very similar, we suggest that melatonin sensitivity is a general feature of the vocal system of songbirds. The binding of the iodo-melatonin in the song system (4, 17, 47) is in general explained by the receptor expression pattern. The area-specific expression of the melatonin receptor types in vocal control nuclei indicates area-specific mechanisms of melatonin action that might affect different song features due to the hierarchical organisation of the song system.

Melatonin affects the firing pattern of vocal neurons

RA and HVC are required for the production of the song pattern. HVC plays a role in the patterning of larger vocal sequences such as syllables whereas RA is thought to control of the motoric details of these syllables (48). In slices of the zebra finch forebrain that include the RA, melatonin application decreases the firing rate of a subpopulation of RA-neurons but does not affect the variability of the firing pattern (22). This effect of melatonin is observed both in slices prepared during the subjective day and night of the animals (Jansen, van der Roest and Gahr, unpubl). The acute direct effect of melatonin on RA neurons (22) is similar to the melatonin-induced reduction in firing rate of neurons of the mouse suprachiasmatic nucleus (23) and of the rat medial vestibular nucleus (33). The latter effects are, however, mediated *via* melatonin action on Mel_{1A} (23, 33). The acute effect of melatonin on vocal neurons is conceptually different from photoperiod-dependent changes of the volume of vocal areas (5); although melatonin was indicated as the mediating endocrine signal (5), this effect is long-term, which makes it difficult to exclude indirect action of melatonin.

Melatonin affects the song pattern of the zebra finch

Adult male zebra finches produce a stereotyped sequence (motif) of syllables, which identifies each male acoustically. Although there is a small temporal drift during long singing bouts (7), song length and song pattern of zebra finches are not circadian, but singing is strictly dependent on light (22). Melatonin levels of zebra finches are high during the night and very low during the day (45). *In vivo*, systemic application of a Mel_{1B} antagonist at the beginning of the night shortens the length of songs uttered the next day (Fig. 1, 22). In these songs, syllables at the end of the strophes are frequently omitted and/or the length of those syllables is modified while the treatment does not affect singing activity. After the following "normal" night, the structure of the songs returns to the pre-treatment pattern, i.e the effect of the melatonin receptor inhibitor is transient (22). The expression of melatonin receptors in vocal areas, the electrophysiological effect of melatonin on vocal neurons, and the behavioral relevance of melatonin receptors suggest that direct action of melatonin on the song control system plays a crucial role in the production of the stereotyped individual song pattern (22). It remains to be seen how the melatonin-dependent firing of RA-neurons relates to melatonin-dependent song pattern (22).

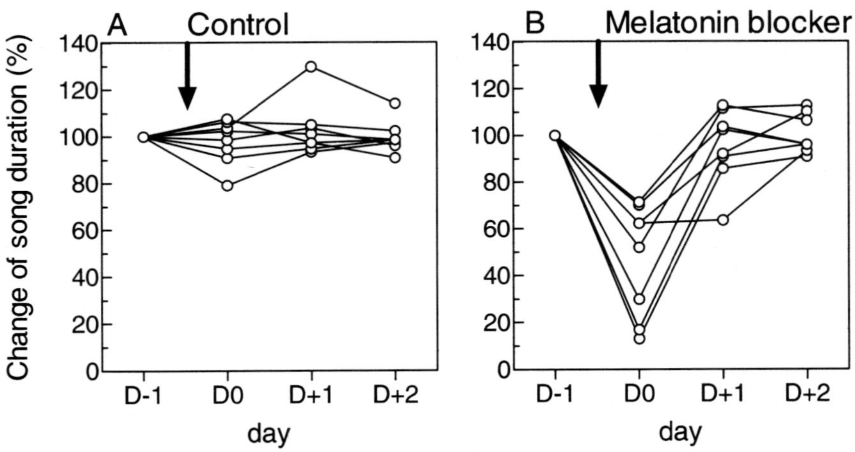

Fig. 1 The song length before (D-1) and after (arrow) application of the control vehicle (A) or the Mel_{1B} antagonist S20928 (B) of each male is shown (mean). The song length is transiently reduced at the day after (day 0, D0) nightly exposure to the antagonist and back to the pre-injection length thereafter (D+1, D+2) (repeated measures ANOVA, $p < 0.001$). Song length at day "D-1" is 100% (Data from 22).

The nightly melatonin surge appears to prime the song control circuit to maintain its stereotyped neural activity during daytime singing episodes (22). This priming might involve the nightly episodes of spontaneous activity of RA neurons that are thought to shape motor output of the daytime singing period (9). Alternatively, the observed changes in song pattern might involve the melatonin dependent expression of androgen or estrogen receptors in the descending motor pathway. In peripheral tissues, such as the uterus of the hamster, melatonin is known to regulate estrogen receptors (8). Although brain areas outside the vocal system that express melatonin receptors, including the preoptic area (Fusani, Metzdorf, Gwinner and Gahr, subm), might be involved in the control of singing activity, such areas are not involved in the control of the song pattern (38). The effect of melatonin on song motor pattern of zebra finches is independent of the circadian singing activity (22). Since melatonin receptors in vocal areas most likely are a general feature of songbirds, it remains to be seen to what extent the song patterns of other species are melatonin sensitive. Since melatonin binding of Area X, a song control nucleus not relevant for acute vocal output, is dynamic but not dependent on the photoperiod (4), melatonin might have further function on singing behavior besides shaping the motor pattern.

Melatonin function: From circadian timing to motor organization

The hormone melatonin is well known for its role in entrainment of circadian and circannual rhythms (10, 19, 32, 36). In view of the widespread distribution of melatonin target sites, melatonin is suggested to link higher brain functions and neural plasticity to circadian (birds, mammals) and circannual (mammals) homeostasis (6, 10, 12, 14, 24, 33, 42). Although melatonin might be involved in some aspects of seasonality of birds, such as seasonal plasticity of vocal areas (5), sexual behaviors including singing are generally thought to be primarily controlled by gonadal hormones, testosterone and 17β-estradiol (3, 15, 16). However, the melatonin-sensitive motor pattern (which is not circadian) of the song of the zebra finch indicates (1) that endocrine systems other than gonadal hormones affect sexual behavior and (2) that melatonin plays a major role in neural functions and behavioral control beside its involvement in circadian rhythms (Fig. 2, 22).

Such a function of melatonin could mediate day-to-day changes in the socio-sexual and physical environment (Fig. 3). Male zebra finches use their individual song to maintain contact with other zebra finches or their mates, but the song is also an integral feature of the male courtship behavior. In the latter, males produce their songs with higher speed and song length is significantly increased (35, 41). Although we

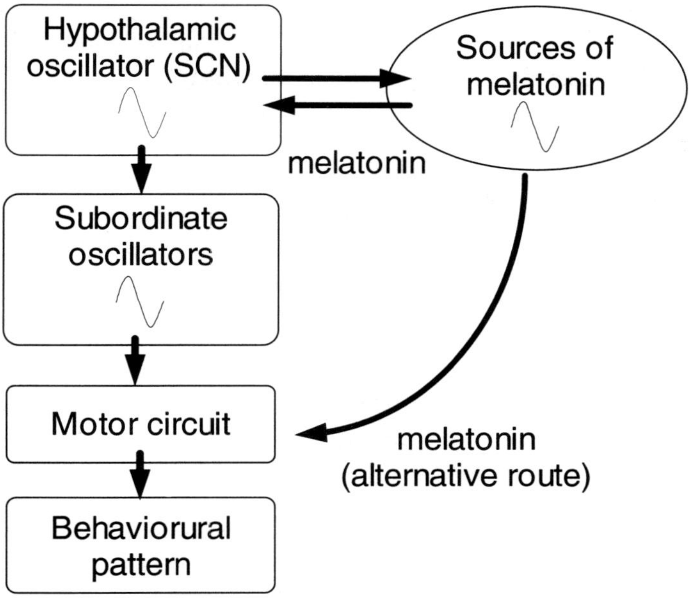

Fig. 2 Schematic model of melatonin-dependent behaviour. The photoperiod dependent production of melatonin of various origins (pineal gland, retina) is thought to entrain hypothalamic oscillators such as the SCN (suprachiasmatic nucleus) *via* an endocrine route. These oscillators might then entrain subordinate oscillators that innervate motor control circuits, which results in circadian occurrence of behaviour. Here we suggest that in addition to this hypothalamic route, melatonin acts directly on premotor and motor areas of a behavioural control circuit (22). This route of melatonin action might modulate the structure of the behaviour such as the song of the zebra finch independent of the circadian behavioural activity.

did not observe many courtship songs in the males treated with a melatonin receptor antagonist, all songs produced by these males were short (22). Thus, on a day-by-day basis, melatonin appears to modulate the males' ability to sing songs that are preferred by females (29, 41). In contrast, modulations of the singing of adult zebra finches that depend on the gonadal production of sex hormones are rather slow and long lasting and do not concern the basic song pattern (1, 20, 35). In keeping with the notion that nightly blockade of melatonin receptors makes the song less attractive, zebra finch males treated with a melatonin receptor antagonist obtained no copulation on the day of treatment but copulated on the days before and after the treatment-day (22; Gahr, unpublished).

The interpretation of melatonin dependent sexual behavior in light of sexual selection theories such as honest signaling remains speculative, due to the lack of detailed knowledge about the control of melatonin production of birds other than by the photoperiod. The photoperiod interacts with an endogenous clock (located in the pineal) so that melatonin production is high in the night and low during the day (42, 49). Cyclic AMP rapidly modulates the activity of the penultimate enzyme of melatonin synthesis, the serotonin N-acetyltransferase (AA-NAT) (42, 49). Thus, next to light, physical (cold stress), neural (adenosine, norepinephrine, vasoactive intestinal peptide) and hormonal (testosterone, histamine) factors are likely to modulate rapidly the secretory activity of the pineal gland *via* the cAMP pathway (13, 21, 27, 28, 30, 34, 42, 43). These factors have been analysed in some detail only in non-songbird species (chicken, quail, pigeon, parakeet). Since there are large species differences in the morphology and innervation of the pineal gland among birds (40), it remains to be seen which factors modulate the synthesis and secretion of melatonin of oscines. Future research shall show to what extent these mechanisms of melatonin synthesis are under sexual selection pressure.

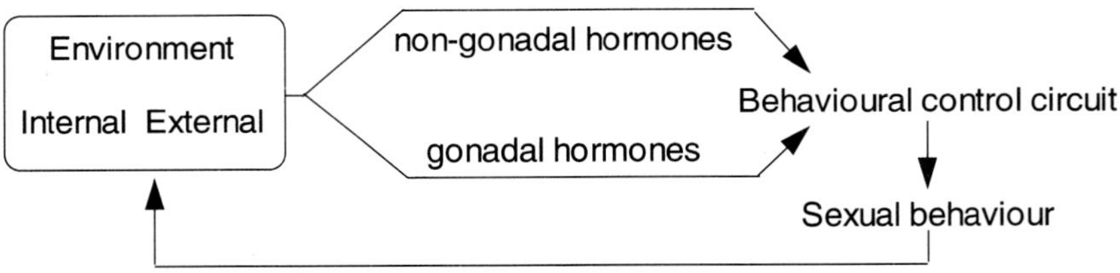

Fig. 3 External (socio-sexual and physical) and internal (body physiology) signals specify the neural phenotype of brain areas controlling sexual behaviour (e.g. the vocal control system) either (1) through acting on the HPG-axis and differential release of gonadal hormones or (2) through non-gonadal hormone systems such as melatonin, both of which have target neurons in behavioural control areas (e.g. the vocal control system).

Conclusion

The example of melatonin-dependent singing clearly suggests proximate control mechanisms for sexual behavior other than by gonadal hormones (22; Fig. 3). Such mechanisms might be widespread in view of the many sites of melatonin receptors in sensory systems, sensory-motor integration areas, and motor systems of birds (2, 4, 17, 22, 37, 47; Fusani, Metzdorf, Gwinner and Gahr, subm). Further, similar to birds, melatonin receptors as well as receptors of other endocrine systems, including the glucocorticoids, the cytokines, the leptins, and the thyroids, are present in neurons of various sensory and motor systems of mammals. Since the external (socio-sexual and physical) and internal (body physiology) environment certainly affects the activity of the hypothalamus-pituitary-gonadal-axis (10), multiple endocrine signals act directly on behavioral control circuits (22; Fig. 3). Because various entities of the body physiology such as the pineal gland are sensitive to gonadal hormones (21, 28) and since the different endocrine signals might interfere with the cerebral expression or activity of receptors of the other endocrine system (11, 31, 50), we need to consider that singing and sexual behavior in general are regulated by a complex network of neuroendocrine systems.

References

1 Arnold AP (1975) The effects of castration and androgen replacement on song, courtship, and aggression in zebra finches (*Poephila guttata*). *Journal of Experimental Zoology* **191** 301-325

2 Aste N, Cozzi B, Stankov B & Panzica G. (2001) Sexual differences and effect of photoperiod on melatonin receptors in avian brain. *Microcopy Research and Technique* **55** 37-47

3 Ball GF & Balthazart J (2004) Hormonal regulation of brain circuits mediating male sexual behavior in birds. *Physiology & Behavior* **83** 329-46

4 Bentley GE & Ball GF (2000) Photoperiod-dependent and -independent regulation of melatonin receptors in the forebrain of songbirds. *Journal of Neuroendocrinology* **12** 745-752

5 Bentley GE, Van't Hof TJ & Ball GF (1999) Seasonal neuroplasticity in the songbird telencephalon: A role for melatonin. *Proceedings of the National Academy of Science USA* **96** 4674-4679

6 Brotto LA, Barr AM & Gorzalka BB (2000) Sex differences in forced-swim and open-field test behaviours after chronic administration of melatonin. *European Journal of Pharmacology* **402** 87-93

7 Chii Z & Margoliash D (2001). Temporal precision and temporal drift in brain and behavior of Zebra Finch song. *Neuron* **32** 899-910

8 Danforth DNJ, Tamarkin L, Do R & Lippman ME. (1983) Melatonin-induced increase in cytoplamic estrogen receptor activity in hamster uteri. *Endocrinology* **113** 81-85

9 Dave AS & Margoliash D (2000) Song replay during sleep and computational rules for sensorimotor vocal learning. *Science* **290** 812-816

10 Dawson A, King VM, Bentley GE & Ball GF (2001) Photoperiodic control of seasonality in birds. *Journal of Biological Rhythms* **16** 365-380

11 Diano S, Kalra SP, Sakamoto H & Horvath TL (1998) Leptin receptors in estrogen receptor-containing neurons of the female rat hypothalamus. *Brain Research* **812** 256-259

12 El-Sherif Y, Tesoriero J, Hogan MV & Wieraszko A (2003) Melatonin regulates neuronal plasticity in the hippocampus. *Journal of Neuroscience Research* **72** 454-460

13 Falcon J, Van Camp G & Collin JP (1995) Adenosine A2 receptor-mediated stimulation of cyclic AMP in cultured chicken pineal cells. *Journal of Pineal Research* **19** 72-78

14 Fraschini F Cesarani A, Alpini D, Esposti D & Stankov BM (1999) Melatonin influences human balance. *Biological Signals and Receptors* **8** 111-119

15 Gahr M. (2001) Distribution of sex steroid hormone receptors in the avian brain: functional implications for neural sex differences and sexual behaviors. *Microscopy Research and Technique* **55** 1-11

16 Gahr M. (2004) Hormone-dependent neural plasticity in the juvenile and adult song system: what makes a successful male? *Annals of the New York Academy of Science* **1016** 684-703

17 Gahr M & Kosar E (1996) Identification, distribution, and developmental changes of a melatonin binding site in the song control system on the zebra finch. *Journal of Comparative Neurology* **367** 308-318

18 Gall von C, Stehle JH & Weaver DR (2002) Mammalian melatonin receptors: molecular biology and signal transduction. *Cell and Tissue Research* **309** 151-162

19 Gwinner E, Hau M & Heigl S (1997). Melatonin: generation and modulation of avian circadian rhythms. *Brain Research Bulletin* **44** 439-444

20 Harding CF, Sheridan K & Walters MJ (1983) Hormonal specificity and activation of sexual behaviour in male zebra finches. *Hormones and Behavior* **17** 111-133

21 Herbute S & Bayle JD (1976) Influence of castration and testosterone on pineal multi-unit activity in quail. *Journal of Physiology* **72** 949-957

22 Jansen R, Metzdorf R, Roest van der M, Fusani L, Maat ter A & Gahr M (in press). Melatonin affects the temporal organization of the song of the zebra finch. *FASEB Journal*

23 Jin X, Gall van C, Pieschl RL, Gribkoff VK, StehleJH, Reppert S & Weaver DR (2003) Targeted Disruption of the Mouse Mel_{1b} Melatonin Receptor. *Molecular And Cellular Biology* **23** 1054-1060

24 Laurido C, Pelissier T, Soto-Moyano R, Valladares L, Flores F & Hernández (2002) Effect of ketamine on spinal cord nociceptive transmission in normal and monoarthritic rats. *NeuroReport* **13** 89-91

25 Leitner S, Voigt C, Garcia-Segura LM, Van't Hof T & Gahr M (2001) Seasonal activation and inactivation of song motor memories in free living canaries is not reflected in neuroanatomical changes of forebrain song areas. *Hormones and Behavior* **40** 160-168

26 Louissaint A Jr, Rao S, Leventhal C & Goldman SA (2002) Coordianted interaction of neurogenesis and angiogenesis in the adult songbird brain. *Neuron* **34** 945-960

27 Lee PP, Allen AE & Pang SF (1990) Cold stress during scotophase elicited differential responses in quail pineal, retinal, and serum melatonin levels. *Acata Endocrinologica (Copenh)* **122** 535-539

28 Maitra SK & Dey M (1994) Castration and testosterone induced changes in the pinealocytes of roseringed parakeet. Psittacula krameri, during different phases of the annual testicular cycle. *Anatomischer Anzeiger* **176** 363-368

29 Neubauer RL (1999) Super-nomal length song preferences of female zebra finches (*Taeniopygia guttata*) and a theory of the evolution of the bird song. *Evolutionary Ecology* **13** 365-380

30 Nowak JZ & Sek B (1994) Stimulatory effect if histamine on cyclic AMP formation in chick pineal gland. *Journal of Neurochemistry* **63** 1338-1345

31 Persengiev SP (1999) Multiple domains of melatonin receptor are involved in the regulation of glucocorticoid receptor-induced gene expression. *Journal of Steroid Biochemistry Molecular Biology* **68** 181-187

32 Pevet J (2003) Melatonin: from seasonal to circadian signal. *Journal of Neuroendocrinology* **15** 422-426

33 Podda MV, Deriu F, Giaconi E, Milia M & Tolu E (2003) Melatonin inhibits rat medial vestibular nucleus neuron activity *in vitro*. *Neuroscience Letters* **341** 209-212

34 Pratt BL & Takahashi JS (1989) Vasoactive intestinal polypeptide and alpha 2-adrenoceptor agonists regulate adenosine 3', 5'-monophosphate accumulation and melatonin release in chick pineal cell cultures. *Endocrinology* **125** 2371-2384

35 Pröve E (1974) Der Einfluss von Kastration und Testosteronsubstitution auf das Sexualverhalten männlicer Zebrafinken (*Taeniopygia guttata castanotis* Gould). *Journal für Ornithologie* **115** 338-347

36 Reppert SM (1997) Melatonin receptors: molecular biology of a new family of G protein coupled receptors. *Journal of Biological Rhythms* **12** 528-531

37 Reppert SM Weaver DR, Cassone VM, Godson C & Kolakowski LF jr (1995) Melatonin receptors are for the birds: molecular analysis of two receptor subtypes differentially expressed in chick brain. *Neuron* **15** 1003-1015

38 Riters LV & Ball GF (1999) Lesions to the medial preoptic area affect singing in the male European starling (*Sturnus vulgaris*). *Hormones and Behavior* **36** 276-86

39 Roemmich JN, Li X, Rogol AD & Rissman EF (1997) Food availability affects neural estrogen receptor immunoreactivity in prepuberal mice. *Endocrinology* **138** 5366-5373

40 Sato T (2001) Sensory and endocrine characteristics of the avian pineal gland. *Microscopy Research and Technique* **53** 2-11

41 Sossinka R & Böhner J (1980) Song types in the zebra finch Poephilia guttata castanotis. *Zeitschrift für Tierpsychologie* **53** 123128

42 Underwood H, Steele C & Zivkovic B (2001) Circadian organization and the role of the pineal in birds. *Microsopy Research and Technique* **53** 48-62

43 Vakkuri O, Rintamaki H & Leppaluoto J (1992) Noradrenergic inhibition and alpha 2-adrnergic stimulation of melatonin secretion of the pigeon. *Journal of Neurochemistry* **59** 915-918

44 Vanecek J (1998) Cellular mechanisms of melatonin Action. *Physiological Reviews* **78** 687-721

45 Van't Hof TJ & Gwinner E (1996) Development of post- hatching melatonin rhythm in zebra finches (*Poephila guttata*). *Experientia* **52** 249-252

46 Welty JC & Baptista L (1990) The Life of Birds. 4ed. Harcourt Brace Jovanovich College Publishers:Fort Worth

47 Whitfield-Rucker MG & Cassone VM (1996) Melatonin binding in the house sparrow song control system: sexual dimorphism and the effect of photoperiod. *Hormones and Behavior* **30** 528-537

48 Yu AC & Margoliash D (1996) Temporal hierarchical control of singing in birds. *Science* **273** 1871-1875

49 Zatz M, Gastel JA, Heath JR 3rd & Klein DC (2000) Chick pineal melatonin synthesis: light and cyclic AMP control abundance of serotonin N-acetyltransferase protein. *Journal of Neurochemistry* **74** 2315-2321

50 Zhu YS, Yen PM, Chin WW & Pfaff DW (1996) Estrogen and thyroid hormone interaction on regulation of gene expression. *Proceedings of the National Academy of Science, USA* **93** 12587-12592

Functional Avian Endocrinology
A Dawson & P J Sharp (Eds)

Melatonin regulation of clock genes in birds

S Ebihara, S Yasuo and T Yoshimura

Division of Biomodeling, Graduate School of Bioagricultural Sciences,
Nagoya University, Nagoya 464-8601, Japan

Introduction

Avian circadian rhythms are regulated by a multioscillatory system consisting of circadian oscillators located in the pineal gland, the eye and the suprachiasmatic nucleus (SCN) (5, 9, 14, 18, 22). These oscillators are integrated into a single circadian pacemaker controlling many physiological rhythms with neural and hormonal interactions. Melatonin synthesized in the pineal and retina shows a distinct circadian rhythm and is though to work as a synchronizer among the multiple oscillators. In fact, daily injections of melatonin entrain circadian rhythms of wheel-running activity in rats with an intact SCN, but not SCN-lesioned rats, suggesting that melatonin entrains circadian rhythms of the SCN (4). Likewise, daily administration of melatonin restores behavioral rhythms in arrhythmic, pinealectomized and/or blinded birds and entrains the free-running rhythm to its administration cycles (6, 18). This seems to show that the SCN of birds is entrained by melatonin rhythms, however this possibility could not be tested until recently because the avian SCN had not been clearly determined (32).

To identify the avian SCN and test this possibility, the most effective way would be to examine the expression of clock genes constituting the core oscillator mechanism in proposed SCN sites. Cloning of avian circadian clock genes, thus has given us a good opportunity to test this possibility and provided a new avenue for molecular understanding of avian circadian system.

Clock gene expression in the pineal and the eye

It is thought that circadian oscillation is generated by a cyclic transcription/translation feedback loop of the clock genes (7). We have cloned several clock genes (*qClock, qPer2, qPer3*) from Japanese quail and characterized these genes (31). All of the genes we cloned were expressed in the circadian oscillators of the eye and the pineal gland (*Bmal1, Cry1, Cry2*, and *E4bp4* were also shown to be expressed in the pineal gland). *qClock* mRNA in the eye was expressed throughout the day, but its relative level was higher during the light than the dark in light and dark (LD) cycles, although this rhythmic change was not detected in the pineal. *qPer2* and *qPer3* mRNA levels were high during the light and low during the dark in LD, which continued in constant darkness (DD) with a reduced amplitude.

In both the eye and the pineal, light induced *qPer2* expression, but neither *qClock* nor *qPer3* were induced by light (31).

Clock gene expression in various brain areas

Clock genes were expressed in various brain areas. As shown in chapter 11 (Yoshimura *et al.*), these genes were expressed in the mediobasal hypothalamus and the pars tuberalis of the pituitary (27, 28). These two structures are important for the control of photoperiodic time measurement of birds and it is likely that clock genes are somehow involved in this phenomena (16). Other areas expressing clock genes were the optic tectum and the cerebellum (26). Of these, the optic tectum seems to be regulated by the circadian system. For example, the response to light pulses in the pigeon is greater during the day than during the night in both LD and DD, as demonstrated by the amplitude of electroretinograms and vision-evoked potentials from the optic tectum (24). Basal metabolic activity in the tectum shows daily rhythmicity, as determined by the uptake of 2-deoxy-[^{14}C]glucose (3). In Japanese quail, all the known circadian clock genes including *Clock, Per2, Per3, Bmal1, Cry1, Cry2*, and *E4bp4* were expressed in the optic tectum, suggesting that the circadian oscillator is localized in this structure (26). *Per2* mRNA expression showed robust oscillation with a peak around at *zeitgeber* time (ZT)0 (ZT0 corresponds to light onset) under LD12:12 cycles. This rhythm continued under constant light conditions (LL) and DD with a similar profile to that in LD. Although the optic tectum receives retinal projections, light did not induce *Per2* expression, indicating that rhythmic expression of *Per2* in the optic tectum is not directly regulated by LD cycles.

Signals of antisense probes of clock genes were detected in the cerebellum, however it is unclear whether these signals represent real mRNA of clock genes, because sense probes of some of the clock genes (*qPer2, qClock* etc.) also showed positive reactions in the cerebellum.

Identification of the SCN

It is clear that the SCN is the dominant biological clock regulating circadian rhythms in mammals (15). In birds, however, the site of the nucleus homologous to the mammalian SCN has been the subject of controversy for a long time and two nuclei, the so-called medial SCN (mSCN) and the visual SCN (vSCN), have been proposed to the avian

SCN (9). The mSCN is located near the angle of the preoptic recess of the third ventricle, and the vSCN is slightly more lateral and caudal to the mSCN. The vSCN receives the retinohypothalamic projection, but this projection to the mSCN is not clear, although a few lesion studies have indicated that the mSCN is necessary for the regulation of circadian rhythms (8, 9).

To identify the avian SCN functioning as a circadian clock, clock gene expression was examined in these areas. As shown in Fig. 1, the signals of *qClock, qPer2*, and *qPer3* were observed in the mSCN, but no positive signals were detected in the vSCN in Japanese quail (32). The results were confirmed in Java sparrow (*Padda oryzivora*), chicken (*Gallus gallus*), and pigeon (*Columba livia*). The temporal pattern of clock genes was extensively studied in Japanese quail (27, 29), and it has been found that many of the clock genes show circadian patterns that are different between long days and short days.

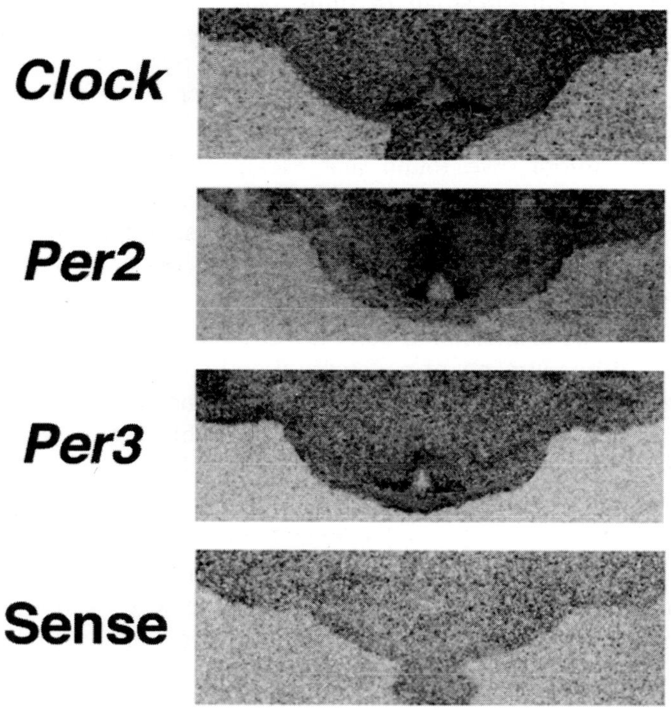

Fig. 1 Clock genes are expressed in the mSCN of Japanese quail. Reproduced from Yoshimura *et al.* (32)

In the pineal and retina, *qPer2* expression was induced by light. Therefore it is expected that light-induced expression of the gene would also occur in the mSCN. In fact, *qPer2* was significantly induced by light in intact birds and this also occurred in the birds whose eyes were covered with black rubber caps (32). The latter indicates that extra-retinal photoreceptors mediate light input to the mSCN, which is consistent with the previous results of entrainment of activity rhythms in blinded birds (10).

Additional evidence supporting the idea that the mSCN functions as a circadian clock, is that small lesions aimed at the mSCN disrupts circadian rhythms of locomotor activity (32). Although there is still disagreement about the site of the avian SCN (2), it is obvious from our results that the mSCN is the primary important hypothalamic circadian oscillator in birds.

Melatonin effects on clock gene expression

There is considerable evidence that melatonin is involved in the regulation of avian circadian rhythms. Since in mammals, melatonin affects circadian rhythms of electrical activity of SCN neurons and melatonin receptors reside in the SCN (17, 20, 23), one would expect that avian SCN is also affected by melatonin. In fact, we found that both Mel_{1a} and Mel_{1c} mRNAs are expressed in the mSCN (but not in the vSCN). In addition, we found that Mel_{1a} mRNA levels show daily rhythms with a peak at ZT3. Mel_{1a} is also expressed in other areas including the optic tectum, the nucleus rotundus, and the ventral lateral geniculatus lateralis (29).

Since melatonin receptors reside in the mSCN, we next examined effects of melatonin administration on clock gene expression. However, both treatments of melatonin (continuously administered high levels of melatonin, which masks endogenous melatonin rhythms and a single injection of melatonin) did not remarkably affect *qPer2, qPer3* and *qClock* gene expression in the mSCN of Japanese quail (Fig. 2). These are somewhat unexpected results because melatonin is thought to be an important factor regulating avian circadian rhythms.

To explain these results, the model proposed by Gwinner at an early stage of avian circadian studies might be worth noting (13). He explained the different effects of pinealectomy on circadian locomotor activity rhythms among avian species on the theory of differences in the coupling strengths of the self-sustaining oscillators. These oscillators are located outside of the pineal, and are strongly affected by the pineal, probably *via* its hormone, melatonin. Thus, when the coupling among the self-sustaining oscillators is too weak to keep them synchronized with each other, as is the case of the pinealectomized house sparrow, overt

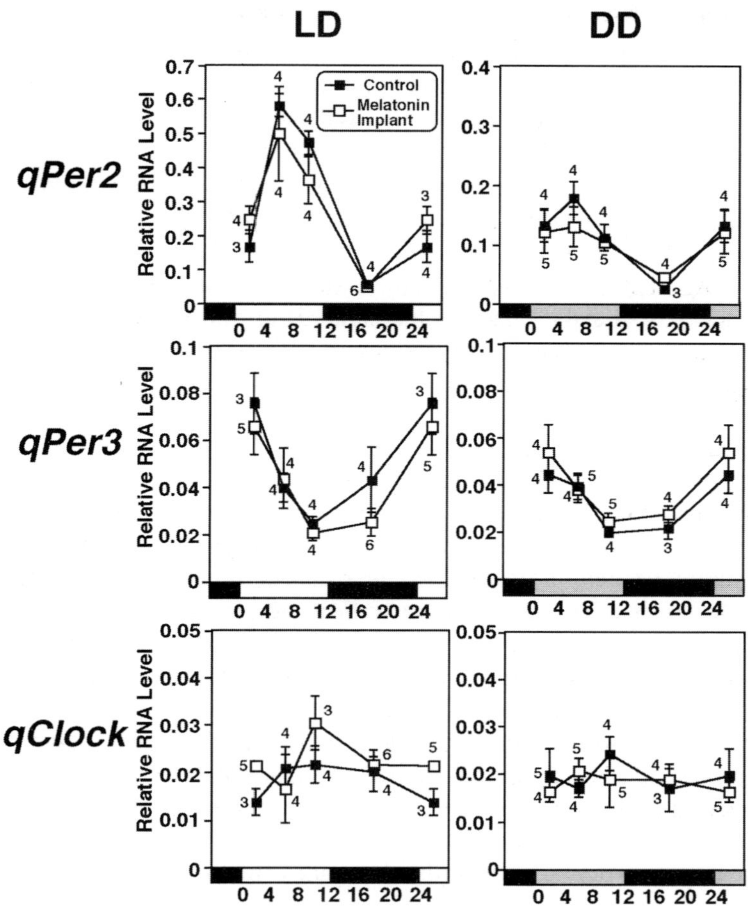

Fig. 2 Temporal changes of *qPer2, qPer3, qClock* mRNA expression in the mSCN of Japanese quail with a melatonin capsule (white square) or an empty capsule (black square). The bar at the bottom of each graph represents LD cycles and those to which the birds were exposed before being transferred to constant conditions. Each value is the mean ± SEM. The number of birds used is shown (n = 3-6). Reproduced from (29)

rhythms damp out and finally disappear. However if the coupling is strong, as in the case of Japanese quail, pinealectomy does not lead to loss of overt circadian rhythms. In this model, melatonin affects the coupling strength among the self-sustaining oscillators but does not impose its rhythmicity on the oscillators. Although Gwinner did not describe the site of the self-sustaining oscillators, it is reasonable to suppose that the SCN corresponds to these self-sustaining oscillators. Our results that the clock gene expression rhythms are not altered by melatonin suggest that melatonin does not affect transcription of clock genes, but may instead strengthen the synchronization among SCN oscillatory cells. It is interesting to note that melatonin receptors are located in astrocytes, which might be important for coupling among SCN cells in mammals (1, 21).

In rats, however, it is reported that a melatonin injection does not immediately affect clock gene expression patterns, but the patterns are changed during the second night after the melatonin injection (19), suggesting that phase-shifting effect of melatonin on the SCN may implicate post-translational mechanisms. Thus further effort is required to understand mechanisms of melatonin effects on the SCN molecular loop in vertebrates.

Conclusion

Although avian circadian rhythms are regulated by multiple oscillator system, intrinsic circadian clocks in mammals have been thought to reside only in the SCN. However, recent studies have revealed that mammalian peripheral tissues also have autonomous circadian oscillators (25). In particular, the demonstration of long-lasting *Per2* gene expression rhythms in cultured peripheral tissues of luciferase-fusion gene transgenic mice suggest that the SCN acts as a master synchronizer, rather a pacemaker in mammals (30). These data indicate that vertebrate circadian system consists of multiple oscillators in nature, and synchronization among these oscillators by the SCN and/or the pineal is a fundamentally important mechanism for maintaining overt circadian rhythms. Melatonin, which seems to be involved in the synchronizing mechanisms particularly in birds, does not affect clock gene expression in the SCN of Japanese quail and mammals, despite its phase-shifting action (11, 12, 29). To understand mechanisms of melatonin effects on the SCN, birds with different circadian responses to melatonin treatment such as passerine birds are particularly interesting, which might contribute to understanding synchronizing mechanism among multiple oscillators in vertebrates.

References

1 Adachi A, Natesan AK, Whitfield-Rucker MG, Weigum SE & Cassone VM (2002) Functional melatonin receptors and metabolic coupling in cultured chick astrocytes. *Glia* **39** 268-278

2 Brandstätter R, Abraham U & Albrecht U (2001) Initial demonstration of rhythmic *Per* gene expression in the hypothalamus of a non-mammalian vertebrate, the house sparrow. *Neuroreport* **12** 1167-1170

3 Cantwell EL & Cassone VM (2002) Daily and circadian fluctuation in 2-deoxy[(14)C]-glucose uptake in circadian and visual system structures of

the chick brain: effects of exogenous melatonin. *Brain Research Bulletin* **57** 603-611

4 Cassone VM (1991) Melatonin and suprachiasmatic nucleus function. *In* DC Klein, RY Moore & SM Reppert (Eds.) *Suprachiasmatic Nucleus: The Mind's Clock.* Oxford University Press, New York pp 309-323

5 Cassone VM & Menaker M. (1984) Is the avian circadian system a neuroendocrine loop? *Journal of Experimental Zoology* **232** 539-49

6 Chabot CC & Menaker M. (1992) Effects of physiological cycles of infused melatonin on circadian rhythmicity in pigeons. *Journal of Comparative Physiology* A **170** 615-622

7 Dunlap JC (1999) Molecular bases for circadian clocks. *Cell* **96** 271-290

8 Ebihara S & Kawamura H. (1981) The role of the pineal organ and the suprachiasmatic nucleus in the control of circadian locomotor rhythms in the Java sparrow, *Padda oryizivora. Journal of Comparative Physiology* A **141** 207-214

9 Ebihara S, Oshima I, Yamada H, Goto M & Sato K. (1987) Circadian organization in the pigeon. *In* T. Hiroshige and K. Homma (Eds.), *Comparative aspects of circadian clocks.* Hokkaido University Press, Sapporo pp 127-145

10 Ebihara S, Uchiyama K & Oshima I. (1984) Circadian organization in the pigeon, *Columba livia*: the role of the pineal organ and the eye. *Journal of Comparative Physiology* A **154** 59-69

11 Gall Cv, Garabette ML, Kell CA, Frenzel S, Dehghani F, Schumm-Draeger PM, Weaver DR, Korf HW, Hastings MH & Stehle JH (2002) Rhythmic gene expression in pituitary depends on heterologous sensitization by the neurohormone melatonin. *Nature Neuroscience* **5** 234-238

12 Gillette MU & McArthur AJ. (1996) Circadian actions of melatonin at the suprachiasmatic nucleus. *Behavior Brain Research* **73** 135-139

13 Gwinner E (1978) Effects of pinealectomy on circadian locomotor activity rhythms in European starlings, *Sturnus vulgaris. Journal of Comparative Physiology* A **126** 123-129

14 Gwinner E (1989) Melatonin in the circadian system of birds: model of internal resonance. *In* T Hiroshige & K Homma (Eds.) *Circadian Clocks and Ecology* Hokkaido University Press, Sapporo pp 88-94

15 Klein DC, Moore RY & Reppert SM (1991) *Suprachiasmatic Nucleus: the Mind's Clock.* Oxford University Press, New York

16 Lincoln GA, Messager S, Andersson H & Hazlerigg DG (2002) Temporal expression of seven clock genes in the suprachiasmatic nucleus and the pars tuberalis of the sheep: evidence for an internal coincidence timer. *Proceedings of the National Academy of Sciences U S A* **99** 13890-13895

17 McArthur AJ, Hunt AE & Gillette MU (1997) Melatonin action and signal transduction in the rat suprachiasmatic circadian clock: activation of protein kinase C at dusk and dawn. *Endocrinology* **138** 627-634

18 Oshima I, Yamada H, Goto M, Sato K & Ebihara S (1989) Pineal and retinal melatonin is involved in the control of circadian locomotor activity and body temperature rhythms in the pigeon. *Journal of Comparative Physiology* A **166** 217-226

19 Poirel VJ, Boggio V Dardente H, Pevet P, Masson-Pevet M & Gauer F (2003) Contrary to other non-photic cues, acute melatonin injection does not induce immediate changes of clock gene mRNA expression in the rat suprachiasmatic nuclei. *Neuroscience* **120** 745-755
20 Reppert SM, Weaver DR & Ebisawa T (1994) Cloning and characterization of a mammalian melatonin receptor that mediates reproductive and circadian responses. *Neuron* **13** 1177-1185
21 Shinohara K, Funabashi T, Mitushima D & Kimura F (2000) Effects of gap junction blocker on vasopressin and vasoactive intestinal polypeptide rhythms in the rat suprachiasmatic nucleus *in vitro. Neuroscience Research* **38** 43-47
22 Underwood H & Siopes T (1984) Circadian organization in Japanese quail. *Journal of Experimental Zoology* **232** 557-566
23 Vanecek J (1988) Melatonin binding sites. *Journal of Neurochemistry* **51** 1436-1440
24 Wu WQ, McGoogan JM & Cassone VM (2000) Circadian regulation of visually evoked potentials in the domestic pigeon, *Columba livia. Journal of Biological Rhythms* **15** 317-328
25 Yamazaki S, Numano R, Abe M, Hida A, Takahashi R, Ueda M, Block GD, Sakaki Y, Menaker M & Tei H (2000) Resetting central and peripheral circadian oscillators in transgenic rats. *Science* **288** 682-685
26 Yasuo S, Ebihara S & Yoshimura T (2004) Circadian expression of clock gene in the optic tectum of Japanese quail. *Brain Research* **1005** 193-196
27 Yasuo S, Watanabe M, Okabayashi N, Ebihara S & Yoshimura T (2003) Circadian clock genes and photoperiodism: Comprehensive analysis of clock gene expression in the mediobasal hypothalamus, the suprachiasmatic nucleus, and the pineal gland of Japanese Quail under various light schedules. *Endocrinology* **144** 3742-3748
28 Yasuo S, Watanabe M, Tsukada A, Takagi T, Iigo M, Shimada K, Ebihara S & Yoshimura T (2004) Photoinducible phase-specific light induction of *Cry1* gene in the pars tuberalis of Japanese quail. *Endocrinology* **145** 1612-1616
29 Yasuo S, Yoshimura T, Bartell PA, Iigo M, Makino E, Okabayashi N & Ebihara S (2002) Effect of melatonin administration on *qPer2, qPer3*, and *qClock* gene expression in the suprachiasmatic nucleus of Japanese quail. *European Journal of Neuroscience* **16** 1541-1546
30 Yoo SH, Yamazaki S, Lowrey PL, Shimomura K, Ko CH, Buhr ED, Siepka SM, Hong HK, Oh WJ, Yoo OJ, Menaker M & Takahashi JS (2004) PERIOD2::LUCIFERASE real-time reporting of circadian dynamics reveals persistent circadian oscillations in mouse peripheral tissues *Proceedings of the National Academy of Sciences U S A* **101** 5339-5346
31 Yoshimura T, Suzuki Y, Makino E, Suzuki T, Kuroiwa A, Matsuda Y, Namikawa T & Ebihara S (2000) Molecular analysis of avian circadian clock genes. *Molecular Brain Research* **78** 207-215
32 Yoshimura T, Yasuo S, Suzuki Y, Makino E, Yokota Y & Ebihara S (2001) Identification of the suprachiasmatic nucleus in birds. *American Journal of Physiology-Regulatory Integrative and Comparative Physiology* **280** R1185-1189

Part six

Stress hormones

White-crowned Sparrow
Pierre Deviche

Functional Avian Endocrinology
A Dawson & P J Sharp (Eds)

Modulation of the adrenocortical response to acute stress in breeding birds

J C Wingfield

Department of Biology, Box 351800, University of Washington, Seattle, Washington 98195

Introduction

The adrenocortical response to stress and other hormones released from the hypothalamo-pituitary-adrenal (HPA) axis allow vertebrates to cope with perturbations of the environment in general. The behavioral and physiological responses to hormones of the HPA axis are thought to be highly adaptive and include stimulatory, suppressive and permissive actions of glucocorticoids (32). These physiological and behavioral responses to elevated glucocorticoid secretion have been collectively termed the emergency life history stage that allows the individual to avoid chronic stress (42, 40). In contrast to diversity of perturbations and other stresses, the emergency life history stage is a remarkably consistent suite of traits among all vertebrates that allows animals to cope with a stochastic environment and maximize lifetime fitness.

The frequency and magnitude of unpredictable perturbations vary along environmental gradients. For example, Kitaysky *et al.* (17, 18) have shown that in sea birds, behavioral responses to variability in food resources reflects their phylogeny and ecology including that of their prey. If individuals rely on continuously available food resources, then a short-term decrease in food availability may increase glucocorticoid secretion and trigger an emergency life history stage. In contrast, if individuals rely on highly variable food resources, then short-term food shortages should not result in immediately elevated glucocorticoid secretion. Only a severe long-term food shortage should trigger an emergency life history stage. Thus, in predictable environments (less stochastic – a low probability of perturbations), the HPA axis should be more responsive to environmental stress resulting in more rapid physiological and behavioral responses such as the emergency life history stage. In less predictable environments (more stochastic – a greater probability of perturbations), the HPA axis will be resistant to the same types of stresses and those responses typical of the emergency life history stage are expected be delayed (17, 18, 40).

Despite the apparent adaptive value of the emergency life history stage, activation of the HPA axis in response to perturbations (stress) may not be adaptive at all stages in the life cycle, and many species of birds show dramatic modulation (suppression and hyper-activation) of the HPA axis while breeding. For example, outside of the reproductive season survival is maximized whereas when breeding, the number of viable offspring produced is critical. Therefore, whether or not an individual modulates its responsiveness of the HPA axis to perturbations of the environment should reflect which specific component of lifetime fitness is currently being maximized (Wingfield and Kitaysky, 40). The ecological bases underlying such modulation have been suggested, although much remains to be done to experimentally test these ideas (36, 44, 42).

More recently, Wingfield and Sapolsky (43) summarized when and why individuals should maintain reproductive function under stressful conditions rather than interrupt breeding in favor of survival. Examples relevant to birds are:

1) Long-lived individuals with decreasing likelihood of future reproductive success as they age (e.g. many seabirds) should suppress their HPA axis response to acute perturbations when breeding. An example comes from a known age population of Leaches storm-petrels, *Oceanodroma castro*, in which the oldest indviduals showed a reduced increase in plasma corticosterone levels following capture, handling and restraint (24, 5).
2) Seasonal breeders in severe environments, when the breeding season is very short and only one breeding attempt is possible per year (e.g. many species especially at high latitudes and altitudes), should temporarily suppress their HPA axis response stress (36). If both members provide parental care that is essential for maximum reproductive success, then loss of the mate (to a predator) means the remaining parent must compensate to raise young successfully (e.g. many species in all environments). The surviving mate should then suppress the HPA axis response to stress so as to maximize the likelihood of raising young (35).
3) In many tournament species, an individual may only have access to mates for a short period (e.g. dominant birds in leks). In these cases, the dominant birds with greatest potential reproductive success should avoid interruptions of breeding by acute stresses and suppress the response to the HPA axis.

There is a growing literature on examples of suppression of the adrenocortical response to acute stress, their ecological bases and possible mechanisms by which they occur (e.g. 44, 34, 42). However, there is also growing evidence that some species may increase responsiveness of the HPA axis to stress at specific times in the breeding season (3, 28, 15). Such dynamic regulation of responsiveness to environmental stress was unexpected and the reasons why remain equivocal. In this paper I will give some specific examples of modulation of the adrenocortical responses to stress in natural situations, and then go on to suggest why this occurs and possible underlying mechanisms.

Examples of modulation of the adrenocortical response to acute stress

Migrant birds in the Arctic

In spring, millions of migrant birds stream north of the Arctic Circle to settle on the tundra and breed. They display and sing despite severe weather, reproduce quickly during the ephemeral summer, and then migrate south to avoid the arctic winter. This biological spectacle appears at first to be a frantic rush into a challenging environment. Closer observation and experiments reveal, however, a highly coordinated process with morphological, physiological and behavioral adaptations that permit great phenotypic flexibility to cope with the arctic weather (47). Capricious weather in spring has been endured by indigenous peoples of the Arctic for millennia. An old proverb from the Saami Reindeer People of Lapland states "spring is the weather's struggletime" (16). This phrase epitomizes spring in the Arctic with dramatic swings between high winds, low temperatures and snow typical of winter to sunshine and rapidly increasing temperatures that herald summer. At our long term field site near Toolik Lake, Alaska (68°N), snow cover is extremely variable in mid-May when many migrant song birds begin arriving (Fig. 1). Snow may melt from the tundra as early as the first week in May, or persist into June. Storms and temperatures down to -20°C may also inundate early migrants (Fig. 1; 13, 22). Furthermore, within a year, snow cover along the migratory route is no indication of conditions on the breeding grounds further north. Nonetheless, arrival dates of migrant songbirds at Toolik Lake are very consistent from year to year with the main influx being May 17-22 (47).

Even after arrival, snow storms with high winds can occur at any time emphasizing the need for birds to seek out micro-habitats that allow them to shelter (37, 47). By the third week in May, day length is virtually 24 hours and insolation is sufficient to melt snow rapidly (22). Depth of snow cover is important in determining which areas of tundra open up first and where many migrant birds will settle (13). Wind action may scour ridges and some valleys reducing snow cover to a few centimeters. These areas typically melt out very quickly (within hours). When late

Fig. 1 Photographs of the tundra near Toolik Field Station, North Slope of Alaska (68°38'N, 149°38'W). All were taken around May 20 but in different years. Snow cover may be 100 percent in some years and less than 5 percent in others. Arriving migrant birds have no way to predict what snow conditions will be. However, even in years with virtually complete snow cover on arrival, some refuges with suitable microhabitat for temporary survival can be found (see 47).

spring snow storms occur, the accumulation of snow on these bare patches is much lower than on existing snow pack suggesting that once microhabitats open up they can persist even during late storms (13). As soon as tundra patches melt out, increases in ground temperature and wind attenuation allow birds to forage in more benign conditions although these areas are patchy and unpredictable both spatially and temporally (47). These patches contain seeds and berries (e.g. *Vaccinium* sp.) from the previous summer. High ground temperatures in melted patches trigger rapid emergence of arthropods. We have observed live Dipteran flies, beetles and wolf spiders, important sources of food, within 2 h of becoming snow free (47). As patches of snow melt, the flow of water washes out seeds and arthropods from the surface of the tundra providing further sources of food. Winter winds form drifts of snow that also act as traps for wind-blown seeds. Water percolating through melting snow concentrates and exposes seeds at the lower edge of the drift.

One of the most severe climatic problems for migrant birds on the tundra is high wind speed. This coupled with low temperature can result in extreme wind chill factors. Wind speed is greatly attenuated close to the ground, especially as patches of tundra open up as snow melts. Even on exposed grassy meadows wind speed is attenuated by about 75% within 5 cm of the ground. In willow patches and tussock tundra, wind speeds may be reduced by over 90% (47). Thus the frozen and snow covered arctic landscape can quickly expose patches of open tundra that provide temperatures higher than the surrounding air, shelter from high winds, and multiple sources of food. However, these resources are also vulnerable to the unpredictable storms that can occur at any time covering refuges, refreezing water (making washed out food items unobtainable), and increasing density of birds in a patch resulting in more competition for food (39, 47).

Clearly, unpredictable features of the arctic environment require behavioral and physiological flexibility as described by Piersma and Drent (26). Migrants arrive under extreme conditions in spring, but must be prepared to retreat during storms if necessary (to the nearest refuge usually, but not always, to the south). Responses to these potentially stressful events must be fine-tuned to the capricious arctic environment and achieve a balance between threats posed to survival as they endure severe conditions, and the risk of missing a single reproductive opportunity if they are overly cautious.

Adrenocortical responses to stress in Arctic breeding birds

Responses to local unpredictable perturbations require rapid, facultative changes in behavior and physiology. The latter have been collectively termed the emergency life history stage (ELHS) diverting the individual into a physiological and behavioral state that will promote survival (46). Although the ELHS is highly adaptive in most conditions, it may actually be detrimental in arriving arctic migrant birds because breeding must begin as soon as conditions allow and not be delayed by severe weather.

Vertebrates activate the hypothalamo-pituitary adrenal (HPA) axis in response to stress (12). Adrenocorticotropin (ACTH) is released from the precursor molecule pro-opiomelanocortin in the anterior pituitary under the control of corticotropin releasing hormone (CRH) from the hypothalamus. ACTH acts primarily on the adrenocortical cells to promote synthesis and secretion of glucocorticoids (corticosterone in birds). Chronic elevation of corticosterone over weeks can have severe debilitating effects such as: inhibition of the reproductive system, suppression of the immune system, promotion of severe protein loss from skeletal muscle, disruption of second cell messengers particularly the arachidonic acid cascade, neuronal cell malfunction, and suppression of growth (e.g. 32). However, in birds and mammals, corticosterone has several behavioral and physiological effects that promote fitness at least in the short term. These include: suppression of reproductive behavior without inhibiting the reproductive system; regulation of the immune system; increased gluconeogenesis; regulation of foraging behavior; promotion of escape (irruptive) behavior during day; promotion of night restfulness; and facilitation of recovery on return to normal life history stage (32, 42). These short-term effects of corticosterone during responses to stress suppress "unnecessary" physiological and behavioral functions, activate alternate behavioral and physiological patterns that promote survival (i.e. temporary emergency behavior) and avoid the long-term, detrimental effects of stress-induced high levels of corticosterone. But, on spring arrival in the Arctic, migrants have to resist the potential for stress despite extreme conditions. Our working hypothesis states " during arrival there are adjustments of hormonal responses to stress that allow individuals some degree of resistance until environmental conditions improve" (46). More specifically, we predicted that the stress-induced increase in corticosterone levels would be inhibited at arrival providing resistance to the potential stress of severe weather. This may be energetically expensive, especially when spring storms occur, but the benefit is enhanced reproductive success.

To measure responsiveness to stress we take advantage of the fact that capture, handling and restraint of wild birds induces a strong activation of the HPA axis. A small blood sample is collected within 2 min of capture and the level of corticosterone in this sample represents the baseline. Further samples are taken at intervals (e.g. 5, 10, 30 and 60 min) to track the increase of corticosterone secretion following the stress of capture, handling and restraint. Most, if not all, stressors have the potential to elicit increases in corticosterone secretion, although the time course and extent of the response may differ (36, 44, 46). The capture stress protocol can be applied equally to all individuals providing measures of the stress response across seasons, habitats, species, ages and genders.

In arctic birds, a reduction in this stress response was found in the most severe environments (36). Field studies of several Arctic breeding birds showed that the adrenocortical response to capture handling and restraint was markedly reduced once onset of nesting begins, and became more suppressed as the parental phase continued (e.g. 44, 25, 15). A corollary of this is the prediction that the sex providing the most parental care should show the greatest suppression of the stress response to the standardized capture stress protocol. O'Reilly and Wingfield (25) used closely related species of sandpipers breeding on the North Slope as Alaska and that have varied degrees of parental care from all male care in some to equal care by males and females and all female care in others. In each case, the sex showing greatest parental effort underwent the greatest suppression of the adrenocortical response to stress.

Thus, reduced responsiveness to acute stress when arctic birds were breeding supports our original hypothesis. Moreover, prolonged severe weather resulted in abandonment of the nest, and the adrenocortical response to capture stress increased as birds formed flocks roaming the snow-covered tundra (3). This observation also suggested that while breeding, responsiveness to stress was suppressed. However, in birds that were arriving on the tundra, such as white-crowned sparrows, *Zonotrichia leucophrys gambelii*, the response to stress in males was actually increased markedly over that seen in winter – prior to migration (28). Similar results were obtained from recently arrived Smith's longspurs, *Calcarius pictus*, in Alaska (21). This elevated response to a standardized stressor was accompanied by decreased sensitivity to glucocorticoid feedback in male *Z.l. gambelii* (2). Although the response to stress was then inhibited as they progressed into the nesting phase (15), on arrival it was greatly enhanced. Females show similar changes but much less pronounced than in males. Note that a non-migratory race of white-crowned sparrow, *Z.l. nuttalli*, breeding at mid-latitudes (central coastal California) did not show this type of modulation of the response to stress (47). Similar enhancement of the HPA response to capture stress at the beginning of the nesting season has now been demonstrated in other avian populations breeding at the northern edge of their range such as the Bush Warbler, *Cettia diphone* in

Hokkaido, Japan (45), the Snow Bunting, *Plectrophenax nivalis*, and Lapland Longspur, *Calcarius lapponicus*, at the northern edge of their range at Thule, Greenland (J C Wingfield *et al.*, unpublished), and in Red Knots, *Calidris canutus* at Alert, Ellesmere Island, Canada (82°N, 27). This remarkable increase in responsiveness to acute stress contradicts our original hypothesis that can be revised as follows. Migrant birds arriving in the Arctic may increase their HPA response to acute stress such as inclement weather until settled on a territory, and then suppress the HPA responses to the same stresses as nesting gets underway.

Mechanisms of hypothalamo-pituitary-adrenal axis modulation

Fig. 2 indicates that the dramatic modulation of the HPA axis responses to standardized stresses in free-living male white-crowned sparrows is completely abolished in captive males. This suggests strongly that environmental cues associated with natural conditions have an important role in the regulation of the HPA axis sensitivity to perturbations. What the mechanisms are underlying modulations of the HPA axis are still unclear, but several promising lines of research are underway in relation to stress modulation to maintain reproductive function. Wingfield and Sapolsky (43) summarized several possible mechanisms underlying resistance of the hypothalamic pituitary gonadal (HPG) axis to stress.

1) Environmental perturbations may not be perceived as stressful, i.e. blockade at the CNS level and before the HPA response to stress is triggered, e.g. lekking species, some arctic songbirds.
2) Stressors do not result in secretion of glucocorticoids, i.e. blockade at the level of the HPA axis itself, e.g. many avian species.
3) One or several components of the hypothalamo-gonadal-axis (HPG) may become resistant to actions of high stress-levels of glucocorticoids. e.g. American tree sparrow, *Spizella arborea.*
4) Inhibition of reproduction by high circulating levels of glucocorticoids may be overcome by compensatory inputs to the HPG axis, e.g. some songbirds.
5) Circulating steroid binding proteins (e.g. CBG), or perhaps even transport proteins in blood/tissue barriers protect the HPG axis from actions of glucocorticoids, e.g. songbirds.

In white-crowned sparrows, the increase in stress response on arrival in the Arctic in spring was accompanied by decreased sensitivity to glucocorticoid feedback (2), and enhanced response of the adrenal cortex to ACTH (29, 30) thus allowing an overall increase in baseline as

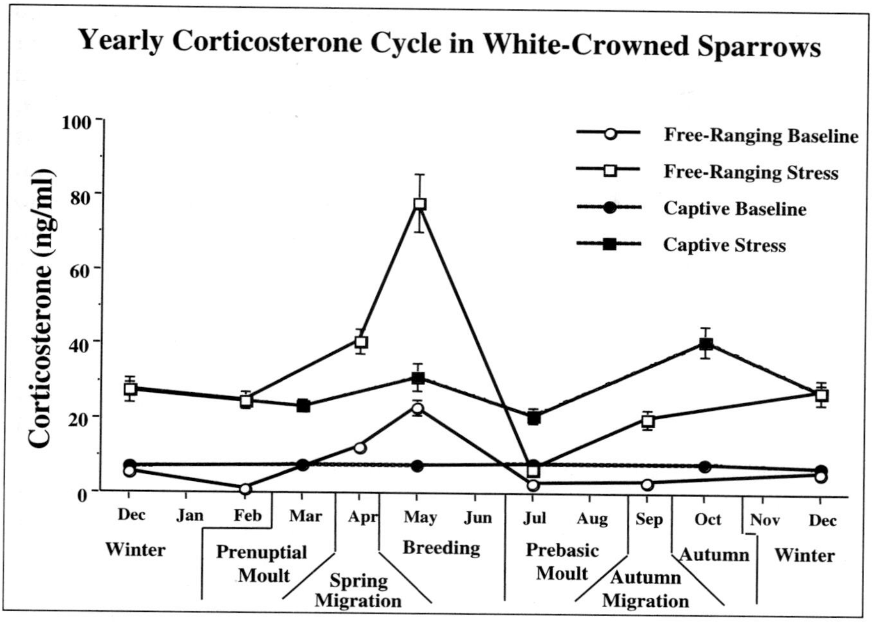

Fig. 2 Seasonal changes in circulating baseline and stress levels of corticosterone (after 30 min of capture, handling and restraint) in free-ranging and captive male white-crowned sparrows, *Zonotrichia leucophrys gambelii*. Note that in free-ranging males, corticosterone titers increase in spring and both baseline and stress levels peak at the time of arrival on the breeding grounds on the North Slope of Alaska. Seasonal lows occur during the annual molt in July with intermediate levels the rest of the year. Note that in captive birds, this dramatic seasonal modulation of the adrenocortical response to stress is almost completely absent. These data suggest that environmental factors associated with the annual cycle in the field are important regulators of responsiveness of the hypothalamo-pituitary-adrenal axis in these birds. From Romero and Wingfield (30), courtesy of Elsevier Press.

well as capture stress induced-plasma levels of corticosterone. It is important to note that circulating glucocorticoids are bound to corticosteroid-binding globulin (CBG) and while bound to this protein, it is thought that they are unable to enter cells and interact with receptors that mediate biological actions (7, 8). If this is the case, then it is possible that CBG levels are modulated to buffer increases in responsiveness to stress (CBG buffer hypothesis). Although the adrenocortical response to acute stress was enhanced at arrival in male white-crowned sparrows, there was also an increase in CBG levels that may buffer stress-induced increases in corticosterone resulting in a reduction of free (unbound) corticosterone able to enter cells and interact with receptors (30, 9). In a comparison of sub-species of white-crowned sparrow, it has been shown in the mid-latitude breeding *Z.l. pugetensis* that does not increase responsiveness to stress in early spring, lower binding capacity of CBG resulted in higher free levels of corticosterone 30 minutes post capture than in *Z.l. gambelii* (9). These data on free corticosterone are actually consistent with our hypothesis that the adrenocortical response to stress is reduced! Furthermore, there is growing evidence that the rise in CBG in early spring is regulated by rising testosterone levels in blood (33).

There is precedent for this kind of regulation in mammals (11) and other avian species. In the House Sparrow, *Passer domesticus*, seasonal changes in adrenocortical responses to capture stress were buffered by changes in CBG binding capacity of plasma so that the free level of corticosterone 30 minutes post-capture did not change (7, 8). So why show seasonal modulation of the adrenocortical response to stress and then buffer it by modulating CBG binding capacity in blood? It is possible that this may maximize flexibility of metabolic and behavioral responses to stress at different times of year. If true, then CBG levels in blood should be regulated as well (8). There is recent evidence that CBG binding capacity may be regulated within hours if a stress persists. In captive male *Z.l. gambelii*, food restriction (to mimic the effects of a storm that covers food resources), increased plasma corticosterone levels by 2 h but had no effect on CBG binding capacity. By 22 h of food restriction, corticosterone levels were similar to controls but CBG levels had declined and free corticosterone titers were higher (19).

The hormone-behavior system associated with arrival of migrant birds in the Arctic is highly flexible, allowing an apparent high adrenocortical response to stress but buffering the early effects with high CBG binding capacity. Although much more work needs to be done to assess how quickly the changes in CBG binding capacity occur and how they change through the rest of the arctic summer, this may be an intriguing system by which migrant birds arriving in the Arctic maintain high activity of the HPA axis intact, but modulate the expression of the CBG gene and thus access of hormone to receptors. Are corticosterone receptors also modulated? In mammals and birds there are two genomic receptors for corticosterone: the high affinity, mineralocorticoid (MR) receptor activated at basal (normal) levels, and low affinity, glucocorticoid (GR) receptor activated at higher "stress" levels (7, 9). *In vitro* binding experiments show that the MR-like receptor in sparrows has high affinity for corticosterone but has different specificity characteristics from mammalian MR (9). The GR-like receptor has a lower binding affinity for corticosterone similar in specificity to that of mammals (7, 9). Because the higher levels of corticosterone in response to stress are of primary interest here (they mediate specific components of the ELHS) we now focus on the GR-like receptor because it is only activated at higher circulating concentrations of corticosterone. There were no significant differences in binding capacity of GR-like receptors in either the liver or brain of *Z.l. gambelii* (arctic-breeding) or *Z.l. pugetensis* (mid-latitude-breeding) males. Using mean levels of free corticosterone and GR-like receptor capacity, it is possible to calculate the receptor occupancy during stress. Although binding capacity of GR-like receptors did not vary between the two taxa of white-crowned sparrow, more GR-like receptors were occupied by corticosterone 30 minutes post capture in both liver and brain of *Z.l. pugetensis* (9). These data strengthen the argument that CBG buffering actually decreases sensitivity to stress in *Z.l. gambelii* arriving in the Arctic.

There is also a putative non-genomic receptor mediating more rapid (within minutes) effects of glucocorticoids (6). Male white-crowned sparrows fed meal worms injected with known doses of corticosterone showed an increase in perch-hopping activity within about 10 minutes compared with controls. This rapid action is not consistent with genomic receptors that require at least 30 min and usually several hours (6, 8, 23). Interestingly, the lower dose of corticosterone (consistent with intermediate plasma levels) was effective but a higher dose (consistent with high stress levels) was not (6). Actions of this putative membrane-type receptor appear to be greatest in spring (when arrival occurs) and less so in winter (10). This suggests that rapid effects of corticosterone on activity may be related to final stages of spring migration in the Arctic when birds arrive on their breeding grounds, locate suitable breeding habitat, and familiarize themselves with immediate food sources should conditions deteriorate and conceal them. Stress levels of corticosterone (as generated by subcutaneous implants of corticosterone) in the absence of food also increase perch hopping activity as well as escape type behavior in male *Z.l. gambelii* (1). This behavior is different from that observed by Breuner *et al.* (6) and is perhaps consistent with leaving the area (irruptive migration) in response to stress (37, 38). Further evidence suggests that irruptive type behavior (and its perch hopping equivalent) is also influenced by CRF injected centrally into the third ventricle of male *Z.l. gambelii* in the laboratory (Maney and Wingfield, 1998), as well as by central injections of CRF into the lateral ventricles of male *Z.l. gambelii* in the field (31). Thus corticosterone action may play a permissive role and secretion of brain peptides, or other substances, provide specificity of behavior for particular environmental scenarios.

Conclusions

Responses of the HPA axis to perturbations of the environment are well known in avian species. However, the ecological bases of modulation of adrenocortical responses, and the mechanisms by which this occurs, are only just beginning to be explored. Work on song birds migrating to arctic breeding grounds shows that males increase baseline levels of corticosterone, and the response to a standardized stressor as they arrive. This increase appears to be buffered by an accompanying elevation of CBG that results in an actual decrease in baseline and stress levels of free cort. Because CBG potentially has this buffering effect means that males (and females to a lesser extent) are able to show a full HPA axis response to a perturbation, without incurring the physiological and behavior effects of high levels of glucocorticoids that would normally shut down migration and inhibit onset of breeding. This may be a mechanism by which birds breeding in severe habitats such as the Arctic may be able to settle and initiate breeding despite a high frequency of inclement weather. If storms persist for longer periods (24 hours or more) then CBG levels begin to decrease thus releasing glucocorticoids that enter cells and then trigger an emergency life history stage and delay of breeding. This might be adaptive because prolonged weather may result in extensive debilitation of birds as a result of declining energy reserves. Thus, more acute perturbations can be resisted and responses to more chronic stress delayed until CBG levels decline. Many of the mechanisms underlying this intriguing CBG buffer hypothesis remain to be determined and probably will await good techniques to manipulate CBG levels in free-living birds. A good start would be to clone the CBG gene in song birds.

Once migrant birds settle and initiate actual nesting, the HPA responsiveness to a standardized stressor declines precipitously in both sexes, suggesting that the birds become even more resistant to environmental perturbations. This again may be adaptive because the breeding season in the Arctic is short (5 weeks) allowing production of one brood and few chances to renest. Therefore, once committed to breeding, birds should resist the potential stressful consequences of storms etc. so as to maximize reproductive success. However, even here chronic storms or other perturbations may result in loss of body condition to a point where continued breeding is no longer possible. At this point birds "reactivate" their responsiveness to stress and abandon the nest. Mechanisms underlying these adjustments are essentially unknown and the role CBG may play has not been explored.

It is entirely possible that these patterns of HPA responsiveness to stress and its modulation may be more widespread, at least in passerine birds. We have accumulating evidence that species from lower latitudes and also high altitude can modulate their responses to standardized stressors. At this point, the hypothesis is that increased responsiveness to stress is buffered by CBG for individuals arriving in, or generally encountering, challenging conditions prior to breeding. Once nesting is actually triggered then responsiveness to stress should be decreased allowing individuals to resist acute perturbations and continue breeding. Furthermore, it is becoming apparent that the adrenocortical response to stress is also modulated during migration – the migration modulation hypothesis of Holberton *et al.* (14). Many avian model systems are available to explore this area of investigation where ecology, behavior and endocrinology meet, and many fascinating mechanisms await discovery.

Acknowledgments

The author is most grateful to the Division of Integrative Biology and Neuroscience, and the Office of Polar Programs, National Science Foundation, for a series of grants that made much of the research summarized in this paper possible. JCW also acknowledges support from the Russell F. Stark University Fellowship, University of Washington, and a John Simon Guggenheim Fellowship.

References

1 Astheimer LB, Buttemer WA, & Wingfield JC (1992) Interactions of corticosterone with feeding, activity and metabolism in passerine birds. *Ornis Scandinavica* **23** 355-365

2 Astheimer LB, Buttemer WA & Wingfield JC (1994) Gender and seasonal differences in the adrenocortical response to ACTH challenge in an arctic passerine, *Zonotrichia leucophrys gambelii. General and Comparative Endocrinology* **94** 33-43

3 Astheimer LB, Buttemer WA & Wingfield JC (1995) Seasonal and acute changes in adrenocortical responsiveness in an arctic-breeding bird. *Hormones and Behavior* **29** 442-457

4 Astheimer LB, Buttemer WA & Wingfield JC (2000) Corticosterone treatment has no effect on reproductive hormones or aggressive behavior in free-living male tree sparrows, *Spizella arborea. Hormones and Behavior* **37** 31-39

5 Baker JG & O'Reilly KM (2000) The effects of age, environmental conditions, and mass on the stress response of Leach's storm-petrels. *Pacific Seabirds* **27** 31

6 Breuner CW, Greenberg AL & Wingfield JC (1998) Non-invasive corticosterone treatment rapidly increases activity in Gambel's White-crowned Sparrows (*Zonotrichia leucophrys gambelii*). *General and Comparative Endocrinology* **111** 386-394

7 Breuner CW & Orchinik M (2001) Seasonal regulation of membrane and intracellular corticosteroid receptors in the house sparrow brain. *Journal of Neuroendocrinology* **13** 412-420

8 Breuner CW & Orchinik M (2002). Plasma binding proteins as mediators of corticosteroid action in vertebrates. *Journal of Neuroendocrinology* **175** 99-112

9 Breuner CW, Orchinik M, Hahn TP, Meddle SL, Moore IT, Owen-Ashley NT, Sperry TS & Wingfield JC (2003) Differential mechanisms for plasticity of the stress response across latitudinal gradients. *American Journal of Physiology: Regulatory Integrative and Comparative Physiology* **285** R594-R600

10 Breuner CW & Wingfield JC (2000) Rapid behavioral response to corticosterone varies with photoperiod and dose. *Hormones and Behavior* **37** 23-30

11 Dallman M & Bhatnagar S (2001) Chronic stress and energy balance: role of the hypothalamo-pituitary-adrenal axis. In *Coping With the Environment: Neural and Endocrine mechanisms. Handbook of Physiology Section 7 Vol. IV* pp. 179-210. Ed BS McEwen. Oxford University Press, Oxford

12 Greenberg N & Wingfield JC (1987) Stress and reproduction: reciprocal relationships. In *Reproductive Endocrinology of Fishes, Amphibians and Reptiles* pp. 389-426. Eds DO Norris & RE Jones, Wiley: New York

13 Hahn TP, Wingfield JC, Deviche P & Mullen R (1995) Spatial and temporal opportunism in Arctic birds. *American Zoologist* **35** 259-273

14 Holberton R, Parrish JD & Wingfield C (1996) Modulation of the adrenocortical stress response in neotropical migrants during autumn migration. *Auk* **113** 558-564

15 Holberton R & Wingfield JC (2003) Modulating the corticosterone stress response: a mechanism for balancing risk and reproductive success in arctic breeding sparrows? *Auk* **120** 1140-1150

16 Järvinen A (1989) Geographical variation in temperature variability and predictability and their implications for the breeding strategy of the pied flycatcher, *Ficedula hypoleuca Oikos* **54** 331-336

17 Kitaysky AS, Piatt JF, Wingfield JC & Romano M (1999a). The adrenocortical stress response of black-legged kittiwake chicks in relation to dietary restrictions. *Journal of Comparative Physiology B* **169** 303-310

18 Kitaysky AS, Wingfield JC & Piatt JF (1999b) Dynamics of food availability, body condition and physiological stress response in breeding black-legged kittiwakes. *Functional Ecology* **13** 577-584

19 Lynn SE, Breuner CW & Wingfield JC (2003) Short-term fasting affects locomotor activity, corticosterone, and corticosterone-binding globulin in a migratory songbird. *Hormones and Behavior* **43**150-157

20 Maney DL & Wingfield JC (1998) Neuroendocrine suppression of female courtship in a wild passerine: corticotropin-releasing factor and endogenous opioids. *Journal of Neuroendocrinology* **10** 593-599

21 Meddle SL, Owen-Ashley NT, Richardson MI & Wingfield JC (2003) Modulation of the hypothalamic-pituitary-adrenal axis of an arctic-breeding polygynandrous songbird, the Smith's longpsur, *Calcarius pictus. Proceedings of the Royal Society, London Series B* **270** 1849-1856

22 Olsson PQ, Sturm M, Racine CH, Romanovsky V & Liston GE (2003) Five stages of the Alaskan Arctic cold season with ecosystem implications. *Arctic, Antarctic Alpine Research* **35** 74-81

23 Orchinik M, Gasser P & Breuner CW (2002) Rapid corticosterone action and behavior: mechanisms and functional consequences. In *Steroids, Brain and Behavior* pp. 567-600. Ed D Pfaff. Academic Press: New York
24 O'Reilly KM, Kurkinen JA, Savage AR (1999) The effect of age and gender on the adrenocortical response to stress in Leach's storm-petrels. *Pacific Seabirds* **26** 42
25 O'Reilly, KM & Wingfield JC (2001) Ecological factors underlying the adrenocortical response to capture stress in Arctic breeding shorebirds. *General and Comparative Endocrinology* **124** 1-11
26 Piersma T & Drent J (2003) Phenotypic flexibility and the evolution of organismal design. *Trends in Ecology and Evolution* **18** 228-233
27 Reneerkens J, Morrison RIG, Ramenofsky M, Piersma T, & Wingfield JC (2002) Baseline and stress-induced levels of corticosterone during different life cycle stages in a shore bird on the high Arctic breeding grounds. *Physiological and Biochemical Zoology* **75** 200-208
28 Romero LM, Ramenofsky M & Wingfield JC (1997) Season and migration alters the corticosterone response to capture and handling in an arctic migrant, the white-crowned sparrow (*Zonotrichia leucophrys gambelii*). *Comparative Biochemistry and Physiology* **116C**171-177
29 Romero LM & Wingfield JC (1998). Seasonal changes in adrenal sensitivity alter corticosterone levels in Gambel's white-crowned sparrows (*Zonotrichia leucophrys gambellii*) *Comparative Biochemistry and Physiology* **119C** 31-36
30 Romero LM & Wingfield JC (1999) Alterations in hypothalamic-pituitary-adrenal function associated with captivity in Gambel's white-crowned sparrows (*Zonotrichia leucophrys gambelii*). *Comparative Biochemistry and Physiology Part B* **122** 13-20
31 Romero LM, Dean SC & Wingfield JC (1998) Neurally active peptide inhibits territorial defense in wild birds. *Hormones and Behaviour* **34** 239-247
32 Sapolsky RM, Romero LM & Munck AU (2000) How do glucocorticosteroids influence stress responses? Integrating permissive, suppressive, stimulatory and preparative actions. *Endocrine Reviews* **21** 55-89
33 Schoech S, Ketterson ED & Nolan V Jr (1999) Exogenous testosterone and the adrenocortical response in the dark-eyed junco, *Junco hyemalis Auk* **11** 64-72
34 Silverin B (1998) Stress responses in birds. *Poultry and Avian Biology Reviews* **9** 153-168
35 Silverin B & Wingfield JC (1998) Adrenocortical responses to stress in breeding Pied Flycatchers, *Ficedula hypoleuca*: relation to latitude, sex and mating status. *Journal of Avian Biology* **29** 228-234
36 Wingfield JC (1994) Modulation of the adrenocortical response to stress in birds. In *Perspectives in Comparative Endocrinology* pp.520-528. Eds Davey KG, Peter RE, Tobe SS. National Research Council: Ottawa, Canada

37 Wingfield JC (2003a) Avian migration: regulation of facultative-type movements. In *Avian Migration* pp. 113-126. Eds P Berthold, E Gwinner, E Sonnenschein. Springer-Verlag: Berlin
38 Wingfield JC (2003b) Control of behavioural strategies for capricious environments. *Animal Behaviour* **66** 807-816
39 Wingfield JC & Hunt K (2002) Arctic spring: hormone-behavior interactions in a severe environment. *Comparative Biochemistry and Physiology B* **132** 275-286
40 Wingfield JC & Kitaysky AS (2002) Endocrine responses to unpredictable environmental events: stress or anti-stress hormones? *Integrative and Comparative Biology* **42** 600-610
41 Wingfield JC & Ramenofsky M (1999) Hormones and the behavioral ecology of stress. In *Stress Physiology in Animals* pp. 1-51. Ed PHM Balm. Sheffield Academic Press, Sheffield, UK
42 Wingfield JC & Romero LM (2000) Adrenocortical responses to stress and their modulation in free-living vertebrates. In *Handbook of Physiology, Section 7: The Endocrine System, Volume 4: Coping With The Environment: Neural and Endocrine Mechanisms* **2000** 211-236. Ed BS McEwen. Oxford University Press: Oxford,
43 Wingfield JC & Sapolsky RM (2003) Reproduction and resistance to stress: when and how. *Journal of Neuroendocrinology* **15** 711-724
44 Wingfield JC, O'Reilly KM & Astheimer LB (1995a) Ecological bases of the modulation of adrenocortical responses to stress in Arctic birds. *American Zoologist* **35** 285-294
45 Wingfield JC, Kubokawa K, Ishida K, Ishii S & Wada M (1995b) The adrenocortical response to stress in male bush warblers, *Cettia diphone*: a comparison of breeding populations in Honshu and Hokkaido. *Japanese Zoological Science* **12** 615-621
46 Wingfield JC, Breuner C, Jacobs J, Lynn S, Maney D, Ramenofsky M & Richardson R (1998) Ecological Bases of Hormone-behavior Interactions: the "Emergency Life History Stage". *American Zoologist* **38** 191-206
47 Wingfield JC, Owen-Ashley NT, Benowitz-Fredericks ZM, Lynn SE, Hahn TP, Wada H, Breuner CM, Meddle SL & Romero LM (2004) Arctic spring: the arrival biology of migrant birds. *Acta Zoologica Sinica* **50** 948-960

Functional Avian Endocrinology
A Dawson & P J Sharp (Eds)

A cardiovascular mechanism for the behavioral effects of AVT during and following acute stress

B C Nephew and L M Romero

Tufts University, Medford, MA 02155, USA

Introduction

Arginine vasotocin (AVT), a nonapeptide neurohormone which is found primarily in birds and amphibians, and its mammalian homologue, arginine vasopressin (AVP), were first characterized and studied for their effects on diuresis and blood pressure. In recent years AVT and AVP have been identified as major components in the vertebrate stress response (2, 11, 21, 32). In addition to being significant secretagogues of adrenocorticotropin hormone (ACTH) in the hypothalamic-pituitary-adrenal (HPA) axis, AVT/AVP have been found to have robust behavioral effects in amphibians 3, 9, 23, 27, mammals (17, 20); (4, 10, 13) and birds, (5, 12, 14). The results of these behavioral studies are difficult to interpret considering the potential physiological and endocrine actions of this nonapeptide hormone. Given the lack of studies that simultaneously examine the behavioral, cardiovascular, and endocrine effects of AVT/AVP, it is unclear whether these behavioral effects are due to central or peripheral actions. While AVP/AVT acts independently of its pressor actions on sexual behaviors in rats (29) and Japanese quail (*Coturnix japonica*) (5), verification of these results in other avian species are needed to distinguish its behavioral effects from its HPA secretagogue actions.

The corticosterone (CORT) stress response is a vital mechanism that allows vertebrates to survive acute or chronic exposure to noxious stimuli, such as starvation or severe weather. While this response is generally adaptive, it can be harmful if expressed at inappropriate times or for extended periods (24). Until recently, much of the research on the stress response had focused on corticosteroid releasing factor (CRF), the other major ACTH secretagogue. A recent review of AVP's role in HPA axis activation concluded that additional studies are needed to fully understand AVP's role in the HPA response to stressors (22). AVT/AVP play a significant role in maintaining the CORT response in both mammals and birds, and over-expression of these hormones may be involved in the deleterious secretion of CORT. Specifically in avian species, AVT is a more potent ACTH secretagogue than CRF (6), and studies in breeding Japanese quail (8) and chickens (19) suggest that AVT may upregulate to allow for greater adaptability during exposure to stressors. Understanding how AVT regulates the CORT response to stress will result in better treatment of stress related illnesses, as well as a better understanding of how animals respond to stress, and how our presence affects this response.

AVT and AVP can act through multiple mechanisms in regulating cardiovascular physiology. These peptides can modulate both blood pressure and heart rate at the central and peripheral levels, and impairments in the AVT/AVP cardiovascular system can result in difficulties maintaining the appropriate ranges of these variables. Over time, this may lead to cardiovascular disease. Studies using an AVP-overexpressing transgenic rat model suggest that AVP plays a significant role in cardiovascular regulation (30). Additional studies investigating the role of AVP in the lateral septal area concluded that AVP may play a role in water intake and the modulation of arterial blood pressure (1). Specifically, V1a receptors in rat cardiac tissue mediate the inotropic effects through an IP3 pathway (7). In birds, AVT can increase (18), decrease (28), or have no effect (16) on heart rate. In chickens, low doses of AVT have vasopressor effects, whereas it acts as a vasodepressor at high doses (28). AVT's diurectic effects are thought to be due to pressor effects on vascular tissues (31). Overall, these studies indicate a clear need for further study of the cardiovascular effects of AVT in avian species.

While the behavioral effects of these two peptides have been studied in a variety of vertebrate species, the exact mechanisms of these behavioral effects have not been elucidated. In their 2001 review of the social behavior effects of AVT/AVP, Goodson and Bass suggest that the mechanisms of AVT/AVP's behavioral effects need to be explored (15). The work presented in the following chapter focuses upon the potential mechanisms of AVT's behavioral effects using the European starling (*Sturnus vulgaris*) as a model system. Starlings were chosen due to their well-characterized behavioral and CORT responses to acute stress (25). Specifically, the experiments discussed in this chapter were investigated how AVT simultaneously modulates the behavior, heart rate (HR), and CORT responses to acute stress. AVT's roles in the regulation of these three responses were examined following exposure to both standard capture and handling stress, as well an ecologically relevant stressor, conspecific crowding, a condition know to cause substantial modulations in behavior, HR, and plasma CORT concentrations. These effects are dependent on the social role of the starling (26). By measuring all three of these potential effects of AVT simultaneously, we were able to evaluate possible interactions between them (see Fig. 1). While no comprehensive measures of fitness were evaluated, heart rate, behavior, and CORT have been shown to have significant impacts on overall fitness in several vertebrate species.

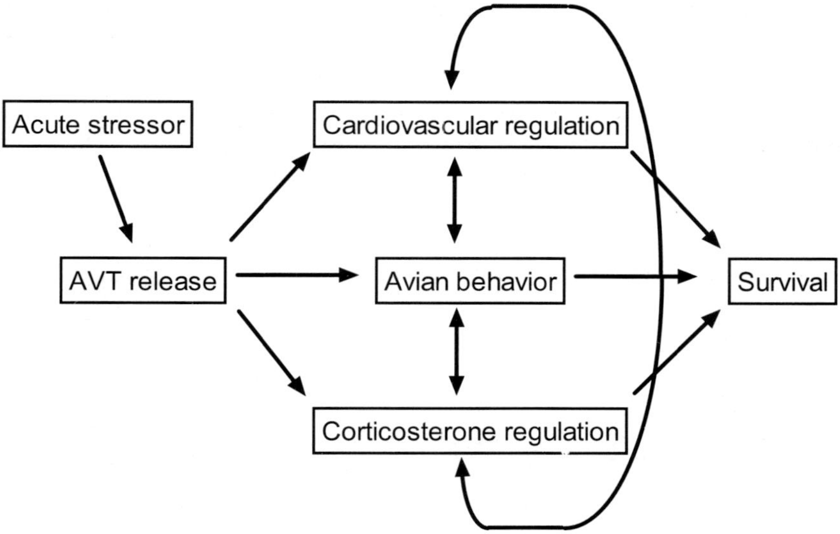

Fig. 1 Diagram of AVT's roles in the behavioral, cardiovascular, and corticosterone responses to stress in birds and the potential interactions between these responses.

Independent analyses of the effects of AVT on the behavioral, cardiovascular, and CORT responses to acute handling stress

Initial experiments investigated the role of subcutaneously administered AVT on the behavioral, cardiovascular, and CORT responses to the acute stress of handling in starlings. The goal was to determine if AVT is behaviorally active in the starling. The first experiment involved injecting several doses of AVT (0.5, 4.0, 8.0μg) and recording several behaviors for 30 min: feeding, drinking, preening, and activity. At the conclusion of the 30 min behavioral observations, blood samples were withdrawn to measure circulating CORT concentrations. The results demonstrated that AVT exerted a general attenuating effect on behavior, and this effect was independent of plasma CORT concentrations (only preening data presented, Fig. 2). This suggests that the behavioral effects of AVT are not dependent on elevated levels of CORT.

Following these experiments, the potential role of cardiovascular physiology in AVT's behavioral effects was investigated. Data Science

International (St. Paul, MN) transmitters implanted subcutaneously in the starlings were used to record HR in the absence of human exposure. These transmitters have two leads which are secured subcutaneously and transmit to a receiver placed on the side of the starling cages. The receiver sends a continuous ECG recording to a computer located outside the testing room. The 60 min protocol consisted of a 15 min basal recording of HR, an initial bleed and treatment injection (lactated ringers solution control, 0.5, 1.0, 2.0μg AVT), a 30 min post-treatment recording of HR, a 30 second acute handling stimulus, and a 15 min post-stimulus HR recording. A final blood sample was taken at the conclusion to assess plasma CORT concentrations during the trial.

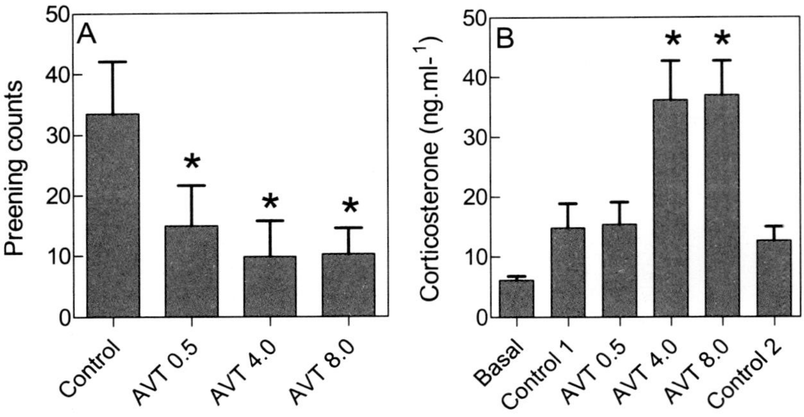

Fig. 2 The effect of AVT on preening and plasma corticosterone concentration. A. Mean ± SEM preening counts after 0.5, 4.0, and 8.0μg subcutaneous AVT or vehicle control. B. Mean ± SEM corticosterone ($ng.ml^{-1}$) following the same treatments. * indicates significant difference compared to control (n=8, $p<0.05$).

The results indicated that the handling and injection of the lactated Ringers solution vehicle control into the starlings elicited significant increases in HR (40-60% above basal, Fig. 3). The AVT treatments resulted in attenuated HR following the subcutaneous administration. AVT had no immediate effect on the increase in HR recorded during the injection procedure, but the 1.0 and 2.0μg doses elicited a lower HR during the post-injection and post-stressor periods (Fig. 3). Since it is possible that these cardiovascular effects result from secondary CORT actions, the pre and post-treatment blood samples were assayed for CORT. Although the 2.0μg dose significantly increased plasma CORT concentrations, the 1.0μg dose had no effect compared to either basal or control injection levels (Fig. 4). These data indicate that

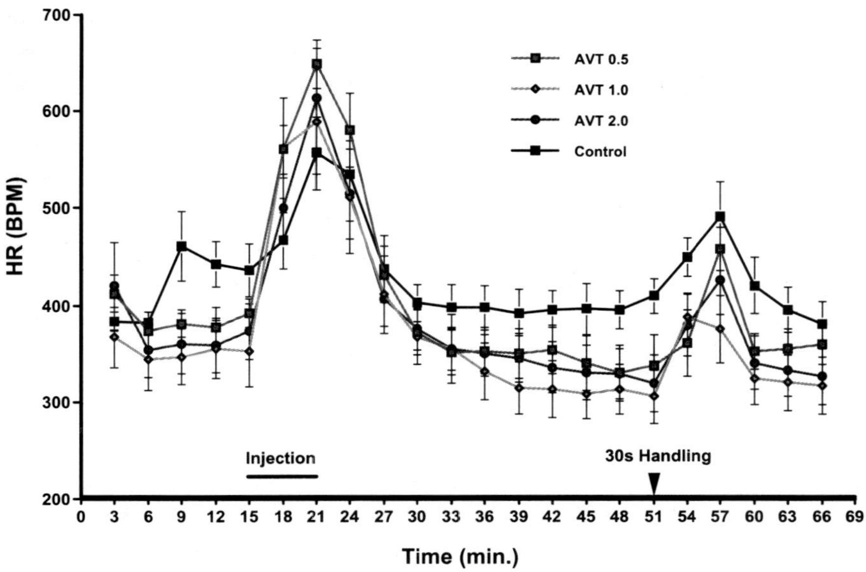

Fig. 3 Effects of AVT and CORT on heart rate. Mean ± SE HR in 3 min intervals before and after subcutaneous injection of vehicle control, 0.5, 1.0, 2.0μg AVT. HR during injections (15-21 min) and 30s handling stimulus is included (n=8). Mean HR after 1.0 and 2.0μg AVT treatments was lower compared to control ($p<0.05$)

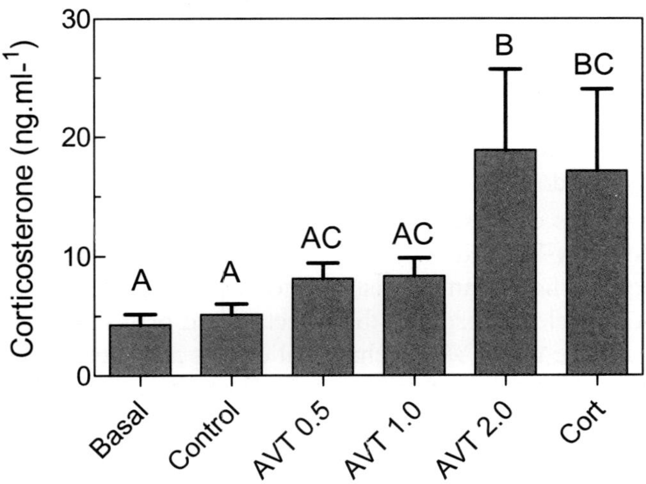

Fig. 4 Effects of AVT and CORT on plasma CORT concentration. Mean ± SE basal and post-treatment CORT concentrations following subcutaneous injection of vehicle control, 0.5, 1.0, 2.0μg AVT or 100μg CORT. Treatments with the same letter are not significantly different ($n=8$, $p<0.05$).

while subcutaneous AVT is capable of increasing plasma CORT, this effect is not necessary for AVT to alter HR. AVT's cardiovascular effects do not require a significant modulation in plasma CORT, similar to its behavioral effects.

Simultaneous analysis of the effects of AVT on the behavioral, cardiovascular, and CORT responses to acute handling

Having demonstrated the effects of AVT on behavior and HR in independent experiments, it was hypothesized that the behavioral attenuations may be due to AVT's effects on cardiovascular physiology. In order to address this possibility we employed an experimental design that allowed for the simultaneous measurement of behavior, HR, and CORT. To monitor the starling's behavior, a digital video camera was placed outside the animal room behind a 2-way glass window. As described previously, the HR transmitters were set up to record ECG data, whereas blood samples taken before and after the trials provided CORT concentrations during the behavioral observations and HR recording.

Since the route of administration can affect the actions of AVT, a set of intracerebroventricular (icv) AVT treatments were included in this protocol. These were given by way of injections through chronic cannulae directed into the lateral ventricle. The AVP V1 receptor antagonist dPTyr(Me)AVP was also included to examine whether AVT exerts its effects through the V1 receptor, a receptor system involved in many of AVT's behavioral effects (see 15). Therefore, the six treatments consisted of an icv infusion of lactated Ringers solution control, 4.0μg subcutaneous AVT, 200 ng of AVT icv, 500 ng of the V1 antagonist icv, the subcutaneous treatment followed immediately by the central antagonist dose, and the central AVT treatment combined with the central antagonist dose.

The results of these treatments on preening behavior are shown in Fig. 5. Both AVTsc and AVTicv depressed preening counts. Central administration of the V1 antagonist did not significantly alter preening, and was not capable of rescuing the effects of sc or icv AVT (Fig. 5). These data indicate that AVT's behavioral effects may not be dependent on a V1-mediated mechanism. Alternative hypotheses are that AVT's behavioral actions are mediated through the V2 receptor pathway, or that the dose of antagonist was too low to block access of available AVP to the V1 receptor.

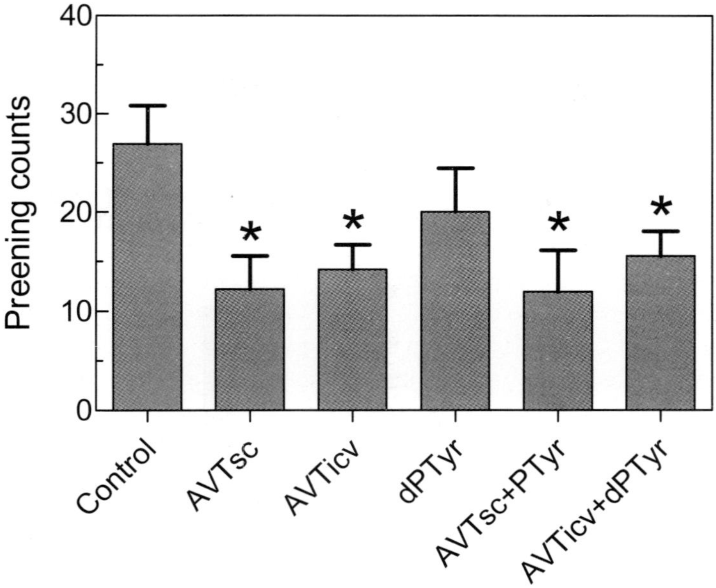

Fig. 5 Mean number ± SEM of preening counts during 45 min observation following icv injection of vehicle control, 4.0μg subcutaneous AVT, the V1 antagonist dPTyr(Me)AVP (denoted dPTyr and administered icv), or a combination of the AVT and antagonist treatments. sc = subcutaneous injection and icv = intracerebroventricular injection. AVTsc = 4.0μg AVT, AVTicv = 200ng. dPTyr = 500ng, and was always injected icv. * indicates significant difference from control (n=8, $p<0.05$).

The HR data obtained during the behavioral recording supported the earlier study on the cardiovascular effects of AVT, with the 4μg AVT causing decreases in post-injection and post-stressor HR. Although the effect was not as robust, intracerebroventricular AVT also attenuated starling HR (Fig. 6). It is likely that central AVT would have had a greater effect if the dose had been administered to a specific cardiovascularly relevant nucleus, such the paraventricular nucleus or preoptic area. This central cardiovascular effect did not significantly alter plasma CORT, providing additional support for the hypothesis that CORT modulation is not involved in the effects of AVT on HR (Fig. 6). In addition, starling HR following 100μg subcutaneous CORT was not significantly different compared to HR following the vehicle control (data not shown). The injection of dPTyr(Me)AVP into the lateral ventricles had no effect on HR, and was not able to completely rescue the effects of either central or peripheral AVT. This suggests that the cardiovascular effects of AVT may not be mediated by a V1 receptor mechanism. Future studies using peripheral and central V1 and V2 antagonists at cardiovascularly relevant nuclei need to be conducted to determine the specific mechanisms involved in AVT's cardiovascular effects.

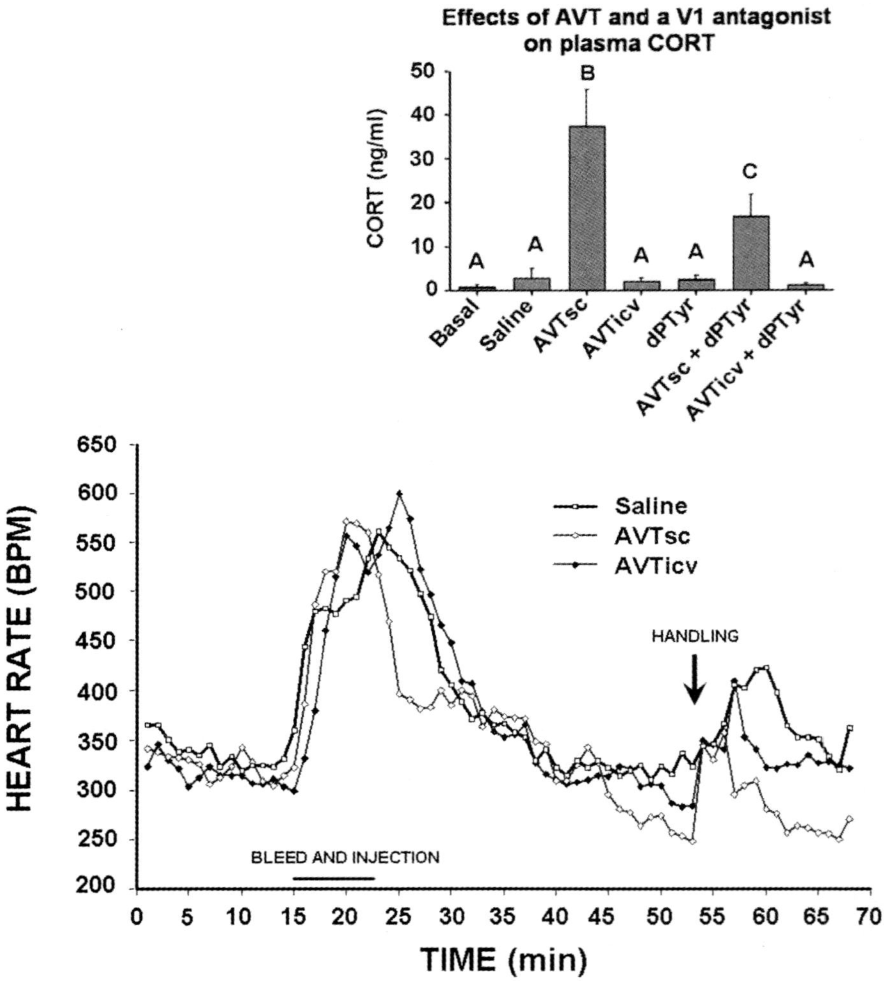

Fig. 6 Mean HR in 1 min intervals before and after icv lactated Ringers saline control, 4.0μg subcutaneous AVT, or 200 ng icv AVT. HR during the injection (15-23 min) and 30 second handling stimulus is included. Both AVTsc and AVTicv decreased HR significantly (n=8, $p<0.05$, error bars omitted for clarity). Insert graph shows plasma CORT concentrations. Only AVTsc significantly increased plasma CORT ($p<0.05$). Reproduced with permission from Hormones and Behavior.

Simultaneous analysis of the effects of AVT on the behavioral, cardiovascular, and CORT responses to acute crowding

To further test the hypothesis that AVT is influencing behavior through a cardiovascular mechanism, the effects of a more ecologically relevant stressor, acute crowding, was evaluated. During the fall and winter months, starlings congregate in flocks that may number in the thousands. Increased conspecific density, or crowding, commonly occurs at feeding sites and roosts during these seasons. Initial studies of acute crowding in wild-caught captive starlings indicated that crowding can act as a significant behavioral, cardiovascular, and CORT stressor (26). Aggressive behavior, HR and CORT all increase during acute crowding in a resident-intruder paradigm. While only the intruder bird significantly increases CORT, HR and aggression increase in both residents and intruders (26).

The crowding protocol involved minimal experimenter interference. Two standard size bird cages were stacked on top of one another, with the door to the bottom cage facing up into the top cage. The resident bird is housed in the top cage, which contains the receiver for the HR transmitter, and the intruder birds are placed in the lower cage prior to the experimental trial. The protocol consists of an initial basal blood sample followed by the injection of lactated Ringers solution (control), or AVT (1.0 or 4.0μg). Pre-crowding HR and behavioral data was recorded for 30 min. The crowding was initiated by remotely opening the door between the two cages with a string. The door was closed once all the intruders (1, 3, or 5 starling intruder treatment groups) had entered the resident's cage (30-90 seconds). Following the intrusion, 30 min of crowding HR and behavioral data were recorded, and a final post-crowding blood sample was obtained.

The results are consistent with previous studies using the handling stressor. Subcutaneous AVT during the crowding treatments decreased feeding, drinking, preening, beak wiping, and activity (preening data shown in Fig. 7A). In addition, the typical increases in the aggressive behaviors upright posture, head feather expansion, and pecking were attenuated by AVT (pecking data shown in Fig. 7B). The effects of AVT on maintenance behaviors and aggression are dose dependent.

The HR data supported our initial investigations on the cardiovascular effects of crowding, as the intrusion of conspecifics elicited a significant increase in resident HR (Fig. 8). This elevation lasted approximately 5 min. The 1.0μg AVT treatment decreased HR throughout the pre- and post-intrusion periods. However, the absolute intrusion induced HR values (400-440 bpm) were the same for the control and AVT treatments. After the intrusion, the 1.0μg AVT caused

HR to return to an attenuated basal level in a similar duration to the control injection (5 min, Fig. 8). In contrast, the 4.0μg dose caused a transient increase in HR for 10 mins following the injection. After 10 min, the HR decreased to a lower basal level, similar to the effects of the 1ug dose. The 4.0μg failed to alter the max HR response to the intrusion. Since only the 4.0μg treatment significantly increased CORT (Fig. 9), we hypothesize that the transient increase in HR is due to sympathetic activation. It is possible that this effect of the dose is due to an interaction between the AVT treatment and the presence of the intruders in the cage below.

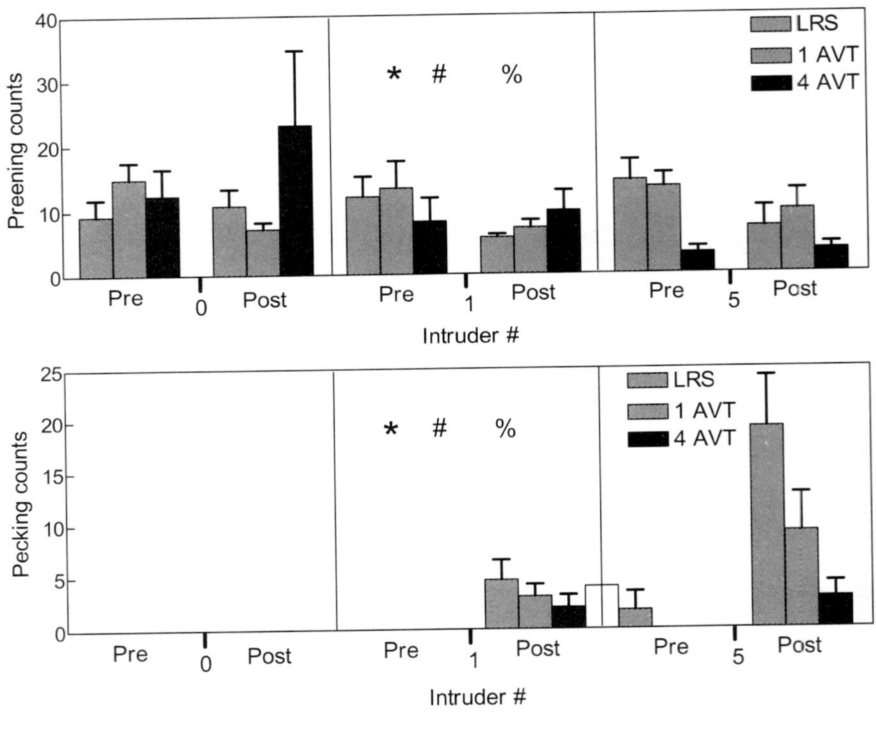

Fig. 7 Mean ± SEM counts of resident starling preening (A) and pecking (B) behavior both pre- and post-intrusion of 0, 1, or 5 intruder starlings following subcutaneous LRS vehicle control, 1.0, or 4.0μg AVT. AVT treatment resulted in significant decreases in preening behavior. The agonistic behavior, pecking, increased due to crowding, and this increase was attenuated by AVT. * indicates significant overall effect of AVT; # indicates significant effect of crowding; **%** indicates significant interaction between AVT and crowding (n=8, $p<0.05$). Reproduced with permission from Hormones and Behavior.

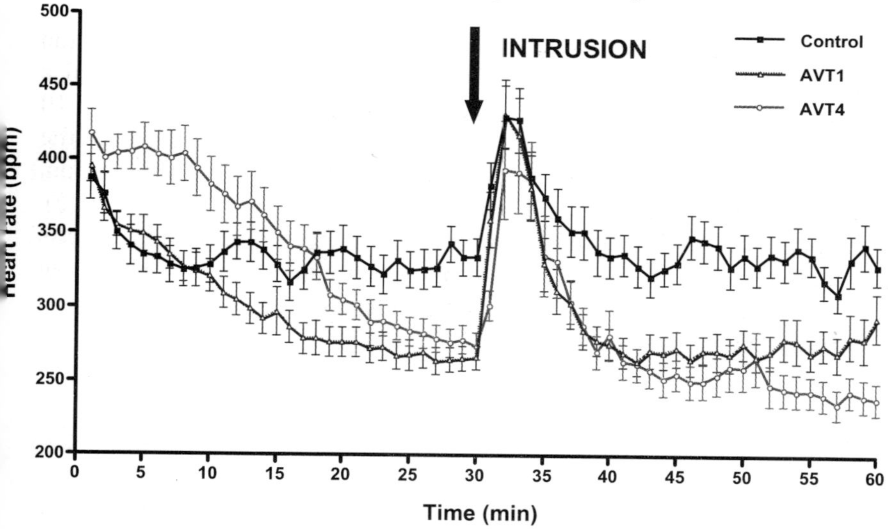

Fig. 8 Effects of AVT on HR during crowding. Mean ± SEM heart rate (HR) per min during the crowding protocol following subcutaneous vehicle control, 1.0, or 4.0μg AVT. All crowding treatments were combined to clearly display the effects of AVT on HR. 1.0 and 4.0μg subcutaneous AVT reduced post-intrusion HR in starlings (n=8, p<0.01). The 4.0μg dose elicited a significant transient increase (0-10min, p<0.01), and the initial response to crowding (31-35min) was statistically similar among all treatments (p<0.05). Arrow indicates intrusion (initiation of crowding). Reproduced with permission from Hormones and Behavior.

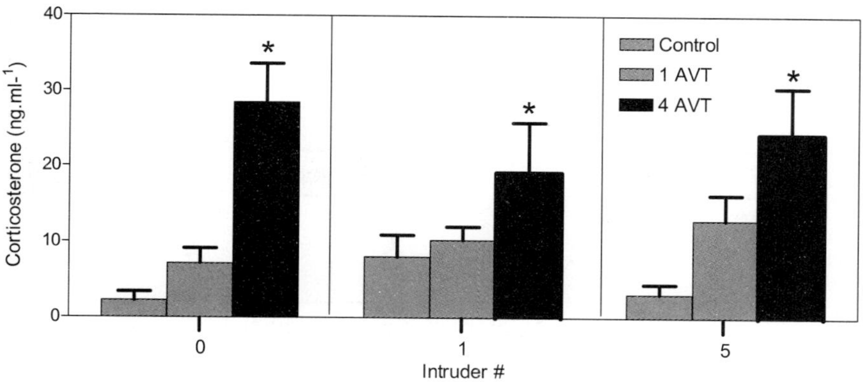

Fig. 9 Mean ± SEM resident plasma CORT concentrations post-intrusion of 0, 1 or 5 intruder starlings following treatment with LRS vehicle control, 1.0 or 4.0μg AVT. Subcutaneous AVT treatment significantly increased resident CORT concentrations post-intrusion compared to the corresponding controls, as well as basal values (n=8, p<0.05). Crowding had no effect on resident CORT. * represents significant effect of 4.0μg AVT.

To summarize AVT's effects on crowding, AVT decreased maintenance behaviors (feeding, drinking, preening), and the attenuated the crowding induced increases in activity and aggression. It simultaneously attenuated the crowding induced increase in HR. Together, these data provide additional support to the hypothesis that the behavioral effects of AVT in starlings are mediated by a CORT-independent cardiovascular mechanism. Although the evidence presented in the previous experiments is not conclusive, it underscores the need to consider AVT's cardiovascular effects when investigating its role in the control of behavior.

Conclusions

In conclusion, our data support the hypothesis that AVT's behavioral effects in avian species are mediated through a CORT-independent cardiovascular mechanism (see Fig. 10). Neither the behavioral nor the cardiovascular effects of AVT are dependent on significant increases in plasma CORT in starlings. Although previous work in birds suggests that AVT may directly affect behavior through centrally mediated mechanism, the present data do not support this pathway. Based on these data, we suggest that investigations of the behavioral effects of AVT or

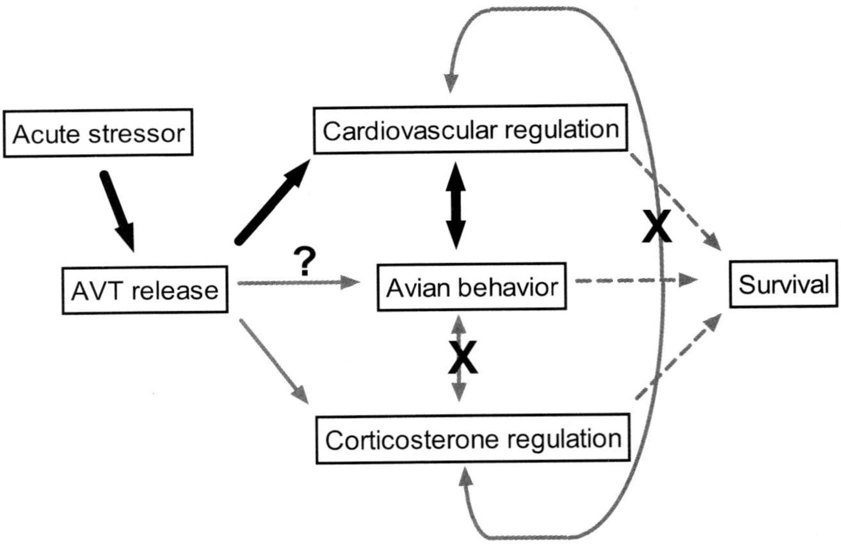

Fig. 10 Diagram of AVT's roles in the behavioral, cardiovascular, and corticosterone responses to stress in starlings. Bold arrows represent the pathway supported by the starling data presented in this chapter. **X** indicates a pathway that was not supported by the present data.

AVP should examine blood pressure and/or heart rate to address the potential cardiovascular effects of these hormones. Future studies investigating AVT/AVP's effects at specific behaviorally active nuclei using additional V1 antagonists, as well as viral-mediated gene transfer, will help to further elucidate the specific mechanisms of both central and peripheral AVT/AVP's effects on behavior.

Acknowledgements

B.C.N and L.M.R were supported by NSF grant IBN-9975502 awarded to L.M.R

References

1 Abrao SW *et al.* (2004) Influence of arginine vasopressin receptors and angiotensin receptor subtypes on the water intake and arterial blood pressure induced by vasopressin injected into the lateral septal area of the rat. *Autonomic Neuroscience-Basic and Clinical* **111** 66-70

2 Aguilera G, et al. (1996) Regulation of pituitary ACTH secretion during stress: role of corticotrophin releasing hormone and vasopressin. In *Stress: Molecular, Genetic, and Neurobiological Advances* pp 385-399 Eds R McCarty, G Aguilera, E Sabban & R Kvetnansky. Gordon and Breach: New York

3 Boyd SK (1991) Effect of vasotocin on locomotor activity in bullfrogs varies with developmental stage and sex. *Hormones and Behavio*r **25** 57-69

4 Buwalda B *et al.* (1993) Neuroendocrine and behavioral effects of vasopressin in resting and mild stress conditions. *Physiology and Behavior* **54** 947-953

5 Castagna C et al. (1998) Systemic and intracerebroventicular injections of vasotocin inhibit appetitive and consummataory components of male sexual behavior in Japanese quail. *Behavioral Neuroscience* **112** 233-250

6 Castro MG, Estivariz FE & Iturriza FC (1986) The regulation of the corticomelanotropic cell activity in Aves - II. Effect of various peptides on the release of ACTH from dispersed, perfused duck pituitary cells. *Comparative Biochemistry and Physiology* **83A** 71-75

7 Chandrashekhar Y et al. (2003) The role of arginine vasopressin and its receptors in the normal and failing rat heart. *Journal of Molecular and Cellular Cardiolog*y **35** 495-504

8 Chaturverdi CM et al. (2000) A sexual dimorphism in hypothalamic arginine vasotocin (AVT) gene expression and AVT plasma levels in the Japanese quail (*Coturnix coturnix japonica)* in response to water deprivation *General and Comparative Endocrinology* **117** 129-137

9 Chu J, Marler CA and Wilczynski W (1998) The effects of arginine vasotocin on the calling behavior of male cricket frogs in changing social contexts. *Hormones and Behavior* **34** 248-261

10 Delville Y. Mansour KM and Ferris CF (1996) Serotonin blocks vasopressin-facilitated offensive aggression: interactions within the

ventrolateral hypothalamus of golden hamsters. *Physiology and Behavior* **59** 813-816

11 de Goeij DCE, Dijkstra H and Tilders FJH (1992) Chronic psycosocial stress enhances vasopressin, but not corticotropin-releasing factor, in the external zone of the median eminence of male rats: Relationship to subordinate status. *Endocrinology* **131** 847-853

12 de Kloet ER, Elands J and Voorhuis DAM (1993) Implication of central neurohypophyseal hormone receptor-mediated action in the timing of reproductive events: evidence from novel observations on the effect of a vasotocin analogue on singing behavior of the canary. *Regulatory Peptides* **45** 85-89

13 Ferris CF et al. (1986) A vasopressin antagonist can reverse dominant/subordinate behavior in hamsters. *Physiology and Behavior* **38** 135-138

14 Goodson JL (1998) Territorial aggression and dawn song are modulated by septal vasotocin and vasoactive intestinal polypeptide in male field sparrows. *Hormones and Behavior* **34** 67-77

15 Goodson JL and Bass AH (2001) Social behavior functions and related anatomical characteristics of vasotocin/vasopressin system in vertebrates. *Brain Research Reviews* **35** 246-265

16 Hassinen E, Pyornila A and Raimo H (1999) Cardiovascular and themoregulatory responses to vasotocin and angiotensin II in the pigeon. *Comparative Biochemistry and Physiology* **123A** 279-285

17 Insel TR and Young LJ (2000) Neuropeptides and the evolution of social behavior. *Current Opinion in Neurobiology*. **10** 784-789

18 John TM and George JC (1992) Effects of arginine vasotocin on cardiorespiratory and thermoregulatory responses in the pigeon. *Comparative Biochemistry and Physiology – C* **10** 353-359.

19 Jurkevich A, Barth SW and Grossmann R (1997) Sexual dimorphism of arg-vasotocin gene expressing neurons in the telencephalon and dorsal diencephalon of the domestic fowl. An immunocytochemical and in situ hybridization study. *Cell and Tissue Research* **287** 69-77

20 Landgraf R et al. (2003) Viral vector-mediated gene transfer of the vole V1a vasopressin receptor in the rat septum: improved discrimination and active social behavior. *European Journal of Neuroscience* **18** 403-411

21 Ma X and Lightman SL (1998) The arginine vasopressin and corticotrophin-releasing hormone gene transcription responses to varied frequencies of repeated stress in rats. *Journal of Physiology* **510.2** 605-614

22 Makara GB, Mergl Z and Zelena D (2004) The role of vasopressin in hypothalmo-pituitary-adrenal axis activation during stress: An assessment of the evidence. *Annals of the New York Academy of Sciences* **1018** 151-161

23 Marler CA, Boyd SK and Wilczynski W (1999) Forebrain arginine vasotocin correlates of alternative mating strategies in cricket frogs. *Hormones and Behavior* **36** 53-61

24 McEwen BS and Wingfield JC (2003) The concept of allostasis in biology and biomedicine. *Hormones and Behaviour* **43** 2-15

25 Nephew BC, Kahn SA and Romero (2003) LM Heart rate and behavior are regulated independently of corticosterone following diverse acute stressors. *General and Comparative Endocrinology* **133** 173-180

26 Nephew BC and Romero LM (2003) Behavioral, physiological, and endocrine responses of starlings to acute increases in density. *Hormones and Behaviour* **44** 222-232

27 Propper CR and Dixon TB (1997) Differential effects of arginine vasotocin and gonadotropin-releasing hormone on sexual behaviors in an anuran amphibian. *Hormones and Behavior* **32** 99-104

28 Robinzon B, Koike TI and Marks PA (1993) At low dose, arginine vasotocin has vasopressor rather than vasodepressor effect in chickens. *General and Comparative Endocrinology* **91** 105-112

29 Sodersten P et al. (1983) Vasopressin alters female sexual behaviour by acting n the brain independently of alterations in blood pressure. *Nature* **301** 608-610

30 Tachikawa K et al. (2003) Altered cardiovascular regulation in arginine vasopressin-overexpressing transgenic rat. *American Journal of Physiology-Endocrinology and Metabolism* **285** E1161-1166

31 Wilson X and West N H (1986) Cardiovascular responses to neurohormones in conscious chickens and ducks. *General and Comparative Endocrinology* **62** 268-280

32 Wotjak CT et al. (1996) Release of vasopressin within the rat paraventricular nucleus in response to emotional stress: A novel mechanism of regulating adrenocorticotropic hormone secretion? *Neuroscience.* **16** 7725-7732

Functional Avian Endocrinology
A Dawson & P J Sharp (Eds)

Corticosterone and memory in birds

V V Pravosudov

Department of Psychology, University of California, Davis, CA 95616, USA

Introduction

The relationship between stress, glucocorticoid hormones and memory has received much attention in mammalian studies (10, 11, 28). It is generally assumed that while short-term elevation of stress hormones could be beneficial for memory, prolonged elevation has deleterious effects on cognitive abilities (10, 11, 28). Avian studies have mostly concentrated on the relationship between adverse environmental conditions and corticosterone, and the effects of stress-induced corticosterone on behavior and physiology (36, 37). The relationship between corticosterone and memory has received little attention. Memory is an important mechanism by which all animals acquire and retain information about the surrounding environment and it is critically important for survival. Therefore, to fully understand the impact of stress hormones on fitness consequences in birds, we need to investigate the effects of naturally occurring variation in corticosterone on memory.

Available data on the relationship between corticosterone and memory in birds are scarce but suggest that corticosterone plays an important role in memory formation (8, 26, 27). In domestic chicks, it has been demonstrated that administration of moderate, but not high, corticosterone doses enhanced memory for a weak aversant in a passive avoidance task (26), whereas blocking glucocorticoid synthesis impaired memory performance (8). However, in a task involving a strong aversant, both moderate and high doses of administered corticosterone impaired memory retention (26). In greylag goose goslings (*Anser anser*), better learning performance was correlated with higher levels of excreted corticosterone (12). Rapid short-term corticosterone elevations enhanced spatial memory performance and increased cache retrieval efficiency in food caching mountain chickadees (*Poecile gambeli*, 25). Unfortunately, it is hard to evaluate the results of this study because it appears that corticosterone was experimentally elevated well above naturally occurring maximum stress-induced corticosterone levels. The authors used a noninvasive corticosterone treatment, used in a study of white-crowned sparrows (*Zonotrichia leucophrys*) (1), which rapidly increased corticosterone to about 100 ng.ml^{-1} (1). Several studies of mountain chickadees showed that they did not reach such high corticosterone levels even during standardized stress-response tests (19, 20, 22). Also, white-crowned sparrows weigh about twice as much as mountain chickadees so the same corticosterone dose is likely to have produced a greater effect in chickadees. Thus, it remains to be examined whether high corticosterone levels within the range experienced by these birds during adrenocortical response to acute stress have an effect on spatial memory. Here I discuss the relationship between prolonged but moderate elevations in plasma corticosterone and spatial memory within the context of natural history in food-caching birds.

Long-term moderate corticosterone elevation

The benefits of short-term corticosterone elevation in response to adverse environmental changes are well known, but prolonged corticosterone elevations are generally considered harmful (10, 11, 28, 36, 37). However, it appears that some birds may maintain moderately elevated corticosterone levels for fairly long periods of time. For example, food-caching willow tits (*Parus montanus*) appear to maintain moderately elevated baseline corticosterone for several months (Fig. 1).

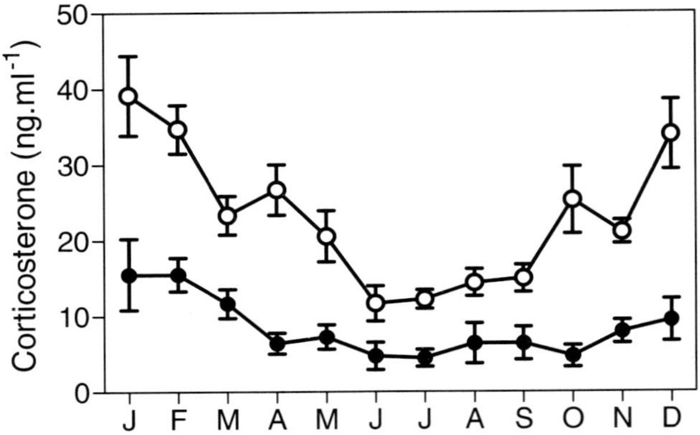

Fig. 1 Seasonal changes in baseline (filled circles) and stress-induced (open circles) plasma corticosterone levels in free-ranging willow tits in Sweden (data from 33).

It seems that the period during which baseline corticosterone was elevated coincided with winter months when birds experience their most challenging environmental conditions due to limited and unpredictable food supply, low ambient temperature, persistent snow cover and short days available for foraging. Elevated corticosterone levels in willow tits were registered for *ca.* 5 months – from November until March. These elevations in baseline corticosterone were moderate –

only 2-3 times lowest summer levels. Stress-induced levels, on the other hand are usually much higher and they can exceed baseline more than 10 times (19, 20, 22, 33). It appears that stress-induced corticosterone levels also varied seasonally in willow tits (Fig. 1). During the summer, however, stress-induced corticosterone levels in willow tits were lowest, lower even than baseline levels experienced during winter (Fig. 1).

Long-term limited unpredictable food and corticosterone

Several environmental factors could potentially cause moderate corticosterone elevation in wintering birds and only experiments can establish the role of each one of them. For example, when food was unlimited, baseline corticosterone levels of mountain chickadees maintained on short, winter-like days in a laboratory were indistinguishable from those of chickadees maintained on long, summer-like days (20). Poor foraging conditions, on the other hand, seem to be sufficient to cause moderate, but significant elevation in baseline corticosterone levels in food-caching mountain chickadees (Fig. 2A) and these changes resemble those observed in willow tits (33).

It is well known that short-term food deprivation elicits corticosterone elevation in birds (7, 9). But in natural conditions, resident birds may experience poor foraging conditions for several months during the winter and thus corticosterone levels may remain elevated over extended periods of time (33). Supporting this view, baseline corticosterone levels of mountain chickadees maintained on limited and unpredictable food supply for over 90 days were significantly higher than those of chickadees maintained on *ad libitum* food (Fig. 2A). Such an increase in corticosterone levels was only moderate and elevated levels were much smaller than stress-induced corticosterone levels (Fig. 2B). The effect of food supply was limited to baseline corticosterone only and there were no differences in a magnitude of the adrenocortical response to acute stress between birds maintained on limited and unpredictable food supply and birds maintained on *ad libitum* food (Fig. 2B). Interestingly, in free-ranging willow tits stress-induced levels also vary seasonally similarly to the baseline values and highest stress-induced levels coincided with the highest baseline corticosterone levels (Fig. 1). Laboratory experiments with mountain chickadees showed that food supply did not affect stress-induced levels and thus, some other factors must be responsible for the seasonal variation in stress-induced corticosterone levels observed in willow tits (33).

Thus, experimental data on mountain chickadees suggest that baseline corticosterone levels may remain elevated for prolonged periods of time in resident birds experiencing poor foraging conditions. Further support for this finding comes from a study showing that baseline plasma corticosterone levels in red knots (*Calidris canutus*) maintained on a variable feeding schedule were significantly higher compared to plasma levels in knots maintained on a constant feeding schedule (23). While the exact timing of the experimental feeding schedules was unfortunately not presented in the paper, it appears that the treatments lasted for about a month. Therefore, this study also suggested that birds might maintain moderately elevated baseline corticosterone levels for extended periods of time during unpredictable foraging conditions. It is widely believed that prolonged or chronic elevation in glucocorticoid levels has harmful effects on animals' health (10, 11, 28). The experiments with mountain chickadees and red knots, on the other hand, suggest that some animals may routinely experience extended periods of moderate corticosterone elevations in adverse natural conditions and such a response may be adaptive (19, 23).

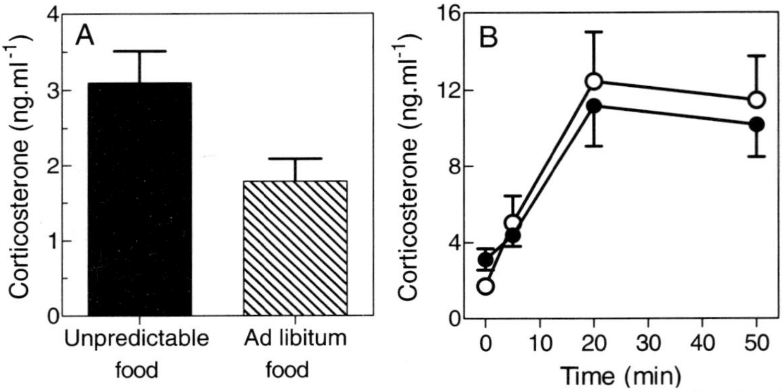

Fig. 2 Plasma corticosterone levels in mountain chickadees maintained on limited and unpredictable food supply (solid bar and filled circles) and on *ad libitum* food supply (hatched bar and open circles). A. Baseline corticosterone levels; B. Adrenocortical response to acute stress. Columns and points represent mean ± SE (data from 19).

The benefits of short-term increases in corticosterone in response to sudden changes in environmental conditions are well described (36, 37) and include re-direction of behavior and physiology towards immediate survival. In particular, elevated corticosterone might trigger opportunistic migration and facilitate the mobilization of stored energy reserves (36, 37). But resident birds, such as food-caching willow tits, usually do not leave their home ranges during the winter and even experimentally increased levels of corticosterone failed to trigger migratory movements in these birds during the winter (32). Instead, it appears that corticosterone elevation might facilitate behaviors that increase survival without the birds moving to more favorable locations, such as exploratory behavior, feeding, food caching, and enhanced spatial memory needed for successful cache retrieval. Unlike short-term strong responses to acute stress, adrenocortical responses to typical harsh winter conditions might involve only moderate elevations in baseline corticosterone, which may last for a relatively long time (weeks or even months). While maintaining even moderately elevated corticosterone levels might have costs (e.g. a suppressed immune response), the benefits of such elevations likely outweigh these costs.

Long-term limited and unpredictable food and memory

During the winter, resident birds have to accumulate enough energy reserves to survive long cold nights and thus foraging proficiency should be crucial for their survival. Food-caching birds, such as willow tits or mountain chickadees, scatter-hoard numerous food caches around their territory during the time when food is available and then rely on these food caches when natural food is scarce (35). Whereas these birds create most of their caches during late summer and autumn when natural food is abundant, they continue caching even during the winter (13). It is not clear whether winter caching consists of finding and storing new food items or whether these birds simply re-cache food stores created previously (13, 16). In either case, these food caches are mostly important during the winter when resident birds face the greatest energetic challenges (16). To benefit from caches, food-caching birds have to successfully recover them and spatial memory has been implicated as one of the mechanisms used for cache retrieval (31).

Fitness consequences of better memory may not be equal throughout the year; forgetting cache locations during periods with plenty of naturally occurring food might not yield the same costs as forgetting locations of food caches during energetically demanding periods such as winter, when natural food is scarce. Indeed, mountain chickadees maintained on limited and unpredictable feeding schedule retrieved their caches more accurately and performed better on spatial memory tests than chickadees maintained on *ad libitum* food (Fig. 3). Thus, chickadees experiencing long-term limited and unpredictable food supply had moderately but significantly elevated corticosterone levels and they also demonstrated better spatial memory performance compared to the birds maintained on *ad libitum* food. These results suggest that elevation in baseline corticosterone could have been responsible for improved spatial memory performance.

Long-term corticosterone elevation and memory

The effect of glucocorticoid stress hormones on memory has received much attention in mammalian studies (10, 11, 28). Generally, they suggest that short-term elevation in these hormones could be highly beneficial for memory while any prolonged, chronic elevation usually results in memory impairments (10, 11, 28).

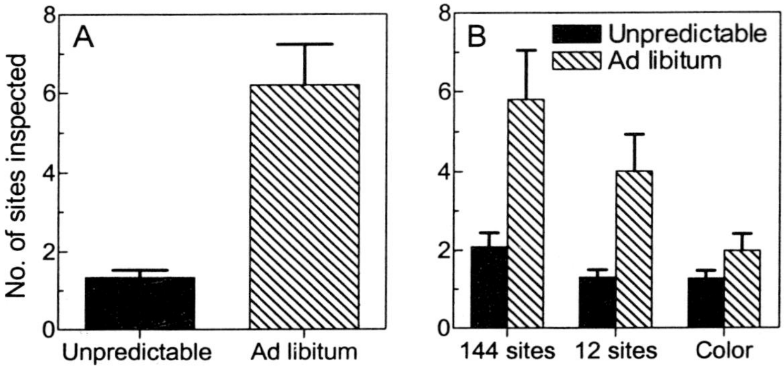

Fig. 3 Performance of mountain chickadees maintained on limited and unpredictable (filled bars) and on *ad libitum* (hatched bars) food supply in a cache-recovery (A) and in a one-trial associative learning task (B). Memory performance was measured as the number of sites inspected in order to find previously hidden caches (cache-recovery task) or the site previously containing food (one-trial associative learning task). During a one-trial associative learning task, birds had to search for the site previously containing food either among 144 or 12 available sites. Fewer sites inspected indicate better memory performance. To investigate whether the differences found between the experimental groups were mainly due to differences in spatial memory and not in motivation to search, a color version of a one-trial associative learning task was performed in which a single site (of 12 available sites) containing food was marked with a unique color. The color task suggested that motivational differences were unlikely because there were no significant differences between the two groups of birds in the number of searches to find the color-marked site. This result suggested that the differences were specific to spatial memory. Columns represent mean ± SE (data from 15).

In food-caching mountain chickadees, it was found that long-term limited and unpredictable food supplies resulted in enhanced spatial memory and in moderate but significant corticosterone elevation (Fig. 2). These data, however, provide only a correlational link between elevated corticosterone and enhanced spatial memory.

To establish a causal link between moderately elevated corticosterone and spatial memory, mountain chickadees were implanted with 90-day continuous time release biodegradable implants containing 0.075 mg of corticosterone and tested on spatial memory tests starting on7th and ending on 48th day after implantation (14). Such implants result in moderate corticosterone elevations resembling those found in chickadees maintained on limited and unpredictable food supply (14, 19).

Corticosterone-implanted birds cached and consumed significantly more food and demonstrated more efficient cache recovery and enhanced spatial memory performance compared with placebo-implanted individuals (Fig. 4). Thus, contrary to prevailing assumptions, prolonged but moderate elevations of corticosterone actually enhanced spatial memory in food-caching mountain chickadees.

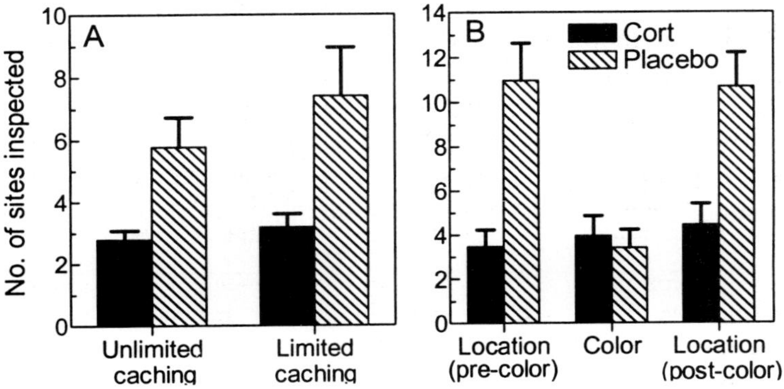

Fig. 4 Performance of corticosterone- (solid bards) and placebo-implanted (hatched bars) mountain chickadees in a cache-retrieval task (A) and in a one-trial associative learning task (B). Columns represent mean ± SE. During unlimited caching trials, birds were not restricted in the amount of caching and corticosterone-implanted birds cached significantly more seeds than placebo-implanted individuals. During limited caching trials caching was restricted to three caches in all birds. Memory performance was measured as the number of sites inspected in order to find previously hidden caches (cache-recovery task) or the site previously containing food (one-trial associative learning task). Fewer sites inspected indicate better memory performance. To investigate whether the differences found between the experimental groups were mainly due to differences in spatial memory and not in motivation to search, a color version of a one-trial associative learning task was performed in which a site containing food was marked with a unique color. The color task suggested that motivational differences were unlikely because there were no significant differences between the two groups of birds in the number of searches to find the color-marked site. The differences between corticosterone- and placebo-implanted birds re-appeared when the color task was followed by a spatial task reinforcing the conclusion that the differences between the groups were specific to spatial memory (data from 14).

Long-term moderate corticosterone elevations and the hippocampus

The hippocampus has been implicated in spatial memory processing in both birds and mammals (29), and in food-caching birds, it has been shown that lesions of the hippocampal formation lead to impaired spatial memory performance (6, 30). It has been frequently suggested that any prolonged, chronic elevation in glucocorticoid stress hormones is damaging for both the hippocampus and its functions such as spatial memory (10, 11, 28).

Experiments with mountain chickadees demonstrated that moderately elevated corticosterone actually correlated with enhanced spatial memory performance in contrast with the expectation that any prolonged elevation of corticosterone should be detrimental for spatial memory. Furthermore, the volume of the hippocampal formation and the total number of the hippocampal neurons did not differ significantly between chickadees maintained on limited and unpredictable food and those on *ad libitum* food (Fig. 5). Thus, plasma corticosterone levels elevated due to long-term limited and unpredictable food supply had no significant effects on hippocampal anatomy in these birds.

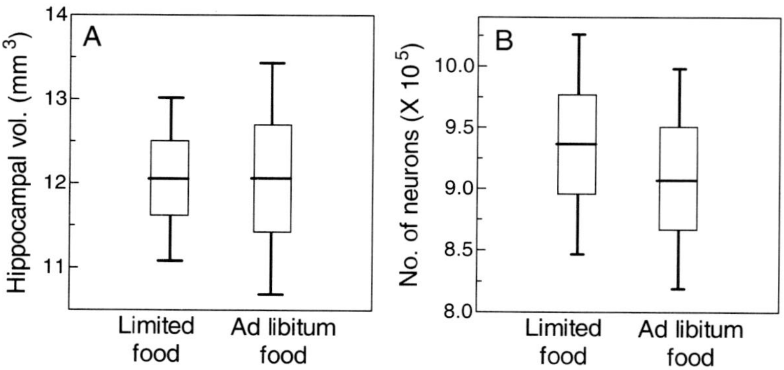

Fig. 5 Hippocampal volume (A) and the total number of hippocampal neurons (B) in mountain chickadees maintained on limited and unpredictable and on *ad libitum* food supply. Boxes represent SE and whiskers – 95% CI (data from 21).

Prolonged corticosterone elevation caused by implants enhanced spatial memory performance in mountain chickadees (14) but it did not have a significant effect on the hippocampal anatomy; hippocampal volume or the total number of hippocampal neurons (Fig. 6).

Mammalian studies have suggested that chronic elevation in glucocorticoids might have a negative effect on the hippocampus by reducing hippocampal neurogenesis rates (5). However, chronic but moderate corticosterone elevation in mountain chickadees had no significant effect on hippocampal cell proliferation rates (Fig 7).

It is still possible that moderate chronic elevations of corticosterone have some negative effects on the avian hippocampus, but these effects did not seem to concern spatial memory since birds with

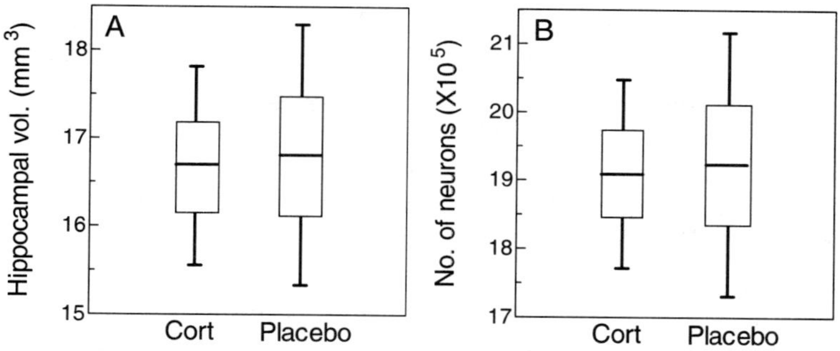

Fig. 6 Hippocampal volume (A) and the total number of hippocampal neurons (B) in corticosterone- and placebo-implanted mountain chickadees. Boxes represent SE and whiskers – 95 CI (data from 18).

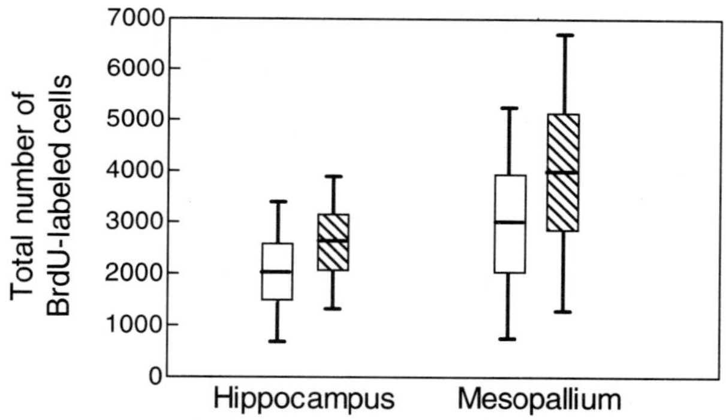

Fig. 7 Cell proliferation rates in the ventricular zone adjacent to either the hippocampus or to the mesopallium in corticosterone- (open bars) and placebo-implanted (hatched bars) mountain chickadees. Boxes represent SE and whiskers – 95% CI (data from 18).

moderately elevated corticosterone actually demonstrated enhanced spatial memory (14). The experiment with limited and unpredictable feeding schedule and the experiment with corticosterone implants both resulted in spatial memory enhancements in mountain chickadees (14, 15). However, these changes in spatial memory were not associated with any detectable changes in the hippocampal anatomy or cell proliferation rates. This suggests that temporary (seasonal) changes in spatial memory are not necessarily mediated by changes in hippocampal volume or neuron numbers as was suggested previously (34), but rather they could be mediated by changes in plasma corticosterone levels.

Mechanism of corticosterone's effects on memory

There are several possible ways stress hormones can enhance memory formation. Elevated corticosterone could increase blood glucose levels and glucose has been implicated in enhancing memory function (4). Elevated corticosterone may also facilitate protein synthesis in the brain, which is also necessary for memory function (2, 24). Finally, there appears to be an inverted-U relationship between corticosterone levels and hippocampal neuron firing rates (3). Thus, moderate elevation in corticosterone levels might increase hippocampal neuron firing rate.

Conclusions

Chronic moderate elevation of corticosterone appears to be an adaptive response to unpredictable environments. Food-caching birds wintering in high latitudes appear to respond to energetically demanding and variable conditions by maintaining moderately elevated corticosterone, which would facilitate more intense foraging, fat accumulation, food caching and more efficient cache retrieval through enhanced spatial memory. Memory is an important mechanism used by food-caching birds and, by enhancing spatial memory, elevated corticosterone may increase survival probability during the winter (17). Such adaptive effects of prolonged but moderate elevations in corticosterone may be prevalent but as yet are not well explored. The fact that corticosterone levels are reduced when conditions are favorable suggests that maintaining even moderately elevated corticosterone has its costs even though these costs do not seem to include previously assumed negative impacts on memory.

Acknowledgments

I was supported by NIH/NIMH career grant. I thank Nicky Clayton, Sasha Kitaysky, Pierre Lavenex, John Wingfield and Alicja Omanska who collaborated with me on various projects, and Marty Morton for critical comments on the manuscript.

References

1 Breuner CW, Greenberg AL & Wingfield JC (1998) Noninvasive corticosterone treatment rapidly increases activity in Gambel's white-crowned sparrows (*Zonotricia leucophrys gaambelii*). *General and Comparative Endocrinology* **111** 386-394

2 Cahill L & McGaugh JL (1996) Modulation of memory storage. *Current Opinion in Neurobiology* **6** 237-242

3 Diamond DM, Bennett MC, Fleshner M & Rose GM (1992) Inverted-U relationship between the level of peripheral corticosterone and the magnitude of hippocampal primed burst potentiation. *Hippocampus* **2** 421-430

4 Gold P (1995) Role of glucose in regulating the brain and cognition. *American Journal of Clinical Nutrition* **61** 987S-995S

5 Gould E & Tanapat P (1999) Stress and hippocampal neurogenesis. *Biological Psychiatry* **46** 1472-1479

6 Hampton RR & Shettleworth SJ (1996) Hippocampal lesions impair memory for location but not for color in passerine birds. *Behavioral Neuroscience* **110** 831-835

7 Harvey S, Phillips PG, Rees A & Hall TR (1984) Stress and adrenal function. *Journal of Experimental Zoology* **232** 633-645

8 Loscertales M, Rose SPR & Sandi C (1997) The corticosteroid synthesis inhibitors metyrapone and aminoglutethimide impair long-term memory for a passive avoidance task in day-old chicks. *Brain Research* **769** 357-361

9 Lynn SE, Breuner CW & Wingfield JC (2003) Short-term fasting affects locomotor activity, corticosterone, and corticosterone binding globulin in a migratory songbird. *Hormones & Behavior* **43** 150-157

10 McEwen BS & Sapolsky RM (1995) Stress and cognitive function. *Current Opinions in Neurobiology* **5** 205-216

11 McEwen BS. (2000) The neurobiology of stress: from serendipity to clinical relevance. *Brain Research* **886** 172-189

12 Pfeffer K, Fritz J & Kotrschal K (2002) Hormonal correlates of being an innovative greylag goose, *Anser anser. Animal Behaviour* **63** 687-695

13 Pravosudov VV (1985) Search for and storage of food by *Parus cinctus lapponicus* and *P. montanus borealis. Zoologicheskiy Zhurnal* **64** 1036-1043

14 Pravosudov VV (2003) Long-term moderate elevation of corticosterone facilitates avian food-caching behaviour and enhances spatial memory. *Proceedings of Royal Society of London Biological Sciences* **270** 2599-2604

15 Pravosudov VV & Clayton NC (2001) Effects of demanding foraging conditions on cache retrieval accuracy in food-caching mountain chickadees (*Poecile gambeli*). *Proceedings of Royal Society of London Biological Sciences* **268** 363-368

16 Pravosudov VV & Grubb TC Jr (1997) Energy management in passerine birds during the non-breeding season: a review. *Current Ornithology* **14** 189-234

17 Pravosudov VV & Lucas JR (2001) A dynamic model of short-term energy management in small food-caching and non-caching birds. *Behavioral Ecology* **12** 207-218

18 Pravosudov VV & Omanska A (2004) Prolonged moderate elevation of corticosterone does not affect hippocampal anatomy or cell proliferation rates in mountain chickadees (*Poecile gambeli*). *Journal of Neurobiology* In Press

19 Pravosudov VV, Kitaysky AS & Wingfield JC (2001) Long-term unpredictable foraging conditions and physiological stress response in mountain chickadees (*Poecile gambeli*). *General and Comparative Endocrinology* **123** 324-331

20 Pravosudov VV, Kitaysky AS, Saldanha CJ, Wingfield JC & Clayton NC (2002) The effect of photoperiod on adrenocortical stress response in mountain chickadees (*Poecile gambeli*). *General and Comparative Endocrinology* **126** 242-248

21 Pravosudov VV, Lavenex P & Clayton NC (2002) Changes in spatial memory mediated by experimental variation in food supply do not affect

hippocampal anatomy in mountain chickadees (*Poecile gambeli*). *Journal of Neurobiology* **51** 142-148
22 Pravosudov VV, Mendoza SP & Clayton NS (2003) The relationship between dominance, corticosterone, memory, and food caching in mountain chickadees (*Poecile gambeli*). *Hormones & Behavior* **44** 93-102
23 Reneerkens J, Piersma T & Ramenofsky M (2002) An experimental test of the relationship between temporal variability of feeding opportunities and baseline levels of corticosterone in a shorebird. *Journal of Experimental Zoology* **293** 81-88
24 Rose SP (2000) God's organism? The chick as a model system for memory studies. *Learning and Memory* **7** 1-17
25 Saldanha CJ, Schlinger BA & Clayton NC (2000) Rapid effects of corticosterone on cache recovery in mountain chickadees (*Parus gambeli*). *Hormones & Behavior* 37 109-115
26 Sandi C, Rose SPR (1997) Training-dependent biphasic effects of corticosterone in memory formation for a passive avoidance task in chicks. *Psychopharmacology* **133** 152-160
27 Sandi C, Rose SPR, Mileusnic R & Lancashire C (1995) Corticosterone facilitates long-term memory formation via enhanced glycoprotein synthesis. *Neuroscience* **4** 1087-1093
28 Sapolsky RM. (1996) Why stress is bad for your brain. *Science* **273** 749-750
29 Sherry DF, Jacobs LF & Gaulin SJC (1992) Spatial memory and adaptive specialization of the hippocampus. *Trends in Neuroscience* **15** 298-303
30 Sherry DF, Vaccarino AL, Buckenham K & Hertz RS (1989) The hippocampal complex of food-storing birds. *Brain Behavior and Evolution* **34** 308-317
31 Shettleworth SJ (1995) Memory in food-storing birds: from the field to the Skinner box. In *Behavioral Brain Research in Naturalistic and Seminaturalistic Settings* 158-179. Eds E Alleva, A Fasolo, HP Lipp, L Nadel & L Ricceri. Kluwer: The Hague
32 Silverin B (1997) The stress response and autumnal dispersal behavior in willow tits. *Animal Behaviour* **53** 451-459
33 Silverin B (1998) Stress responses in birds. *Poultry and Avian Biology Reviews* **9** 153-168
34 Smulders TV, Shiflett MW, Sperling AJ, DeVoodg TJ (2000) Seasonal changes in neuron numbers in the hippocampal formation of a food-hoarding bird: the black-capped chickadee. *Journal of Neurobiology* **44** 414-422
35 Vander Wall SB (1990) *Food hoarding in animals*. University of Chicago Press: Chicago
36 Wingfield JC, Breuner C & Jacobs J (1997) Corticosterone and behavioral responses to unpredictable events. In *Perspectives in avian andocrinology* 267-278. Eds S Harvey & RJ Etches. J Endocrinology Ltd: Bristol
37 Wingfield JC, Maney DL, Breuner CW, Jacobs JD, Lynn S, Ramenofsky M & Richardson RD (1998) Ecological bases of hormone-behavior interactions: the "emergency life history stage". *American Zoologist* **38** 191-206.

Part seven

Endocrine disrupting chemicals

Burrowing Owl
Pierre Deviche

Functional Avian Endocrinology
A Dawson & P J Sharp (Eds)

Neuroendocrine and behavioral consequences of embryonic exposure to endocrine disrupting chemicals

M A Ottinger, M J Quinn Jr, E Lavoie, M A Abdelnabi, N Thompson, J Hazelton, M McKernan, J Wu, P Henry,[1] C Viglietti-Panzica[2], and G Panzica[2]

Department of Animal and Avian Sciences, University of Maryland, College Park, MD 20742, USA
[1]Patuxent Wildlife Research Center, USGS, Laurel, MD, USA
[2]Dept. of Anatomy, Pharmacology and Forensic Medicine, University of Turin, Torino, Italy

Introduction

Endocrine disrupting chemicals (EDCs) have been defined functionally due to their activity on endocrine systems. Much of the attention given to these chemicals arose from the recognition that prenatal exposure to diethylstilbestrol (DES) had long term consequences in women. As more information was collected, it became clear that EDCs had the potential to impact animals across many phyla (8, 9, 10, 38). Furthermore, the exposure of wildlife at field relevant exposures is often at sublethal concentrations, making detection of effects of EDCs challenging, especially when attempting to separate other interacting factors in the animal's environment. These additional factors may include environmental conditions, food availability, disease, or confounds such as simultaneous exposure to several chemicals. It is therefore important to develop reliable and sensitive measurement end points that are appropriate for the endocrine or organ system that is the target of the EDC. A number of workshops and symposia have addressed these issues, generating publications and reports that raised awareness to the potential effects of EDCs on wildlife and humans (15, 23, 32). These and other publications have also provided a summary of the data available on the impact of EDCs in wild populations and the incidence of exposure (15, 4, 14). In 1998, the Endocrine Disruptor Screening and Testing Advisory Committee (EDSTAC) submitted a report that recommended a tiered approach to testing suspected EDCs, which involved screening followed by selected bioassays, and subsequent single and multigenerational tests when indicated. In birds, a two-generation testing

paradigm was selected for evaluation. Development and validation of these testing paradigms have continued (see Endocrine Disruptor Screening Program:
(http://www.epa.gov/oppfead1/cb/csb_page/updates/endocrin.htm)
for progress in the area of regulatory activities and for updates on the status of validation of specific tests. A parallel activity has been ongoing through the Organization for Economic Co-operation and Development, which established a Working Group on Avian Reproduction Toxicity Testing. This group has been considering the testing parameters and measurement end points that are reliable and sensitive indices of EDC exposure in birds and generated a discussion document (6). This Working Group has revised the Avian Reproduction Toxicity Test guideline, which is under review and is drafting a protocol for a Multigenerational Reproduction Toxicity Test guideline, to be reviewed by the OECD member nations.

In birds and mammals, early attention had been focused on the estrogenic activity of many of the environmental chemicals (30). Recognition that many of the suspected EDCs were likely to have estrogenic activity also led to development of *in vitro* screens for rapid detection of these chemicals (4). Simultaneously, there was growing concern about the effects of environmental chemicals of species living near water, such as the Great Lakes or Chesapeake Bay, resulting in biomagnification of chemicals for fish eating birds (25, 43). Because some of the EDCs also interact with the thyroid and adrenal endocrine systems, they have the ability to affect metabolic endocrine responses, as well as reaction to stressors and immune function. Overall, this field has a great deal of complexity due to the number of EDCs in the environment, temporal and spatial variation in somewhat sporadic distribution of specific EDCs, and their potential interaction with other contaminants. Fortunately, there are several databases available that integrate the available data into an accessible format for determining potential risk for populations in selected regions (for example, see 44). Some of the basic issues and an overview of the data collected in birds is discussed below. Clearly this is an area that requires an integrated approach amongst avian biologists that are working in the areas of endocrinology, neurobiology, field biology, and toxicology in order to address these questions in a responsible and scientific manner.

Issues associated with the identification and regulation of EDCs in avian species

Relative potency and effects compared to other environmental factors

A major confound in the issue of EDC effects in birds is distinguishing the impact of other environmental and health impacts from those associated with EDC exposure. In fact, it is likely that these factors all interact with one another to affect the overall viability and fitness of the individual. Furthermore, unless a site is highly contaminated in a homogeneous manner, it is likely that an individual may not be consistently exposed to EDCs. Therefore, it is difficult to clearly assess the impact of EDCs in the field because the level of exposure is often variable and the bird's contact with the EDC(s) may be inconsistent according to the distribution of the compounds in their habitat. Therefore, in order to evaluate EDC impact for the purpose of risk assessment, it is critical to determine the availability of the EDCs for species that are likely to be exposed and if there are metals and other contaminants in the environment that will also affect the birds (19). Mode of exposure is also an issue, especially when considering sensitive stages in the life cycle. For example, pesticides sprayed in an orchard are likely to coat eggs in nests located in the trees that are sprayed. Moreover, contaminants deposited on the fur or feathers may be ingested during grooming or occur on/in prey items that are ingested or fed to the young. These embryos are then exposed to EDCs during a critical time in development. However, passerines appear to be less sensitive to sublethal doses of EDCs during embryonic development compared to precocial species potentially due to the timing and mechanisms of sexual differentiation as well as to the mode of exposure to EDCs associated with diet and habitat (2, 34, 35, 38). Therefore, these factors must all be considered, especially when considering the risk of EDCs to field species. Clearly, more characterization of the effects of EDCs to precocial and altricial must be available along with the pharmacokinetics of the EDCs in the environment in order to generate realistic risk assessment models. Finally, these models must consider avian biology rather than being based on mammalian, fish, or invertebrate models because of phylogenetic differences in metabolism, reproductive strategies, habitat considerations, and life history.

Regulatory issues and risk assessment

Toxicity testing with end points that detect endocrine disruption

A basic issue is the development of an appropriate testing paradigm, including reliable end points. One source of confusion has been that avian toxicology tests currently in use generally include reproductive end points as a measure of the fitness of the adults. Because these reproductive end points, such as fertility and hatching success, are included in the testing paradigm, it has been assumed that these measures are sufficient. However, these reproductive end points may not necessarily be appropriate, reliable, or sensitive for detecting an endocrine disrupting chemical. Therefore, it is important to distinguish

the toxicology test from a test designed for detecting an endocrine disrupting chemical. A number of studies are underway in laboratories around the world; many of these laboratories have participation in the OECD Expert Group on Avian Reproduction Testing. At this time, the measurement end points for consideration in a testing paradigm include neuroendocrine, behavioral, endocrine, and whole organism measures which relate to the health of the bird. Deliberations continue and information and progress in revisions, including consideration of EDC impact in the testing paradigms used in preparing applications for registration of pesticides and other chemicals for agricultural or residential may be found at the following website:
http://www.epa.gov/pesticides/regulating/index.htm

Avian species selected for monitoring and testing

In most of the toxicological applications for regulatory monitoring, species indigenous to the United States have been used for testing. This means that the species of choice have been the mallard duck and northern bobwhite quail (6, 37). Although the toxicological responses of these species have been tested with a variety of compounds, few background or baseline data exist for the reproductive endocrinology of these birds especially relative to EDC impact. Conversely, there are data available on the reproductive and metabolic endocrinology and on the neural regulation of reproductive behavior in Japanese quail, zebra finches, and doves. However, these species were not considered for toxicological testing until recently. Japanese quail have been under intense study recently with the recognition that multigenerational approaches are necessary to detect some of the more subtle effects of EDCs. Moreover, since field exposures are often at low levels or in an unpredictable frequency, the impact may be more significant over several generations. As such, multigenerational testing is necessary to detect this type of low exposure, long term outcome. In order to do multigeneration tests, slowly maturing species, including northern bobwhite quail and mallard ducks require long term studies. Our experience in conducting a two-generation bobwhite quail dietary study was that, beginning with hatching the parent generation, at least 24 months was required to complete the in life portion of the study. This becomes a prohibitive time frame for use in regulatory applications.

Baseline on Japanese quail neuroendocrinology and rationale for choice of endpoints used in studies discussed below

Studies in Japanese quail take advantage of a great deal of information available on the biology, neurobiology, and endocrinology of this species. In addition, Japanese quail have a short incubation period (17 days), mature quickly (8 weeks), and have a long peak production period (3-12 months). EDCs impact development and sexual differentiation in Japanese quail because they are hormonally active at a time that native estrogens and androgens exert long-term effects on endocrine and behavioral responses (7, 22). Further, estradiol is a critical element in sexual differentiation in quail. During embryonic development, females have high relative concentrations of estradiol/androgen whereas males have low relative estradiol/androgen. Moreover, there are sex related differences in the patterns of changes in these steroids. In females embryos, plasma E_2 rose until hatch and decreased, post hatch (34, 35). In males, plasma androgen peaked at embryonic days 14-17 (E14-E17) and declined post hatch (34). These steroid hormones are primarily produced by the embryonic gonads, with some contribution from the adrenal glands (1). In addition, yolk steroid hormone content reflects embryonic steroid hormone levels, suggesting that the yolk serves as a hormone depot for the embryo (Abdelnabi and Ottinger, unpublished data). This developmental pattern allows for precise experimental manipulations at timed phases in ontogeny as well as selection for suitable sampling paradigms that can be used for range finding tests and other manipulations.

Sex differences were observed in hypothalamic catecholamine and indolamine concentrations in untreated quail hatchlings (1, 34, 35). Females had higher levels of norepinephrine (NE) than males over embryonic development, whereas males had higher dopamine (DA) levels (34, 35). Our current data provide evidence that estrogenic EDC exposure can alter these monoamines in hatchling and adult quail, suggesting potential consequences for neuroendocrine regulation of reproductive endocrine responses.

Finally, maternal transfer of steroids is a likely mode of embryonic exposure; female quail given E_2 implant transferred estradiol to offspring *via* the yolk (3). Therefore, embryonic steroid exposure from endogenous and exogenous sources appears to be critical in the sexual differentiation of hypothalamic neurotransmitter systems (35). Altering the relative exposure to the gonadal steroids, even with weakly hormonally active compounds therefore has the potential to affect sexual differentiation if exposure occurs during sensitive periods in development.

Testing paradigms and measurement end points

Chemicals chosen for testing in our laboratories and effects of these chemicals

Estrogenic EDCs

Estradiol benzoate: Administration of exogenous gonadal steroids to embryonic quail alters sexual differentiation, resulting in abnormalities in the female reproductive tract and impaired male sexual behavior (3, 45). The action of estradiol may differ in males and females in that both have sex-specific endogenous steroid hormone patterns during embryonic development; specifically males have relatively higher androgen to estradiol levels and vice versa in females (34). Relatively low concentrations (20µg/egg estradiol benzoate or 2ppm estradiol) injected at embryonic day 11 (ED11), decreased male sexual behavior and decreased egg production and fertility in females (35). Fifty µg/egg (5ppm estradiol) at ED11 caused uterine prolapse, egg bound females and low egg production, while 500µg/egg (50ppm) estradiol at ED 11 caused uterine prolapse, egg bound females and low egg production, while 500µg/egg (50ppm estradiol) at ED11 resulted in no egg production and both 50 and 500µg/egg caused persistent right oviducts (Ottinger *et al.*, unpublished data).

Methoxychlor (MXC): The deleterious effects of EDCs have long been recognized in wildlife, with effects of DDT on the avian egg shell as one such example. Methoxychlor is an organochlorine pesticide and was developed to replace DDT. It is used as a pesticide for fruit, vegetables, trees, home gardens, forage crops, and livestock. Highest intensity of use is in the northwest and in eastern seaboard states, with concern for human exposure (17, 36). In mammals, MXC has been linked to reproductive dysfunction.

In laboratory experiments MXC was maternally transferred into eggs when hens were fed 2.5mg/day in a capsule over five days (Ottinger *et al.* unpublished data). Exposure of Japanese quail embryos to 1.5 or 3.0mg/egg MXC resulted in a significant increase in gonadotropin-releasing hormone-I (GnRH-I) in female hatchlings and a significant decrease of GnRH-I in female adults (39). Adult males exposed to 1.5 or 3.0mg/egg MXC had less sexual behavior (35). Exposure of Japanese quail at ED 0 to 5mg/egg MXC decreased or completely inhibited adult male sexual behavior (12). Therefore, if embryonic exposure to MXC has negative effects on female fitness, also, then hatchling GnRH-I will be a sensitive biomarker.

Soy Phytoestrogens: The phytoestrogens in soy, especially genistein are hormonally active. The primary biologically active isoflavone in soybeans is genistein, which is approximately 1000 times less active than estradiol *in vitro* (13, 46, 26, 47). The phytoestrogens are found primarily in the soy protein, which in feeds, may have relatively high levels of genistein or other phytoestrogens in their aglycone, glycosylated, and acylated forms (27). Interestingly we also found that of the phytoestrogens in soy, genistein was the primary isoflavone transferred into the yolks when fed to quail hens in a dose dependent manner. Therefore, the soy protein content in feeds is noteworthy as there is the potential for is provides evidence that this biologically active phytoestrogen is a potential for maternal deposition into the yolk and subsequent effects on the developing embryo.

Studies in which quail eggs were injected with 10, 100, or 1000 µg of genistein showed that this phytoestrogen has bioactivity in birds. Results showed that the highest dose of genistein inhibited sexual behavior in adult males (Panzica *et al.*, unpublished data). Because of the association of vasotocin and male sexual behavior, this system was further investigated in phytoestrogen treated males (40). Phytoestrogen treated males had less vasotocin-immunreactive cells stained in areas that regulate reproductive behavior, suggesting that the genistein may have interfered with sexual differentiation of male behavior. It is intriguing to note that there is evidence from mammalian studies that phytoestrogens affected aromatase enzyme (41). Therefore, it is is possible that the EDCs act via several ways to impact reproductive endocrine and behavioral responses.

Androgen or thyroid active EDCs

V*inclozolin:* Vinclozolin is a fungicide that blocks androgen receptors (24, 21). In quail, we found that egg injections with vinclozolin at relatively low levels resulted reduced male sexual behavior and GnRH-I in hatchlings (29).

p,p'-DDE: DDE is a persistent, bioaccumulative environmental contaminant, exposure to which was associated with egg shell thinning in many bird species. Although DDT (the parent compound of DDE) is no longer used in the United States and the European Union, DDE is persistent in many ecosystems and is a chemical of concern in many risk assessments. Exposure of Japanese quail embryos to 40µg/egg DDE injected at embryonic day 4 decreased male sexual behavior and early onset of puberty in females, but did not affect hatchability compared to controls. Egg injection with DDE (low=20µg/egg and high=40 µg/egg) also impacted reproductive end points. Females showed delayed sexual maturation, at the higher dose, and males treated with the higher dose showed impaired mating behavior ($p<0.05$; Quinn and Ottinger, unpublished data).

Polychlorinated biphenyls (PCBs): PCBs have been the subject of intense investigation for many years as they have been identified as contaminants of concern, especially in fish eating birds (25, 28, 32, 48). The PCBs are especially complex due to the number in the environment and their persistence. Further, depending on the structure of the compound, there are estrogenic, androgen, thyroidgenic, or a combination of these effects from PCBs. PCBs are also a problem for carnivorous birds due to relatively higher concentrations in the fatty tissue of their prey, which resulted in decreased thyroxine and estradiol levels in species such as the kestrel (42). Studies conducted in our laboratory with a primarily estrogenic PCB have shown that fertility and reproductive performance were reduced by PCB exposure during embryonic development (see Table 1). As with many of the pesticides, sexual behavior proved to be a sensitive index of embryonic exposure in Japanese quail.

Table 1. Exposure at E4 by egg injection resulted in decreased reproductive performance (fertility and egg production), due in part to reduced male reproductive behavior (Henry *et al.*, unpublished data).

Generation	Treatment	n	Latency (sec)	Attempts	Cloacal Contacts
P1	Control PCB	13 8	35 ± 10.05 77.8 ± 19.45	5.5 ± 1.98 5.9 ± 0.97	2.16 ± 0.45 2.0 ± 0.44
F1	Control PCB	11 6	42.6 ± 14.44 107.0±14.97	8.4 ± 1.95 2.8 ± 0.82	2.3 ± 0.47 0.3 ± 0.20
F2	Control PCB	16 12	49.8 ± 9.96 77.0 ±13.25	6.0 ± 0.86 5.3 ± 2.6	2.2 ± 4.0 0.9 ± 0.33

PCB = 2',4',6 –trichloro-4-biphenylol
P1 = parent generation
F1 = first generation
F2 = second generation
n = number of tested males/day

Effects of EDCs in birds and validating appropriate measures

Population level effects: measurement end points and assessment end points

Ultimately, it is critical to ascertain the potential of EDCs to affect wild birds, especially at a population level. As discussed above, many of the mechanisms that regulate reproductive endocrine and behavioral responses have been extensively investigated in both laboratory models and in wild birds, particularly those that are likely to be impacted by EDCs. Once the appropriate measurement end points have been established, they very often are also used as assessment endpoints to determine population level effects.

Measurement end points: reliability and use in laboratory studies:

Fitness end points: provide an indication of the individual's fitness, including health, reproduction, and vigor. These measures are crucial to survival and production of offspring, but may underestimate long-term or transgenerational effects of the EDCs. Our studies have shown that it takes relatively high doses of weakly estrogenic compounds to exert measurable effects on these types of end points. Conversely, some EDCs also have toxicological effects and will be detected in both toxicology tests as well as tests designed for detecting EDCs. These compounds present less of an issue due to their potency. As such, regulatory decisions and risk assessment for these compounds are much less complicated.

Endocrine, behavioral, and neuroendocrine end points: As mentioned earlier, a primary route of exposure is via maternal deposition of the EDC into the yolk (3). This is the case for methoxychlor as well as for genistein and presumably other lipid soluble chemicals. Although EDCs are generally weaker in action than endogenous steroids, including exogenous estradiol, the estrogenic pesticide methoxychlor (MXC) slowed sexual maturation in both males and females. Treated males also had impaired sexual behavior, similar to the effects of embryonic estradiol (35).

Similarly, androgenic EDCs, such as vinclozolin, a fungicide that blocks androgen receptors also affected GnRH in hatchlings and sexual behavior in adult males that had been exposed as chicks (5). These adults also showed slower sexual maturation and impaired behavior during maturation. Although in many cases, the birds appear to "catch up", a critical question would be the impact of delayed maturation and impaired sexual behavior for bird populations in the field.

Therefore, in our studies male sexual behavior has consistently been a sensitive and responsive measurement end point. Aromatase enzyme has been clearly shown to modulate male sexual behavior (5, 11). However, in our studies, we have not observed a dose dependent effect of embryonic EDCs on aromatase enzyme activitiy in adults. Arginine vasotocin (VT), which is localized in brain regions that modulate reproductive behavior is also sexually dimorphic and responsive to gonadal steroids (40). In fact VT may prove to be a more sensitive dose dependent index of EDC effects. This is based on the findings from enbryonic genistein exposure in which a dose dependent decrease occurred in immunostaining that correlated with the loss of male sexual behavior (Panzica *et al.*, unpublished data).

Immune end points: Administration of many different types of androgens, including androsterone, androstene-3, 17-dione, methylandrostene diol, 5α-dihydrotestosterone (5α-DHT), testosterone propionate (TP), and 19-nortestosterone, disrupts embryonic development of the bursa (18, 33). DHT treatments resulted in reduced numbers of proliferating bursal cells in broiler chicks (31). Bursas from chicken embryos treated with TP at E3 were considerably smaller than controls on E10 (33).

Some laboratory studies have shown specific immunotoxic effects of individual EDCs. Planar halogenated aromatic hydrocarbons, such as polychlorinated biphenyls (PCBs) and 2,3,7,8-tetrachlorodibenzo-*p*-dioxin have been shown to have immunotoxic effects (20), including developmental effects on the bursa. PCB 126 caused decreases in bursa weight and numbers of developing B lymphocytes in the bursas of chicken embryos (16). PCBs and dioxins have also been linked to thymic atrophy and suppression of mitogen-induced proliferation, mixed lymphocyte responses, cell-mediated cytotoxicity, and humoral responses (reviewed in 20).

Interpretation of field studies is difficult when trying to establish links between individual chemicals and individual effects. Organisms are most often exposed to multiple types of EDCs at any one period of time in the environment. Exposure to multiple EDCs makes it impossible to link any immunotoxic effect to any one chemical. Further, many of these studies are observational, and as such, causal relationships cannot be established. The matter is further complicated by the fact that many of the mechanisms behind EDC-induced immunosuppression remain unknown.

Concluding remarks

The issue of EDCs and their impact on avian species has gained visibility in recent years due the evidence of endocrine activity. The extent of hormone-like activity is the subject of extensive investigation. Results from these studies will provide the basis for regulatory guidelines and tests that should serve to protect field birds from exposure to deleterious levels of EDCs. It is not clear, however, if EDCs interact with other contaminants or predispose an individual to effects from other environmental challenges or disease. As such, more information on the extent of EDCs impact and stages of the life cycle that are most sensitive will be extremely useful. Furthermore, the traditional toxicological tests, even those including reproductive end points may not be sufficient to reliably detect and characterize the effects of EDCs. It is important to establish reliable indices that are appropriate for the mode and target of action for each class of EDC and for regulation to be based upon these tests.

Aknowledgements

Supported by EPA #R826134010 (Star Grant), NSF#9817024, and EPA R-2877801(MAO). The two generation diet studies were possible only with our collaborators, Dr. Joanne Beavers and Mark Jaber, Wildlife International, Ltd., Easton, Maryland. In addition, the authors thank our gifted and diligent students, Kameka Henry, Elizabeth Humphries, Erin

Quigley, Elizabeth Reed, Maie Abdelnabi, Ashley Barton, and Laurie Mann who worked with us throughout all phases of these studies.

References

1 Abdelnabi MA & Ottinger MA (2003) Hypothalamic indolamines during embryonic development and effects of steroid exposure. *General and Comparative Endocrinology* **130** 13-19

2 Adkins-Regan E, Abdelnabi M, Mobarak M & Ottinger MA (1990) Sex steroid levels in developing and adult male and female zebra finches (*Poephila guttata*) *General and Comparative Endocrinology* **78** 93-109

3 Adkins-Regan EK, Ottinger MA & Park J (1995) Maternal transfer of estradiol to egg yolks alters sexual differentiation of avian offspring. *Journal of Experimental Zoology* **271** 466-470

4 Ankley G, Mihaich E, Stahl R, Tillitt D, Colborn T, McMaster S, Miller R, Bantle J, Campbell P, Denslow N, Dickerson R, Folmar L, Fry M, Giesy J, Gray E, Guiney P, Hutchinson T, Kennedy S, Kramer V, LeBlanc G, Mayes M, Nimrod A, Patino R, Peterson R, Purdy R, Ringer, R, Thomas P, Touart L, Van der Kraak G & Zacharewski T (1997) Overview of a workshop on screening methods for detecting potential (anti-) estrogenic/androgenic chemicals in wildlife. *Environmental Toxicology and Chemistry* **17** 68-87

5. Aste N, Panzica GC, Aimar P, Viglietti-Panzica C, Foidart N & Balthazart J (1996) Morphometric studies demonstrate that aromatas-immunoreactive cells are the main target of androgens and estrogens in the quail medial preoptic nucleus. *Experimental Brain Research* **101** 241-252

6 Bennett RS, Bentley R, Shiroyama T & Bennett JK (1990) Effects of the duration and timing of dietary methyl parathion exposure on bobwhite reproduction. *Environmental Toxicology and Chemistry* **7** 1473-1480

7 Berg C, Halldin K, Fridolfsson AK, Brandt I & Brunstrom B (1999) The avian egg as a test *system* for endocrine disrupters: effects of dietylstilbestrol and ethynylestradiol on sex organ development. *The Science of the Total Environment* **233** 57-66

8 Bern HA (1992) The fragile fetus. In *Chemically Induced Alterations in Sexual and Functional Development: The Wildlife/Human Connection,* pp. 9-16. Eds T Colborn & C Clement. Princeton: Princeton Scientific Publishing

9 Colborn T, vom Saal FS & Soto AM (1993) Developmental effects of endocrine-disrupting chemicals in wildlife and humans. *Environmental Health Perspectives* **101** 378-384

10 Crews D, Willingham E & Skipper JK (2000) Endocrine disruptors: present issues, future directions. *Quarterly Review of Biology* **75** 243-260

11 Dellovade TL, Rissman EF, Thompson N, Harada N & Ottinger MA (1995) Co-localization of aromatase enzyme and estrogen receptor immunoreactivity in the preoptic area during reproductive aging. *Brain Research* **674** 181-187

12 Eroschenko VP, Amstislavsky SY, Schwabl H & Ingermann RL (2002) Altered behaviors in male mice, male quail, and salamander larvae

following early exposures to the estrogenic pesticide methoxychlor. *Neurotoxicology and Teratology* **24** 29-36

13 Farmakalidis E, Hathcock JN & Murphy PA (1985) Oestrogenic potency of genistin and daidzin in mice. *Food and Chemical Toxicology* **23** 741-45

14 Fairbrother A, Landis WG, Dominquez S, Shiroyama T, Buchholz P, Roze MJ & Matthews GB (1998) A novel nonmetric multivariate approach to the evaluation of biomarkers in terrestrial field studies. *Ecotoxicology* **7** 1-10

15 Fox GA (1992) Epidemiological and pathobiological evidence of contaminant-induced alterations in sexual development in free-living wildlife In *Chemically-induced alterations in sexual and functional development: the wildlife / human connection*. pp.147-158. Eds T Colborn & C Clement. Princeton: Princeton Scientific Publishing Co. Inc.

16 Fox LI & Grasman KA (1999) Effects of PCB 126 on primary immune organ development in chicken embryos. *Journal of Toxicology and Environmental Health* **58** 233-244

17 Fry M (1995) Reproductive effects in birds exposed to pesticides and industrial chemicals. *Environmental Health Perspectives* **103**(Suppl 7) 165-171

18 Glick B (1991) How it all began: The continuing story of the bursa of Fabricius In *Basis and Practice.* Eds A Toivanen & P Toivanen. Boca Raton, Florida: CRC Press

19 Golden NH, Rattner BA, McGowan PC, Parsons KC & Ottinger MA (2003) Concentrations of metals in feathers and blood of nestling black-crowned night-herons (Nycticorax nycticorax) in Chesapeake and Delaware Bays. *Bulletin of Environmental Contamination and Toxicology* **70** 385-393

20 Grasman KA & Fox GA (2001) Associations between altered immune function and organochlorine contamination in young Caspian terns (Sterna caspia) from Lake Huron, 1997-1999. *Ecotoxicolog*y **10** 101-114

21 Gray Jr LE, Ostby J, Monosson E & Kelce WR (1999) Environmental antiandrogens: low doses of the fungicide vinclozolin alter sexual differentiation of the male rat. *Toxicology and Industrial Health* **15** 48-64

22 Halldin K, Berg C, Brandt I & Brunstrom B (1999) Sexual behavior in Japanese quail as a test end point for endocrine disruption: Effects of in Ovo exposure to ethinylestradiol and diethylstilbestrol. *Environmental Health Perspectives* **107** 861-866

23 Kavlock RJ, Daston GP, DeRosa C, Fenner-Crisp P, Gray LE, Kaatari S, Lucier G, Luster M, Mac MJ, Maczka C, Miller R, Moore J, Rolland R, Scott G, Sheehan DM, Sinks T & Tilson H (1996) Research needs for the risk assessment of health and environmental effects of endocrine disruptors: a report of the U.S. EPA-sponsored workshop. *Environmental Health Perspectives* **104**(Suppl 4) 715-740

24 Kelce WR, Monosson E, Gamesik MP, Laws SC & Gray Jr LE (1994) Environmental hormone disruptors: evidence that vinclozolin developmental toxicity is mediated by anti-androgenic metabolites. *Toxicology and Applied Pharmacology* **126** 276-285

25 Kubiak TJ, Harris HJ, Smith LM, Schwartz TR, Stalling DL, Trick, JA, Sileo L, Docherty DE & Erdman TC (1989) Microcontaminants and reproductive impairment of the Forster's Tern on Green Bay, Lake

Michigan - 1983. *Archives of Environmental Contamination and Toxicology* **18** 706-727
26 Lephart, ED, Rhees, RW, Setchell KD, Bu LH & Lund TD (2003) Estrogens and phytoestrogens: brain plasticity of sexually dimorphic brain volumes. *Journal of Steroid Biochemistry and Molecular Biology* **85** 299-309
27 Lin F, Wu J, Abdelnabi MA, Ottinger MA & Guisti MM (2004) Effects of dose and glycosylation on the transfer of genistein into the eggs of Japanese quail (Coturnix japonica). *Journal of Agricultural and Food Chemistry* **52** 2397-2403
28 McCarty JP & Secord AL (1999) Reproductive ecology of tree swallows (Tachycineta bicolor) with high levels of polychlorinated biphenyl contamination. *Environmental Toxicology and Chemistry* **18** 1433-1439
29 McGary S, Henry PF & Ottinger MA (2001) Impact of vinclozolin exposure on Japanese quail reproduction. *Environmental Toxicology and Chemistry* **20** 2487-2493
30 McLachlan JA & Korach KS (1995) Symposium on estrogens in the environment, III. *Environmental Health Perspectives* **103**(Suppl 7) 3-5
31 Novotny EA, Raveche ES, Sharrow S, Ottinger MA & Steinberg AD (1983) Analysis of thymocyte subpopulations following treatment with sex hormones. *Clinical Immunology and Immunopathology* **28** 205-217
32 NRC (1999). Hormonally Active Agents in the Environment. Washington, DC: National Academy Press
33 Olah I, Glick B & Toro I (1986) Bursal development in normal and testosterone-treated chick embryos. *Poultry Science* **65** 574-588
34 Ottinger MA, Pitts S & Abdelnabi MA (2001a) Steroid hormones during embryonic development in Japanese quail: Plasma, gonadal, and adrenal levels. *Poultry Science* **80** 795-799
35 Ottinger MA, Abdelnabi MA, Henry PFP, McGary S, Thompson N & Wu J (2001b) Neuroendocrine and behavioral implications of endocrine disrupting chemicals in quail. *Hormones and Behavior* **40** 234-247
36 Ottinger MA, Abdelnabi MA, Quinn M, Golden N, Wu J & Thompson N (2002) Reproductive consequences of EDCs in birds: What do laboratory effects mean in field species? *Neurotoxicology Teratology* **24** 17-29
37 Ottinger MA & Rattner B (1999) Husbandry and care of quail. *Poultry and Avian Biology Reviews* **10** 117-120
38 Ottinger MA & vom Saal FS (2002) Impact of environmental endocrine disruptors on sexual differentiation in birds and mammals. In *Hormones and Behavior in Higher Vertebrates. Vol 4*, pp 325-383. Eds D Pfaff, A Arnold, A Etgen & R Rubin. Academic Press
39 Ottinger MA, Wu J, Hazelton J, Abdelnabi M, Thompson N, Quinn Jr M, Donoghue D, Schenck F, Ruscio M, Beavers J & Jaber M (2004) Assessing the consequences of the pesticide methoxychlor: Neuroendocrine and behavioral measures as indicators of biological impact of an estrogenic environmental chemical. *Brain Research Bulletin* In press
40 Panzica GC, Garcia-Ojeda E, Viglietti-Panzica C, Thompson NE & Ottinger MA (1995) Testosterone effects on vasotocinergic innervation of sexually dimorphic medial preoptic nucleus and lateral septum during aging

in male quail. *Brain Research* **712** 190-198

41 Pelissero C, Lenczowski MJP, Chinzi D, Davail-Cuisset B, Sumpter JP & Fostier A (1996) Effects of flavonoids on aromatase activity, an in vitro study. *Journal of Steroid Biochemistry and Molecular Biology* **57** 215-223

42 Quinn M, French JB, McNabb AF & Ottinger MA (2002) The effects of polychorinated biphenyls (Aroclor 1242) on thyroxine, estradiol, molt, and plumage characteristics in the American kestrel (*Falco sparverious*). *Environmental Toxicology and Chemistry* **21** 1417-1422

43 Rattner BA, Melancon MJ, Custer TW, Hothem RL, King KA, LeCaptain LJ, Spann JW, Woodin BR & Stegeman JJ (1993) Biomonitoring environmental contamination with pipping black-crowned night heron embryos:induction of cytochrome P450. *Environmental Toxicology and Chemistry* **12** 1719

44 Rattner BA, Pearson JL, Golden NH, Cohen JB, Erwin RM & Ottinger MA (2000) Contaminant exposure and effects in a terrestrial vertebrates database: Trends and data gaps for Atlantic coast estuaries. *Environmental Monitoring and Assessment* **63** 131-142

45 Rissman EF, Ascenzi M, Johnson P & Adkins-Regan E (1984) Effect of embryonic treatment with oestradiol benzoate on reproductive morphology, ovulation, and oviposition and plasma LH concentrations in female quail (*Coturnix coturnix japonica*). *Journal of Reproduction and Fertility* **71** 411-417

46 Setchell KDR (1998) Phytoestrogens: the biochemistry, physiology, and implications for human health of soy isoflavones. *American Journal of Clinical Nutrition* **68** 1333S-1346S

47 Sheehan DM (1998) Herbal medicines, phytoestrogens and toxicity: risk: benefit considerations. *Proceeding of the Society for Experiemental Biology and Medicine* **217** 379-385

48 Stanton B, DeWitt J, Henshel D, Watkins S & Lasley B (2003) Fatty acid metabolism in neonatal chickens (Gallus domesticus) treated with 2,3,7,8-tetrachlorodibenzo-p-dioxin (TCDD) or 3,3',4,4',5-pentachlorobiphenyl (PCB-126) in ovo. *Comparative Biochemistry and Physiology Part C* **136** 73-84

Functional Avian Endocrinology
A Dawson & P J Sharp (Eds)

Endocrine disruption in a songbird, the Zebra finch

J R Millam

Department of Animal Science, University of California, Davis, CA 95616, USA

Introduction

Estrogen plays key roles in the sexual differentiation of birds, particularly as a morphogen for the development of ovaries. Thus, perturbing estrogen metabolism during embryonic development produces intersexual or sex-reversed genetic females (8). Behavior can also be affected. Treatment of quail embryos with estradiol alters adult reproductive performance (39) and estradiol-treated male quail embryos (5) or estradiol-treated male Zebra finch chicks (24) show altered mate preference and/or demasculinized copulatory and other sexual behaviors as adults (e.g., 1, 3, 5; for review see 8).

Development of the passerine song control system reflects another aspect of estrogen metabolism that was poorly understood for many years. Estradiol-treated female songbird chicks develop masculinized song control nuclei and can produce song, when administered testosterone as adults (e.g., 3, 19). However, gonadal steroids are probably not involved in the normal development of the song system in males (6). Rather, normal masculinization of song nuclei likely results from steroids synthesized *de novo* in brain tissue (20, 36, 41). Regardless, female brain is subject to the masculinizing effects of estrogens from whatever source, so female chicks remain vulnerable to exposure to estrogens in the environment.

Recurring reports of sexual disruption by environmental contaminants (11) suggest that environmental estrogens – estrogenic substances in the environment of either manmade (i.e., xenoestrogens) or plant origin (phytoestrogens) – might adversely affect normal development, leading to impairments in adult reproduction. Whether environmental estrogens act hormonally to impair reproduction in birds has been questioned (12), but abnormalities in wild bird populations exposed to putative xenoestrogens suggest an endocrine mode of action (16), and many industrial chemicals and pesticides are estrogenic in *in vitro* test systems (7). These issues prompted us to ask whether oral exposure to environmental estrogens might disrupt normal sexual differentiation of the song control system and reproductive performance in an Australian grass finch, the Zebra finch (*Taeniopygia guttata*).

Oral estrogens and masculinization of brain

The extreme dimorphism of the song control system of the Zebra finch brain provides an excellent system in which to study sexual differentiation. Four song control nuclei, area X of the parolfactory lobe, HVC, nucleus robustus of the archistriatum (RA), and lateral magnocellular nucleus of the anterior neostriatum (LMAN), are highly developed in the male Zebra finch, whereas HVC, RA and LMAN are greatly reduced in the female brain, and area X is absent. The size of these song control nuclei correlates with singing behavior: male Zebra finches sing and females do not (30).

Previous experiments used parenteral routes of administration to study mechanisms of sexual differentiation of the song system. However, as chicks in the wild would likely face dietary exposure to xenobiotics, e.g., by being fed insects or seeds tainted with pesticides, we administered test substances orally. Additionally, we used equimolar dosing of putative xenoestrogens to permit the characterization of estrogenicity in comparison to a prototypical estrogen, estradiol benzoate. As the sensitive period for masculinization of the female brain includes a period of rapid growth (approximately days 5 through 11 post-hatch), we dosed chicks daily on a per body mass basis to reflect the changing amounts of feed consumed during development. As adults, birds were perfused and brains were sectioned and Nissl-stained to reveal song nuclei. Song nuclei were then traced, re-constructed and their volumes calculated using Neurolucida (Microbrightfield, Williston, VT USA) (34).

Using this approach, we tested: canola oil, the vehicle for all test substances; estradiol benzoate (EB), a prototypical estrogen; methoxychlor (MXC), an organochlorine pesticide used on a variety of food crops, particularly apples and also licensed for veterinary use to control parasites on some livestock species (10, 18, 35, 38); dicofol (D), another organochlorine pesticide used on a variety of food and fiber crops, particularly fruits and cotton (35); and octylphenol (OP), a non-ionic surfactant used in a variety of industrial and agricultural applications (29, 35, 37, 39), OP is found in occasional abundance in such substrates as sewage treatment plant sludge, which is used as a soil amendment (9). All of these substances, except canola oil, have shown estrogenic activity in other test systems (13, 21, 28, 33, 43).

No significant differences were found in the song nuclei of male brains, though LMAN was excluded from analysis because of high and unexplained variability of LMAN in all treatment groups. Also, no significant differences were found in the volumes of song control nuclei in female finches between a no vehicle control group (no treatment of any kind) and canola oil-treated controls. Canola oil thus showed no detectable estrogenicity in this system. In contrast, all doses of EB tested (1, 10, 100 and 1000 nmol/g body mass) increased area X volume and the three highest doses significantly increased nuclear area of RA and HVC (Fig. 1). The greatest increase in song control nuclear volume was observed in the 100 nmol/g body mass dose of EB. None of the xenoestrogens induced significant growth of song control areas at the doses tested, although only the highest doses (1000 nmol/g body mass) of octylphenol and methoxychlor had sufficient replication to have any confidence in these negative findings.

In sum, we found that an oral route of administration of EB was effective in masculinizing female song control nuclei of Zebra finch chicks, but we found no evidence that octylphenol, methoxychlor or dicofol masculinized female brain.

Oral estrogens and impact on reproduction

Although exogenous estrogen has been known for some 25 years to alter developing Zebra finch brain, whether such treatment impairs adult fecundity in males or females has received little attention. Adkins-Regan and Ascenzi (2) found that early treatment with EB de-masculinized male sexual behavior, suggesting that environmental estrogens might generally impair reproductive performance. Indeed, reproduction of California quail was impaired when they were fed extracts of phytoestrogenic plants (23). As we found that oral exposure to estrogen masculinizes female Zebra finch brain and confers the ability to sing (34), we asked whether such treatment would impair reproductive performance in either males or females.

We tested this idea by administering EB in the same manner, as above: daily, oral gavage of the estrogenic or putatively estrogenic compounds (EB, octylphenol or methoxychlor) dissolved or suspended in canola oil and dosed on a per gram body mass basis, from day 5 through 11. Birds were then grown to somatic maturity (over 130 days of age), and provided with nest boxes and shredded burlap to stimulate breeding.

Reproductive performance was initially assessed in canola-treated control birds in social housing (five randomly chosen males and females housed communally [repeated once, N=10 potential pairs], with access to a surplus of nest boxes [8]), and in individual breeding cages, in which males and females were force-paired (N=10). Number of hatched eggs per next box was significantly reduced (P<0.05) and candled fertility was lower (P<0.06) in force-paired, individually housed birds, though other measures of performance (number of eggs per nest box, incidence of cracked/broken or missing eggs, and percentage of dead embryos) did not differ. Communal housing was chosen for subsequent experiments.

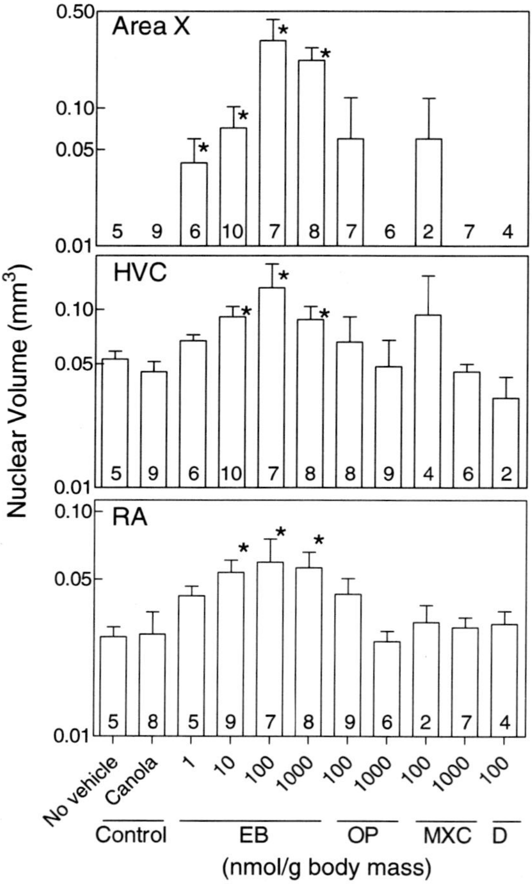

Fig. 1 Volumes (mm^3; mean ± SE; log plot; N noted on bar of each treatment group) of female Zebra finch song nuclei area X, HVC and nucleus robustus of the archistriatum (RA). Finches were treated as chicks with substances denoted under the bars: Canola was used as vehicle/control; estradiol benzoate (EB); octylphenol (OP); methoxychlor (MXC); dicofol (D). Asterisks denote groups that are significantly greater than canola-treated controls. Area X was not detectable in groups that lack error bars. (Redrawn from 34)

To assess male- and female-only effects of early estrogen treatment, reproductive performance was tested in trials in which only females were estrogen-treated (communally housed with canola-treated males), only males were estrogen-treated (communally housed with canola-treated females), as well as trials in which both genders were estrogen-treated and communally housed.

Inspection of Fig. 2 shows that egg production was impaired in females treated with 100 nmol/g EB, whether they were paired with canola-treated or EB-treated males. No other groups were affected (Fig. 2A). Percentage of dead embryos was also significantly elevated in the females-only 100 nmol/g EB dose-treatment. Percent eggs candled fertile showed an analogous male-associated impairment; fertility was decreased in both groups receiving 100 nmol/g EB, but not in other groups (Fig. 2B). These sex-specific impairments appeared additive in that while a lower dose of EB (10 nmol/g) did not impair either egg production or candled fertility, this dose did reduce number of eggs hatched (Fig. 2C); percentage of missing eggs was significantly elevated in the same group. A separate preliminary experiment showed that EB-treated females (100 nmol/g dose) have thinner eggshells, as measured by scanning electron microscopy.

Further trials evaluated the effects of treatment with methoxychlor, which has been shown to alter normal male sexual differentiation in precocial species, such as the chicken and Japanese quail when treated *in ovo* (14, 31). Results showed that Zebra finches orally dosed with MXC at doses calculated to be within the range that songbirds could be exposed to in field conditions following licensed application rates, as predicted by Kenaga's nomogram (17), had significantly decreased reproductive performance due to impaired egg and gamete quality. Though latency to lay (Fig. 3A) and number of eggs (Fig. 3B) did not vary between MXC-dosed and control groups, the 1000 nmol/g dose-group had fewer hatchlings relative to controls (Fig. 3F, $P<0.05$, Bonferroni). The number of eggs candled fertile was only borderline in significance at $P=0.06$ (Fig. 3C), so multiple means tests were not done. Lowered fertility, if present, may have been either due to male or female infertility. The incidence of reproductive failure due to missing eggs and egg breakage was greater in the high dose group, compared to controls (Fig. 3D, $P<0.01$, Bonferroni). No other differences were significant.

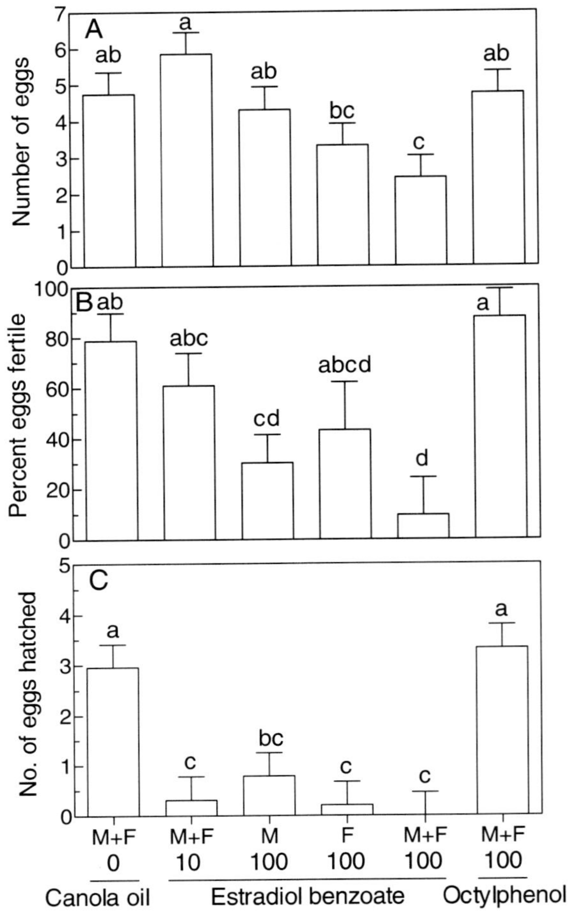

Fig. 2 A, mean number of eggs per nest box, B, mean percent eggs candled fertile and C, mean number of eggs hatched, of males (M) and/or females (F) housed communally and treated orally, once per day, with 10 or 100 nmol/g body mass on days of age 5 through 11 with estradiol benzoate or octylphenol in canola oil or canola oil alone, as a control. N = 10 per treatment. Error bars are pooled estimates. Means with unlike superscripts are significantly different ($P<0.05$). (Redrawn from 26)

Oral estrogen stimulated oviduct growth

The role of ovarian hormones in maintaining oviduct mass after ovariectomy was first demonstrated by Adler (4). The response of the oviduct to estrogen was later exploited in adult chickens, as an assay for placental hormone (22), and in chicks, as an assay for a variety of synthetic estrogens (27). The chick oviduct was also used to investigate

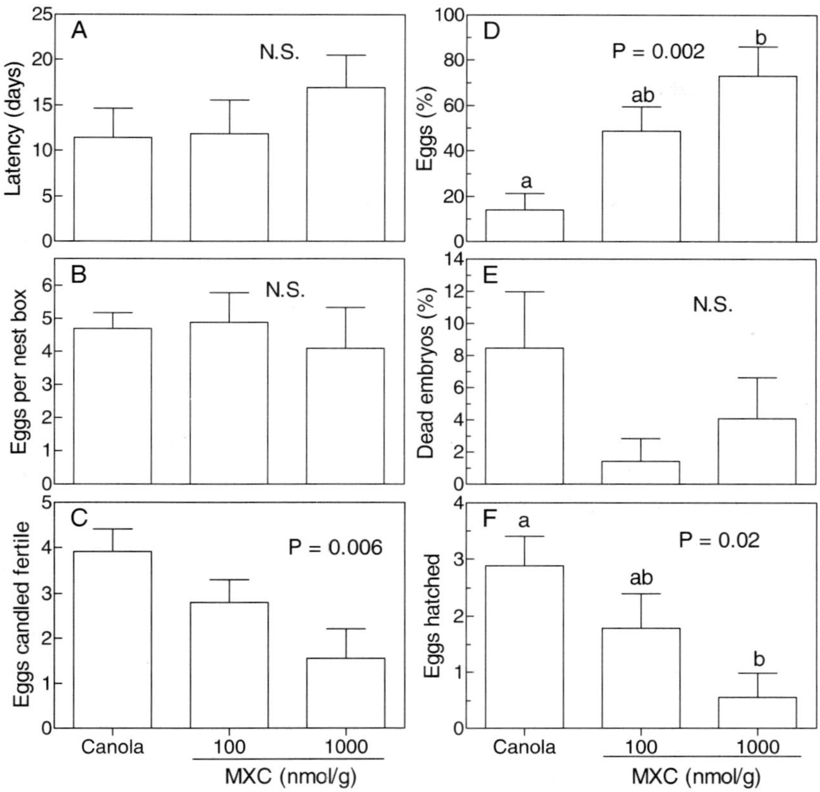

Fig. 3 Reproductive performance of Zebra finch pairs treated orally as chicks with canola oil (control) or 100 or 1000 nmol/g methoxychlor (MXC): A. Latency to egg laying after nest box presentation; B. Number of eggs per nest box; C. Number of eggs candled fertile in each nest box; D. Percent cracked/broken and missing eggs per nest box; E. Percent dead embryos; and F. Number of eggs hatched. Unlike superscripts denote significant differences at indicated levels using Bonferroni's multiple means test, when main effect is significant (ANOVA). (Redrawn from 17)

steroid hormone action (32). But estrogen-induced oviduct growth has been studied in only a few precocial, species (e.g., Japanese quail, guinea fowl [*Numida meleagris*], duck [*Anas platyrhyncus*], and a smaller number of altricial species (e.g., ring dove [*Streptopelia risoria*], English sparrow [*Passer domesticus*], European starling [*Sturnus vulgaris*] and Zebra finch) (see 25). Most of these studies have involved only adult birds; little is known about the estrogen sensitivity of oviduct growth in altricial songbird chicks, in which estrogen-sensitive sexual differentiation occurs after, as well as before hatch.

We asked whether an estrogen-induced oviduct growth response similar to that occurring in rats, mice and chickens may also occur in female Zebra finch chicks. If so, the Zebra finch could provide a useful model for characterizing the estrogenicity of putative environmental estrogens and their potential ability to alter normal development of brain and behavior and impair adult reproductive performance. We dosed birds, as above, with EB and putative environmental estrogens (methoxychlor, octylphenol, and genistein, an estrogenic isoflavone found in soy products), then compared oviduct growth on day of age 12, one day after the last oral gavage (25).

We found that oral dosing with EB produced dose-dependent increases in oviduct mass ranging from approximately 4.2 mg (mean) at the 0.1 nmol/g body weight dose to over 108 mg at the 1000 nmol/g dose (Fig. 4), a sixty-fold induction in mass over canola-treated controls (1.7 mg). The sensitivity of the assay is comparable to the rat uterotrophic assay. Methoxychlor failed to induced significant oviduct growth at the lower of two tested doses (100 nmol/g) but did at a ten-fold higher dose (1000 nmol/g), with average mass of 14.6 mg, an over eight-fold increase in mass over controls. The 100 nmol/g genistein dose induced significant oviduct growth to 7.8 mg, a nearly five-fold induction; genistein appeared about 0.01 as estrogenic as EB. The 1000 nmol/g octylphenol dose induced a slight but significant oviduct growth to 3.6 mg, an approximately two-fold induction (Fig. 4).

Two of six finches treated with 100 nmol/g EB and two of twelve finches treated with 1000 nmol/g EB developed right as well as left oviducts (24.1 mg ±2.6 [mean ± standard error]; 18.0 to 33.0 mg [range]), with no apparent relation to dose of EB. Right oviducts were not noticed in 68 females treated with canola oil, lower doses of EB or any of the environmental estrogens. Gonads of treated chicks appeared normal by gross inspection and were not examined histologically.

Histologic analysis of magnal portions of oviducts were categorized into four groups, A, B, C or D based on the extent of growth and differentiation (Fig. 5). In general, growth of oviduct correlated with differentiation and EB treatment, i.e., compare oviduct categories A (showing small, dense sheets of stromal cells with no mucosal tubular glands and a uniform columnar ciliated epithelium [Fig. 5A]; treated with canola and low doses of EB), B (showing rare to light numbers of tubular glands in the mucosal stroma [Fig. 5B]; treated with 1 to 100 nmol/g EB or 100 nmol/g genistein), C (showing pseudostratified columnar epithelium lining the mucosal surface and multi-focal to densely packed, tubular glands [Fig. 5C]; treated with intermediate doses of EB or 1000 nmol/g methoxychlor) and D (showing luminal pseudostratified, ciliated, columnar epithelium and a markedly thickened mucosa densely packed with tubular glands [Fig. 5D]; treated with either 10 or 100 nmol/g EB). Methoxychlor, however, stimulated relatively more differentiation than growth, i.e., the same oviductal mass was associated with greater differentiation in methoxychlor-treated birds than in EB-treated birds, suggesting a mechanism of action that differs from EB.

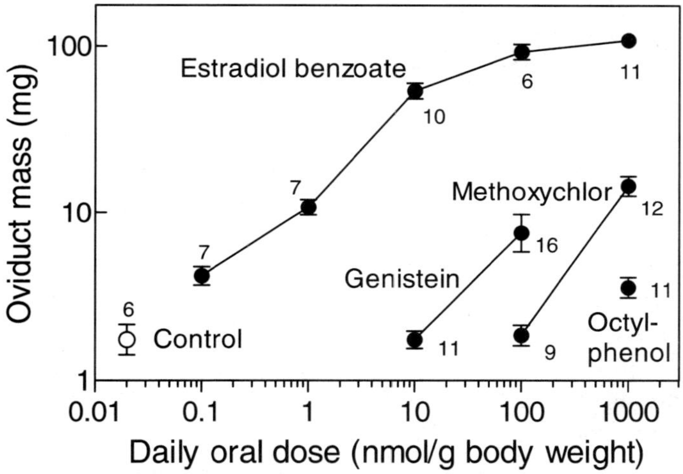

Fig. 4 Oviduct mass (log plot; mean ± SE) of 12-day old Zebra finch chicks dosed daily from days 5 through 11 with canola oil (Control), estradiol benzoate, genistein, methoxychlor or octylphenol at doses indicated. N of animals treated is indicated next to each data point. (Redrawn from 25)

Conclusions

These findings demonstrate, not unexpectedly, that oral exposure of Zebra finch chicks to estrogens act largely the same as estrogens delivered parenterally (after adjusting dose for route of administration) to masculinize female brains. This opens up the possibility that substances in the environment possessing estrogenic activity, e.g., xenoestrogens and phytoestrogens, may produce effects similar to estradiol or EB. However, at the doses tested, we found no adverse effects of either octylphenol or methoxychlor on the development of song control nuclei.

However, oral dosing of both genders with doses of EB (10 or 100 nmol/g body mass) sufficient to masculinize HVC and RA in females severely impaired reproduction in communally housed finches. Similar effects on reproduction were produced by 1000 nmol/g methoxychlor. Because methoxychlor also stimulated oviduct growth, as did EB, the impairment is consistent with an endocrine mechanism of action. Finally, as oral estrogens induced both oviduct growth and impaired reproductive performance, it may be possible to use chick oviduct growth to monitor exposure to environmental estrogens in the wild and predict impairments in adult reproduction.

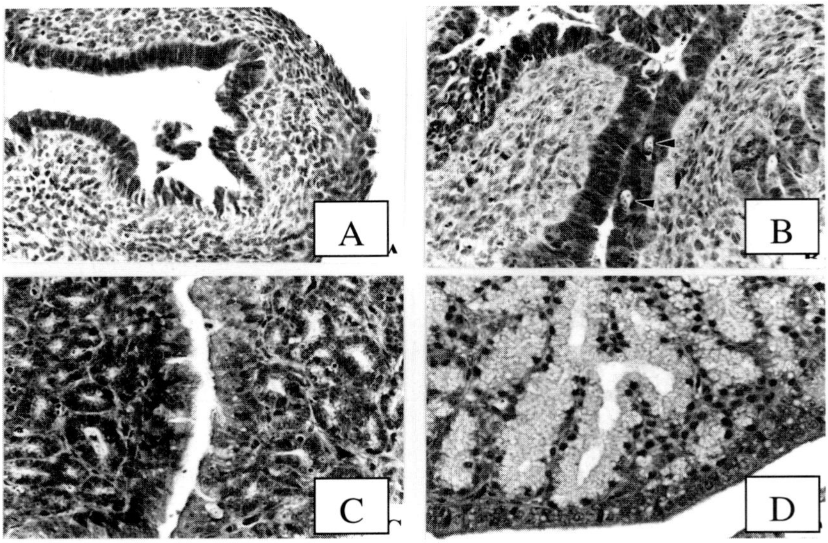

Fig. 5 Representative photomicrographs of oviducts treated with canola oil, EB or environmental estrogens and categorized according to extent of growth and differentiation: (A) least, (D) most (H&E-staining; X370 magnification). (From 25)

Acknowledgements

This work was supported by EPA grant R-855294-01 and NSF #0314510. I would like to thank Ms. Johanna Rochester for assistance in preparing this manuscript.

References

1 Adkins EK (1975) Hormonal basis of sexual differentiation in the Japanese quail. *Journal of Comparative Physiolology and Psychology* **89** 61-71

2 Adkins-Regan E & Ascenzi M (1987) Social and sexual behavior of male and female zebra finches treated with estradiol during the nestling period. *Animal Behaviour* **35** 1100-1112

3 Adkins-Regan E, Mansukhani V, Seiwert C & Thompson R (1994) Sexual differentiation of brain and behavior in the zebra finch: critical periods for effect of early treatment. *Journal of Neurobiology* **25** 865-877

4 Adler L (1911) Zur physiologie and pathologie der ovarialfunktion. *Archiv fuer Gynaekologie* (Berlin) **95** 349-424

5 Arnold AP (1997) Sexual differentiation of the zebra finch song system: Positive evidence, negative evidence, null hypothesis, and paradigm shift. *Journal of Neurobiology* **33** 572-584

6 Aste N, Panzica GC, Viglietti-Panzica C & Balthazart J (1991) Effects of *in ovo* estradiol benzoate treatments on sexual behavior and size of neurons in the sexually dimorphic medial preoptic nucleus of Japanese quail. *Brain Research Bulletin* **27** 713-720

7 Andersen HR, Andersson A-M, Arnold SF, Autrup H, Barfoed M, Beresford NA, Bjerregaard P, Christiansen LB, Gissel B, Hummel R, Jørgensen EB, Korsgaard B, Le Guevel R, Leffers H, McLachlan J, Møller A, Nielsen JB, Olea N, Oles-Karasko A, Pakdel F, Pederson KL, Perez P, Skakkeboek NE, Sonnenschein C, Soto AM, Sumpter J P, Thorpe SM & Grandjean P (1999) Comparison of short-term estrogenicity tests for identification of hormone-disrupting chemicals. *Environmental Health Perspectives* **107** (Suppl. 1) 89-108

8 Balthazart J, Tlemcani O & Ball GF (1996) Do sex differences in the brain explain sex differences in hormonal induction of reproductive behavior? What 25 years of research on Japanese quail tells us. *Hormones and Behavior* **30** 627-661

9 Bennie DT (1999) Review of the environmental occurrence of alkylphenols and alkylphenol ethoxylates. *Water Quality Research Journal of Canada* **34** 79-122

10 Bulger WH, Muccitelli RM & Kupfer D (1978) Studies on the in vivo and in vitro estrogenic activities of methoxychlor and its metabolites. Role of hepatic mono-oxygenase in methoxychlor activation. *Biochemical Pharmacology* **27** 2417-2423

11 Colborn T (1995) Environmental estrogens: health implications for humans and wildlife. *Environmental Health Perspectives* **103** (Suppl. 7) 135-136

12 Dawson A (2000) Mechanisms of endocrine disruption with particular reference to occurrence in avian wildlife: A review. *Ecotoxicology* **9** 59-69

13 Desbrow C, Routledge EJ, Brighty GC, Sumpter JP & Waldock M (1998) Identification of estrogenic chemicals in STW effluent. Chemical fractionation and *in vitro* biological screening. *Environmental Science and Technology* **32** 1549-1558

14 Eroschenko VP, Amstaslavsky SYY, Schwabel H & Ingermall RL (2002) Altered behaviors in male mice, male quail, and salamander larvae following early exposures to the estrogenic pesticide methoxychlor. *Neurotoxicology and Teratology* **24** 29-36

15 Flavin DF (1982) The effects of soybean trypsin inhibitors on the pancreas of animals and man: A review. *Veterinary and Human Toxicology* **24** 25-28

16 Fry D M (1995) Reproductive effects in birds exposed to pesticides and industrial chemicals. *Environmental Health Perspectives* **103** (Suppl. 7) 165-171

17 Gee JM, Craig-Veit CB & Millam JR. Post-hatch methoxychlor exposure adversely affects reproduction of adult zebra finches, *Taeniopygia guttata. Bulletin Environmental Contamination and Toxicology* **73** 607-612

18 Gray LE Jr, Ferrell JM & Ostby JS (1985) Alteration of behavioral sex differentiation by exposure to estrogenic compounds during a critical neonatal period: effects of zearalenone, methoxychlor, and estradiol in hamsters. *Toxicology and Applied Pharmacology* **80** 127-136

19 Gurney ME & Konishi M (1980) Hormone-induced sexual differentiation of brain and behavior in zebra finches. *Science* **208** 1380-1382

20 Holloway CC & Clayton DF (2001) Estrogen synthesis in the male brain triggers development of the avian song control pathway *in vitro. Nature Neuroscience* **4** 170-175

21 Jobling S & Sumpter JP (1993) Detergent components in sewage effluent are weakly oestrogenic to fish: an *in vitro* study rainbow trout (*Oncorhynchus mykiss*) hepatocytes. *Aquatic Toxicology* **27** 361-372.

22 Juhn M & Gustavson RG (1930) The production of female genital subsidiary characters and plumage sex characters by injection of human placental hormone in fowls. *Journal of Experimental Biology* **56** 31-60

23 Leopold AS, Erwin M & Browning B (1976) Phytoestrogens: adverse effects on reproduction in California quail. *Science* **191** 98-100

24 Mansukhani V, Adkins-Regan E & Yang S (1996) Sexual partner preference in female zebra finches: The role of early hormones and social environment. *Hormones and Behavior* **30** 506-513

25 Millam JR, Craig-Veit CB, Batchelder ME, Viant MR, Herbeck TM & Woods LW (2002) An avian bioassay for environmental estrogens: the growth response of zebra finch (*Taeniopygia guttata*) chick oviduct to oral estrogens. *Environmental Toxicology and Chemistry* **21** 2663-68

26 Millam JR, Craig-Veit CB, Quaglino AE, Erichsen AL, Famula TR & Fry DM (2001) Post-hatch oral estrogen exposure impairs adult reproductive performance of zebra finch in a sex-specific manner. *Hormones and Behavior* **40** 542-549

27 Munro SS & Kosin IL (1940) The relative potency of several estrogenic compounds tested on baby chicks of both sexes. *Endocrinology* **27** 687-692

28 Naylor CG (1992) Environmental fate of alkylphenol ethoxylate. *Soap Cosmetics Chemical Specialties* **68** 27-72

29 Nimrod AC & Benson WH (1996) Environmental estrogenic effects of alkylphenol ethoxylates. *Critical Reviews in Toxicology* **26** 335-364

30 Nottebohm F & Arnold AP (1976) Sexual dimorphism in vocal control areas of the songbird brain. *Science* **194** 211-213

31 Ottinger MA, Abdelnabi MA, Henry P, McGary S, Thompson N, Wu JM (2001) Neuroendocrine and behavioral implications of endocrine disrupting chemicals in quail. *Hormones and Behavior* **40** 234-47

32 Palmiter RD, Mulvihill ER, McKnight GS & Senear AW (1978) Regulation of gene expression in the chick oviduct by steroid hormones. *Cold Spring Harbor Symposium on Quantitative Biology* **42 (Pt 2)** 639-47

33 Purdom CE, Hardiman PA, Bye VJ, Eno NC, Tyler CR & Sumpter JP (1994) Estrogenic effects of effluents from sewage treatment works. *Chemistry and Ecology* **8** 275-285

34 Quaglino AE, Craig-Veit CB, Viant MR, Erichsen AL, Fry DM & Millam JR (2002) Oral estrogen masculinizes female zebra finch song system. *Hormones and Behavior* **41** 236-241

35 Roy D, Palangat M, Chen C, Thomas RD, Colerangle J, Atkinson A & Yan Z (1997) Biochemical and molecular changes at the cellular level in response to exposure to environmental estrogen-like chemicals. *Journal of Toxicology and Environmental Health* **50** 1-29

36 Schlinger BA & Arnold AP (1993) Estrogen synthesis *in vivo* in the adult zebra finch: Additional evidence that circulating estrogens can originate in brain. *Endocrinology* **133** 2610-2616

37 Servos MR (1999) Review of the aquatic toxicology, estrogenic responses and bioaccumulation of alkylphenols and alkylphenol polyethoxylates. *Water Quality Research Journal of Canada* **34** 123-177

38 Shelby MD, Newbold RR, Tully DB, Chae K & Davis VL (1997) Assessing environmental chemicals for oestrogenicity using a combination of *in vitro* and *in vivo* assays. *Environmental Health Perspectives* **104** 1296-1300

39 Staples CA, Weeks J, Hall JF & Naylor CG (1998) Evaluation of aquatic toxicity and bioaccumulation of C8- and C9- alkylphenol ethoxylates. *Environmental Toxicology and Chemistry* **17** 2470-2480

40 Vaillant S, Guémené D, Dorizzi M, Pieau C, Richard-Mercier N & Brillard J-P (2003) Degree of sex reversal as related to plasma steroid levels in genetic female chickens (*Gallus domesticus*) treated with fadrozole. *Molecular Reproduction and Development* **65** 420-428

41 Wade J &Arnold AP (2004) Sexual differentiation of the zebra finch song system. *Annals of the New York Academy of Science* **1016** 540-559

42 Wade J, Gong A & Arnold AP (1997) Effects of embryonic estrogen on differentiation of the gonads and secondary sexual characteristics of male zebra finches. *Journal of Experimental Zoology* **278** 405-411

43 White R, Jobling S, Hoare SA, Sumpter JP & Parker MG (1994) Environmentally persistent alkylphenolic compounds are estrogenic. *Endocrinology* **135** 175-182

44 Whitsett JM, Irvin EW, Edens FW & Thaxton JP (1977) Demasculinization of male Japanese quail by prenatal estrogen treatment. *Hormones and Behavior* **8** 254-263

45 Williams TD & Martyniuk CJ (2000) Tissue mass dynamics during egg-production in female zebra finches *Taeniopygia guttata*: Dietary and hormonal manipulations. *Journal of Avian Biology* **31** 87-95

46 Wynne-Edwards KE (2001) Evolutionary biology of plant defenses against herbivory and their predictive implications for endocrine disruptor susceptibility in vertebrates. Environmental Health Perspectives **109** 443-448

Functional Avian Endocrinology
A Dawson & P J Sharp (Eds)

Impact of endocrine disrupting chemicals on sexual differentiation in Japanese quail

K Halldin*

Department of Environmental Toxicology, Uppsala University,
Norbyvägen 18A, SE-752 36 Uppsala, Sweden

* Present address; Institute of Environmental Medicine, Karolinska Instititutet,
P.O. Box 210, SE-171 77 Stockholm, Sweden

Introduction

Avian wildlife is present in terrestrial, marine and freshwater habitats and birds attract the attention not only of ornithologists but also of the general public. Population declines in various avian species have historically functioned as a warning signal of adverse effects of environmental pollutants in wildlife. However, in endocrine disruption research, studies on mammals and fish are presently dominating. A search on selected endocrine disruptors using PubMed clearly shows that the numbers of studies using birds are scarce (Table 1). Whether this is due to priorities from the regulatory authorities and funding sources, or reflects the general interest among researchers for studying certain animal classes is left unsaid. There are however laboratories that have used birds as model organisms for investigating disturbances of reproductive function caused by exposure to endocrine disrupters. The Japanese quail (*Coturnix japonica*) has, for us and others, been the species of choice for these studies. In this chapter, data from studies carried out in our laboratory are reviewed with focus on reproductive disturbances in adult males caused by embryonic exposure to exogenous substances.

Table 1. Number of publications on selected endocrine disruptors published between 1994 and 2004, and indexed in PubMed. The search phrase used was; Ethinylestradiol OR bisphenol OR nonylphenol OR methoxychlor AND (birds, fish or mammals). The search was conducted on the 29th of October 2004.

class	Number of indexed publications
Birds	28
Fish	275
Mammals	1048

For studies of exogenous chemicals that disturb endocrine function, the Japanese quail offers several advantages, some of which are shared with other avian species. One advantage is that exposure *in ovo* can be carried out under well-controlled conditions, and at different stages of embryonic development. The egg offers an enclosed environment for the embryo and controlled doses of chemicals are easily administered *via* egg injections. Other advantages to using quail include the relatively small sizes of the adults and eggs. Also, in spite of its small size, the embryo is large enough for experimental studies. Domestication has lead to high egg production and an availability of eggs throughout the year. The quail becomes sexually mature at approximately seven to eight weeks, allowing studies of reproductive function of the adult following embryonic exposure within a relatively short time frame. The process of sexual differentiation, taking place during the embryonic period in galliform species such as the Japanese quail, is sensitive to disturbances in the hormonal milieu and a transient exposure during this period often results in irreversible changes of the endocrine system. The Japanese quail has for long been used as a model species for investigating the hormonal basis of sexual differentiation and the resulting dimorphism in physiology and behavior. Since our studies on chemicals interfering with this process are based on the knowledge collected from this research a short description is presented below.

Sexual differentiation in Japanese quail

During the embryonic stage, the brain is exposed to hormones organizing the neural tissue to react in a specific way to hormones later in life. Studies in various animal species have shown that hormonal manipulation during the organizational period of the brain can lead to changes in adult sexual behavior (3, 10, 17). Sexual differentiation of brain and behavior has been studied in several avian species. Differentiation of singing behavior in oscine passerines such as the zebra finch has been extensively studied, and one of the most studied gallinaceous species with regard to differentiation of the brain and the reproductive behavior is the Japanese quail.

The relation between embryonic hormones and adult behavioral phenotype is obvious in the Japanese quail. Normally the Japanese quail male is sexually vigorous and its behavior is easily studied in the laboratory. The sexual behavior of Japanese quail has been thoroughly described by others (2, 31), but a short description is provided here (Fig. 1). When encountering a female, a sexually mature male displays a defined sequence of behaviors. Initially, the male grabs the feathers on the back of the female's head (neck grab) and after that tries to mount the female. Succeeding with the mount attempt leads to a completed mount, when the male has both his legs on the back of the female while grabbing the neck. The male thereafter bends his tail under the tail of the female and the cloacas come into contact (cloacal contact movement). During cloacal contact movement, copulation may occur. In addition to the copulatory sequence, male-specific behaviors include strutting, when the male stretches his body while walking on stiff legs, and the male vocalization (crowing). Crowing and strutting are normally not observed in females. However, they can be induced in adult females by treatment with testosterone and are therefore not strictly organized (8).

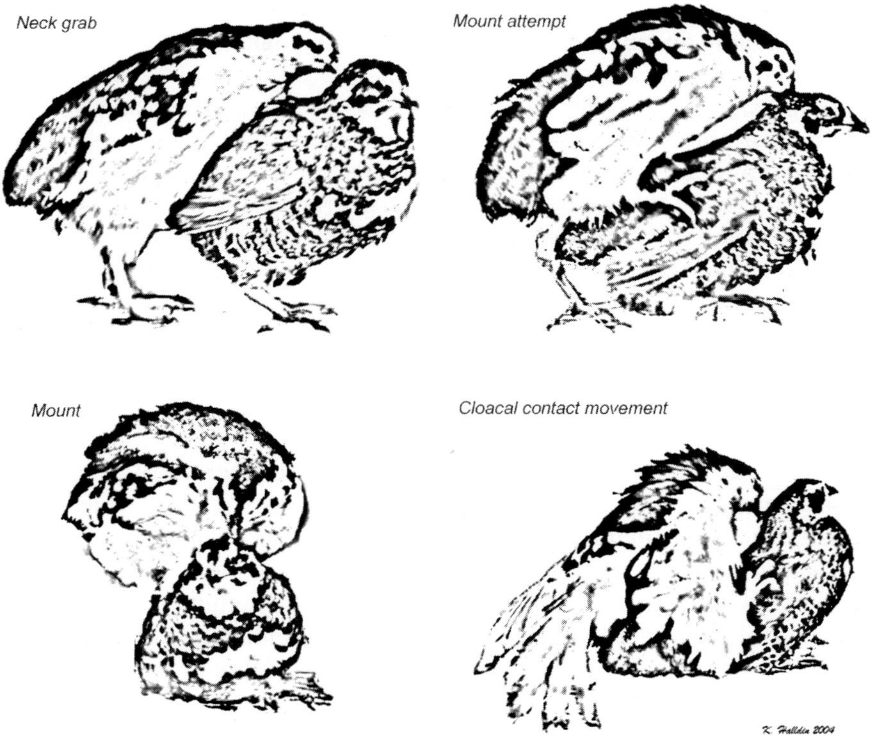

Fig. 1 Illustration of a typical sequence of copulatory behaviors in Japanese quail. Initially the male grabs the feathers on the back of the female's head (neck grab) and then tries to mount the female (mount attempt). The mount attempt is followed by a completed mount. The male thereafter bends his tail under the female and the cloacas come in contact (cloacal contact movement).

Through hormonal manipulations it has been shown that in Japanese quail, estrogens produced during embryonic development demasculinize the female brain (4, 7). This embryonic exposure to estrogens organizes the brain in a female manner and the female will not respond in the same way as a male to hormones in adulthood. Demasculinization can also be produced in the genetic male by exposing the brain to estrogens during differentiation. Demasculinized males will not display male sexual behavior as adults regardless of the hormonal milieu (5, 47).

It is clear that behavioral sex differences reflect neural sex differentiation, but the neurological differences between male and female quail are not yet fully understood. Comprehensive reviews on sexual differentiation in quail are available (8, 9, 40) and it is therefore not described in this chapter. However, an interesting finding on sexual dimorphism in quail that has recently also been suggested as a neural marker for embryonic exposure to xeno-estrogens (42) is the vasotocin-innervation in areas known to be important for the expression of sexual behavior. The POM, the bed nucleus of the stria terminalis, pars medialis (BSTm), and the lateral septum (SL) are innervated by a denser network of vasotocin-immunoreactive (VT-ir) fibers in male quail vis-à-vis females (32, 48, 49). The VT-ir structures of these areas are organized by the embryonic hormonal milieu, and male quail embryos treated with estradiol benzoate on day 9 lost the VT-ir structures of the POM, BSTm, and SL, simultaneously losing the capacity to display copulatory behavior as adults (41).

The sexual apparatus of quail is differentiated early in embryonic life. Before differentiation, the structures of males and females are indistinguishable and both sexes have bilateral, bipotential gonadal anlagen consisting of a medulla and a cortex. In males, the medulla, with its primary sex cords, develops into seminiferous tubules, with sertoli cells and spermatocytes. The differentiation of the gonads and ducts are sensitive to hormones. Estrogen exposure of male embryos results in formation of an ovotestis with a thickened cortex and oocyte-like germ cells (11, 43, 45). The ovotestis does however not seem to persist until adulthood in Japanese quail but asymmetry in testicular weight at adulthood has been observed in males exposed to ethinylestradiol *in ovo* (25).

The androgen-dependent cloacal gland, posterior to the cloaca is well developed in sexually mature males but only rudimentary in females. The cloacal gland produces foam that is transferred to the female at copulation. The foam has been suggested to increase the fertilization capacity of the sperm. Growth of the cloacal gland can be induced in females by treatment with testosterone but the female gland never reaches the same size as in males (1). The cloacal gland becomes demasculinized by embryonic exposure to estrogen and is reduced in size in adult males after *in ovo* treatment with estrogen (3, 25). Sensitivity of the cloacal gland to androgens is therefore organized during differentiation.

Effects of environmental pollutants on sexual differentiation in quail

There are few studies on the effects of environmental estrogens on brain development and reproductive behavior in birds. It is well established however from studies using synthetic estrogens that estrogen-like compounds can have dramatic effects on the differentiation of brain and sexual behavior. In Japanese quail, environmental contaminants suspected to have estrogen-like effects have been administered during critical periods of differentiation. Some of these studies have confirmed the ability of pollutants to interfere in an estrogen-like manner with sexual differentiation of brain and behavior in birds. Below, selected results from studies carried out in our laboratory are presented and the results are also summarized in Table 2.

Ethinylestradiol is used as a human contraceptive and consequently has also been identified as an environmental pollutant released *via* effluent water from sewage treatment works (16, 18, 33). In Halldin *et al*. (25) ethinylestradiol was injected into the yolk of embryonated Japanese quail eggs. After the male birds had been raised to sexual maturity, their sexual behavior, plasma testosterone concentrations and testis morphology were examined. Depressed male sexual behavior was the most sensitive of the end points studied and the lowest dose resulting in a significant effect was 6 ng/g egg.

The plastic monomer Bisphenol A is extensively used as a component in polycarbonate plastics and also as a component of epoxy resins. Conjugated bisphenol A has been detected in fish exposed to sewage treatment effluent (33). Furthermore, it has been found to contaminate food and to leak from plastic dental sealants (12, 36). Bisphenol A has been shown to have estrogenic properties in several *in vivo* studies on rodents (6, 21, 28, 29), and following embryonic exposure (200 μg/g egg), we examined sexually mature male quail for reproductive behavior and testis morphology. Bisphenol A at this dose did not cause any significant estrogen-like effects on the variables studied (26).

One of the most well known environmental estrogens is o,p´-DDT. DDT (dichloro diphenyl trichloroethane) was first synthesized in 1874 and its insecticidal properties were discovered in 1939. DDT has been extensively used as an insecticide worldwide and although banned in most western countries in the 1970s, DDT is still used for mosquito control in Asia, Africa and South America. Commercial DDT consists mainly of the p,p´- isomer but 15–20% is the o,p´-isomer. Several mammalian *in vivo* studies have shown estrogenic effects of o,p´-DDT (13, 20, 39) and Bryan et al. (14) reported decreased sexual behavior in male quail after injection of 6.25 mg /egg of this compound on day 1 of incubation. In Halldin *et al.* (27) quail embryos were exposed *in ovo* to o,p´-DDT at a dose of 150 µg/g egg (2 mg/egg). In adult males the number of positive tests for mount attempt and the number of positive tests for cloacal contact movement were significantly reduced following the treatment. Also the time until first sexual interaction in males being sexually active at each test occasion was significantly longer in the o,p´-DDT -treated birds vis-à-vis the controls.

Plasma testosterone concentration and cloacal gland area were significantly reduced in the treated males. Neither body weight, gonado-somatic index, testis weight asymmetry, nor tubule diameter were significantly affected by treatment with o,p´-DDT.

Methoxychlor is an organochlorine insecticide with relatively low toxicity used as a substitute for DDT. The reported LD50 for methoxychlor is greater than 2.000 mg/kg in the mallard duck, sharp-tailed grouse, and California quail (30). Methoxychlor has little or no affinity for the estrogen receptor in vitro (15, 38) but has in several studies been shown to have estrogenic activity *in vivo* (22, 23, 34, 35, 46). Methoxychlor has also been shown to affect sexual behavior in quail after *in ovo* exposure. Eroshenko et al. (19) found a significantly prolonged latency to mount after injecting 5 mg methoxychlor/egg, and similar effects were reported by Ottinger et al. (37). In Halldin et al. (24) methoxychlor was administered at 150 µg/g egg (2 mg/egg), the same dose as previously used for o,p´-DDT. No significant effects on male sexual behavior were seen at this dose. However, when methoxychlor was co-administered with a technical mixture of polychlorinated biphenyls (PCBs), which is known to induce biotransformation enzymes, a significant reduction of sexual behavior was found. The conclusion of that study was that the induction of metabolizing enzymes with PCBs probably increased the demethylation of methoxychlor into more estrogenic metabolites. These results emphasize the fact that exposure to mixtures of pollutants with different mechanisms of action, which normally occurs in free living birds, is important to consider when assessing the risk for developmental disturbances caused by maternal transfer of contaminants.

Table 2. Significant ($p<0.05$) changes in male reproductive variables in adult quail after *in ovo* treatment with Ethinylestradiol (EE_2), *o,p*´-DDT, or a combination of Methoxychlor (MXC) and Clophen A50 (CA50) as compared to their respective control group. (▼ = decreased, ▲ = increased, × = no effects, - = not measured).

Compound	Dose µg/g egg	Sexual behavior[1]	GSI[2]	Testis assymmetry[3]	T conc.[4]	CGA[5]	Ref.
EE_2	0.002	×	×	×	×	-	(24)
	0.006	▼	×	▲	×	-	(24)
o,p´-DDT	150	▼	×	×	▼	▼	(26)
MXC + CA50	150+40	▼	×	×	×	×	(23)

[1] Behavioral activity was scored as the number of tests, given 1 per day for 5 consecutive days, during which at least one mount attempt (MA) occurred.
[2] Gonado-somatic index
[3] Weight of left testis/weight of right testis
[4] Plasma testosterone concentration
[5] Cloacal gland area

Conclusions

The effects on male sexual behavior after exposure to environmental estrogens are consistent with results from previous studies on estradiol. Injection of estradiol before day 12 of incubation irreversibly demasculinized male embryos and the adult males lacked male-typical sexual behavior in response to testosterone (3, 44). From the perspective of hazard assessment of environmental pollutants, our studies and those carried out in other labs have shown that birds offer good models for studying endocrine disrupting chemicals with estrogenic effects. The crucial role of estrogens during sexual differentiation of the Japanese quail and the diversity of effects in the adult animal offers several endpoints that can be included in test systems. Future research should hopefully provide us with insight into the developmental changes in the brain portending adverse effects on reproductive function in adult birds. If the organizational events leading to the differentiated adult behavior can be identified, it may in a longer perspective be possible to test the endocrine disrupting potential of chemicals in an avian *in vivo* system that only requires use of embryos.

In birds, maternal transfer of contaminants to the eggs can be high and especially lipophilic compounds are often present in high concentrations in eggs of wild-living birds. Hence, the growing embryo is exposed from early development until after hatching when the remaining yolk has been resorbed. Effects on sexual behavior, induced during the organizational period, are subtle and such effects may remain undetected in wild-living avian populations. Experimental studies on endocrine effects of chemicals are therefore important since they may help us to foresee potential effects on avian wildlife.

Acknowledgements

Financial support was provided by the Swedish Research Council for Environment, Agricultural Sciences and Spatial Planning and by the Swedish Environmental Protection Agency (ReproSafe).

References

1 Adkins EK & Adler NT (1972) Hormonal control of behavior in the Japanese quail. *Journal of Comparative and Physiological Psychology* **81** 27-36

2 Adkins EK (1975) Hormonal basis of sexual differentiation in the Japanese quail. *Journal of Comparative and Physiological Psychology* **89** 61-71

3 Adkins EK (1976) Embryonic exposure to an antiestrogen masculinizes behavior of female quail. *Physiology and Behaviour* **17** 357-359

4 Adkins EK (1979) Effect of embryonic treatment with estradiol or testosterone on sexual differentiation of the quail brain. *Neuroendocrinology* **29** 178-185

5 Ashby J & Tinwell H (1998) Uterotrophic activity of bisphenol A in the immature rat. *Environmental Health Perspectives* **106** 719-720

6 Balthazart J, De Clerck A & Foidart A (1992) Behavioral demasculinization of female quail is induced by estrogens: studies with the new aromatase inhibitor, R76713. *Hormones and Behaviour* **26** 179-203

7 Balthazart J & Foidart A (1993) *Neural bases of behavioral sex differences in quail.* Kluwer Academic Publishers

8 Balthazart J, Tlemcani O & Ball GF (1996) Do sex differences in the brain explain sex differences in the hormonal induction of reproductive behavior? What 25 years of research on the Japanese quail tells us. *Hormones and Behaviour* **30** 627-661

9 Beach FA & Inman NG (1965) Effects of castration and androgen replacement on mating in male quail. *Proceedings of the National Academy of Sciences USA* **54** 1426-1431

10 Berg C, Halldin K, Fridolfsson AK, Brandt I & Brunström B (1999) The avian egg as a test system for endocrine disrupters: effects of diethylstilbestrol and ethynylestradiol on sex organ development. *Science of the Total Environment* **233** 57-66

11 Brotons JA, Olea-Serrano MF, Villalobos M, Pedraza V & Olea N (1995) Xenoestrogens released from lacquer coatings in food cans. *Environmental Health Perspectives* **103** 608-612

12 Brown NM & Lamartiniere CA (1995) Xenoestrogens alter mammary gland differentiation and cell proliferation in the rat. *Environmental Health Perspectives* **103** 708-713

13 Bryan TE, Gildersleeve RP & Wiard RP (1989) Exposure of Japanese quail embryos to o,p'-DDT has long-term effects on reproductive behaviors, hematology and feather morphology. *Teratology* **39** 525-536

14 Bulger WH, Muccitelli RM & Kupfer D (1978) Interactions of methoxychlor, methoxychlor base-soluble contaminant, and 2,2-bis(p-hydroxyphenyl)-1,1,1-trichloroethane with rat uterine estrogen receptor. *Journal of Toxicology and Environmental Health* **4** 881-893

15 Cargouet M, Perdiz D, Mouatassim-Souali A, Tamisier-Karolak S & Levi Y (2004) Assessment of river contamination by estrogenic compounds in Paris area (France). *Science of the Total Environment* **324** 55-66

16 Clemens LG, Hiroi M & Gorski RA (1969) Induction and facilitation of female mating behavior in rats treated neonatally with low doses of testosterone propionate. *Endocrinology* **84** 1430-1438

17 Desbrow C, Routledge EJ, Brighty GC, Sumpter JP & Waldock M (1998) Identification of estrogenic chemicals in STW effluent. 1. Fractionation and in vitro ciological screening. *Environental Science and Technology* **32** 1549-1558

18 Eroschenko VP, Amstislavsky SY, Schwabel H & Ingermann RL (2002) Altered behaviors in male mice, male quail, and salamander larvae following early exposures to the estrogenic pesticide methoxychlor. *Neurotoxicology Teratology* **24** 29-36

19 Faber KA, Basham K & Hughes CL, Jr (1991) The effect of neonatal exposure to DES and o,p'-DDT on pituitary responsiveness to GnRH in adult castrated rats. *Reproductive Toxicoloyg* **5** 363-369

20 Farabollini F, Porrini S & Dessi-Fulgherit F (1999) Perinatal exposure to the estrogenic pollutant bisphenol A affects behavior in male and female rats. *Pharmacology Biochemistry & Behaviour* **64** 687-694

21 Golub MS, Hogrefe CE, Germann SL, Lasley BL, Natarajan K & Tarantal AF (2003) Effects of exogenous estrogenic agents on pubertal growth and reproductive system maturation in female rhesus monkeys. *Toxicological Sciences* **74** 103-113

22 Gray LE, Jr., Ostby JS, Ferrell JM, Sigmon ER & Goldman JM (1988) Methoxychlor induces estrogen-like alterations of behavior and the reproductive tract in the female rat and hamster: effects on sex behavior, running wheel activity, and uterine morphology. *Toxicology and Applied Pharmacology* **96** 525-540

23 Halldin K, Axelsson J & Brunström B. Embryonic co-exposure to methoxychlor and Clophen A50 alters sexual behavior in adult male quail. *Archives of toxicology*. In Press

24 Halldin K, Berg C, Brandt I & Brunström B (1999) Sexual behavior in Japanese quail as a test end point for endocrine disruption: effects of in ovo exposure to ethinylestradiol and diethylstilbestrol. *Environmental Health Perspectives* **107** 861-866

25 Halldin K, Berg C, Bergman A, Brandt I & Brunström B (2001) Distribution of bisphenol A and tetrabromobisphenol A in quail eggs, embryos and laying birds and studies on reproduction variables in adults following in ovo exposure. *Archives of Toxicology* **75** 597-603

26 Halldin K, Holm L, Ridderstråle Y & Brunström B (2003) Reproductive impairment in Japanese quail (Coturnix japonica) after in ovo exposure to o,p`-DDT. *Archives of Toxicology* **77** 116-122

27 Honma S, Suzuki A, Buchanan DL, Katsu Y, Watanabe H & Iguchi T (2002) Low dose effect of in utero exposure to bisphenol A and diethylstilbestrol on female mouse reproduction. *Reproductive Toxicology* **16** 117-122

28 Howdeshell KL, Hotchkiss AK, Thayer KA, Vandenbergh JG & vom Saal FS (1999) Exposure to bisphenol A advances puberty. *Nature* **401** 763-764

29 Hudson R, Tucker R & Haegele (1984) Handbook of Acute Toxicity of Pesticides to Wildlife, Resource Publication 153. U.S. Department of Interior, Fish and Wildlife Service, Washington, DC, 6-54

30 Hutchison RE (1978) Hormonal differentiation of sexual behavior in Japanese quail. *Hormones and Behaviour* **11** 363-387

31 Jurkevich A, Barth SW, Aste N, Panzica G & Grossmann R (1996) Intracerebral sex differences in the vasotocin system in birds: Possible implication in behavioral and autonomic functions. *Hormones and Behavior* **30** 673-681

32 Larsson DGJ, Adolfsson-Erici M & Parkkonen J et al (1999) Ethinyloestradiol-an undesired fish contraceptive? *Aquatic Toxicology* **45** 91-97

33 Laws SC, Carey SA, Ferrell JM, Bodman GJ & Cooper RL (2000) Estrogenic activity of octylphenol, nonylphenol, bisphenol A and methoxychlor in rats. *Toxicological Sciences* **54** 154-167

34 Millam JR, Craig-Veit CB, Batchelder ME, Viant MR, Herbeck TM & Woods LW (2002) An avian bioassay for environmental estrogens: the growth response of zebra finch (Taeniopygia guttata) chick oviduct to oral estrogens. *Environmental Toxicology and Chemistry* **21** 2663-2668

35 Olea N, Pulgar R & Perez P *et al* (1996) Estrogenicity of resin-based composites and sealants used in dentistry. *Environmental Health Perspectives* **104** 298-305

36 Ottinger MA, Abdelnabi MA, Henry P, McGary S, Thompson N & Wu JM (2001) Neuroendocrine and behavioral implications of endocrine disrupting chemicals in quail. *Hormones and Behaviour* **40** 234-247

37 Ousterhout J, Struck RF & Nelson JA (1981) Estrogenic activities on methoxychlor metabolites. *Biochemical Pharmacology* **30** 2869-2871

38 Palanza P, Parmigiani S, Liu H & vom Saal FS (1999) Prenatal exposure to low doses of the estrogenic chemicals diethylstilbestrol and o,p'-DDT alters aggressive behavior of male and female house mice. *Pharmacology Biochemistry Behavior* **64** 665-672

39 Panzica GC, Viglietti Panzica C & Balthazart J (1996) The sexually dimorphic medial preoptic nucleus of quail: a key brain area mediating steroid action on male sexual behavior. *Frontiers in Neuroendocrinology* **17** 51-125

40 Panzica GC, Castagna C, Viglietti-Panzica C, Russo C, Tlemcani O & Balthazart J (1998) Organisational effects of estrogens on brain vasotocin and sexual behavior in quail. *Journal of Neurobiology* **37** 684-699

41 Panzica GC, Bakthazart J, Pessatti M & Viglietti-Panzica C (2002) The parvocellular vasotocin system of Japanese quail: a developmental and adult model for the study of influences of gonadal hormones on sexually

differentiated and behaviorally relevant neural circuits. *Environmental Health Perspectives* **110 Suppl 3** 423-428

42 Scheib D (1983) Effects and role of estrogens in avian gonadal differentiation. *Differentiation* **23 (Suppl.)** S87-S92

43 Schumacher M, Hendrick JC & Balthazart J (1989) Sexual differentiation in quail: Critical period and hormonal specificity. *Hormones and Behavior* **23** 130-149.

44 Teng CS & Teng CT (1979) Prenatal effect of the estrogenic hormone on embryonic genital organ differentiation. In *Ontogeny of Receptors and Reproductive Hormone Action* 421-440. Eds TH Hamilton, JH Clark & WA Sadler. Raven Press: New York

45 Thorpe KL, Hutchinson TH, Hetheridge MJ, Scholze M, Sumpter JP & Tyler CR (2001) Assessing the biological potency of binary mixtures of environmental estrogens using vitellogenin induction in juvenile rainbow trout (*Oncorhynchus mykiss*). *Environmental Science* **35** 2476-2481

46 Whitsett JM, Irvin EW, Edens FW, Thaxton JP (1977) Demasculinization of male Japanese quail by prenatal estrogen treatment. *Hormones and Behavior* **8** 254-263

47 Viglietti Panzica C, Anselmetti G, Balthazart J, Aste N & Panzica G (1992) Vasotocinergic innervation of the septal region in the japanese quail: sexual differences and the influence of testosterone. *Cell and Tissue Research* **267** 261-265

48 Viglietti Panzica C, Aste N, Balthazart J & Panzica GC (1994) Vasotocinergic innervation of sexually dimorphic medial preoptic nucleus of the male Japanese quail: influence of testosterone. *Brain Research* **657** 171-184

Part eight

Ovarian growth and differentiation factors

Long-billed Dowitcher
Pierre Deviche

Functional Avian Endocrinology
A Dawson & P J Sharp (Eds)

Growth differentiation factor 9: An oocyte factor regulating ovarian follicle development

P A Johnson, M J Dickens, T R Kent and J R Giles

Dept of Animal Science, Cornell University, Ithaca, NY, USA

Introduction

The avian ovarian follicle consists of the germ cell or oocyte, and the surrounding somatic cell layers, the granulosa and theca cell layers. The oocyte in avian species is remarkable for its size and accumulated yolk. As follicle development proceeds, steroidogenic and maturational changes in the follicle are coordinated with yolk uptake by the oocyte. Until recently, the oocyte has not received very much attention from reproductive biologists because interest has focused on the growth and steroid producing capacity of the surrounding granulosa and theca cell layers. In avian species the growth of the oocyte is profound because of the accumulation of increasing amounts of yellow yolk.

Recent work has indicated that the oocyte is far from passive and signals emanating from it are crucial to follicle development. Growth differentiation factor 9 (GDF9) is a growth factor that was first identified in rodents and has been subsequently characterized in many other mammals (4, 5, 6, 7, 14, 16, 17, 21). It is produced primarily by the oocyte although there is evidence for granulosa production in primates (6, 7, 27). In a mouse model with the GDF9 gene inactivated by transgenesis (GDF9 knockout), folliculogenesis is blocked at the primary stage and sterility ensues (5). The ovaries of the GDF9 knock-out mouse lack normal ovarian structures. Although follicle development does not proceed past the primary (1 layer of granulosa cells) stage, oocytes grow larger than normal. Furthermore, granulosa cell proliferation is reduced and recruitment of the theca cell layer is inhibited. Many other functions have been attributed to GDF9 in the developing follicle. It stimulates an increase in follicle size in rats (10) and increases inhibin B production in human granulosa cells (14). GDF9 also attenuates steroid production (25, 27) and this has led to the suggestion that GDF9 may stimulate granulosa proliferation but modulate differentiation (24).

Avian ovarian follicle development

Follicle development in the laying hen (Fig. 1) proceeds from the very small follicles embedded within the ovary to the large, preovulatory stage (that is, approximately 35 mm). Significant growth of the follicle occurs up to 2 mm, with no accumulation of yellow yolk (9). After this size threshold is reached, white yolk is incorporated into the growing oocyte which has less lipid content than the yellow yolk. Gradually, yellow yolk is taken up by the growing oocyte so that by the 6 mm stage, the oocyte has entered a phase of rapid growth and the follicle is growing rapidly as the oocyte incorporates large amounts of yellow yolk. The yellow yolk is synthesized in the liver under the influence of estrogen and is taken into the growing follicles by a receptor mediated process (3). Associated with the profound increase in size of the oocyte is an increase in the vascularity of the growing follicles, necessary to facilitate the transport of a large increase in yolk components into the oocyte.

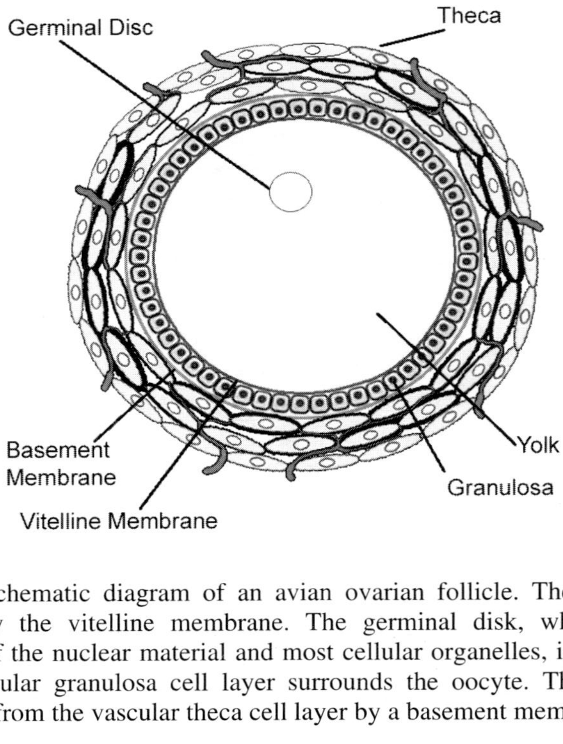

Fig. 1 Schematic diagram of an avian ovarian follicle. The oocyte is limited by the vitelline membrane. The germinal disk, which is the location of the nuclear material and most cellular organelles, is indicated. The avascular granulosa cell layer surrounds the oocyte. This layer is separated from the vascular theca cell layer by a basement membrane.

Concomitant with the increase in follicle size, due mainly to the increase in yolk incorporation by the oocyte, maturational changes also occur in the surrounding follicular layers. In the small, growing white follicles, estrogen is the main steroid secreted (19). The granulosa layer is steroidogenically inactive at this early stage due in large part, to the absence of cholesterol side-chain cleavage enzyme activity (22). As the granulosa layer becomes competent to secrete steroids (11; approximately >8 mm), the main product is progesterone, although only low levels of progesterone are produced by the granulosa layer surrounding small follicles. The granulosa layer of the large, pre-ovulatory follicles is the primary site of production of progesterone with increasing secretory capacity correlated with increasing follicle size (2). The largest, F1 follicle produces the greatest amount of progesterone and progesterone is responsible for eliciting the preovulatory LH surge by stimulatory feedback on the neuroendocrine system (8). In contrast, the small white follicles (< 1 mm) are the predominant source of plasma estradiol (19).

Polarity is conferred on the oocyte by the germinal disk. This structure, a pale white disk on the surface of the oocyte, is the location of the nuclear material as well as most cellular organelles. The germinal disk interacts with the overlying granulosa cells through specialized granulosa cell interdigitations (18). Granulosa cells overlying the germinal disk divide more rapidly (18, 23) and produce less progesterone than more peripheral granulosa cells (15). In addition, germinal disk-associated granulosa cells have fewer FSH receptors and produce an epidermal growth factor (EGF)-like molecule (26, 28, 29). These features of the germinal disc have led to it being identified as the proliferative center of the oocyte from which a paracrine gradient emanates to regulate the development of the rest of the oocyte (28).

GDF9 in hens

In order to investigate the chicken homologues of mammalian GDF9 we utilized a partial sequence for chicken GDF9 cDNA (EST, accession no, BM439817; http://udgenome.ags.udel.edu/-cogburn/index.html) deposited by Dr L Cogburn (Univ. of Delaware) in GenBank. This EST was reported to be most similar to mammalian GDF9 sequences. We generated a cDNA which was used in Northern blot analysis, to demonstrate a single band of RNA of approximately 1.7 kb, only in RNA extracted from ovary and not from other tissues. On the basis of this ovarian-specific localization, we utilized repeated rounds of PCR amplification to obtain the full length cDNA deposited in GenBank (Accession no, AY672110). The full length cDNA sequence is approximately 65% similar to mammalian GDF9 sequences and approximately 80% similar at the C-terminal, biologically active region (12).

Quantitative RNA analysis demonstrated that GDF9 mRNA is more abundant in very small (< 1mm) follicles than in larger follicles (1-3 mm) (12). There is a smaller amount present in RNA extracted from the whole ovary and very low, although detectable RNA for GDF9 in the granulosa layer of 6 mm follicles. Moreover, there is no difference in

GDF9 expression between the granulosa layer from the germinal disk area as compared to that from the non-germinal disk area. There is negligible GDF9 mRNA in the granulosa layer isolated from larger follicles, including the preovulatory follicles. These data suggested that GDF9 is important in the early stages of follicle development. Moreover, our finding of a low level of GDF9 expression in the granulosa layer was verified by immunocytochemistry and is consistent with findings in primates (6, 7, 27). The function of granulosa GDF9 is not known but it may serve to augment oocyte derived GDF9 and/or provide another source closer to the theca layer, where it also has effects (5).

Immunocytochemical staining for GDF9 in the hen ovary is shown in Fig. 2. GDF9 is localized throughout the oocyte in the small follicles embedded within the ovary (12 and Fig. 2A). In contrast,

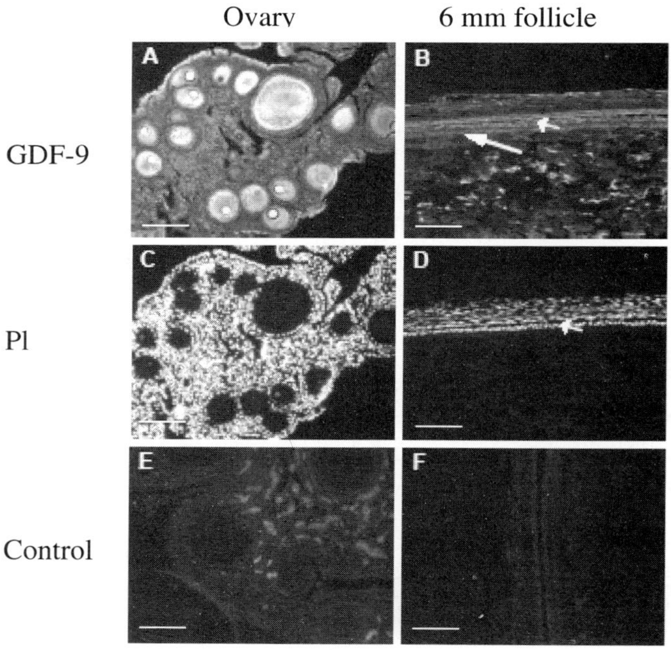

Fig. 2 Immunocytochemical localization of growth differentiation factor 9 (GDF9) in the chicken ovary (A) and large yellow yolky follicles (B). Labelling with the fluorescent nuclear marker, propidium iodide (PI) demonstrates the location of cells (C, D). Control sections were not labeled (E, F). The primary antiserum against GDF-9 was made in a rabbit against mouse GDF-9 (JH131; obtained from Dr. S.-J. Lee). The secondary antiserum was Alexa Fluor 488® goat anti-rabbit IgG. The small arrow (B, D) indicates the granulosa cell layer whereas the larger arrow (B) indicates cytoplasmic GDF9 localized at the periphery of the oocyte. Control slides (E, F) were incubated in the presence of normal rabbit serum. The scale bar equals 50 microns.

in oocytes of larger follicles (6 mm; Fig. 2B), GDF9 is localized under the vitelline membrane, apparently pushed to the periphery by accumulating yolk (20). This location is unlikely to be random because of the effects that GDF9 exerts on granulosa cell function (12). Rather, the location may indicate the importance of close contact necessary for exchange of paracrine factors secreted by the oocyte and the adjacent granulosa cells.

We have reported that oocyte conditioned medium (OCM, presumably a source of GDF9) increases granulosa cell proliferation (12), an effect that is neutralized by pre-treatment of the conditioned medium with GDF9 antiserum. As shown in Fig. 3, OCM inhibits granulosa cell progesterone production. Progesterone production from cultured granulosa cells increases with stage of follicle development (2) and is markedly decreased when the granulosa cells are cultured with 25% or 50% of OCM in the culture medium.

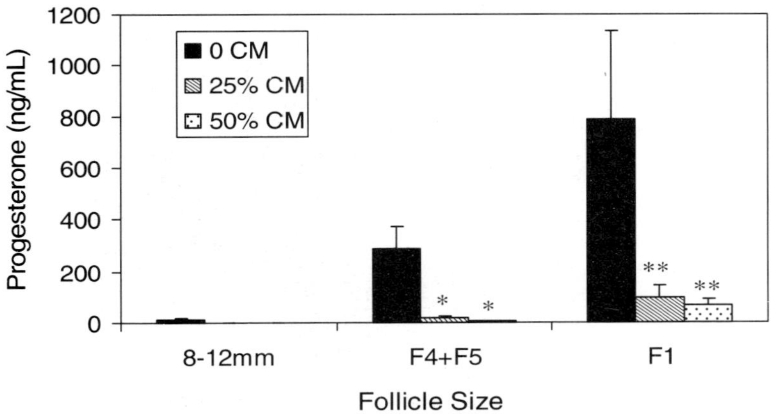

Fig. 3 Effect of oocyte conditioned medium (CM) on progesterone production by granulosa cells of different maturity. Granulosa cells were cultured in the presence of control medium or 25% or 50% medium previously conditioned by oocyte culture. Progesterone production was significantly decreased by oocyte conditioned medium incubated with granulosa cells from the F4+F5 and F1 follicles. Limited progesterone is secreted from granulosa cells of smaller 8-12 mm follicles.

We measured the concentration of estradiol in OCM because the small white follicles are the source of most estradiol in the hen (19) and a previous study indicated that estradiol inhibits progesterone production from the granulosa layer of hens (13). Estradiol was found to be present at an approximate concentration of 100 pg/ml in the OCM, a value less than 1/1000 of the dose previously found necessary to inhibit progesterone production by the granulosa cell layer (13). This suggests that an oocyte-secreted factor, most likely GDF9, modulates avian steroidogenesis as has been reported in mammalian species (21, 24, 27).

Recombinant mammalian GDF9 has not been tested in our system so we do not know whether it is bioactive in avian granulosa cells. The antiserum (JH131 Dr Lee, Johns Hopkins University) used in our studies was raised against mouse GDF9 and was found to recognize GDF9 in immunocytochemical studies and to inhibit OCM bioactivity. We conclude from this that at least some of the biologically significant sites have been evolutionarily conserved, in agreement with DNA sequence analysis.

GDF9 is unusual in having no obvious counterpart in males. Many hormones, while having a well established function in one gender, are often found to be co-opted by the opposite gender for a related function. Surprisingly, GDF9 is not present in the male (17), suggesting a unique function/s in oocyte development and survival.

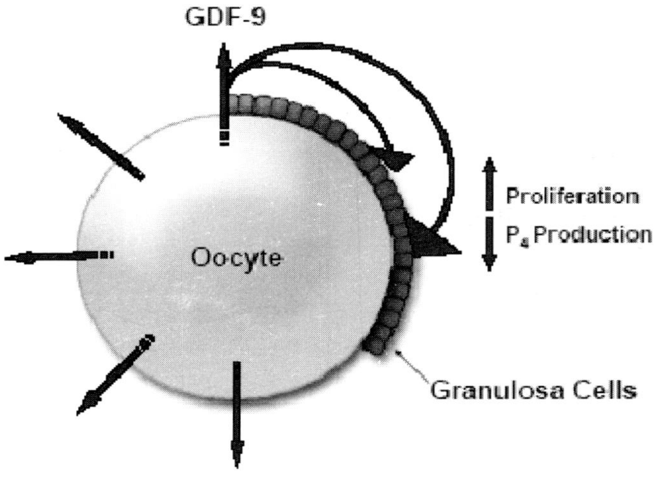

Fig. 4 Schematic diagram of some effects of GDF9 on adjacent granulosa cells. Although not indicated in this figure, it is likely that GDF9 also affects theca cell function.

Conclusions - role of the oocyte

In Fig. 4 we present a model of the proposed paracrine effect of GDF9 from the oocyte on the surrounding granulosa cells. Other factors produced by the oocyte have been described such as GDF9b (1) and EGF (26, 28, 29). The role of the oocyte as a driving force in follicle development has not been adequately considered. Our current studies indicate that oocyte-derived GDF9 affects granulosa cell proliferation and steroidogenesis. A fuller understanding of follicle maturation requires knowledge of the mechanism of yolk uptake. The growth of follicles is coordinated with availability of yolk precursors and expression of lipoprotein receptors on the growing oocytes (20). Just as steroid production and gonadotropin receptors are regulated in a coordinated manner in granulosa and theca cells, the uptake of yolk and therefore, the growth of the oocyte, is also tightly controlled. It is evident two-way communication between oocyte and surrounding somatic cells is necessary for production of a mature avian oocyte.

Since GDF-9 decreases progesterone secretion from the granulosa cells *in vitro*, it is likely that it functions in this way *in vivo*. The timing of this effect must be modulated because it is critical that the granulosa cell layer surrounding maturing follicles secretes progesterone to generate the preovulatory LH surge. We have not yet examined GDF-9 expression in the oocyte of the large hierarchical follicles, which produce the most progesterone. In mammals, GDF9 is expressed throughout follicle development and is present in ovulated oocytes (16). While mouse GDF9 gene knockout studies demonstrate the critical role of GDF9 in early follicle development (5), assessment of the role in later stages could not be determined because follicles never develop to advanced stages without GDF9. With the wide range of functions demonstrated for GDF-9 (5, 10, 14, 21, 24, 25, 27), it is likely that granulosa and/or theca cell responsiveness or access to GDF9 is regulated during follicle development.

Acknowledgements

These studies were supported by USDA 00-35203-9116, 03-35203-13397. We thank Dr. S.-J. Lee of Johns Hopkins University, Baltimore, MD for the gift of the GDF9 antiserum.

References

1 Aaltonen J, Laitinen MP, Vuojolainen K, Jaatinen R, Horelli-Kuitunen N, Seppa L, Louhio H, Tuuri T, Sjoberg J, Butzow R, Hovatta O, Dale L & Ritvos O (1999) Human growth differentiation factor 9 (GDF-9) and its novel homolog GDF-9B are expressed in oocytes during early folliculogenesis. *Journal of Clinical Endocrinology and Metabolism* **84** 2744-2750

2 Bahr JM, Wang S-C, Huang M & Calvo F (1983) Steroid concentration in isolated theca and granulosa layers of the domestic hen. *Biology of Reproduction* **29** 326-334

3 Barber DL, Sanders EJ, Aebersold R & Schneider WJ (1991) The receptor for yolk lipoprotein deposition in the chicken oocyte. *Journal of Biological Chemistry* **266** 18761-18770

4 Bodensteiner KJ, Clay CM, Moeller CL & Sawyer HR (1999) Molecular cloning of the ovine growth/differentiation factor-9 gene and expression of

growth/differentiation factor-9 in ovine and bovine ovaries. *Biology of Reproduction* **60** 381-386

5 Dong J, Albertini DF, Nishimori K, Kumar TR, Lu N & Matzuk MM (1996) Growth differentiation factor-9 is required during early ovarian folliculogenesis. *Nature* **383** 531-535

6 Duffy DM (2003) Growth differentiation factor-9 is expressed by the primate follicle throughout the periovulatory interval. *Biology of Reproduction* **69** 725-732

7 Duffy DM & Stouffer RL (2001) Growth differentiation factor-9 (GDF-9) is expressed by the primate follicle throughout the periovulatory interval. *Biology of Reproduction* **64** 296 (Abstract)

8 Etches RJ & Cunningham FJ (1976) The interrelationship between progesterone and luteinizing hormone during the ovulation cycle of the hen (*Gallus domesticus*). *Journal of Endocrinology* **71** 51-58

9 Gilbert, A. B (1971) The Ovary. In *Physiology and Biochemistry of the Domestic Fowl.* Vol. III. pp.1225-1233. Eds DJ Bell & BM Freeman. New York and London: Academic Press

10 Hayashi M, McGee EA, Min G, Klein C, Rose UM, van Duin M & Hsueh AJW (1999) Recombinant growth differentiation factor-9 (GDF-9) enhances growth and differentiation of cultured early ovarian follicles. *Endocrinolog*y **140** 1236-1244

11 Johnson AL (1996) The avian ovarian hierarchy: a balance between follicle differentiation and atresia. *Poultry and Avian Biology Reviews* **7(2/3)** 99-110

12 Johnson PA, Dickens MJ, Kent TR & Giles JR Expression and function of GDF9 in an oviparous species, *Gallus domesticus. Biology of Reproduction* published 29 Dec 2004; 10.1095/biolreprod.104.036822

13 Johnson PA, Green C, Lee HT & Bahr JM (1988) Inhibition of progesterone secretion from granulosa cells by estradiol and androgens in the domestic hen. *Endocrinology* **123** 473-477

14 Kaivo-Oja N, Bondestam J, Kamarainen M, Koskimies J, Vitt U, Cranfield M, Vuojolainen K, Kallio JP, Olkkonen VM, Hayashi M, Moustakas A, Groome NP, ten Dijke P, Hsueh AJW & Ritvos O (2003) Growth differentiation factor-9 induces smad2 activation and inhibin B production in cultured human granulosa-luteal cells. *Journal of Clinical Endocrinology and Metabolism* **88** 755-762

15 Marrone BL, Jamaluddin M & Hertelendy F (1990) Regional pattern of cell maturation and progesterone biosynthesis in the avian granulosa cell layer. *Biology of Reproduction* **42** 405-412

16 McGrath SA, Esquela AF, & Lee S-J (1995) Oocyte-specific expression of growth/differentiation factor-9. *Molecular Endocrinology* **9** 131-136

17 McPherron AC & Lee S-J (1993) GDF-3 and GDF-9: Two new members of the transforming growth factor- superfamily containing a novel pattern of cysteines. *Journal of Biological Chemistry* **268** 3444-3449

18 Perry MM, Gilbert AB & Evans AJ (1978) The structure of the germinal disc region of the hen's ovarian follicle during the rapid growth phase. *Journal of Anatomy* **127** 379-392

19 Robinson, FE & Etches RJ (1986) Ovarian steroidogenesis during follicular

maturation in the domestic fowl (*Gallus domesticus*). *Biology of Reproduction* **35** 1096-1105

20 Shen X, Steyrer E, Retzek H, Sanders EJ & Schneider WJ (1993) Chicken oocyte growth: receptor-mediated yolk deposition. *Cell and Tissue Research* **272** 459-471.

21 Solovyeva EV, Hayashi M, Margi K, Barkats C, Klein C, Amsterdam A, Hsueh AJW & Tsafriri A (2000) Growth differentiation factor-9 stimulates rat theca-interstitial cell androgen biosynthesis. *Biology of Reproduction* **63** 1214-1218

22 Tilly JL, Kowalski KI & Johnson AL (1991) Stage of ovarian follicular development associated with the initiation of steroidogenic competence in avian granulosa cells. *Biology of Reproduction* **44** 305-314.

23 Tischkau SA, Neitzel LR, Walsh JA & Bahr JM (1997) Characterization of the growth center of the avian preovulatory follicle. *Biology of Reproduction* **56** 469-474.

24 Vitt, UA, Hayashi M, Klein C & Hsueh AJW (2000) Growth differentiation factor-9 stimulates proliferation but suppresses the follicle-stimulating hormone-induced differentiation of cultured granulosa cells from small antral and preovulatory rat follicles. *Biology of Reproduction* **62** 370-377

25 Vitt UA, McGee EA, Hayashi M & Hsueh AJH (2000) In vivo treatment with GDF-9 stimulates primordial and primary follicle progression and theca cell marker CYP17 in ovaries of immature rats. *Endocrinology* **141** 3814-3820

26 Volentine KK, Yao HHC & Bahr JM (1998) Epidermal growth factor in the germinal disc and its potential role in follicular development in the chicken. *Biology of Reproduction* **59** 522-526

27 Yamamoto N, Christenson LK, McAllister JM & Strauss III JF (2002) Growth differentiation factor-9 inhibits 3'5'-adenosine monophosphate-stimulated steroidogenesis in human granulosa and theca cells. *Journal of Clinical Endocrinology and Metabolism* **87** 2849-2856

28 Yao HHC & Bahr JM (2001) Germinal disc-derived epidermal growth factor: a paracrine factor to stimulate proliferation of granulosa cells. *Biology of Reproduction* **64** 390-395

29 Yao HHC & Bahr JM (2001) Chicken granulosa cells show differential expression of epidermal growth factor (EGF) and luteinizing hormone (LH) receptor messenger RNA and differential responsiveness to EGF and LH dependent upon location of granulosa cells to the germinal disc. *Biology of Reproduction* **64** 1790-1796

Functional Avian Endocrinology
A Dawson & P J Sharp (Eds)

The inhibin-activin system and ovarian folliculogenesis in the chicken

P G Knight, R T Gladwell and T M Lovell

School of Animal and Microbial Sciences, The University of Reading, Whiteknights, Reading RG6 6AJ, UK

Introduction

Evidence has accumulated over the last 10-15 years to support the functional involvement of the inhibin-activin system in the regulation of ovarian folliculogenesis in birds. As in mammals, inhibins of gonadal origin are implicated as classical endocrine feedback regulators of pituitary gonadotrophin secretion. Both activins and inhibins appear to play intragonadal autocrine/paracrine roles but there is no evidence that gonadally-derived activins act on the pituitary (17, 40).

Inhibins and activins are members of the structurally conserved but functionally diverse transforming growth factor β (TGFβ) superfamily of extracellular signaling molecules. Inhibins are dimeric glycoproteins consisting of an α subunit disulphide-linked to one of two distinct β subunits (βA or βB), to generate inhibin-A and inhibin-B, respectively. Activins are disulphide-linked homo- or hetero-dimers of βA and/or βB subunits, termed activin-A (βA-βA), activin-B (βB-βB) and activin-AB (βA-βB). As with other TGFβ superfamily members the mature forms of inhibin and activin comprise the carboxy-terminal regions of larger precursor molecules that are processed post-translationally. The genes encoding the full-length precursor molecules of the chicken inhibin/activin α (42), βA (4) and βB (11) subunits have been cloned and characterised; they each share a high degree of homology with their mammalian counterparts. Messenger RNAs encoding inhibin/activin α, βA and βB subunits (3, 5, 16), activin receptors (ActR) (38, this article), activin-binding protein (follistatin) (6) and putative inhibin co-receptor (betaglycan) (this article) are expressed in chicken ovarian follicles consistent with a fully operational intraovarian inhibin-activin system.

Activin and inhibin receptors and signaling mechanisms

Generally inhibins and activins exert antagonistic effects on their target cells. The receptors and downstream signal transduction cascade through which activins exert their biological effects are well-characterised (45). This is not the case for inhibins; indeed conclusive evidence for the existence of a cognate inhibin-specific signaling receptor has yet to emerge (34). As with other ligands of the TGFβ superfamily, activins bind to and form a hetero-oligomeric complex with two types of cell-surface receptor, designated type-I and type-II. Several type-I and -II activin receptors (ActRIA, ActRIB, ActRIIA, ActRIIB) have been cloned and characterized in mammals (1). Three receptor sub-types (ActRI, ActRIIA, ActRIIB) have been cloned in avian species (19, 32). Each receptor is a monomeric glycoprotein with an amino-terminal extracellular ligand-binding domain, a single transmembrane domain and an intracellular serine/threonine kinase domain (1, 45). The association of activin with type-I receptor requires the presence of type-II receptors and both receptor types are required to mediate ligand signaling. The type-II receptors are constitutively phosphorylated and, following ligand binding, transphosphorylate the type-I receptors. These, in turn, phosphorylate downstream substrates called receptor-Smads which associate with co-Smads before translocating to the nucleus to modulate gene transcription (1, 45).

Inhibin can also bind to activin type-II receptors although with an affinity ~20 times lower than activin (45). The binding of inhibin to type-II receptors prevents the recruitment of type-I receptors to the complex, and this is the likely mechanism through which inhibin antagonizes the effects of activin in most systems (9, 34, 45). Association of inhibin with type-III TGFβ receptor (aka betaglycan) enhances inhibin binding to activin type-II receptors and so renders inhibin a more effective activin antagonist (9, 34). Thus, betaglycan can be considered as a non-signaling inhibin co-receptor.

Technical issues associated with detecting and measuring inhibins, activins and follistatin

Examination of ovarian inhibin/activin subunit expression at the mRNA level alone does not reveal the extent to which the different subunit precursors are translated and processed posttranslationally into functionally active inhibin/activin protein dimers. Initial studies of plasma inhibin levels in the chicken (17, 39, 40, 41) utilized α subunit-directed radioimmunoassays which do not discriminate between monomeric α subunit(s), inhibin-A and inhibin-B. Since the chicken ovary expresses α, βA and βB inhibin/activin subunits (3, 5, 16) it may be presumed that the tissue has the capacity to produce free inhibin/activin subunits, dimeric inhibin-A and inhibin-B, in addition to the three forms of activin (A, AB and B). Given the likelihood of differential follicular expression and assembly of the different subunits, it is evident that dimer-specific assays are essential tools to investigate the spatio-temporal distribution of the two inhibin forms and three activin forms in tissues (ovary, blood) of birds at different physiological stages. The design of such immunoassays is further complicated by the presence in biological samples of inhibin/activin binding proteins such as α_2 macroglobulin and follistatin that can interfere with antibody binding. Likewise, activin and other ligands with follistatin-binding properties can potentially interfere in follistatin immunoassays.

Using specific 2-site immuoassays that were designed to overcome these pitfalls (10, 21, 24) we sought to quantify inhibin-A, inhibin-B, activin-A and follistatin in: (i) theca and granulosa tissue of developing follicles; (ii) the circulation of hens during sexual maturation, throughout the ovulatory cycle and after induced cessation of lay; and (iii) conditioned media from cultured granulosa cells exposed to gonadotrophins and other factors. We have yet to develop immunoassays for chicken activin-B or-AB. More recently we used quantitative-PCR (Taqman) to examine thecal and granulosal expression of mRNAs encoding activin receptors (ActRI, -IIA and -IIB) and betaglycan (putative inhibin co-receptor) throughout folliculogenesis allowing us to relate these to protein and mRNA expression of inhibins, activins and follistatin. The main findings of these studies are discussed below.

Circulating inhibins during sexual maturation and in laying hens

To investigate the potential role of inhibins in the modulation of gonadotrophin secretion Lovell *et al* (25) measured plasma concentrations of inhibins A and B in growing pullets from juvenile status to sexual maturity and related these to plasma gonadotrophins and progesterone. Inhibin-A and progesterone concentrations were positively correlated and both increased progressively prior to the onset of lay; plasma inhibin-B remained undetectable throughout. Plasma FSH was negatively correlated with inhibin-A during sexual maturation; FSH decreased particularly sharply over the 2 weeks prior to the onset of lay, coincident with the time that plasma inhibin-A peaked. This supports the possibility of a negative feedback loop between inhibin-A and FSH. However, given that the concentrations of progesterone and oestradiol are also rising at this time, a negative feedback action of these hormones in the regulation of FSH secretion cannot be discounted.

A potential endocrine role of inhibin-A in the adult hen is suggested by the finding of cyclic changes in plasma inhibin-A during the ovulatory cycle (26). Inhibin-A concentrations peaked around the time of the preovulatory LH surge, approximately 6h before ovulation. However, no correlation was found between inhibin-A and FSH; plasma FSH levels were uniform throughout. Plasma inhibin-A and FSH were negatively correlated during a period of ovarian regression induced by food restriction (26). Falling plasma inhibin-A level was accompanied by a rise in plasma FSH, indicative of a functional negative feedback relationship.

The finding that plasma inhibin-A levels are maximal around the time of the preovulatory LH surge and that the decline in plasma inhibin-A levels in feed-restricted hens was preceded by a fall in LH (26), raise the possibility that pituitary LH has a positive role in maintaining inhibin secretion by preovulatory follicles. This notion is supported by the ability of LH, as well as FSH, to promote inhibin-A secretion by cultured hen granulosa cells from F1, F2 and F3 follicles (22, 23)

Effect of inhibin immunization in female birds

Active immunization of pullets against inhibin using a synthetic chicken inhibin α-subunit peptide-conjugate was associated with an increased incidence of double ovipositions and an increased number of recent postovulatory follicles per ovary (25). There was no detectable effect on plasma FSH or LH levels, nor on the onset of lay. Active immunization of juvenile Japanese quail with an inhibin α-subunit fusion protein vaccine resulted in a significant advance in the onset of lay combined with an enhanced egg production over a 30 week period (30). More recently Satterlee *et al* (36) reported that a similar vaccine applied to 20 week-old broiler breeder pullets advanced the onset of lay and increased egg laying performance. This evidence from active immunization studies is consistent with the observation that plasma ir-inhibin and ovarian inhibin α subunit mRNA levels are higher in hens with a low rate of lay (42).

It was also noted by Lovell *et al* (25) that inhibin immunization doubled the number of follicles in the 8-10mm diameter size-class. Importantly, these follicles are at the bottom of the preovulatory heirarchy having only recently been selected from the prehierarchical 6-8mm follicle pool. It was subsequently discovered that follicles of this size-class produce high levels of inhibin-B (24). In the absence of significant changes in circulating LH or FSH concentrations, this suggests an intraovarian role for inhibin-A and/or -B in regulating the entry of follicles into the preovulatory hierarchy.

Irrespective of the mechanism through which inhibin immunization brings about an increased rate of lay, the technique shows some promise as a practicable way of increasing poultry egg production.

Changes in inhibin-A, inhibin-B, activin-A and follistatin production during follicle development

Lovell *et al* (24) used a panel of specific two-site immunoassays to quantify inhibin-A, inhibin-B, activin-A and follistatin contents in granulosa and theca layers and thus investigate the potential involvement of these proteins during follicular development and, in particular, in the recruitment of growing follicles to the preovulatory hierarchy. The study revealed a complete dissociation between follicular inhibin-B and inhibin-A expression during follicle development (Fig. 1).

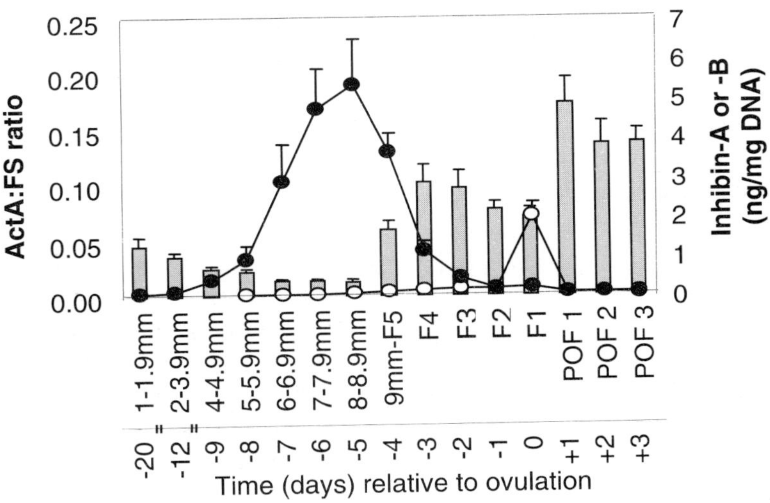

Fig. 1 Follicular development in the laying hen is associated with major shifts in the apparent balance between 'activin tone' and 'inhibin tone'. Bars represent the intrafollicular ratio of activin-A to follistatin (ActA:FS ratio) calculated for *combined* theca and granulosa layers of chicken ovarian follicles ranging from 1-2mm to F1 position. POF*n* represents postovulatory follicles with POF1 being the most recent. Total contents (i.e. theca + granulosa) of inhibin-B (closed circles) and inhibin-A (open circles) are plotted for comparison. Values are means ± SEM (n=11). The lower timescale provides an estimate of the time taken for follicles of a given size category to reach the point of ovulation (data replotted from 24).

Inhibin-B content increased dramatically from 1-2mm, to peak in 7-9mm follicles before falling steeply as follicles entered the hierarchy and advanced towards F1. In complete contrast to inhibin-B, inhibin-A was undetectable in the granulosa cell layer of follicles <9mm but increased ~40-fold from 9mm to F1 with the sharpest increase during the F2 to F1 transition. These changes reflect follicle size-dependant change in βA/βB subunit mRNA expression in avian granulosa cells (5, 11; see

Fig. 2). It is tempting to speculate that the dramatic increase in production of inhibin-B in 6-9mm follicles is somehow involved in the modulation of granulosa cell acquisition of steroidogenic competence and/or increased gonadotrophin responsiveness, prerequisites for recruitment to the preovulatory hierarchy. 'Activin tone' appears to decrease progressively as small follicles grow from 1-2mm to ~9mm, the approximate point of recruitment into the preovulatory hierarchy (24). Indeed the apparently high 'activin tone' of small follicles (1-2mm) might explain how such follicle acquires FSH responsiveness and subsequently progresses to an LH-responsive stage of development.

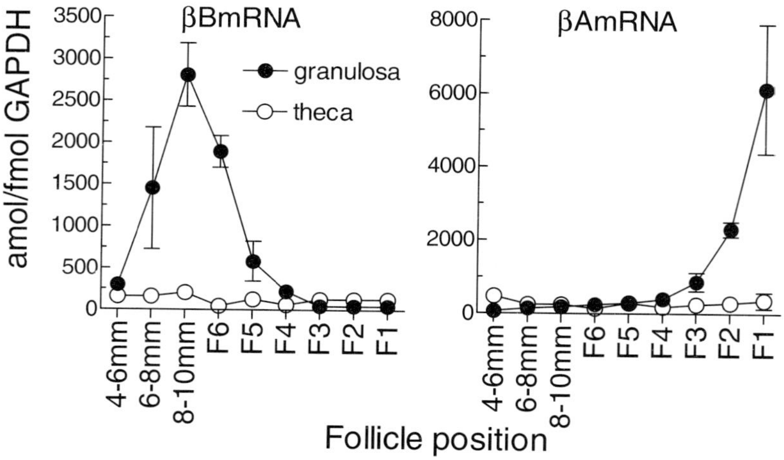

Fig. 2 The pattern of expression of mRNAs for inhibin/activin βB and βA subunits in granuosa and theca layers accords well with observed changes in inhibin-B and inhibin-A protein content of developing hen follicles. Values are means ± sem (n = 4).

On the assumption that the intrafollicular balance between inhibin, activin and follistatin is crucial in determining the follicular response to these proteins (8, 18), the proposed action of activin to promote gonadotrophin receptor expression in small follicles would operate most successfully when follistatin levels are low. Similarly, since inhibins antagonise the actions of activin, low levels of inhibin should also favour activin action. Follicular activin-A contents were relatively uniform in follicles ranging from 1mm to 9mm while follistatin contents increased progressively from 1mm to 7mm suggesting a gradually attenuation of 'activin tone'(24). The subsequent increase in inhibin-B levels between 5mm and 8mm would further diminish 'activin tone' relative to 'inhibin tone'. It is not known whether these changes merely reflect, rather than contribute to, the process(es) which drive follicles through specific developmental checkpoints during folliculogenesis. Nonetheless, the observation of a rise in the follistatin content of both theca and granulosa cell layers of 6-9mm follicles, unaccompanied by a rise in its ligand activin-A, supports the notion of a marked though transient increase in 'inhibin tone' (and corresponding decrease in 'activin tone') in follicles approaching entry into the preovulatory hierarchy (see Fig. 1).

It is, of course, possible that the dramatic though transient inhibin-B rise in 6-9mm follicles reflects not transition into, but delayed entry into the preovulatory hierarchy; after all, only one follicle of this cohort is recruited to the preovulatory hierarchy during a typical ovulatory cycle. The recent finding that activin treatment *in vitro* can upregulate gonadotrophin receptor expression by prehierarchical 6-8mm hen follicles (15) accords with this view as the exogenous activin applied would clearly raise 'activin tone' and allow gonadotrophin-dependent follicle progression. Likewise, inhibin immunization would raise relative 'activin tone' in 6-8mm follicles (which express high levels of inhibin-B) and this could explain the doubling in the number of 8-10mm follicles observed earlier by Lovell *et al* (25). This leads us to hypothesise that only one 'privileged' follicle of the growing SYF cohort is promoted to the preovulatory hierarchy each cycle by virtue of it having an intrinsically high 'activin tone' (or low 'inhibin tone'). Preliminary attempts in our laboratory to identify such 'privileged' follicles amongst the 6-9mm population by examining basal and gonadotrophin-induced hormone secretion by individual follicle wall explants *in vitro* have yielded inconclusive results.

Recruitment to the preovulatory hierarchy was associated with a marked rise in activin-A (~6-fold rise; 9mm to F4), predominantly in the theca layer but also evident in the granulosa cell layer. This was accompanied by a ~4-fold decline in the follistatin content of the granulosa cell layer and a dramatic reduction in the inhibin-B content of both granulosa and theca layers from 9mm to F4 (24). Such conditions would enhance the interaction of theca-derived activin-A with receptors on granulosa cells. We propose that this further shift in the balance between 'inhibin tone' and 'activin tone' might facilitate the rapid growth of follicles as they move up the preovulatory hierarchy. *In vitro* studies with mammalian cells showed that activin can promote granulosa cell proliferation (20, 27) and upregulate basal and gonadotrophin-stimulated P450aromatase activity and oestrogen secretion (12, 28, 37, 38).

As selected follicles moved up the preovulatory hierarchy, and particularly during the final F2-F1 transition when progesterone production increases dramatically, their relative 'activin tone' evidently decreases again as reflected by a large rise in inhibin-A, a small decline in activin-A and a relatively constant follistatin level (24). Various factors have been shown to modulate secretion of inhibin-A and progesterone from cultured chicken F1-F3 granulosa cells including FSH, LH, TGFα, IGF-1 and activin-A itself (22, 23, 33) but precisely how such divergent signals interact within the context of a developing preovulatory follicle *in vivo* remains obscure. The ability of thecal activin-A to promote granulosal inhibin-A production is intriguing given the antagonistic properties of these two proteins. It is tempting to speculate that the dramatic rise in inhibin-A production by the F1 granulosa layer plays a role in the generation of the preovulatory progesterone surge. However, a preliminary *in vitro* experiment in which we attempted to immunoneutralize endogenous inhibin-A produced by F1 follicle explants revealed no effect of the inhibin antibody on basal or LH-induced progesterone secretion (authors unpublished data).

Spatio-temporal pattern of expression of activin receptors and inhibin co-receptor during follicle development

Using real-time quantitative PCR, the expression of ActRI, -IIA, -IIB and betaglycan mRNA was detected in both theca and granulosa layers with ActRI and betaglycan mRNA being much more abundant in theca than granulosa at all stages of follicle development examined (Fig. 3).

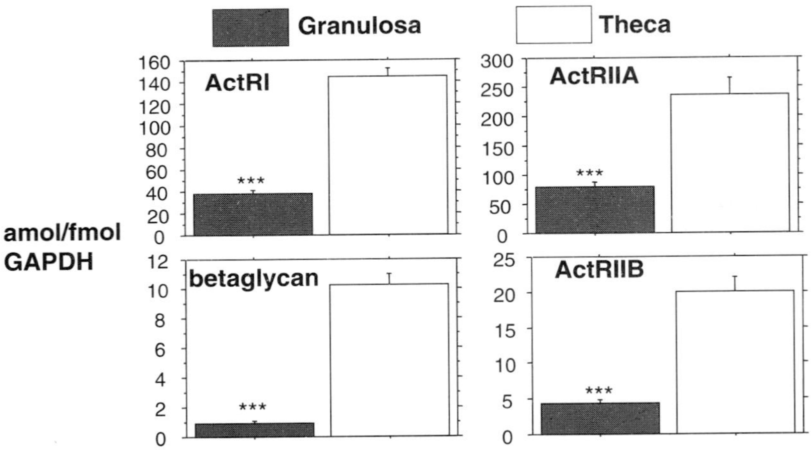

Fig. 3 Comparison of expression of mRNAs encoding activin receptors (ActRI, -IIA, -IIB) and inhibin co-receptor (betaglycan) in granulosa and theca layers of hen follicles. Values are means ± sem (n = 36) and represent combined data for all follicle stages analysed (4mm-F1).*** P<0.001)

Moreover ActRIIA mRNA was much more abundant than ActRIIB mRNA in both cell types. Granulosal ActRI, -IIA and -IIB mRNA levels varied significantly and followed a similar expression pattern during follicle development, being low in 4-6mm follicles, increasing to a peak in 8-10mm follicles before falling progressively as follicles moved up the preovulatory hierarchy to the F1 position (Fig. 4). Thecal ActRI, -IIA and -IIB expression followed a very different pattern, falling to a nadir in F4 follicles before rising to their maximal levels in F1 ($P<0.01$). Granulosal betaglycan mRNA levels were quite variable and differences were not statistically significant. Thecal betaglycan mRNA expression tended to fall ($P<0.1$) as follicles moved into the hierarchy (F6) but then increased significantly to a peak in F2. Taken together with the dynamic changes observed in expression of inhibin/activin subunit mRNAs, assembled protein dimers and activin binding-protein (follistatin), these findings further support local intrafollicular actions of inhibin and activin in the hen.

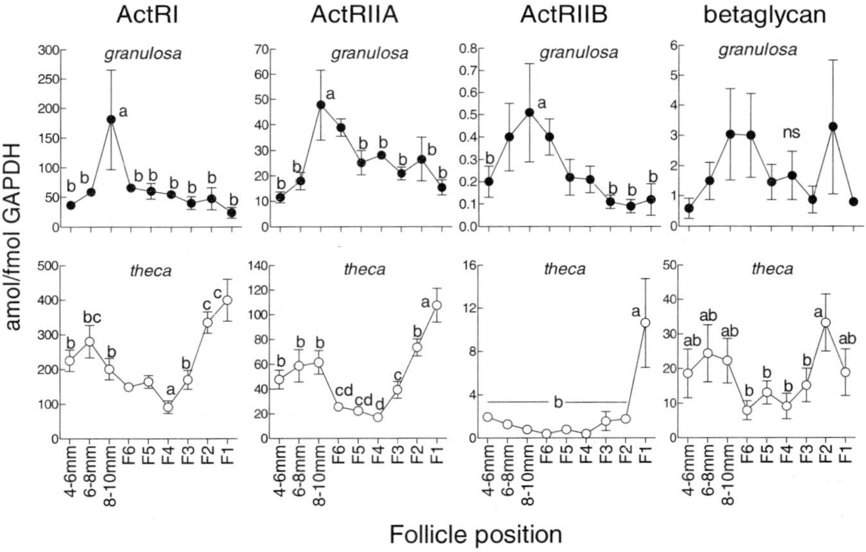

Fig. 4 Spatio-temporal mRNA expression pattern for activin receptors (ActRI, ActRIIA, ActRIIB) and inhibin co-receptor (betaglycan) during follicle development in the laying hen. Values are means ± sem (n = 4) and means without a common letter are significantly different.

Evidence from functional *in vitro* studies supporting intraovarian actions of activin and inhibin in the hen

In mammalian ovarian follicles (reviews:18, 43), in contrast to avian follicles (21, 24), activins and inhibins are both produced almost exclusively by granulosa cells rather than theca cells. *In vitro* studies in rodents have shown that activin-A can up-regulate FSH receptor expression, promote cell proliferation and enhance FSH-dependent processes (e.g. inhibin production, steroidogenesis) by granulosa cells (20, 28, 31, 44). In addition, activin-A has been shown to suppress, while inhibin can enhance basal and/or LH-induced androgen production by mammalian theca cells (18).

Is there any evidence to support such roles for activin in birds? Using mixed ovarian cell cultures derived from chicken embryos Rombauts *et al.* (35) observed an inhibitory effect of activin-A on the conversion of pregnenolone to androgens. This action was reversed by follistatin while inhibin alone raised androgen production consistent with the above findings for mammalian theca cells.

Recently Johnson *et al* (15) reported that treatment of hen granulosa cells from prehierarchical (6-8mm) follicles with activin upregulated expression LH receptor mRNA; combined treatment with activin and FSH upregulated FSH receptor expression. While activin alone did not promote progesterone production, pretreatment of cells with activin potentiated LH/FSH induced StAR expression and progesterone production.

Lovell *et al* (23) compared the effects of activin-A on basal and gonadotrophin-induced secretion of inhibin-A, inhibin-B and progesterone by cultured hen granulosa cells from F1, F2 and F3 follicles (Fig. 5). Treatment with activin-A alone markedly increased inhibin-A secretion with the greatest response in F1 follicles. In contrast, activin-A alone had no effect on progesterone output. Furthermore, activin-A greatly enhanced FSH-induced secretion of both inhibin-A and progesterone the relative response being greatest in F1 follicles and smallest in F3 follicles. In terms of LH-induced inhibin-A and progesterone secretion, cells from F1, F2 and F3 follicles showed similar responses. Co-treatment with activin-A markedly enhanced LH-induced secretion of inhibin-A but had only a marginal effect on LH-induced progesterone secretion the latter response being most pronounced in F1 follicles. Secretion of inhibin-B was below detection limit under all conditions. Immunocytochemistry (23) confirmed that type-I and -II activin receptors are expressed by the cells, supporting the hypothesis that theca-derived activin-A contributes to the regulation of preovulatory follicle development in the hen by exerting a local paracrine action on granulosa cells leading to enhance output of inhibin-A and progesterone. Indeed, it is possible that activin-A dependent upregulation of inhibin-A production by F1 granulosa contributes to the positive feedback mechanism that generates the preovulatory progesterone/LH surge.

Evidence regarding potential autocrine/paracrine actions of inhibin on mammalian granulosa cell steroidogenesis remains contradictory (2, 14) and further experiments are required to examine this possibility in birds. The observation (25) that hens actively immunized against inhibin α subunit had higher levels of activin-A in the theca layer of F1 and F2 follicles despite having unchanged circulating gonadotrophin concentrations implies an intraovarian paracrine action of inhibin on theca cell function. Unfortunately, the study (25) did not examine thecal activin-A content in prehierarchical follicles. Since activin can upregulate gonadotrophin responsiveness of granulosa cells from both preovulatory (23) and prehierarchical (15) follicles, this could provide a 'local' mechanism to account for the positive effect of inhibin immunization on ovulation rate discussed above.

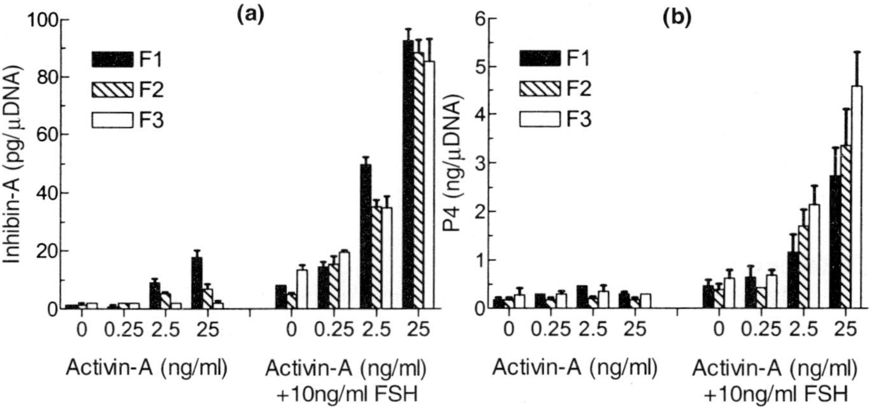

Fig. 5 Effect of activin-A on 'basal' and FSH-induced secretion of (a) inhibin-A and (b) progesterone by chicken granulosa cells from F1, F2 and F3 follicles (data from Lovell *et al*, 23)

Conclusions

On the basis of the sequence of changes observed in follicular expression of inhibin-A, inhibin-B, activin-A and follistatin proteins (see Fig. 1), we suggest that relative 'activin tone' decreases progressively from 1-2mm (the smallest follicle category analysed) to a nadir in 7-9mm follicles on the threshold of entry into the preovulatory hierarchy, increases again as the selected follicle enter the preovulatory hierarchy (F5/F4) before falling again as the follicle approaches the F1 position. These changes are accompanied by alterations in the granulosal and thecal expression of activin type-I and -II receptors and inhibin co-receptor, betaglycan during follicle development. While the interpretation of such spatio-temporal differences in ligand/receptor expression pattern is complex, the observation that sequential shifts in the balance between 'activin tone' and 'inhibin tone' occur over a timescale of about 20 days of follicle development, together with the finding of a striking divergence in the pattern of inhibin-A and inhibin-B expression, imply important functional role(s) of the inhibin-activin system in avian folliculogenesis that await definition. Evidence so far indicates that activin-induced upregulation of gonadotrophin responsiveness, and its antagonism by inhibin and/or follistatin, is a key aspect affecting follicle progression to preovulatory status. We hypothesize that the marked though transient rise in inhibin-B production by prehierarchical follicles has a 'gate-keeper' role in delaying follicle recruitment to the preovulatory hierarchy, perhaps by antagonizing a stimulatory action of thecal activin on cell proliferation, gonadotrophin receptor expression and steroidogenesis. Accordingly, only follicles that are deficient in inhibin-B, inhibin co-receptor (betaglycan) and/or follistatin would proceed beyond this developmental checkpoint to ovulate about one week later.

Acknowledgements
The authors are grateful to Professor Nigel Groome, Professor Frank Cunningham, Dr John Proudman and Dr Al Parlow for providing valuable research reagents and to the BBSRC for financial support.

References
1 Abe Y, Minegishi T & Leung PC (2004) Activin receptor signaling. *Growth Factors* **22**:105-10
2 Campbell BK & Baird DT (2001) Inhibin A is a follicle stimulating hormone-responsive marker of granulosa cell differentiation, which has both autocrine and paracrine actions in sheep. *Journal of Endocrinology* **169** 333-345
3 Chen CC & Johnson PA (1996) Expression of inhibin α and inhibin/activin β_A subunits in the granulosa layer of the large preovulatory follicles of the hen. *Biology of Reproduction* **55** 450-454
4 Chen CC & Johnson PA (1996) Molecular cloning of inhibin/activin beta A-subunit complementary deoxyribonucleic acid and expression of inhibin/activin alpha- and beta A-subunits in the domestic hen. *Biology of Reproduction* **54** 429-435
5 Davis AJ & Johnson PA (1998) Expression pattern of messenger ribonucleic acid for follistatin and the inhibin/activin subunits during follicular and testicular development in *Gallus domesticus*. *Biology of Reproduction* **59** 271-277
6 Davis AJ, Brooks CF & Johnson PA (2000) Estradiol regulation of follistatin and inhibin α- and β_B-subunit mRNA in avian granulosa cells. *General and Comparative Endocrinology* **119** 308-316
7 Davis AJ, Brooks CF & Johnson PA (2001) Follicle-stimulating hormone regulation of inhibin α- and β_B-subunit and follistatin messenger ribonucleic acid in cultured avian granulosa cells. *Biology of Reproduction* **64** 100-106

8 Findlay JK, Drummond AE, Dyson M, Baillie AJ, Robertson DM & Ethier JF (2001) Production and actions of inhibin and activin during folliculogenesis in the rat. *Molecular and Cellular Endocrinology* **180** 139-144
9 Gray PC, Bilezikjian LM & Vale W (2002) Antagonism of activin by inhibin and inhibin receptors: a functional role for betaglycan. *Molecular and Cellular Endocrinology* **188** 254-260
10 Groome NP, Tsigou A, Cranfield M, Knight PG & Robertson DM (2001) Enzyme immunoassays for inhibins, activins and follistatins. *Molecular and Cellular Endocrinology* **180** 73-77
11 Hecht DJ, Davis AJ, Brooks CF & Johnson PA (2000) Molecular cloning and expression analysis of the complementary deoxyribonucleic acid for chicken inhibin/activin β_B subunit. *Biology of Reproduction* **62** 1128-1134
12 Hillier SG & Miro F (1993) Local regulation of primate granulosa cell aromatase activity. *Journal of Steroid Biochemistry and Molecular Biology* **44** 435-439
13 Hillier SG, Whitelaw PF & Smyth CD (1994) Follicular oestrogen synthesis: the 'two-cell, two-gonadotrophin' model revisited. *Molecular and Cellular Endocrinology* **100** 51-54
14 Jimenez-Krassel F, Winn ME, Burns D, Ireland JL & Ireland JJ.(2003) Evidence for a negative intrafollicular role for inhibin in regulation of estradiol production by granulosa cells. *Endocrinology* **144** 1876-1886
15 Johnson AL, Bridgham JT & Woods DC (2004) Cellular mechanisms and modulation of activin A- and transforming growth factor {beta}-mediated differentiation in cultured hen granulosa cells. *Biology of Reproduction* **16** 1844-1851
16 Johnson PA & Wang SY (1993) Characterization and quantitation of mRNA for the inhibin alpha-subunit in the granulosa layer of the domestic hen. *General and Comparative Endocrinology* **90** 43-50
17 Johnson PA, Wang SY & Brooks C (1993) Characterization of a source and levels of plasma immunoreactive inhibin during the ovulatory cycle of the domestic hen. *Biology of Reproduction* **48** 262-267
18 Knight PG & Glister C (2001) Potential local regulatory functions of inhibins, activins and follistatin in the ovary. *Reproduction* **121** 503-512
19 Lai YT, Beason KB, Brames GP, Desgrosellier JS, Cleggett MC, Shaw MV, Brown CB & Barnett JV (2000) Activin receptor-like kinase 2 can mediate atrioventricular cushion transformation. *Developmental Biology* **222** 1-11
20 Li R, Phillips DM & Mather JP (1995) Activin promotes ovarian follicle development *in vitro*. *Endocrinology* **125** 2134-2140
21 Lovell TM, Gladwell RT, Cunningham FJ, Groome NP & Knight PG (1998) Differential changes in inhibin A, activin A, and total α-subunit levels in granulosa and thecal layers of developing preovulatory follicles in the chicken. *Endocrinology* **139** 1164-1171
22 Lovell TM, Gladwell RT, Groome NP & Knight PG (2002) Modulatory effects of gonadotrophins and insulin-like growth factor on the secretion of inhibin A and progesterone by granulosa cells from chicken preovulatory (F1-F3) follicles. *Reproduction* **123** 291-300

23 Lovell TM, Gladwell RT, Groome NP & Knight PG (2002) Activin exerts differential effects on basal and gonadotrophin-induced secretion of inhibin A and progesterone by granulosa cells from chicken preovulatory (F1-F3) follicles. *Reproduction* **124** 649-657

24 Lovell TM, Gladwell RT, Groome NP & Knight PG (2003) Ovarian follicle development in the laying hen is accompanied by divergent changes in inhibin A, inhibin B, activin A and follistatin production in granulosa and theca layers. *Journal of Endocrinology* **177** 45-55

25 Lovell TM, Knight PG, Groome NP & Gladwell RT (2001) Changes in plasma inhibin A levels during sexual maturation in the female chicken and the effects of active immunization against inhibin α-subunit on reproductive hormone profiles and ovarian function. *Biology of Reproduction* **64** 188-196

26 Lovell TM, Vanmontfort D, Bruggeman V, Decuypere E, Groome NP, Knight PG & Gladwell RT (2000) Circulating concentrations of inhibin-related proteins during the ovulatory cycle of the domestic fowl *(Gallus domesticus)* and after induced cessation of egg laying. *Journal of Reproduction and Fertility* **119** 232-238

27 Miro F & Hillier SG (1996) Modulation of granulosa cell deoxyribonucleic acid synthesis and differentiation by activin. *Endocrinology* **137** 464-468

28 Miro F, Smyth CD & Hillier SG (1991) Development-related effects of recombinant activin on steroid synthesis in rat granulosa cells *Endocrinology* **129** 3388-3394

29 Miro F, Smyth CD, Whitelaw PF, Milne M & Hillier SG (1995) Regulation of 3β-hydroxysteroid Δ5/Δ4-isomerase and cholesterol side-chain cleavage cytochrome P450 by activin in rat granulosa cells. *Endocrinology* **136** 3247-3252

30 Moreau JD, Satterlee DG, Rejman JJ, Cadd GG, Kousoulas KG & Fioretti WC (1998) Active immunization of Japanese quail hens with a recombinant chicken inhibin fusion protein enhances production performance. *Poultry Science* **77** 894-901

31 Nakamura M, Minegishi T, Hasegawa Y, Nakamura K, Igarashi S, Ito I, Shinozaki H, Miyamoto K, Eto Y & Ibuki Y (1993) Effect of activin A on follicle-stimulating hormone (FSH) receptor messenger ribonucleic acid levels and FSH receptor expressions in cultured rat granulosa cells *Endocrinology* **133** 538-544

32 Nohno T, Noji S, Koyama E, Myokai F, Ohuchi H, Nishikawa K, Sumitomo S, Taniguchi S & Saito T.(1993) Expression patterns of the activin receptor IIA and IIB genes during chick limb development. *Progress in Clinical Biology Research.* **383B** 705-714

33 Onagbesan OM & Peddie MJ (1995) Effects of insulin-like growth factor I and interactions with transforming growth factor alpha and LH on proliferation of chicken granulosa cells and production of progesterone in culture. *Journal of Reproduction and Fertility* **104** 259-265

34 Phillips DJ & Woodruff TK (2004) Inhibin: actions and signalling. *Growth Factors* **22** 13-18

35 Rombauts L, Vanmontfort D, Decuypere E & Verhoeven G (1996) Inhibin and activin have antagonistic paracrine effects on gonadal steroidogenesis

during the development of the chicken embryo. *Biology of Reproduction* **54** 1229-1237

36 Satterlee DG, Cadd GG & Fioretti WC (2002) Active immunization of broiler breeder hens with a recombinant chicken inhibin fusion protein enhances egg lay. *Poultry Science* **81** 519-528

37 Shukovski L, Dyson M & Findlay JK (1993) The effect of follistatin, activin and inhibin on steroidogenesis by bovine thecal cells. *Molecular and Cellular Endocrinology* **97** 19-27

38 Shukovski L, Findlay JK & Robertson DM (1991) The effect of follicle-stimulating hormone-suppressing protein or follistatin on luteinizing bovine granulosa cells *in vitro* and its antagonistic effect on the action of activin. *Endocrinology* **129** 3395-3402

39 Slappey SN & Davis AJ (2003) Expression pattern of messenger ribonucleic acid for the activin type II receptors and the inhibin/activin subunits during follicular development in broiler breeder hens. *Poultry Science* **82** 338-344

40 Vanmontfort D, Berghman LR, Rombauts L, Verhoeven G, Decuypere E (1994) Changes of immunoreactive inhibin, follicle-stimulating hormone, luteinizing hormone, and progesterone in plasma after short-term food deprivation and during the ovulatory cycle of the domestic hen. *General and Comparative Endocrinology* **95** 117-124

41 Vanmontfort D, Berghman LR, Rombauts L, Verhoeven G & Decuypere E (1995) Developmental changes in immunoreactive inhibin and FSH in plasma of chickens from hatch to sexual maturity. *British Poultry Science* **36** 779-790

42 Wang SY & Johnson PA (1993) Complementary deoxyribonucleic acid cloning and sequence analysis of the alpha-subunit of inhibin from chicken ovarian granulosa cells. *Biology of Reproduction* **49** 453-458

43 Woodruff TK & Mather JP (1995) Inhibin, activin and the female reproductive axis. *Annual Reviews of Physiology* **57** 219-244

44 Xiao S, Robertson DM & Findlay JK (1992) Effects of activin and follicle-stimulating hormone (FSH)-suppressing protein/follistatin on FSH receptors and differentiation of cultured rat granulosa cells. *Endocrinology* **131** 1009-1016

45 Zimmerman CM & Mathews LS (1996) Activin receptors: cellular signalling by receptor serine kinases. *Biochemical Society Symposium* **62** 25-38

Part nine

Behaviour and reproductive strategies

Great-tailed Grackle
Eyal Shochat

Functional Avian Endocrinology
A Dawson & P J Sharp (Eds)

Female mate choice

E Adkins-Regan

Department of Psychology and Department of Neurobiology and Behavior, Cornell University, Ithaca, NY 14853-7601 USA

Introduction

An active and exciting research area has emerged at the interface between behavioral endocrinology and animal behavior. At this hybrid zone, mate choice, mating strategies, and tactical decisions about reproduction are the focus of a cutting edge of efforts to understand behavioral evolution. At this hybrid zone, discovering the mechanisms enabling animals to enact these choices, strategies, and decisions in a fitness-enhancing way is a major priority. Birds are central to this effort, as they are to the study of ecologically relevant behavior more generally.

In comparing the approaches of behavioral endocrinology and animal behavior as applied to avian mate choice, one is struck by two contrasts. First, the study of mate choice emphasizes females: what they are choosing in males and why. This is understandable in light of theoretical predictions, often confirmed, that females should be more discriminating than males in their choices of mating partners (5, 25, 27). Avian behavioral endocrinology, however, emphasizes male behavior. Second, avian behavioral endocrinology focuses on the execution of motor patterns like singing, courting, and copulating. Mate choice, however, is revealed not simply by the behavior executed but by the targets for that behavior. Which other individuals has the animal decided to approach or try to attract? This is a different level of behavioral organization, and might be based on different (or at least not entirely overlapping) mechanisms and brain systems.

Nonetheless, we can ask if some of the concepts and principles that have emerged in the study of singing, courting and copulating also apply to mate choice. One of the goals of the author's research program has been to see if the activation/organization theoretical framework that has been so valuable for understanding the actions of sex steroids in the creation of sex differences in behavior can be extended to the level of mate choice. The distinction between activation and organization was first made decades ago (22). Briefly put, activation refers to hormone effects that normally occur at puberty and in adulthood, that are temporary and reversible (ceasing if the hormone level drops), and that permit the expression of behavior of which the animal is capable. Organization refers to hormone effects that most commonly occur earlier in development, during a critical period, and that permanently establish the sexual phenotype, including the animal's behavioral capabilities. These capabilities may then require activational hormones for expression, and thus complete sexual differentiation of behavior often results from a combination of organizational and activational hormone actions. A detailed summary of the application of these concepts to avian behavior and brains can be found in Balthazart and Adkins-Regan (6).

By distinguishing between the expression of mate choice and the nature of the choice that is expressed, it is possible to generate hypotheses that predict the contributions of these two kinds of sex steroid influences – activation and organization – to mate choice. The focus to date has been on two dimensions of mate choice: (a) choice of sex of partner (sexual partner preference), and (b) choice of opposite-sex partners with particular qualities or attributes. The first dimension is a fairly obvious one with a direct impact on fitness, but its hormonal basis has been largely ignored in birds and other animals that form socially monogamous pairbonds. The second dimension has been a major focus of theory and research by behavioral ecologists seeking to understand the benefits to females of choosing males with extravagant but seemingly useless ornaments such as long tails or elaborate songs. Laboratory studies testing hypotheses about hormonal mechanisms typically measure these two kinds of mating preferences by giving the females a choice of two conspecifics: a male and a female for dimension (a), and two males differing in the trait of interest for dimension (b). Because females don't necessarily display or show other distinctive behavior toward the preferred stimulus, physical proximity (time spent near each choice option) is the most commonly used measure of preference. Such two-choice tests can be followed by more naturalistic tests in which birds freely interact in an aviary over a longer period of time with an array of choices, and are allowed to pair.

Activational hormone effects on the expression of mate choice

When would an activational effect of sex steroids on mate choice be predicted? There are three likely possibilities. First, such effects would be expected when breeding is seasonal and mate preferences are expressed only when breeding. For example, some temperate zone socially monogamous birds form a new pairbond each spring and dissolve it in the fall, pairing with a new partner the following spring. Here an on/off mechanism is needed both for the pair formation and for the partner preference. Obviously seasonal hormonal cycles would be correlated with seasonality of mate choice, but correlation is not evidence of causation. An experimental manipulation is required. An interesting avian example for study would be birds such as California and Gambel's quail (*Callipepla californica* and *C. gambelii*). These are non-migratory and resident year round. During the non-breeding season, birds live in mixed-sex groups (coveys). As spring arrives, the birds form pairs that separate from the group to breed, expressing preferences based, among other qualities, on covey membership and the behavior of the males (11, 13, 15). One could hypothesize that rising sex steroid levels resulting from increasing daylengths stimulate mate choice, including the interest of females in those male behaviors. The right experiments could determine whether that is correct and whether there is a sex difference in which steroids are important. Some decades ago Emlen and Lorenz (12) found that non-breeding quail given high dose steroid implants formed pairs, but there have been no further studies of this kind.

If breeding is aseasonal (the birds breed year round, or they are opportunistic breeders with reproductive systems that are always "on"), however, one would not predict a major role for activational hormone effects on mate choice. The zebra finch (*Taeniopygia guttata*, Estrildidae) provides a test case. Not only are these birds opportunistic and aseasonal, but they pair at a young age and remain continuously paired regardless of whether and when rainfall permits breeding activity (28). There would not seem to be any need for an "on/off" mechanism for either dimension of mate choice during their adult lives (except perhaps when a female is deciding whether to engage in an extra-pair mating with a particular male). Consistent with this prediction, experimental manipulation of adult sex steroid levels (via implants of estradiol benzoate or testosterone propionate) has no effect on the strength of sexual partner preference of females in this species, even though it has very significant effects on the frequency of behaviors such as aggressive chasing (1, 3).

A second time to predict an activational effect is when mate choice is first exhibited at the onset of sexual maturity. A mechanism to turn mate choice "on" at that time would be needed even in birds like zebra finches that do not need such a mechanism thereafter. The onset of mate choice is particularly interesting in the estrildid finches because their close social relationships are expressed through direct physical contact behaviors such as clumping (sitting in contact with another bird) and allopreening (preening another bird). Juvenile zebra finches undergo a transition from an attachment to the family members to an interest in opposite-sex age unpaired birds and then to an attachment to the adult pairing partner, changing the targets for their clumping and allopreening from the parents and siblings to possible partners and finally to a single partner. It is not the behavior that changes, but the targets for the behavior. Hormonal changes of juvenile development are candidate mechanisms to activate this transition in both sexes, a hypothesis that can be tested experimentally.

Note that in the case of the zebra finch, sex steroids are hypothesized to be involved in the initial expression of mate choice and formation of a pairbond, but not in the continued maintenance of that choice and pair relationship. This important conceptual distinction between the formation and maintenance of a social relationship has precedents in the study of the hormonal basis of parental behavior.

Organizational hormone effects on the nature of mate choice

When would an organizational effect on mate choice be predicted? Prime candidates are sex differences in mate choice, when females and males have different preferences, especially if activational hormone manipulations fail to eliminate or reverse these sex differences.

The most robust and species general sex difference in mate choice is sexual partner preference, the preference for opposite-sex partners. As a socially monogamous bird that is continually paired, the zebra finch is an excellent case for studying the mechanisms that produce sexual partner preference. Experiments in which hormone levels of very young finches were manipulated support the hypothesis that organizational hormone actions contribute to the adult female's preference for a male pairing partner.

In one experiment, eggs were injected on day 5 of incubation with 20 μg fadrozole, an estrogen synthesis inhibitor, or with the saline vehicle (4). Fadrozole greatly reduces estrogen synthesis in the brain as well as the gonads, and in genetic females causes testes or ovotestes to develop instead of an ovary. When the females were tested for sexual partner preference as adults, they were masculinized, preferring other females as partners instead of males (Fig. 1). The presence of testes and male-typical androgen levels in adulthood (that is, activational hormone actions) is not likely to account for the same-sex preference of the treated females, because as already discussed giving normal adult females testosterone does not alter their sexual partner preference.

In another experiment, females were injected daily with 20 μg estradiol benzoate (EB) for the first two weeks after hatching (20). One group of EB treated females was then housed in an all-female (unisex) aviary during juvenile and young adult development, while a second group of EB treated females was housed in a mixed-sex aviary. Control groups were given vehicle injections and were housed either in a unisex or a mixed-sex aviary. When tested as adults, the EB treated females housed in a unisex aviary preferred to pair with other females (Fig. 2), whereas none of the other groups showed such a preference, instead pairing with males at the same rate as controls. Here early estrogen exposure seems to be masculinizing the females' preferences, supporting the hypothesis of an organizational hormone contribution to sexual partner preference, but the effect is dependent on a certain kind of social environment (social experience) during the juvenile or young adult period.

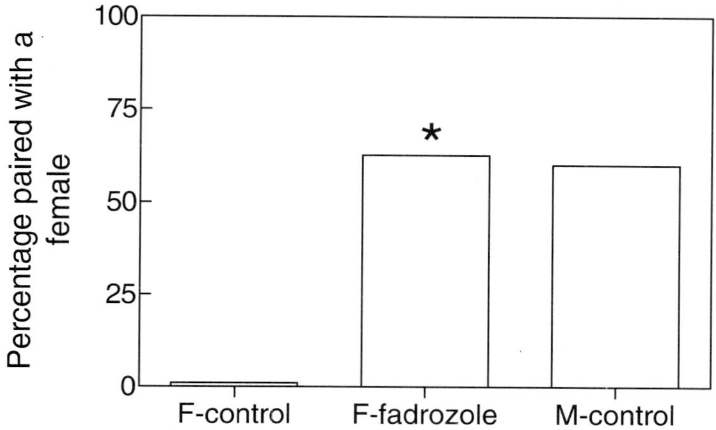

Fig. 1 Female zebra finches (F) hatched from eggs injected with fadrozole resembled males (M), not control females, in their sexual partner preference, as indicated by the percentage pairing with females in free interaction aviary tests. * $P < 0.05$ compared with F-control. Source (4).

This second experiment raises two challenging questions. First, what model of sexual differentiation of mate preference can account for the fact that both fadrozole and estrogen treatment caused females to be masculinized? Masculinization by estrogen has many precedents in the literature on mammalian sexual differentiation, and male zebra finches produce plenty of estrogen at all ages (2). But why do fadrozole and estradiol produce the same, not the opposite, effect? This is one of many puzzles and paradoxes in zebra finch sexual differentiation (6, 26). One possibility is that during early female development an ovarian hormone that is not an estrogen inhibits masculinization. By producing testes, fadrozole eliminates this factor, and by suppressing ovarian hormone production (through negative feedback) so does exogenous estrogen.

The other challenging question is why the combination of early estrogen treatment and all-female living was required for masculinization. One possibility comes from research with this species

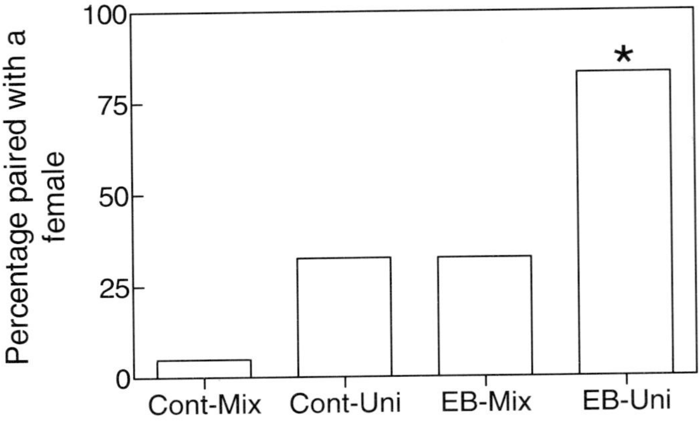

Fig. 2 Female zebra finches treated with estradiol benzoate (EB) as nestlings and housed in an all-female (Uni) aviary as juveniles paired with other females, unlike control females (Cont-Mix and Cont-Uni) or EB females housed in a mixed-sex aviary (EB-Mix). * $P < 0.05$ compared with Cont-Mix. Source (20).

on sexual imprinting and the development of species mating preferences (another dimension of mate choice). This research has revealed a second sensitive period (stage II) later in development that is required for consolidation of what was learned during the earlier sensitive period (stage I) for imprinting (9, 17, 18). It can be hypothesized that early estrogen treatment masculinizes what the females learn during stage I so that they acquire a male-typical, not female-typical, preference. If they are then housed in an all-female group until testing, their only tactile contact during stage II is with other females, which is consistent with and consolidates their stage I preference. (In this species pairing requires tactile contact between birds: 24). If they are housed in a mixed-sex group, however, the males attempt to pair with them, providing social experience that is inconsistent with what was learned in stage I, preventing a stable preference for other females from forming. If this hypothesis is correct, then females that are given estrogen as nestlings but are housed in an all-female aviary that prevents tactile contact should not develop a stable preference for other females. They should be more similar to control females than estrogen treated females that are allowed to have tactile contact with other females during stage II.

This hypothesis is supported by an experiment in which two groups of females treated with 10-20 μg EB daily as nestlings were housed in all-female aviaries prior to testing (Adkins-Regan, unpublished). A third, control, group received vehicle only and was housed in a mixed-sex aviary. For one EB group the aviary was constructed so that females could see and hear each other but had no tactile contact (EB-NTC). The other EB group was allowed to have tactile contact (EB-TC). Although group EB-NTC had the most abnormal social experience, it was predicted to be less affected by the early hormone treatment. The results were as predicted, in that EB-NTC females never differed from controls whereas EB-TC females did on several measures (Fig. 3). (Note that because the EB dosage was lower than in the previous experiment, EB-TC females were not actually masculinized to the point of preferring other females, but only to the neutral point of showing no preference.)

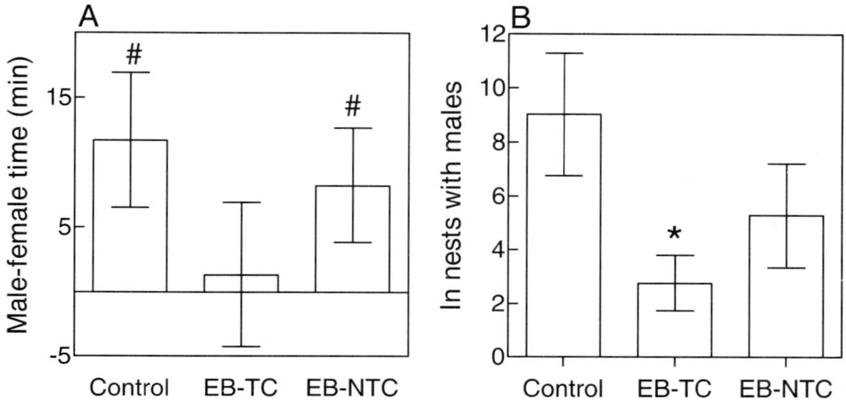

Fig. 3 Nestling estrogen (EB) treatment altered sexual partner preference in female zebra finches only if they had tactile contact (TC) with other females during juvenile development (NTC = no tactile contact). **A**: Mean ± SE difference in time spent near males vs. females in two-choice tests. # $P < 0.05$ compared with zero (no difference), indicating a significant preference for males. **B:** Frequency of being in a nest box with a male in aviary tests (one measure of pairing preference). * $P < 0.05$ compared with control group.

Are sex differences in the preferred qualities or attributes – mate choice dimension (2) – also hormonally organized? There are certainly many such sex differences in mate choice. In zebra finches, females but not males prefer to pair with birds that sing at high rates, sing normal syllables, and have red beaks (10, 16, 19). Here the sex differences take the form of preferences that are present in one sex and absent in the other. An organizational hypothesis is a very attractive possibility. It is not yet known, however, whether early hormone actions are responsible for the development of these sex differences.

A sex difference in preference has been discovered recently in Japanese quail (*Coturnix japonica*) that takes the form of opposite preferences in females and males rather than presence of a preference in one sex and absence in the other. Males spent much less time near females after they had seen them brooding chicks; that is, such females were avoided (23). When females were the subjects in a similar experiment, however, they showed the reverse preference, spending more time near males they had just seen brooding chicks (Fig. 4) (Zhou, Leung, and Adkins-Regan, unpublished). It will be interesting to see if this sex difference in how attracted the birds are to potential mating partners they have observed brooding is an activational or organizational phenomenon. When measured by how interested birds are in looking through a window at a female, sexual partner preference in this species is primarily organizational (7).

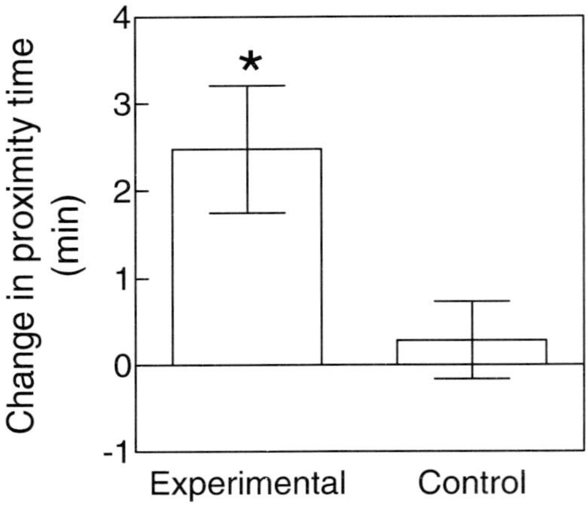

Fig. 4 Mean ± SE change in proximity time in two-choice tests with female Japanese quail choosing between male stimuli. Females spent more time near males after they had seen them brooding chicks (experimental trials). In control trials (no male was seen brooding) the same females' proximity times did not change. * $P < 0.05$, change in time compared with control trials.

In many species there are also sex differences in how discriminating the animals are, with females often showing greater selectivity in mating partners. There are two reasons to think that this sex difference is probably not an organizational phenomenon, however. First, degree of choosiness is not always a fixed species-typical character, but instead can be quite malleable within individuals in response, for example, to changes in the sex ratio of a mating group (8). Females become less choosy and males become choosier if there are more females than males in the group. Second, female dark-eyed juncos with experimentally elevated testosterone become less discriminating in mate choice tests, as if they have been masculinized by activational hormones (21).

Future directions

The activational and organizational sex steroid hypotheses that have been discussed here by no means exhaust the possible hormonal mechanisms underlying female mate choice. For example, oxytocin family peptides and their brain receptors have been shown to play important roles in the mating systems and partner preferences of some mammals, but testing similar hypotheses in birds has only just begun (14).

Acknowledgements

EA-R's research was supported by NSF grants BNS-8204462, IBN-9514088, and IBN-0130986.

References

1 Adkins-Regan E (1999) Testosterone increases singing and aggression but not male-typical sexual partner preference in early estrogen treated female zebra finches. *Hormones and Behavior* **35** 63-70

2 Adkins-Regan E, Abdelnabi M, Mobarak M & Ottinger MA (1990) Sex steroid levels in developing and adult male and female zebra finches (*Poephila guttata*). *General and Comparative Endocrinology* **78** 93-109

3 Adkins-Regan E & Ascenzi M (1987) Social and sexual behaviour of male and female zebra finches treated with oestradiol during the nestling period. *Animal Behaviour* **35** 1100-1112

4 Adkins-Regan E & Wade J (2001) Masculinized sexual partner preference in female zebra finches with sex-reversed gonads. *Hormones and Behavior* **39** 22-28

5 Andersson M (1994) *Sexual Selection*. Princeton University Press: Princeton

6 Balthazart J & Adkins-Regan E (2002) Sexual differentiation of brain and behavior in birds. In *Hormones, Brain and Behavior* **4** 223-301. Eds D Pfaff et al. Academic Press: San Diego

7 Balthazart J, Castagna C & Ball GF (1997) Aromatase inhibition blocks the activation and sexual differentiation of appetitive male sexual behavior in Japanese quail. *Behavioral Neuroscience* **111** 381-397

8 Berglund A & Rosenqvist G (2003) Sex role reversal in pipefish. In *Advances in the Study of Behavior* **32** 131-167. Eds PJB Slater, JS Rosenblatt, CT Snowdon & TJ Roper. Academic Press (Elsevier): Amsterdam

9 Bischof H-J & Clayton N (1991) Stabilization of sexual preferences by sexual experience in male zebra finches *Taeniopygia guttata castanotis*. *Behaviour* **118** 144-155

10 Burley N & Coopersmith CB (1987) Bill color preferences of zebra finches. *Ethology* **76** 133-151

11 Calkins JD, Hagelin JC & Lott DF (1999) California Quail (*Callipepla californica*). In *The Birds of North America*, No. 473. Eds A Poole & F Gill. The Birds of North America, Inc.: Philadelphia

12 Emlen JT & Lorenz, FW (1942) Pairing responses of free-living Valley Quail to sex-hormone pellet implants. *Auk* **59** 369-378
13 Gee JM (2003) How a hybrid zone is maintained: Behavioral mechanisms of interbreeding between California and Gambel's quail (*Callipepla californica* and *C. gambelii*). *Evolution* **57** 2407-2415
14 Goodson JL, Lindberg L & Johnson P (2004) Effects of central vasotocin and mesotocin manipulations on social behavior in male and female zebra finches. *Hormones and Behavior* **45** 136-143
15 Hagelin JC & Ligon JD (2001) Female quail prefer testosterone-mediated traits, rather than the ornate plumage of males. *Animal Behaviour* **61** 465-476
16 Houtman AM (1992) Female zebra finches choose extra-pair copulations with genetically attractive males. *Proceedings of the Royal Society London* B **249** 3-6
17 Immelmann K, Pröve R, Lassek R & Bischof H-J (1991) Influence of adult courtship experience on the development of sexual preferences in zebra finch males. *Animal Behaviour* **42** 83-89
18 Kruijt JP & Meeuwissen GB (1991) Sexual preferences of male zebra finches: effects of early and adult experience. *Animal Behaviour* **42** 91-102
19 Lauay C, Gerlach NM, Adkins-Regan E & DeVoogd TJ (2005) Female zebra finches require early song exposure to prefer high quality song as adults. *Animal Behaviour* in press.
20 Mansukhani V, Adkins-Regan E & Yang, S (1996) Sexual partner preference in female zebra finches: the role of early hormones and social environment. *Hormones and Behaviour* **30** 506-513
21 McGlothlin J, Neudorf DL, Casto JM, Nolan V Jr. & Ketterson ED (2004) Elevated testosterone reduces choosiness in female dark-eyed juncos (*Junco hyemalis*): evidence for a hormonal constraint on sexual selection? *Proceedings of the Royal Society London* B **271** 1377-1384
22 Phoenix CH, Goy RW, Gerall AA & Young WC (1959) Organizational action of prenatally administered testosterone propionate on the tissues mediating behavior in the female guinea pig. *Endocrinology* **65** 369-382
23 Ruscio MG & Adkins-Regan E (2003) Effect of female brooding behavior on male mate choice in Japanese quail, *Coturnix japonica. Animal Behaviour* **65** 397-403
24 Silcox AP & Evans SM (1982) Factors affecting the formation and maintenance of pair bonds in the zebra finch, *Taeniopygia guttata. Animal Behaviour* **30** 1237-1243
25 Trivers RL (1972) Parental investment and sexual selection. In *Sexual Selection and the Descent of Man 1871-1971* 136-179. Ed. B. Campbell. Aldine: Chicago
26 Wade J & Arnold AP (2004) Sexual differentiation of the zebra finch song system. *Annals of the New York Academy of Science* **1016** 540-559
27 Williams GC (1966) *Adaptation and Natural Selection: A Critique of Some Current Evolutionary Thought.* Princeton University Press: Princeton
28 Zann RA (1996) *The zebra finch: a synthesis of field and laboratory studies.* Oxford University Press: Oxford UK

Functional Avian Endocrinology
A Dawson & P J Sharp (Eds)

Activity-dependent regulation of seasonal neuroplasticity in the song system in songbirds

J J Sartor and G F Ball

Department of Psychological and Brain Sciences, Johns Hopkins University, Baltimore, MD 21218 USA

Singing behavior: Neural underpinnings and modulation in a seasonal context

In temperate zone birds, including oscine songbirds, the frequency of reproductive behaviors changes seasonally, being more common during periods of breeding than non-breeding. Most temperate species breed in the spring in order to synchronize the hatching of young with increased food availability (49, 10, 11). A proximate mechanism whereby birds can initiate successful reproduction at the appropriate time of year involves the use of photoperiod (10, 11). Given this, it follows that the occurrence of courtship behaviors, like singing in songbirds, would correlate positively with increasing daylengths and the onset of the breeding season. Physiological and behavioral responses to photoperiod are mediated by a biological system that includes both a neural and an endocrine component (4, 10, 11). In male songbirds, the onset of long daylengths in spring causes a suite of responses including testicular growth, testosterone (T) release, and a subsequent change in many hormone-controlled behaviors (24). Coordinated with changing daylength, T concentrations and changes in song output, temperate zone oscines exhibit seasonal changes in the volumes of many of the nuclei involved in song production, known as song control nuclei (2, 27, 43).

Studies of canaries (*Serinus canaria*) and zebra finches (*Taeniopygia guttata*) have elucidated the interconnected nuclei that comprise the song control circuit (9, 19, 27, 29). The motor pathway involved in song production consists of projections from HVC to the robust nucleus of the arcopallium (RA) which in turn projects to both the intercollicular nucleus and the tracheosyringeal division of the 12th cranial nerve. Projections from this circuit innervate the muscles of the syrinx and nuclei involved in coordination of song production with respiration (48). Projections from HVC also form part of a less direct route – the anterior forebrain pathway (9). Here, HVC projects to Area X of the medial striatum which itself projects to the medial region of the dorsolateral nucleus of the anterior hypothalamus (DLM). DLM projects to the lateral portion of the magnocellular nucleus of the anterior nidopallium (lMAN), which projects to RA. This neuroanatomical specialization known as the song control system has been accompanied by distinct neurochemical specializations, one of which being that several song nuclei contain steroid receptors (5). HVC, RA and MAN appear to contain androgen receptors (2, 6, 12). In addition, all songbirds examined have displayed significant expression of estrogen receptor-α in the caudal nidopallium, including in some cases HVC and the dorsolateral area surrounding RA (5). In HVC, both androgen and estrogen receptors are present in many cells but the occurrence of colocalization of the two receptor types is exceedingly rare (12).

Changes in singing behavior in canaries are accompanied by prominent seasonal changes in the song control system (26). Increases in singing are correlated with increases in the volumes of the song control nuclei in this species, and T is a major mediator of this adult neuroplasticity. Laboratory studies of canaries have demonstrated that volumes of the song control nuclei are large when T concentrations in the blood are high, and smaller when T concentrations are decreased (28). In wild European starlings, HVc, Area X and RA all change seasonally in correlation with circulating testosterone concentrations (34). Further examination of the role of T in altering volumes in the song circuit in other species has confirmed its importance in seasonal neuroplasticity (for example, 3, 25, 40).

How can T affect seasonal changes in song behavior and song system morphology?

Seasonal changes in the volume of HVC and RA involve different factors at the cellular level. In HVC there is evidence for a change in neuron number while for RA changes in cell size and spacing seem to be important (40, 41, 44). Testosterone could affect the volumes of the song control nuclei in several ways, which need not be mutually exclusive (5). One action of T could be a direct affect on HVC. Androgenic or estrogenic metabolites of T usually bind to their cognate receptor and induce gene transcription. Thus T could act either on neurons already within the nucleus or on new neurons migrating to the nucleus, which could result in an increase in singing behavior. In this sense, T is the main regulator of the seasonal response *via* an increase in the recruitment and/or survival of motor-pathway neurons in HVC (32). Secondly, it is possible that neurotrophins mediate the effects of T on HVC in an activity-dependent manner. Recently the activity-dependent neurotrophin brain-derived neurotrophic factor (BDNF) has been implicated as a mediator of the effects of T on HVC (31), and has been shown to be expressed in relation to amount of singing (18). Finally, T could be regulating the seasonal change in HVC size by acting elsewhere in the brain to increase the amount of time birds spend singing, and this increase in singing behavior could feed back to lead to an increase in HVC size. In this manner the activity of producing song could be driving at least some of the changes in HVC size independently of T actions in HVC. Each of the last two possibilities requires an activity-dependent mechanism whereby the act of singing feeds back on HVC to alter its volume seasonally. Activity-dependent changes in the brains of birds and other animals have recently been elucidated across various taxa.

Activity-dependent changes in brains of songbirds and other taxa

Seasonal changes in the brains of songbirds that result from changes in on-going behavior require a cellular mechanism whereby activity-dependent brain changes may occur. One possible class of proteins is the neurotrophins, which often show activity-dependent release and function. Throughout development, as well as into adulthood, there is a coupling of synaptogenesis resulting in morphological plasticity with the sensory input an organism experiences (47). BDNF is a trophic factor that appears to be closely regulated by activity both under physiological and pathological circumstances (7). In adulthood, BDNF and its receptor, TrkB, are involved in plasticity such that BDNF is synthesized and secreted in an activity-dependent manner (47).

The neurotrophin hypothesis proposes that repetitive neuronal activity enhances expression, secretion, and/or actions of neurotrophins at the synapse to modify synaptic transmission and connectivity (30, 39). This hypothesis can provide a connection between neuronal activity and synaptic plasticity, and has been extended to provide a molecular mechanism whereby alteration of an organism's behavior could drive changes in brain morphology, *via* regulation of BDNF secretion. Neurotrophins like BDNF could play an instructive or permissive role in activity-dependent synaptic modification of developing and adult brains (39). In the instructive role, modification results from action of BDNF at the synapse to modify neurotransmitter release, postsynaptic sensitivity, or synapse morphology, so leading to a persistent synaptic modification. In the permissive role, modification is induced by other factors that are associated with neuronal activity, whereas BDNF carries out housekeeping functions necessary for modification of the synapse (39).

Regulation of neurotrophin mRNA expression by electrical and synaptic activity was first demonstrated in the hippocampus, where high basal levels of neurotrophins are present. For example, BDNF mRNA levels are rapidly and potently up-regulated by epileptiform activity in the hippocampus and cerebral cortex (20). Although our knowledge of how neurotrophin expression and release are regulated is sparse, there is evidence suggesting that central neurons and muscle cells may release neurotrophins through activity-regulated release pathways (8, 46). BDNF secretion from BDNF-over-expressing dissociated hippocampal neurons is increased fivefold by depolarization (14). In general, there is evidence at the cellular level that BDNF secretion is influenced by neuronal activity. Interestingly, this secretion can be tied not only to activity of presynaptic neurons, but also to organismal activity and behavior.

BDNF mRNA expression in the rat visual cortex of young animals is developmentally regulated and influenced by visual experience (44). This is manifest as a tightly-regulated dissociation between the presence of mRNA versus protein in the visual cortex at different stages of development. Before eye opening, in all cortical layers a large number of visual cortical neurons contain BDNF mRNA with no detectable amount of protein. At later ages, the number of BDNF-immunostained cells increases; this cellular increase of BDNF protein labelling is blocked in dark-reared animals. Exposure of dark-reared rats to a brief 2h period of light restored the typical match between BDNF mRNA and protein, implying that visual experience controls the neuronal content of both BDNF mRNA as well as protein in developing visual cortex (44).

In adulthood, there is further evidence that alterations in behavior can regulate BDNF messages in the brain. In spontaneously hypertensive rats, which wheel-run up to 20 km/night, abrupt deprivation of habitual running leads to long-lasting decreases in BDNF and TrkB expression in hippocampus in a running length-dependent manner (47). Repetitive exercise also affects the magnitude and stability of these BDNF responses. Rats exercised on a treadmill for 5 consecutive days and euthanized at different intervals after the last bout of exercise showed a significant increase in BDNF mRNA in the spinal cord by 2h post-training (13). Voluntary wheel running has also been shown to lead to increased BDNF mRNA in the hippocampus and caudal neocortex of the rat (35).

Recently a relationship between behavior and neurotrophin expression has been demonstrated in canaries in the context of song production (18). Interestingly, this increase in BDNF and cell number is correlated with a change in song production characteristic of the breeding season, such that neuronal recruitment at this time is temporally related to changes in song (17), and neuronal recruitment, as well as BDNF expression, appear to mediated by testosterone (31, 32). This system provides a compelling model for the idea that changes in behavior can be driving changes in brain morphology, and that these changes may be mediated by the neurotrophin BDNF.

Related to the idea that aspects of a behavior can drive changes in the brain is the notion that not only behavioral, but also cognitive, activity can lead to regulation of BDNF expression. Support for this idea has recently been demonstrated in primates (42) as well as in rats (23). In monkeys, visual long-term memory of objects is stored in the inferior temporal (IT) cortex. During the formation of visual pair-association memory in split-brain monkeys, BDNF was observed to be up-regulated selectively in area 36 of IT cortex, but not in areas involved in earlier stages of visual processing (42). These results suggest that BDNF contributes to the reorganization of these neural circuits after visual long-term memory formation in primates. Furthermore, BDNF has been shown to have a role in the formation of spatial reference and working memory in rats learning a radial arm maze test (23). Radial arm maze training resulted in a significant increase in BDNF mRNA expression in the hippocampus, whereas infusion of antisense BDNF resulted in an impairment of spatial learning. Given this, the idea of "activity-dependent" effects can be expanded to include those changes in the brain that one can see after a change in the behavior of an organism.

Labelling of immediate early genes (IEGs), e.g. *zenk*, which respond rapidly to various signals associated with neural activity and growth (22), has been used in songbirds as indicators of areas and/or circuits in the brain involved in many responses and behaviors. IEG expression patterns can therefore be used in experiments where song output was manipulated as an indicator of activity in brain nuclei. There is evidence that both *zenk* mRNA and ZENK protein are regulated not only in response to song presentation in auditory areas involved in song perception [e.g. the caudomedial nidopallium (22) and the caudomedial mesopallium ventrale (21)] but that zenk can also be induced by singing in the motor pathway. In canaries, the act of singing, but not of hearing song, induced a rapid increase in expression of ZENK in HVC and other song nuclei (15). Results from this study indicate that singing induced ZENK expression in the brain of the singer, rather than the other way around (15), suggesting that ZENK could be a good marker for activity-dependent changes in relevant nuclei. Furthermore, studies of *zenk* mRNA expression in young and adult zebra finches have suggested that IEG expression in RA could be related to motor plasticity (16).

Evidence for singing-dependent seasonal changes in HVC volume

Recently there have been several lines of evidence indicating that seasonal changes in HVC size can be driven by changes in song output. These experiments have largely involved the manipulation of singing behavior and T concentrations in male canaries and starlings, followed by an examination of the effects of these manipulations on new neuron number and/or HVC volume, as well as on BDNF expression in some cases. In male canaries, BDNF mRNA expression is correlated with the number of songs a bird produces, and this induction of BDNF by song is found primarily in the RA-projecting neurons of the motor pathway (18). In castrated male canaries, the act of singing can rescue RA-projecting neurons from death (1). In male European starlings, singing has been experimentally manipulated in several ways and evidence has converged on the notion that seasonal changes in HVC volume can be activity-dependent, and that the effects of T and singing behavior on HVC size are independent. First, male starlings housed in a social situation where some males possessed a nestbox and sang, while others did not, had larger volumes of HVC, despite similar T concentrations between the two groups (36). This difference was not attributable to nestbox possession, but was related to song output. The data from this experiment is summarized in Fig. 1 (36). Furthermore, male starlings that were castrated and given lesions in the medial preoptic area (POM; known to decrease song in starlings, 33) showed smaller HVC volumes compared to males with lesions outside of the POM, regardless of whether or not they experienced T replacement (38). Lastly, castrated male starlings were devocalized using lesions of the syrinx and underwent either T replacement or not. Males with T who received sham surgeries had the largest volumes of HVC. Devocalized males with T had HVC volumes that were similar to those for sham-operated males without T replacement. Devocalized males with no hormone treatment had the smallest volumes of HVC. Interestingly, BDNF mRNA expression in this nucleus showed a similar pattern to the volume data (37). In this manner, multiple methods of manipulating song output all result in decreased volumes of HVC that occur independently of T concentrations, implying that some of the seasonal change in HVC volume must be activity-dependent, resulting from song production.

Conclusion

Given the investigations presented above regarding behavior-driven changes in the brain nuclei involved in a particular behavior, it appears that singing has effects on the size of nuclei involved in the motor production of song that occur independently of the actions of T and are therefore activity-dependent. Whereas correlational data have supported the notion of activity-dependent changes in the songbird brain, recent studies are the first to demonstrate this experimentally. These data therefore challenge the traditional idea of one-way actions of testosterone first on song nucleus size and then on singing behavior, and suggest a feedback loop whereby singing drives changes in the brain and vice versa. Seasonal control of singing behavior in seasonally-breeding songbirds is thus likely a mechanism whereby a change in a courtship behavior can lead to regulation of brain areas involved in that courtship behavior. In this way, seasonal patterns of song output regulate the size of song nuclei and allow the organism to fine-tune his reproductive efforts to that time of the year in which he would be most likely to achieve successful rearing of his offspring.

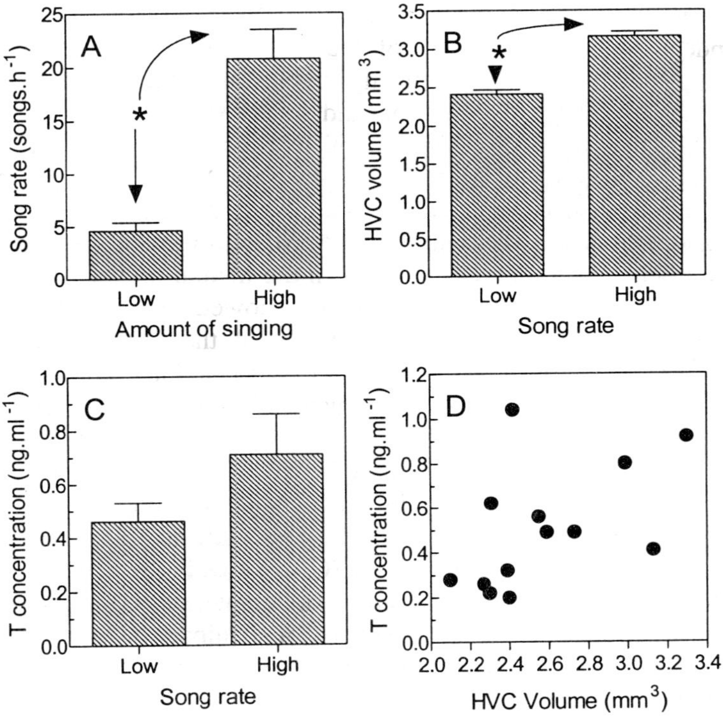

Fig 1 Nestbox manipulation results in differential rates of singing in European starlings (A). Males who sang at a high rate had larger volumes of HVC (B), despite similar concentrations of testosterone (C). There was no correlation between T concentrations and HVC volumes (D). Asterisks represent significance at $p<0.05$. [Adapted from (36).]

Acknowledgements

The authors would like to thank Jacques Balthazart, Eric Fortune, and Lauren Riters for stimulating discussions of the ideas presented here. This work was supported by NIH/NINDS R01 35467 to GFB and NSERC PGS-B to JJS.

References

1 Alvarez-Borda B & Nottebohm F (2002) Gonads and singing play separate, additive roles in new neuron recruitment in adult canary brain. *Journal of Neuroscience* **22** 8684-8690

2 Arnold AP, Nottebohm F & Pfaff DW (1976) Hormone concentrating cells in vocal control areas of the brain of the zebra finch (*Poephila guttata*). *Journal of Comparative Neurology* **165** 487-512

3 Ball GF, Auger CJ, Bernard DJ, Charlier TD, Sartor JJ, Riters LV & Balthazart J (2004) Seasonal plasticity in the song control system: Multiple brain sites of steroid hormone action and the importance of variation in song behavior. *Annals of the New York Academy of Sciences* **1016** 586-610

4 Ball GF & Balthazart J (2002) Neuroendocrine mechanisms regulating reproductive cycles and reproductive behavior in birds. In *Hormones, Brain and Behavior Vol. 2* 649-748. Ed DW Pfaff. Academic Press: New York

5 Ball GF, Riters LV & Balthazart J (2002) Neuroendocrinology of song behavior and avian brain plasticity: Multiple sites of action of sex steroid hormones. *Frontiers in Neuroendocrinology* **23** 137-178

6 Balthazart J, Foidart A, Wilson EA & Ball GF (1992) Immunocytochemical localization of androgen receptors in the male songbird and quail brain. *Journal of Comparative Neurology* **317** 407-420

7 Barde Y (1994) Neurotrophins: A family of proteins supporting the survival of neurons. *Progress in Clinical and Biological Research* **390** 45-46

8 Blochl A & Theonen H (1995) Characterization of nerve growth factor (NGF) release from hippocampal neurons: Evidence for a constitutive and an unconventional sodium-dependent regulated pathway. *European Journal of Neuroscience* **7** 1220-1228

9 Bottjer SW, Halsema KA, Brown SA & Meisner EA (1989) Axonal connections of a forebrain nucleus involved with vocal learning in zebra finches. *Journal of Comparative Neurology* **279** 312-326

10 Dawson A, King VM, Bentley GE & Ball GF (2001) Photoperiodic control of seasonality in birds. *Journal of Biological Rhythms* **116** 365-380

11 Follett BK (1984) Birds. In *Marshall's Physiology of Reproduction* 283-350. Ed GE Lamming. Longman Greene Press: London

12 Gahr, M (1990) Localization of androgen receptors and estrogen receptors in the same cells of the songbird brain. *Proceedings of the National Academy of Sciences of the USA* **87** 8445-9448

13 Gomez-Pinilla F, Ying Z, Opazo P, Roy RR & Edgerton VR (2001) Differential regulation by exercise of BDNF and NT-3 in rat spinal cord and skeletal muscle. *European Journal of Neuroscience* **13** 1078-1084

14 Goodman LJ, Valverde J, Lim F, Geschwind MD & Federov HJ (1996) Regulated release and polarized localization of brain-derived neurotrophic factor in hippocampal neurons. *Molecular and Cellular Neuroscience* **7** 222-238

15 Jarvis ED & Nottebohm F (1997) Motor-driven gene expression. *Proceedings of the National Academy of Sciences of the USA* **94** 4097-4102

16 Jin H & Clayton DF (1997) Localized changes in immediate-early gene regulation during sensory and motor learning in zebra finches. *Neuron* **21** 1049-1059

17 Kirn JR, O'Loughlin B, Kasparian S & Nottebohm F (1994) Cell death and neuronal recruitment in the high vocal center of adult male canaries are temporally related to changes in song. *Proceedings of the National Academy of Sciences of the USA* **91** 7844-7848

18 Li XC, Jarvis ED, Alvarez-Borda B, Lim DA & Nottebohm F (2000) A relationship between behavior, neurotrophin expression, and new neuron survival. *Proceedings of the National Academy of Sciences of the USA* **97** 8584-8589

19 Margoliash D (1997) Functional organization of forebrain pathways for song production and perception. *Journal of Neurobiology* **33** 671-693

20 McAllister AK, Katz LC & Lo DC (1999) Neurotrophins and synaptic plasticity. *Annual Review of Neuroscience* **22** 295-318

21 Mello CV & Clayton DF (1994) Song-induced ZENK gene expression in auditory pathways of songbird brain and its relation to the song control system. *Journal of Neuroscience* **14** 6652-6666

22 Mello CV, Vicario DS & Clayton DF (1992) Song presentation induces gene expression in the songbird forebrain. *Proceedings of the National Academy of Sciences of the USA* **89** 6818-6822

23 Mizuno M, Yamada K, Olariu A, Nawa H & Nabeshima T (2000) Involvement of brain-derived neurotrophic factor in spatial memory formation and maintenance in a radial arm maze test in rats. *Journal of Neuroscience* **20** 7116-7121

24 Nicholls TJ, Goldsmith AR & Dawson A (1988) Photorefractoriness in birds and comparison with mammals. *Physiological Reviews* **68** 133-176

25 Nottebohm F (1980) Testosterone triggers growth of brain vocal control nuclei in adult female canaries. *Brain Research* **189** 429-436

26 Nottebohm F (1981) A brain for all seasons: Cyclical anatomical changes in song-control nuclei of the canary brain. *Science* **214** 1368-1370

27 Nottebohm F, Kelley DB & Paton JA (1982) Connections of vocal control nuclei in the canary telencephalon. *Journal of Comparative Neurology* **207** 344-357

28 Nottebohm F, Nottebohm ME, Crane LA & Wingfield JC (1987) Seasonal changes in gonadal hormone levels of adult male canaries and their relation to song. *Behavioral and Neural Biology* **47** 197-211

29 Nottebohm F, Stokes TM & Leonard CM (1976) Central control of song in the canary, *Serinus canaria. Journal of Comparative Neurology* **165** 457-486

30 Poo MM (2001) Neurotrophins as synaptic modulators. *Nature Reviews Neuroscience* **2** 24-32

31 Rasika S, Alvarez-Buylla A & Nottebohm F (1999) BDNF mediates the effects of testosterone on the survival of new neurons in an adult brain. *Neuron* **22** 53-62

32 Rasika S, Nottebohm F & Alvarez-Buylla A (1994) Testosterone increases the recruitment and/or survival of new high vocal center neurons in adult female canary brains. *Proceedings of the National Academy of Sciences of the USA* **91** 7854-7858

33 Riters LV & Ball GF (1999) Lesions to the medial preoptic area affect singing in the male European starling *(Sturnus vulgaris). Hormones and Behavior* **36** 276-286

34 Riters LV, Eens M, Pinxten R & Ball GF (2002) Seasonal changes in the densities of alpha(2) noradrenergic receptors are inversely related to

changes in testosterone and the volumes of song control nuclei in male European starlings. *Journal of Comparative Neurology* **444** 63-74

35 Russo-Neustadt A, Ha T, Ramirez R & Kesslak JP (2001) Physical activity-antidepressant treatment combination: Impact on brain-derived neurotrophic factor and behavior in an animal model. *Behavioral Brain Research* **120** 87-95

36 Sartor JJ & Ball GF (in press) Social suppression of song is associated with a reduction in volume of a song control nucleus in European starlings (*Sturnus vulgaris*). *Behavioral Neuroscience*

37 Sartor JJ, Charlier TD, Pytte CL, Balthazart J & Ball GF (2002) Converging evidence that song performance modulates seasonal changes in the avian song control system. *Society for Neuroscience Abstracts* **2002** Program No. 781.10

38 Sartor JJ, Riters LV, Alger SJ & Ball GF (2004) Song suppression via POM lesion decreases the volume of song nucleus HVC in European starlings. *Society for Neuroscience Abstracts* **2004**

39 Schinder AF & Poo M (2000) The neurotrophin hypothesis for synaptic plasticity. *Trends in Neurosceinces* **23** 639-645

40 Smith GT, Brenowitz EA, Beecher MD & Wingfield JC (1997) Seasonal changes in testosterone, neural attributes of song control nuclei, and song structure in wild songbirds. *Journal of Neuroscience* **17** 6001-6010

41 Smith GT, Brenowitz EA & Wingfield JC (1997) Roles of photoperiod and testosterone in seasonal plasticity of the avian song control system. *Journal of Neurobiology* **32** 426-442

42 Tokuyama W, Okuno H, Hashimoto T, Li YX & Miyashita Y (2000) BDNF upregulation during declarative memory formation in monkey inferior temporal cortex. *Nature Neuroscience* **3** 1134-1142

43 Tramontin AD & Brenowitz EA (2000) Seasonal plasticity in the adult brain. *Trends in Neurosciences* **23** 251-258

44 Tramontin AD, Smith GT, Breuner CW & Brenowitz EA (1998) Seasonal plasticity and sexual dimorphism in the avian song control system: Stereological measurement of neuron density. *Journal of Comparative Neurology* **396** 186-192

45 Tropea D, Capsoni S, Tongiorgi E, Giannotta S, Cattaneo A & Domenici L (2001) Mismatch between BDNF mRNA and protein expression in the developing visual cortex: The role of visual experience. *European Journal of Neuroscience* **13** 709-721

46 Wang X & Poo M (1997) Potentiation of developing synapses by postsynaptic release of neurotrophin-4. *Neuron* **19** 825-835

47 Widenfalk J, Olson L & Thoren P (1999) Deprived of habitual running, rats downregulate BDNF and TrkB messages in the brain. *Neuroscience Research* **34** 125-132

48 Wild JM (1997) Neural pathways for the control of birdsong production. *Journal of Neurobiology* **33** 653-670

49 Wingfield JC (1983) Environmental and endocrine control of reproduction: An ecological approach. In Avian Endocrinology: Environmental and Ecological Perspectives 265-288. Eds SI Mikami, K Homma & M Wada. Springer-Verlag: Berlin

Functional Avian Endocrinology
A Dawson & P J Sharp (Eds)

Modulation of steroid-dependent male sexual behavior and neural gene expression: A role for steroid receptor co-activators

T D Charlier, G F Ball[1] and J Balthazart

University of Liège, Center for Cellular and Molecular Neurobiology, Research Group in Behavioral Neuroendocrinology, Liège, Belgium
[1]Department of Psychological and Brain Sciences, Johns Hopkins University, Baltimore, MD 21218, USA

Introduction

The neuroendocrine mechanisms that mediate the activation and differentiation of reproductive behavior have been the focus of intensive research during the last 30 years (e.g. 11, 27, 35). In most vertebrate species, male sexual behavior is activated by a localized action in the brain of the gonadal steroid testosterone (T) and the preoptic area (POA) appears to be a key site for the action of this androgen (reviewed in (20). In Japanese quail (*Coturnix japonica*) specifically, male copulatory behavior can be activated in castrates by stereotaxic T implants in the medial preoptic nucleus (POM), which constitutes the most medial part of the POA (10). The POM is a sexually differentiated nucleus (31) that appears to be a necessary and sufficient site for the activation of male sexual behavior by T. T action in the POM is associated with a variety of morphological and neurochemical modifications of this structure. For example, T produces a significant increase in the total volume of the POM (32, 33), induces a marked reorganization of its cellular and sub-cellular organization (4, 13, 30) and regulates the distribution, density and/or activity of several enzymes, neuropeptides, neurotransmitters and their receptors in the POM (5).

Interestingly, male and female quail show a sexually differentiated response to a similar treatment with T. In particular, female quail never show male-typical copulatory behavior even after ovariectomy and exposure to doses of T that are behaviorally effective in males (1, 9). In the majority of avian and mammalian species that have been studied to date, the transformation of androgens into estrogens, a metabolic step catalyzed by the enzyme aromatase, plays a key role in the control of many physiological processes related to reproduction (22, 24). In quail in particular, T aromatization into 17β-estradiol (E2) has been recognized as a limiting step in the activation by T of male sexual behavior (31, 2). Preoptic aromatase activity is sexually differentiated in quail but this neuroendocrine difference alone is not sufficient to explain the behavioral sex dimorphism since injections of estrogens, which should by-pass the enzymatic difference, are not sufficient to activate male-typical copulatory behavior in females (see reference 1 for additional discussion).

The biological effects of steroids result largely from specific interactions with intracellular receptors located in the target tissues. Steroid hormone receptors are members of the nuclear receptor superfamily of ligand-inducible transcription factors that include receptors for estrogens, androgens, progestins, glucocorticoids, thyroid hormones, vitamin D3, retinoic acids and 9-cis retinoic acids along with numerous orphan receptors. When exposed to their ligand, the steroid-receptor complexes dimerize, bind to hormone response elements located on the DNA and regulate gene transcription (review in 23). Recent studies reveal that receptors undergo a characteristic conformational change upon hormone activation that allows the recruitment of several proteins implicated in the transcriptional activation (see Fig. 1; review in 21, 26). The function of these proteins is to increase (coactivators) or decrease (corepressors) the binding to the response elements and thus the transcriptional activity of the receptors. Coactivators enhance transcription by acting as a bridge to the basal transcriptional machinery and via histone acetyltransferease activity (review in 25).

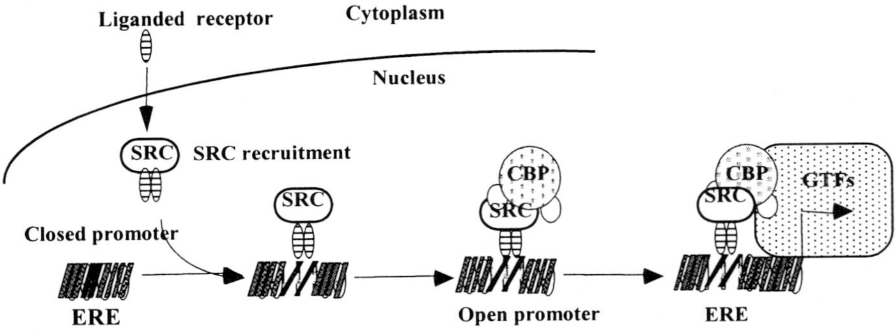

Fig. 1 General dynamic model of coactivator recruitment and chromatin remodeling during estrogen receptor transactivation. **ERE**: Estrogen Receptor Element; **CBP/p300:** CREB-binding protein/ p300; **p/CAF**: p300/CBP associated factor; **SRC**: Steroid Receptor Coactivator;. **GTFs**: General Transcription Factors.

The steroid receptor coactivator (SRC) family consist of three newly identified coactivators designed SRC-1, SRC-2 (TIF2/ GRIP1) and SRC-3 (p/CIP / RAC3 / ACTR / TRAM1 / AIB1) (review in 25). Most studies so far have used *in vitro* assay techniques to elucidate the interactions between nuclear receptors and their coactivators but the importance of these coactivators under physiological conditions is still poorly understood. However, Auger and collaborators (6) showed during *in vivo* experiments that the organizing effects of estrogens on the brain and on sexual behavior can be decreased or even suppressed by SRC-1 antisense oligonucleotide injection, clearly indicating that estradiol action during sexual differentiation is critically associated with the expression of SRC-1. Moreover, antisense oligodeoxynucleotides to SRC-1 infused into the VMN interfere with the expression of progesterone-facilitated lordosis behavior in female rats (29) confirming the physiological relevance of SRC-1 in hormone-mediated processes.

In quail, the neural and behavioral effects of estrogens both during ontogeny and in adulthood have been particularly well characterized (11, 8). The large magnitude of the effects of estrogens that have been described and the detailed understanding of the endocrine control of the sexual differentiation of brain and behavior in this species make it a useful model to analyze the mechanisms of steroid action and in particular the functional significance of steroid receptor coactivators and repressors. We postulated that coactivators may be responsible, at least in part, for the sexually differentiated responses to sex steroids that have been reported in male and female quail. The aim of this study was thus to acquire some insight into the physiological relevance of one of these coactivators, SRC-1.

Characterization of SRC-1 in quail and control of expression

The first part of our study was dedicated to the cloning of the quail SRC-1, the analysis of the neuroanatomical localization of its mRNA and the study of the potential regulation of its expression, in particular by steroids. The full-length cDNA of the quail SRC-1 was obtained by PCR using degenerate primers derived from the mouse and human protein sequences. The quail SRC-1 has all the characteristic features of the p160 coactivator family. The open reading frame encodes a 1508 amino acids protein with a predicted molecular weight of 161.6 kDa (Fig. 2). The presence of bHLH-PAS domains and of multiple NR boxes ("LXXLL" motif) in the quail sequence is an additional hallmark confirming its SRC-1 homology. In mammals, several variants of SRC-1 (SRC-1a-e) have been reported. The longest of these variants (SRC-1a) includes an extreme c-terminal "LXXLL" motif and is the most closely related to our isolated cDNA. No other isoform was detected in our study; this is similar to what has been reported in the chicken brain (3). Further studies should, however, be carried out to determine whether this absence of isoforms truly represents a species difference and is a general characteristic that differentiates birds from mammals.

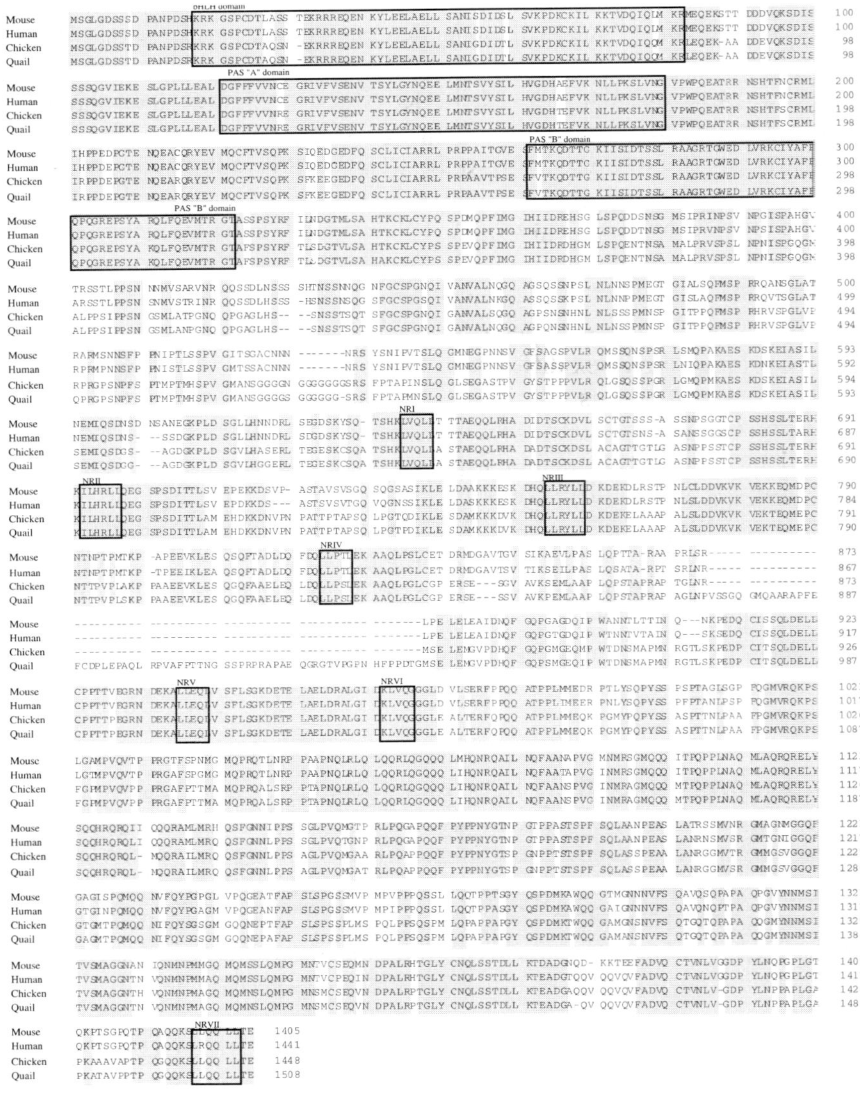

Fig. 2 Multiple alignment of the deduced SRC-1 amino acid sequences in mouse, human, chicken and quail. The amino sequence is shown by the single-letter code. Identical residues in three species are shaded and conserved domains such as bHLH-PAS (basic Helix Loop Helix-Per Arnt Sim) and LXXLL domains are boxed.

The tissue distribution of the SRC-1 mRNA was then identified by semi-quantitative RT-PCR and was found to be reminiscent of the distribution previously reported by northern blot in rats (28). We also analyzed with this study in more detail the relative expression of the coactivator in specific brain regions such as the telencephalon, diencephalon, optic lobes and brain stem in the male and female brain (comparison based on several cycles of amplification). As expected for a relatively ubiquitous coactivator, and in agreement with the results obtained in rodents, all the quail tissues that were studied showed a detectable level of SRC-1 expression (see Fig. 3). Lower levels of SRC-1 transcripts were observed in muscles and in the cerebellum in contrast to what had been observed in rat. The tissue differences in SRC-1 expression observed in this study by RT-PCR suggests that, although SRC-1 is ubiquitously expressed throughout the quail body, its relative level of expression is tissue-dependent. The SRC-1 transcript was expressed in the same tissues of male and female quail suggesting that no major qualitative sex difference is present in SRC-1 expression. Some quantitative differences seem however present.

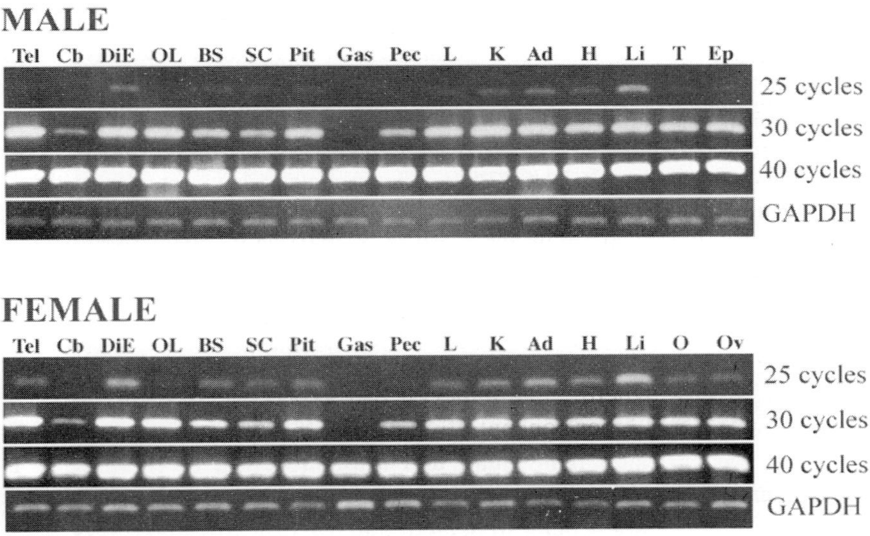

Fig. 3 SRC-1 mRNA tissue distribution as observed by semi-quantitative RT-PCR (25-30-40 amplification cycles) using GAPDH as internal standard. **Tel**: telencephalon; **Cb** Cerebellum; **DiE**: diencephalon; **OL**: optic lobe; **BS**: brain stem; **SP**: spinal cord; **Pit**: pituitary gland; **Gas**: gastrocnemius muscle; **Pec**: pectoral muscle; **Li**: liver; **K**: kidney; **Ad**: adrenal gland; **H**: heart; **L**: lung; **T**: testis; **Ep**: epididymis; **O**: ovary; **Ov**: oviduct

Additional studies were therefore completed using real-time quantitative PCR (qPCR) procedures to quantify precisely the SRC-1 mRNA content in particular brain areas. These studies indicated that SRC-1 expression was higher in the optic lobes of females as compared to males (although this difference was not statistically significant due to a large intra-sex heterogeneity) but was definitely higher in the preoptic area-hypothalamus (HPOA) of males compared to females (17 see Fig. 4a).

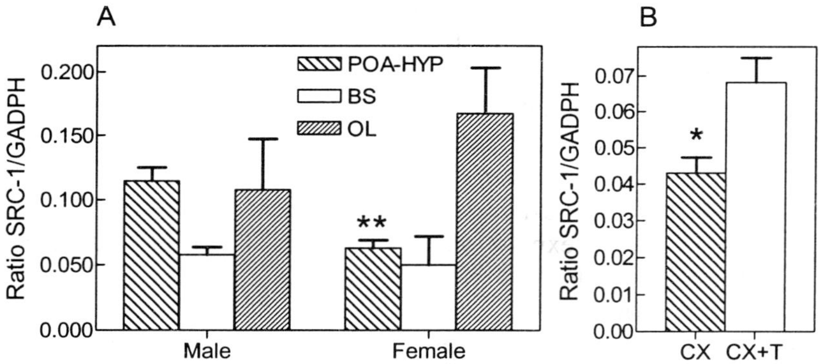

Fig. 4 Bar graphs illustrating the SRC-1/GAPDH ratio as measured by real time quantitative PCR. GAPDH was used to normalize the samples. **A.** Comparison of the SRC-1/GAPDH ratio in males versus females in the preoptic area-hypothalamus (POA-HYP), brain stem (BS) and optic lobes (OL). Data presented are means ± S.E.M. $^{**}p<0.005$ by comparison with the same region in males. **B.** Comparison of the SRC-1/GAPDH ratio in preoptic area hypothalamus in castrated males (CX) and castrated male treated with testosterone (CX+T). $^{*}p<0.05$.

To test whether the higher SRC-1 expression in males was due to their relatively higher concentration of T in the plasma as compared to females, we castrated male quail and implanted half of them with T. SRC-1 expression was then analyzed in the preoptic area-hypothalamus by qPCR. This experiment indicated that SRC-1 mRNA levels were higher in the T-treated males than in untreated castrates clearly suggesting that SRC-1 expression is not constitutive and is regulated by steroids at least in the preoptic area-hypothalamus (See Fig. 4b). SRC-1 expression was higher in T-treated birds but stress (one-hour restraint at 4°C) or behavioral manipulations tended to inhibit these stimulatory effects of the steroid (15), confirming the environmental modulation of SRC-1 expression observed in rat (12). Moreover, subsequent experiments analyzing SRC-1 expression in males and females by qPCR and Western blotting unexpectedly showed that sexually mature females had higher concentrations of SRC-1 mRNA and protein in the HPOA than males. One potential explanation to this discrepancy between results of successive experiments is that the short ovulatory cycle and its associated rapid endocrine changes could regulate SRC-1 expression in females as T does in males. This hypothesis however remains to be tested.

Together, these data support the idea that SRC-1 is not constitutively expressed but regulated by steroids, stress and possibly other unidentified factors. Differential controls that are brain regions specific might also occur but such controls could only be identified by quantitative immunohistochemistry and *in situ* hybridization.

Neuroanatomical localization of SRC-1

In situ hybridization histochemistry procedures were therefore developed to determine the precise neuroanatomical localization of SRC-1 transcripts in quail. SRC-1 mRNA was broadly distributed throughout the quail brain but its expression was found to be particularly dense in several diencephalic, limbic and mesencephalic areas that contain high densities of androgen and estrogen receptors. For example, relatively dense staining was observed in the POM and the BST (see Fig. 5), in the periventricular region of the hypothalamus, including the paraventricular (PVN) and ventromedial (VMN) nuclei, and in the nucleus intercollicularis (17). All these brain regions are known to be steroid targets that are involved in the control of reproductive behavior (reviewed for example in 7).

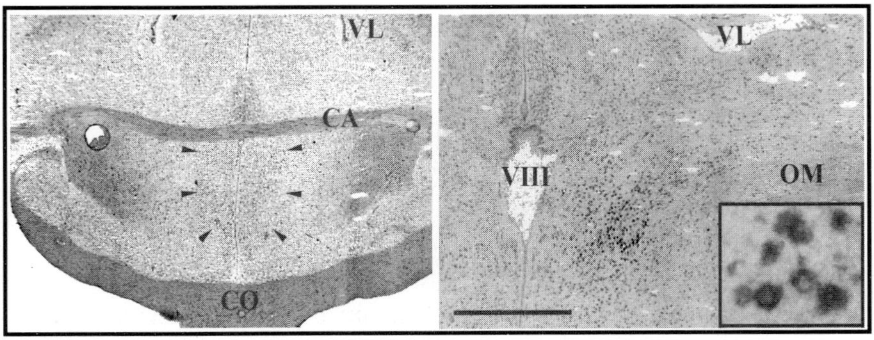

Fig. 5 Photomicrographs illustrating the distribution of the SRC-1 transcript in the medial preoptic area (indicated by arrows in the left panel) and in the nucleus of the stria terminalis as identified by non radioactive in situ hybridization. The insert shows positive cells at high magnification. **CA**: anterior commissure; **CO**: Optic chiasma; **OM**: occipito-mesencephalicus tract; **VL**: Lateral ventricule; **VIII**: 3rd ventricle. Magnification bar corresponds respectively to 1000μm, 250μm and 80μm from left to right.

The distribution of SRC-1 transcripts was thus consistent with the notion that the coactivator could be important for the modulation of the action of sex steroid receptors. To test how general the distribution observed in quail was amongst other avian species and to investigate whether the specialized steroid-regulated vocal control system present in the telencephalon of songbirds is associated with a particular distribution pattern for, we also localized SRC-1 transcripts in the canary brain. Consistent with the results previously obtained in quail, these studies revealed that SRC-1 is broadly expressed throughout the brain of both male and female canaries and that the coactivator expression is particularly high in steroid-sensitive brain regions (16). Interestingly, several telencephalic song control nuclei such as HVC, area X of the medial striatum, the lateral part of the magnocellular nucleus of the nidopallium (lMAN) and the Interfacial nucleus of the nidopallium (Nif) also expressed a high densities of SRC-1 transcripts (See 36 for discussion of song system nomenclature). The analysis of the surface and volume defined by the high density of SRC-1 transcripts identified a marked sex difference in HVC and area X for these measures (see Fig. 6). Moreover, the boundaries of area X could be outlined by the dense expression of SRC-1 in males but not in females. In these females, SRC-1 expression was equivalent within area X and in the surrounding medial striatum (mST).

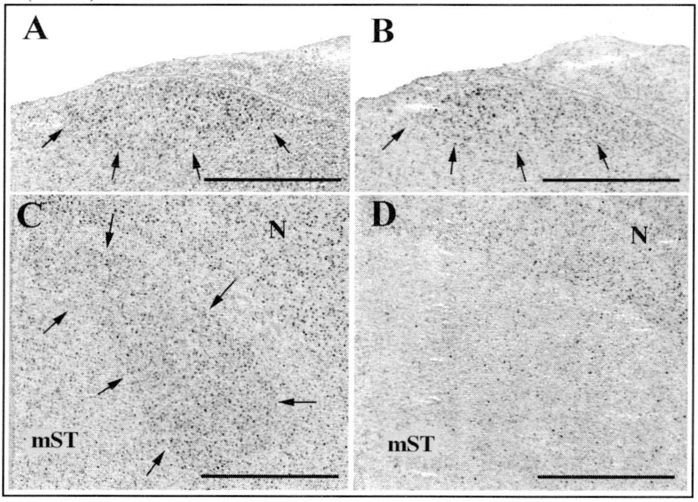

Fig. 6 A-D Photomicrographs illustrating the dense expression of SRC-1 transcripts in the song control nuclei HVC (A-B) and area X (C-D) of male (A and C) and female (B and D) canaries. Magnification bar is 1000 µm. E: Volume of HVC and Area X as defined by the dense SRC-1 signal in male and female canaries. Data presented are means ± S.E.M. *$p<0.05$ for comparison by unpaired Student *t*-tests.

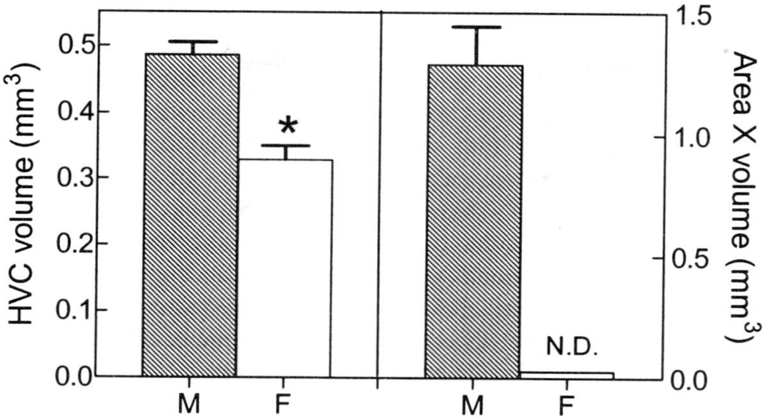

Fig. 6E Volume of HVC and Area X as defined by the dense SRC-1 signal in male (M) and female (F) canaries. Data presented are means ± S.E.M. *$p<0.05$ for comparison by unpaired Student *t*-tests.

Taken together, these results show that the SRC-1 distribution is highly conserved in the brain of different avian species and resembles the distribution identified in rodents. It is also noteworthy that the distribution of coactivators and steroid receptors are to a large extent correlated. SRC1 is densely expressed in classical steroid targets such as the preoptic area and hypothalamus but also in the newly evolved steroid-sensitive areas of the songbird telencephalon.

Physiological function of SRC-1

These anatomical data describing the expression of SRC-1 suggested that this coactivator could play a prominent role in the control of the activation by steroids of male–typical traits in adult. To test this notion, the antisense technique was recently used to specifically down regulate the expression of SRC-1 at the level of the preoptic area (14). Castrated male Japanese quail were stereotaxically implanted with an injection cannula into the third ventricle to allow the central administration of the antisense oligodeoxynucleotide (ODNs) that consisted in chimers of locked nucleic acid (LNA) and deoxyribonucleotides. Birds were injected daily with 1 μg of antisense LNA or two different control substances, either scrambled LNA or the vehicle alone (saline). Injections of LNA antisense ODNs directed against the mRNA for SRC-1 resulted in a significant knock-down of the corresponding protein as measured by Western blots. The knock-down of the coactivator induced in turn a profound inhibition of the testosterone-dependent male sexual behaviors, including neck grabs (NG), mount attempts (MA), mounts

(M) and cloacal contact movements (CCM) as well as struts, a pre- and post-copulatory display. These effects of the ODNs seemed to be limited to the brain given that no effect was observed on peripheral steroid-dependent responses such as the cloacal gland growth. In quail, it has been clearly established that some of these behaviors (e.g., crowing, strutting) are dependent on androgenic metabolites of testosterone while others (e.g., NG, M and CCM) depend mainly on estrogenic metabolites (2, 37). The present results therefore support the notion that SRC-1 facilitates the transcriptional actions of both androgen and estrogen receptors on male-typical behaviors. Adult testosterone action also regulates male-biased sex differences in the volume of the POM in quail, as well as the number of cells expressing immunoreactive aromatase and fibers immunoreactive for the peptide vasotocin in this brain area. Accordingly, the antisense ODN treatment decreased the volume of this nucleus and the density of immunoreactivity for aromatase and vasotocin, so that all these measures were closer after treatment to the female-typical than the male-typical pattern. Together with the Western blot and behavioral results, these histological data therefore confirm that SRC-1 indeed regulates steroid action in the POA. Finally, we also assessed in these studies the recovery of male-typical behaviors and neural attributes in a sub-set of animals that had been injected daily with the antisense LNA and in which these injections were ceased and replaced for two additional days by injections of scrambled LNA. After two days, the behavioral measures as well as the neural measures were similar or even greater in these subjects than the control values. These data thus provide further support for the conclusion that SRC-1 enhances in a reversible manner the transcriptional action of ligand-activated androgen receptors and estrogen receptors.

Conclusion

Steroid hormone effects on brain and behavior are largely mediated by the hormone binding to an intra-cellular receptor that results in an up-regulation of gene transcription, which is ultimately responsible for the implementation of the biological responses (34). However in several situations, variation in peripheral concentrations of steroid hormones do not explain variation in the behavioral response to hormones (e.g., individual and sex differences; (11, 19). In order to understand the actions of sex steroid hormones on reproductive behaviors one must take into account variations in the responsiveness of the target tissue to the hormone. Investigations of the distribution of sex steroid receptors have identified brain sites related to the activation of reproductive behaviors. Few studies have however been able to correlate the behavioral effectiveness of steroid hormones with the concentration of their receptors in a particular brain area (e.g. 11, 18). Co-activators such as SRC-1 represent another class of proteins that appears to be important for regulating steroid action in adulthood on brain and behavior. Our data indicate that SRC-1 functions as a critical regulatory molecule that modulates steroid-dependent behavior and neural gene expression. We know that its expression is not constitutive but rather is regulated by a variety of factors including sex steroid hormones and stress. It remains to be seen to what extent natural variation in this protein can explain intra-specific variation in the effects of sex steroids on sexual behavior.

Acknowledgements

Our work on quail and canaries is supported by grants from the NIH/NIMH (MH50388), NIH/NINDS (NS35467) to GFB and JB and grants from the Belgian FRFC (2.4555.01) and the French Community of Belgium (ARC 99/04-241) to JB. TDC is a FRIA Grant recipient.

References

1 Adkins EK (1975) Hormonal basis of sexual differentiation in the Japanese quail. *Journal of Comparative & Physiological Psychology* **89** 61-71

2 Adkins EK, Boop JJ, Koutnik DL, Morris JB & Pnieswski EE (1980) Further evidence that androgen aromatization is essential for the activation of copulation in male quail. *Physiology and Behavior* **24** 441-446

3 Arai S, Ogawa K, Yamachika S Nishihara T & Nishikawa J (2001) Cloning and functional characterization of chicken p160 coactivator family members. *Biochimica et Biophysica Acta* **1518** 7-18

4 Aste N, Panzica GC, Aimar P, Viglietti-Panzica C, Foidart A & Balthazart J (1993) Implication of testosterone metabolism in the control of the sexually dimorphic nucleus of the quail preoptic area. *Brain Research Bulletin* **31** 601-611

5 Aste N, Viglietti-Panzica C, Balthazart J & Panzica GC (1997) Testosterone modulation of peptidergic pathways in the septo-preoptic region of male Japanese quail. *Poultry and Avian Biology Reviews* **8** 77-93

6 Auger AP, Tetel MJ & McCarthy MM (2000) Steroid receptor coactivator-1 (SRC-1) mediates the developmen of sex-specific brain morphology and behavior. *Proceedings of the National Academy of Science USA* **97** 7551-7555

8 Ball GF & Balthazart J (2002) Neuroendocrine mechanisms regulating reproductive cycles and reproductive behavior in birds. In *Hormones, Brain and Behavior* pp 649-798. Eds DW Pfaff, AP Arnold, AM Etgen, SE Fahrbach & RT Rubin. Academic Press: San Diego, CA

8 Balthazart J & Ball GF (1998) The Japanese quail as a model system for the investigation of steroid-catecholamine interactions mediating appetitive and consummatory aspects of male sexual behavior. *Annual Review of Sex Research* **9** 96-176

9 Balthazart J, Schumacher M & Ottinger MA (1983) Sexual differences in the Japanese quail: behavior, morphology and intracellular metabolism of testosterone. *General and Comparative Endocrinology* **51** 191-207

10 Balthazart J, Surlemont C & Harada N (1992) Aromatase as a cellular marker of testosterone action in the preoptic area. *Physiology and Behavior* **51** 395-409

11 Balthazart J, Tlemçani O & Ball GF (1996) Do sex differences in the brain explain sex differences in the hormonal induction of reproductive behavior? What 25 years of research on the Japanese quail tells us. *Hormones and Behavior* **30** 627-661

12 Bousios S, Karandrea D, Kittas C & Kitraki E (2001) Effects of gender and stress on the regulation of steroid receptor coactivator-1 expression in the rat brain and pituitary. *Journal of Steroid Biochemistry and Molecular Biology* **78** 401-407.

13 Castagna C, Obole A, Viglietti-Panzica C, Balthazart J & Panzica GC (1999) Effects of testosterone on the synaptology of the medial preoptic nucleus of male Japanese quail. *Brain Research Bulletin* **50** 241-249

14 Charlier TD, Ball G & Balthazart J (2004) Gene knock down via antisense oligonucleotides to the steroid receptor coactivator SRC-1 modulates testosterone-dependent male sexual behavior and neural gene expression. *Hormones and Behavior* **46** 86

15 Charlier TD, Ball G & Balthazart J (2004) Plasticity in the expression of the steroid receptor coactivator SRC-1 in quail. *Itinerary planner of The Society for Neurosciences* abstract N°440.4

16 Charlier TD, Balthazart J & Ball G (2003) Sex difference in the distribution of the steroid receptor coactivator SRC-1 in the song control nuclei of male and female canaries. *Brain Research* **959** 263-274

17 Charlier TD, Lakaye B, Ball G & Balthazart J (2004) The steroid receptor coactivator SRC-1 exhibits high expression in steroid-sensitive brain areas regulating reproductive behaviors in the quail brain. *Neuroendocrinology* **76** 297-315

18 Clark AS, Davis LA & Roy EJ (1985) A possible physiological basis for the dud-stud phenomenon. *Hormones and Behavior* **19** 227-230

19 Grunt JA & Young WC (1952) Differential reactivity of individuals and the response of the male guinea pig to testosterone propionate. *Endocrinology* **51** 237-248

20 Hull EM, Meisel RL & Sachs BD (2002) Male sexual behavior. In *Hormones, Brain and Behavior* pp 1-137. Eds DW Pfaff AP Arnold, AM Etgen, SE Fahrbach, RT Rubin. Academic Press: San Diego, CA

21 Leo C & Don Chen J (2000) The SRC family of nuclear receptors coactivators. *Gene* **245** 1-11

22 MacLusky NJ & Naftolin F (1981) Sexual differentiation of the central nervous system. *Science* **211** 1294-1303

23 McEwen BS & Alves SE (1999) Estrogen actions in the central nervous system. *Endocrine Reviews* **20** 279-307

24 McEwen BS & Krey LC (1984) Properties of estrogen-sensitive neurons: aromatization, progestin receptor induction and neuroendocrine effects. In

Metabolism of hormonal steroids in the neuroendocrine structures pp 117-128. Eds F Celotti, F Naftolin, & L Martini. Raven Press: New York
25 McKenna NJ, Lanz B & O'Malley BW (1999) Nuclear receptor coregulators: cellular and molecular biology. *Endocrine Reviews* **20** 321-344
26 McKenna NJ & O'Malley BW (2002) Combinatorial control of gene expression by nuclear receptors and coregulators. *Cell* **108** 465-474
27 Meisel RL & Sachs BD (1994) The physiology of male sexual behavior. In *The physiology of reproduction*, pp 3-105. Eds E Knobil & JD Neill. Raven Press: New York
28 Misiti S, Shomburg L, Yen PM & Chin WW (1998) Expression and hormonal regulation of coactivator and corepressor genes. *Endocrinology* **139** 2493-2500
29 Molenda HA, Griffin AL, Auger AP, McCarthy MM & Tetel MJ (2002) Nuclear receptor coactivators modulate hormone-dependent gene expression in brain and female reproductive behavior in rats. *Endocrinology* **143** 436-444
30 Panzica GC, Castagna C, Aste N, Viglietti-Panzica C & Balthazart J (1996) Testosterone effects on the neuronal ultrastructure in the medial preoptic nucleus of male Japanese quail. *Brain Research Bulletin* **39** 281-292
31 Panzica GC, Viglietti-Panzica C & Balthazart J (1996) The sexually dimorphic medial preoptic nucleus of quail: a key brain area mediating steroid action on male sexual behavior. *Frontiers in Neuroendocrinology* **17** 51-125
32 Panzica GC, Viglietti-Panzica C, Calcagni M, Anselmetti GC, Schumacher M & Balthazart J (1987) Sexual differentiation and hormonal control of the sexually dimorphic preoptic medial nucleus in quail. *Brain Research* **416** 59-68
33 Panzica GC, Viglietti-Panzica C, Sanchez F, Sante P & Balthazart J (1991) Effects of testosterone on a selected neuronal population within the preoptic sexually dimorphic nucleus of the Japanese quail. *Journal of Comparative Neurology* **303** 443-456
34 Pfaff D (1999) Drive: Neural and molecular mechanisms for sexual motivation. The MIT Press: Cambridge, MA
35 Pfaff DW, Schwartz-Giblin S, McCarthy MM & Kow LM (1994) Cellular and molecular mechanisms of female reproductive behaviors. In *The physiology of reproduction* pp 107-220. Eds E Knobil & JD Neill. Raven Press: New York
36 Reiner A, Perkel DJ, Mello CV, & Jarvis E (2004) Songbirds and the revised avian brain nomenclature. *Annals of the New York Academy of Sciences* **1016** 77-109
37 Schumacher M & Balthazart J (1983) The effects of testosterone and its metabolites on sexual behavior and morphology in male and female Japanese quail. *Physiology and Behavior* **30** 335-339

Functional Avian Endocrinology
A Dawson & P J Sharp (Eds)

Evolution of plasticity in hormonally-integrated parental tactics: An example with the house finch

A V Badyaev and R A Duckworth[1]

Department of Ecology & Evolutionary Biology, University of Arizona, Tucson, AZ 85721, USA
[1]Biology Department, Duke University, Durham, NC 27708, USA

Introduction

The evolution of behavioral plasticity, particularly the evolution of context-dependent (i.e., conditional) mating tactics, is one of the most controversial and least understood phenomena in evolutionary theory (18, 44, 50). On the one hand, a growing number of behavioral studies document the benefits of alternative mating tactics where individuals adjust their reproductive behaviors and morphologies to a particular breeding context which ultimately enables them to achieve high fitness even in the most rare of environmental circumstances (8, 23, 37, 47). For example, the costs and benefits of female choice of a particular male sexual trait, as well as male investment into a sexual trait, may depend on the circumstances in which mating occurs (8). It is often assumed explicitly that because such context-dependency is enabled by individual phenotypic plasticity, no genetic polymorphism for distinct mating behaviors and morphologies is needed (2, 24). On the other hand, population genetics theory shows that these assumptions would actually preclude the evolution of context-dependency, which requires inheritance of reliable context-recognition even for the most rare contexts, as well as strong and predictably frequent selection favoring the formation a particular mating strategy (25, 26, 44, 48). Thus, conditions required for the evolution of context-dependency in mating tactics are not likely to extend to fluctuating and infrequent contexts and to all individual morphologies (3, 44).

In theoretical studies, the resolution of this paradox depends on explicit consideration of the demographic dynamics of a population; the lack of such consideration in the past has prevented addressing the basic issue of the importance of population fluctuations as a source of selective pressures promoting genetic polymorphism and/or phenotypic plasticity (27, 28, 34, 36). For example, whereas theoretical modeling suggests that plastic parental strategies will evolve under ecological conditions that promote genetic differentiation and the maintenance of genetic polymorphism, no study of parental strategies so far has investigated the evolution of genetic polymorphism and/or phenotypic plasticity in a unified framework that incorporates ecological, behavioral, and demographic considerations. In empirical studies, resolution of this paradox requires understanding the mechanisms that "switch" phenotypes from one ontogenetic trajectory to the other (1, 21, 36, 49). Such switches are variably called "modifier genes for behavior expression", "developmental switchpoints", or "social status-dependent switchpoints" which betrays the uncertainty about their nature – some assume that these switches are genetically determined, some that they are induced by population dynamics, and some that they are created during ontogeny (13, 32, 35, 38, 46).

Especially puzzling and most revealing are cases when the same individuals employ distinct mating tactics across different breeding contexts. The distinct and demographically structured parental strategies of socially monogamous house finches (*Carpodacus mexicanus*) (5) provide an opportunity to investigate experimentally the developmental and behavioral components of conditional mating tactics. In this species males use distinct parental tactics depending on elaboration of their sexual ornamentation. The fitness consequences of each parental tactic depend on the demographic composition of the population, and across their lifespan, males' parental tactics closely correspond to annual changes in sexual ornamentation (4, 5). In this review, we first address the proximate hormonal mechanisms behind dichotomous parental strategies indicated by continuous sexual ornamentation. Second, we examine how these hormonal mechanisms integrate proximately the expression of sexual ornamentation and parental behaviors. Finally, we discuss how the sensitivity to behavioral and ecological cues for switching between alternative parental tactics can evolve.

Distinct parental tactics in relation to sexual ornamentation

House finch males molt into new sexual ornamentation once a year after the breeding season and based on the elaboration of plumage ornamentation males can be separated into two groups – from orange to purple ("red" hereafter) and from uncolored to bright yellow ("yellow" hereafter) (4). Red and yellow colors in these groups are produced by different carotenoid catabolisms and elaboration within each color category is distinct biochemically (6, 11, 29).

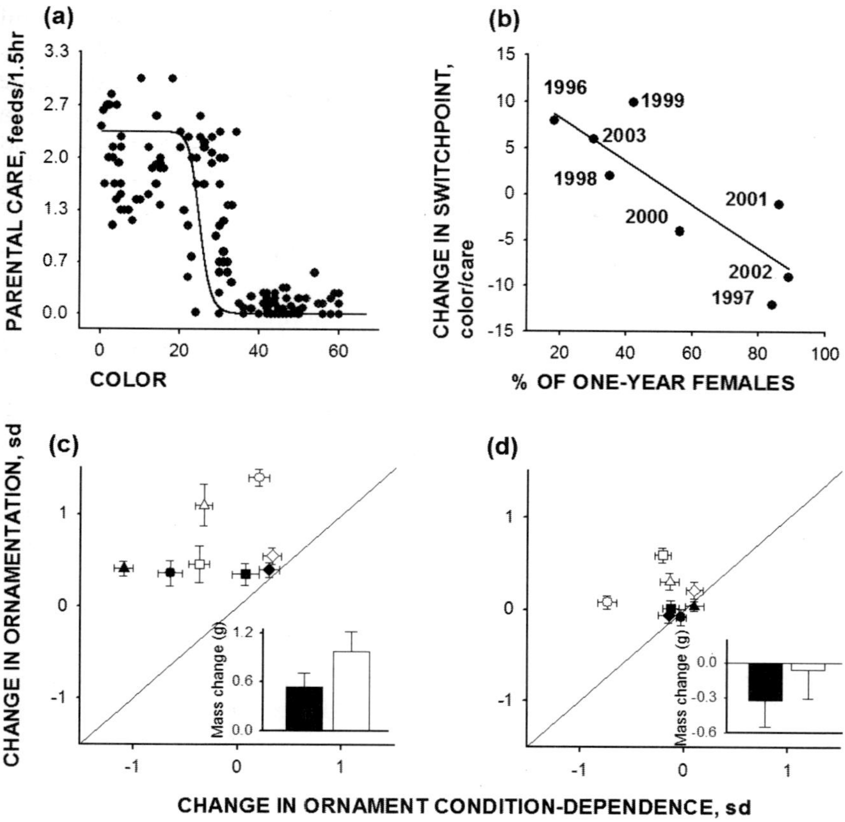

Fig. 1 **(a)** Red (color score >20) and yellow males have distinct parental tactics in Montana house finches, 1995-2003 (5), **(b)** Switchpoint between parental tactics (1st derivative of male color by male care shown in a) varies with the female age ratio in breeding population; **(c-d)** Summary of the changes in sexual ornamentation (in standard deviations) between subsequent molts and changes in ornament condition-dependence (in standard deviations) between subsequent molts in **(c)** after first breeding season (i.e., between postjuvenile and the first postbreding molts) and **(d)** after the second breeding season (i.e, between the first and the second postbreeding molts). Points above diagonal indicate that change in sexual ornamentation are greater than expected from the corresponding change in condition-dependence (hue: ●; saturation: ■; area: ▲; hue asymmetry: ♦). Black symbols indicate change in males that were paired and white symbols indicate males that were single during breeding season preceding molt. Inserts show the change in body mass between molts in males that were paired (black) and single (white) between the molts (4).

A male sexual ornamentation is closely linked to his mating tactic (Fig. 1a). Red males pair earlier, but do not provision incubating females and nestlings, whereas yellow males pair later, but provide most of the food for female and nestlings (5). Consequently, for red males offspring recruitment is primarily affected by earlier nest initiation and number of mates, while in yellow males it is affected most by male provisioning of nestlings (5).

Males change both sexual ornamentation and mating tactics in both directions across their lifespan. Males can live up to 6 years undergoing repeated and reversible changes in ornamentation and mating tactics between subsequent molts (4). During the initial (postjuvenile) molt, the expression of ornamentation is dependent on condition; however the condition-dependence of ornamentation decreases sharply throughout a male's lifetime, and in older males, switching between plumage morphs is largely independent of condition during molt (Fig. 1cd, 4, 11). The fitness consequences of male mating tactics depend strongly on the ratio of "old" (i.e., older than a year) and "young" (i.e., second-year) females in the population (Fig. 1b, 5). Red males are strongly preferred by young females, whereas older females avoid red males and prefer yellow males. This pattern of preference is due to overinvestment into parental care by females that initially mate with red non-parental males (5). The experience-mediated female choice of male plumage morph results in close concordance between the ratio of males' morph and the demographic composition of a population (Fig. 1b).

Empirical evidence for hormonal integration of parental tactics: Natural patterns

Paternal care in the house finch is closely linked with the plasma level of prolactin – a pituitary hormone that is associated with paternal care in many passerines (12, 22, 30, 40, 41). Free-living parental (yellow) males had the highest prolactin levels during provisioning of incubating females, and especially nestlings and fledglings (Fig. 2a, 20). Among these parental males, the level of plasma prolactin was positively correlated with the intensity of provisioning (Fig. 2d, 20). Parental males with higher levels of prolactin had lower hematocrit counts during the provisioning period and postbreeding molt (Fig. 2c, 20), and suffered higher overwinter mortality compared to red (non-parental) males and yellow males that provisioned less (7, 20). During nestling provisioning period, yellow and red males have nonoverlapping hormonal profiles of plasma prolactin (Fig. 2ab): red (non-parental) males have much lower levels of prolactin compared to yellow (parental) males (20).

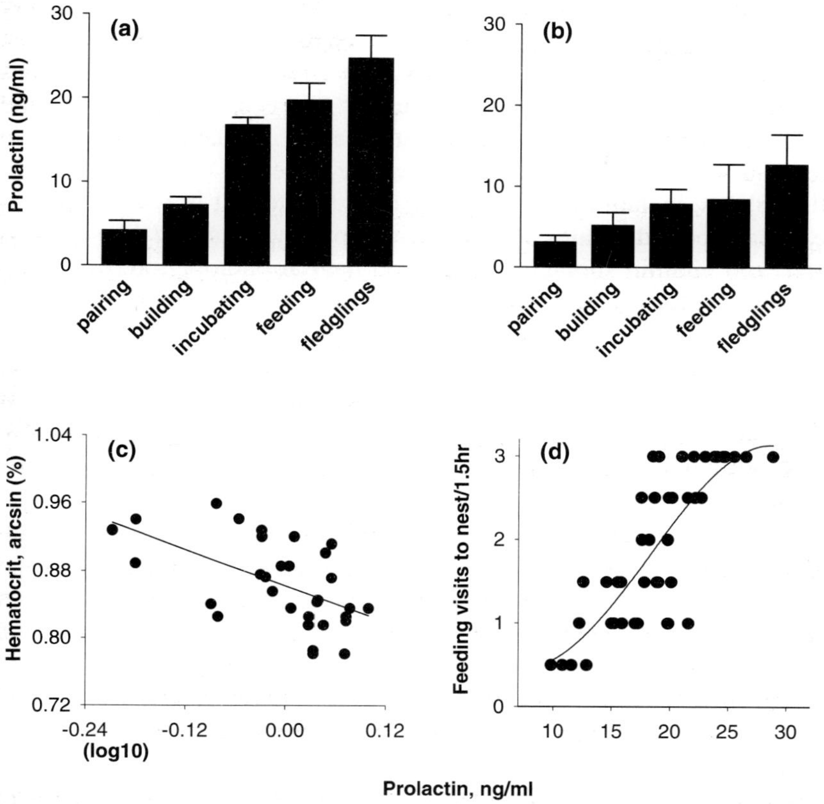

Fig. 2 **(a-b)** Plasma prolactin levels in relation to breeding cycle stage in **(a)** parental (yellow) males, and **(b)** non-parental (red) males (5, 20, including data from 2002-2003); **(c)** Plasma prolactin level (log10) in relation to hematocrit level (arsin) during nestling provisioning period (20); **(d)** Plasma prolactin levels in relation to nestling provisioning frequency immediately preceding prolactin sampling (20, including data from 2002-2003).

Experimental evidence for hormonal integration of parental tactics: Creating *reversed* mating tactics with hormonal implants

Long-term collection of hormonal profile data from breeding finches at our study populations (9, 20) enabled us to calibrate and test on a captive population delayed-release, 2mm-diameter, subcutaneous hormonal implants of vasoactive intestinal peptide (VIP hereafter; that stimulates the pituitary to increase a male's own secretion of prolactin), and hormonal implants of bromocryptine (that blocks the production of prolactin in some male vertebrates) (31, 33, 39, 40, 42).

To validate the performance of these implants in house finches, we captured nine males from the study population and measured their baseline prolactin levels (Fig. 3a). We transported the birds to an aviary facility and assigned them randomly to a treatment group. Each male was implanted subcutaneously with either one bromocryptine (released 0.24 mg/day for 21 days), one VIP (released 7 μg/day for 14 days), or one control pellet (vehicle only; Innovative Research of America, USA). All implants, including placebos, are designed to last only to the end of the release period. Males were kept in captivity on an *ad lib* seed diet, water, and vitamin supplements, and on natural photoperiod. Each male was sampled for prolactin on days 1, 3 and 5 after implantation with the exception of two males that we bled on days 1, 2, and 5 instead. All males were released at the site of capture on day 6. We subsequently recaptured and sampled for prolactin all of the VIP and placebo implanted males but did not recapture the bromocryptine implanted males, although these males were repeatedly re-sighted at the feeder stations.

Both VIP and bromocryptine implants, but not placebo implants, had a significant effect on the level of plasma prolactin in males by day 3 post-implantation (Fig. 3a). In males with baseline levels of prolactin, VIP strongly increased plasma prolactin to a level typically seen in intensively provisioning males (20) (Fig 3a). Bromocryptine implants sharply lowered the level of plasma prolactin beyond detectable levels.

Following this validation experiment, in 2001-2003, we implanted 24 red (non-parental) males with VIP implants and 37 yellow (parental) males with bromocryptine implants (Fig. 3b). For all males, we conducted a pre-implantation nestling provisioning watch on day 3-5 of the nestling stage. Males were subsequently captured, either at the nests (parental yellow males) or at the feeding stations (nonparental, red males) sampled for plasma prolactin (20) and implanted with VIP, bromocryptine, or placebo pellet as describe above. On day 7-9 of the nestling stage, we conducted a post-implantation observation of male provisioning behavior and obtained plasma prolactin samples from implanted males that were captured opportunistically during daily captures. Both pre- and post-implantation behavioral observations were conducted by the same technician who was unaware of the experimental objectives and also did not know which males were assigned to which group.

VIP implants of red (nonparental) males and bromocryptine implants of yellow (parental) males reversed males' parental behavior within a breeding season (Fig. 3c). VIP-implanted red males became parental and fed their nestlings, whereas bromocryptine-implanted yellow males stopped feeding their nestlings (Fig. 3c). There were no other changes in males' behavior either towards their females or other birds. Control males did not change their provisioning frequency (Fig. 3c). Placebo implants did not affect male behaviors in other studies of this species (45).

Taken together, these results show that (i) house finch males have two parental tactics invariantly indicated by two states of sexual ornamentation, (ii) males switch between the plumage morphs and parental tactics throughout their lifetime, (iii) male's switching of mating tactics between years is linked with the female age ratio in a population, and (iv) males with distinct paternal tactics have non-overlapping profiles of plasma prolactin that proximately integrates ornament elaboration and parental behavior.

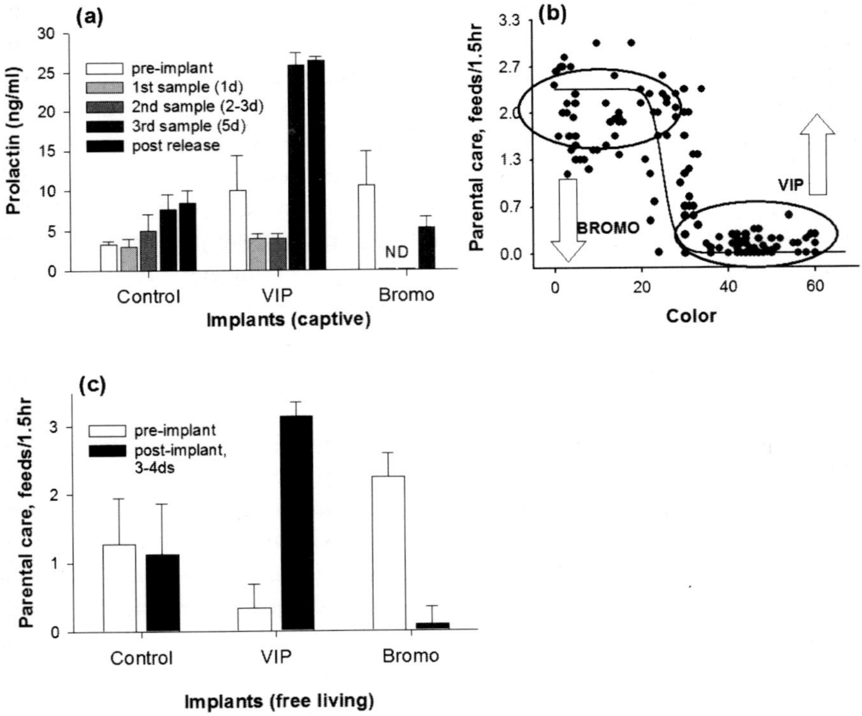

Fig. 3 **(a)** Plasma prolactin levels in control-, VIP, and bromocryptine (Bromo) implanted male house finches. ND indicates levels below detectable. Males were sampled for prolactin three times within five days post-implantation and within a week after release (post-release); **(b)** Experimental design to reverse parental behavior in relation to sexual ornamentation by subcutaneous implants of VIP to red (non-parental) males and bromocryptine to yellow (parental) males. **(c)** Pre- and post-implantation nestling provisioning frequency in Control males, red males implanted with VIP, and yellow males implanted with bromocryptine.

Evolution of hormone-mediated behavioral plasticity in parental tactics: Unresolved questions

Two questions remain unresolved. First, what are the cues for switching between elaboration morphs/parental tactics? Second, how does the ability of males to detect and implement these cues evolve? We propose three general alternative hypotheses for the evolution of hormonally-mediated plasticity in parental tactics in this system.

First, integration of sexual morph and parental tactic can be established simultaneously, under shared developmental cues, during the color acquisition in the fall ("*developmental integration*" of color and behavior). In this case, the cue could include demographic composition of the population or it can be a passive consequence of the overlap between parental care and molt. Male finches undergo postbreeding molt in fall flocks with the same birds as will be breeding at the study site the following year (10), so that the demographic composition of the population in the fall may serve as a cue for development of male's mating tactic. The female age ratio changes strongly among years (Fig. 1b). Thus, immediate developmental cues for changes in tactic can include age-specific food competition (including competition for carotenoid-rich food) or some form of cognitive assessment of general age-composition in the fall (19). Alternatively, simultaneous developmental integration of color and behavior can be due to shared hormonal signaling associated with production of prolactin (14, 15). For example, parental males have higher circulating prolactin late in the summer and this can, potentially, delay their molt onset and affect molt duration (16, 17). If shorter molt results in yellow ornamentation, then this could produce an association between parental behaviors, molt duration, and subsequent color (A. Dawson, pers. comm.).

Second, integration can be established sequentially with color being established in the autumn and the behavioral tactic, independently, during the following breeding season ("*behavioral integration*"). In this case, the cues can include speed of pairing. Color can be determined in the autumn molt (as a result of variation in the access to carotenoid-rich diet, duration of molt, health, or prior breeding season behaviors, whereas the parental behavior can be a consequence of a lag in female choice in the spring – later selected males might be more willing to provide care (43).

Finally, yearly variation in parental tactics can be due to annual population fluctuation of the age ratio of males with particular life history characteristics (e.g., non-parental red males might be younger than males of other tactic) and the parental tactic can be due to heritable association of color and behavior ("*genetic integration*") (1, 3).

Further experiments in this study system should examine, in particular, the effect of plasma prolactin on molt onset and duration (and resulting color morph), the effect of demographic composition of the population at the time of molt on the sex- and age-specific aggression and competition for carotenoid-rich foods as well as hormonal profiles. Further examination of endocrinological mechanisms behind both the integration of sexual ornamentation and parental care and ecological and behavioral cues for implementation of a particular parental tactic is a necessary next step to understand the evolution of behavioral plasticity.

Acknowledgements

AVB was supported by the National Science Foundation (NSF) grants (DEB-0075388, IBN-0218313, DEB-0077804) and the University of Arizona, RAD was supported by NSF (DDIG 0407952). We thank A.F. Parlow for collaborations on prolactin assays for studies cited here, many field assistants for help in the field, and K. Oh, D. Seaman, and R. Young for comments that improved this manuscript.

References

1 Badyaev AV (2004) Integration and modularity in the evolution of sexual ornaments: An overlooked perspective. In *Phenotypic Integration: The Evolutionary Biology of Complex Phenotypes,* pp 50-79. Eds M. Pigliucci and K. Preston. Oxford: Oxford University Press

2 Badyaev AV (2004) Rigorous science for exaggerated traits. *Ecology* **85** 1472–1474

3 Badyaev AV (2005) Colorful phenotypes of colorless genotypes: Towards a new evolutionary synthesis of animal color displays. In *Animal Coloration: Proximate and Ultimate Mechanisms*, pp 1-31. Harvard: Harvard University Press

4 Badyaev AV & Duckworth RA (2003) Context-dependent sexual advertisement: plasticity in development of sexual ornamentation throughout the lifetime of a passerine bird. *Journal of Evolutionary Biology* **16** 1065-1076

5 Badyaev AV & Hill GE (2002) Parental care as a conditional strategy: Distinct reproductive tactics associated with elaboration of plumage ornamentation in the house finch. *Behavioral Ecology* **13** 591-597

6 Badyaev AV, Hill GE, Dunn PO & Glen JC (2001) Plumage color as a composite trait: Developmental and functional integration of sexual ornamentation. *American Naturalist* **158** 221-235

7 Badyaev AV & Martin TE (2000) Sexual dimorphism in relation to current selection in the house finch. *Evolution* **54** 987-997

8 Badyaev AV & Qvarnström A (2002) Putting sexual traits into the context of an organism: a life-history perspective in studies of sexual selection. *Auk* **119** 301-310

9 Badyaev AV et al (2004) Adaptive sex differences in growth of pre-ovulation oocytes in a passerine bird. *Proceedings of the Royal Society Biological Sciences Series B*: In review

10 Badyaev AV, Whittingham LA & Hill GE (2001) The evolution of sexual size dimorphism in the house finch: III. Developmental basis. *Evolution* **55** 176-189

11 Badyaev AV & Young RL (2004) Complexity and integration in sexual ornamentation: An example with carotenoid and melanin plumage pigmentation. *Journal of evolutionary biology* **17** 1317-1327

12 Buntin JD (1996) Neural and hormonal control of parental behavior in birds. *Advances in the Study of Behavior* **25** 161-212

13 Caro TM & Bateson P (1986) Organization and ontogeny of alternative tactics. *Animal Behaviour* **34** 1483-1499

14 Dawson A (1997) Plasma-lutenizing hormone and prolactin during circannual rhythms of gonadal maturation and molt in male and female European starlings. *Journal of Biological Rhythms* **12** 371-377

15 Dawson A, King VM, Bentley GE & Ball GF (2001) Photoperiodic control of seasonality in birds. *Journal of Biological Rhythms* **16** 365-380

16 Dawson A & Sharp PJ (1998) The role of prolactin in the development of reproductive photorefractoriness and postnuptial molt in the European Starling (*Sturnus vulgaris*). *Endocrinology* **139** 485-490

17 Deviche P, Wingfield JC & Sharp PJ (2000) Year-class differences in the reproductive system, plasma prolactin, and corticosterone concentrations, and onset of prebasic molt in male dark-eyed juncos (*Junco hyemalis*) during the breeding season. *General and Comparative Endocrinology* **118** 425-435

18 DeWitt TJ & Scheiner SM (2004) Phenotypic Plasticity: Functional and Conceptual Approaches. New York: Oxford University Press.

19 Doligez B, Danchin E & Clobert J (2002) Public information and breeding habitat selection in a wild bird population. *Science* **297** 1168-1170

20 Duckworth RA, Badyaev AV & Parlow AF (2003) Males with more elaborated sexual ornaments avoid costly parental care in a passerine bird. *Behavioral Ecology & Sociobiology* **55** 176-183

21 Emlen DJ & Nijhout HF (2000) The development and evolution of exaggerated morphologies in insects. *Annual Reviews in Entomology* **45** 661-708

22 Goldsmith AR (1982) Plasma concentrations of prolactin during incubation and parental feeding throughout repeated breeding cycles in canaries (*Serinus canaria*). *Journal of Endocrinology* **94** 51-59

23 Griffith SC (2000) A trade-off between reproduction and a condition-dependent sexually selected ornament in the house sparrow *Passer domesticus*. *Proceedings of Royal Society, Lond. Series B*. **267** 1115-1119

24 Gross MR (1996) Alternative reproductive strategies and tactics: diversity within sexes. *Trends in Ecology and Evolution* **11** 92-98

25 Gross MR & Repka J (1998) Stability with inheritance in the conditional strategy. *Journal of Theoretical Biology* **192** 445-453

26 Hazel W & Smock R (1998) Inheritance in the conditional strategy revisited. *Journal of Theoretical Biology* **204** 307-309

27 Houston AI & McNamara JM (1992) Phenotypic plasticity as a state-dependent life-history decision. *Evolutionary Ecology* **6** 243-253

28 Houston AI & McNamara JM (2002) A self-consistent approach to paternity and parental effort. *Philosophical Transactions of the Royal Society of London B Biological Sciences* **357** 351-362

29 Inouye CY, Hill GE, Stradi RD & Montgomerie R (2001) Carotenoid pigments in male house finch plumage in relation to age, subspecies, and ornamental coloration. *Auk* **118** 900-915

30 Khan MZ, McNabb FMA, Walters JR & Sharp PJ (2001) Patterns of testosterone and prolactin concentrations and reproductive behavior of helpers and breeders in the cooperatively breeding red-cockaded woodpecker (*Picoides borealis*). *Hormones and Behavior* **40** 1-13

31 Li C, Kelly PA & Buntin JD (1995) Inhibitory effects of anti-prolactin receptor antibodies on prolactin binding in brain and prolactin-induced feeding behavior in ring doves. *Neuroendocrinology* **61** 125-135

32 Luttbeg B & Warner RR (1999) Reproductive decision-making by female peacock wrasses: flexible versus fixed behavioral rules in variable environments. *Behavioral Ecology* **10** 666-674

33 Maney DL, Schoech SJ, Sharp PJ & Wingfield JC (1999) Effects of vasoactive intestinal peptide on plasma prolactin in passerines. *General and Comparative Endocrinology* **113** 323-330

34 Maynard Smith, J (1977) Parental investment: a prospective analysis. *Animal Behaviour* **25** 1-9

35 Moore MC (1991) Application of organization-activation theory of alternative male reproductive strategies: A review. *Hormones and Behavior* **25** 154-179

36 Moran NA (1992) The evolutionary maintenance of alternative phenotypes. *American Naturalist* **139** 971-989

37 Qvarnström A (2001) Context-dependent genetic benefits from mate choice. *Trends in Ecology and Evolution* **16** 5-7

38 Radwan J, Unrug J & Tomkins JL (2002) Status-dependence and morphological trade-offs in the expression of a sexually selected character in the mite, *Sancassania berlesei*. *Journal of Evolutionary Biology* **15** 744-752

39 Rozenboim I, Tabibzadeh C, Silsby JL & El Halawani ME (1993) Effect of ovine prolactin administration on hypothalamic vasoactive intestinal peptide (VIP), gonadotropin releasing hormone I and II content, and anterior pituitary VIP receptors in laying turkey hens. *Biology of Reproduction* **48** 1246-1250

40 Scanes CG (2000) Introduction to Endocrinology: Pituitary gland. In *Sturkie's Avian Physiology*, pp 437-460. Ed G. Whittow. San Diego: Academic Press

41 Schoech SJ, Mumme RL & Wingfield JC (1996) Prolactin and helping behaviour in the cooperatively breeding Florida scrub-jay, *Aphelocoma c. coerulescens*. *Animal Behaviour* **52** 445-456

42 Sharp PJ, Dawson A & Lea RW (1998) Control of lutenizing hormone and prolactin secretion in birds. *Comparative Biochemistry and Physiology C* **119** 275-282

43 Sheldon BC (2000) Differential allocation: tests, mechanisms and implications. *Trends in Ecology and Evolution* **15** 397-402
44 Shuster SM & Wade MJ (2003) Mating systems and strategies. Princeton: Princeton University Press
45 Stoehr AM & Hill GE (2000) Testosterone and the allocation of reproductive effort in male house finches (*Carpodacus mexicanus*). *Behavioral Ecology and Sociobiology* **48** 407-411
46 Tomkins JL & Brown GS (2004) Population density drives the local evolution of a threshold dimorphism. *Nature* **431** 1099-1103
47 Vaclav R & Hoi H (2002) Different reproductive tactics in house sparrows signaled by badge size: Is there a benefit to being average? *Ethology* **108** 569-582
48 Wade M & Shuster SM (2002) The evolution of parental care in the context of sexual selection: A critical reassessment of parental investment theory. *American Naturalist* **160** 285-292
49 West-Eberhard MJ (1989) Phenotypic plasticity and the origins of diversity. *Annual Reviews in Ecology and Systematics* **20** 249-278
50 West-Eberhard MJ (2003) *Developmental Plasticity and Evolution*. Oxford: Oxford University Press

Part ten

Hormonal interactions during development

Mourning Dove
John Alcock

Functional Avian Endocrinology
A Dawson & P J Sharp (Eds)

Growth hormone: a pituitary and extrapituitary chameleon during development

K L Hull, M Luna[1], S Harvey[2] and C Arámburo[1]

Department of Biology, Bishop´s University, Lennoxville, Quebec, Canada, JIM IZ7
[1]Department of Cellular and Molecular Neurobiology, Institute of Neurobiology, Universidad Nacional Autónoma de México, Querétaro, México, 76230
[2]Department of Physiology, University of Alberta, Edmonton, Canada, T6G 2H7.

Introduction

Growth hormone (GH) has classically been considered as a pituitary endocrine, released from somatotrophs into peripheral circulation to stimulate the growth and differentiation of distant target sites, directly or through the mediation of hepatic (insulin-like) growth factors (IGFs). It is, however, now known that GH is not a single entity, but a family of size and charge variants produced both in the pituitary and in many extrapituitary tissues. In addition to its endocrine actions, GH acts locally as an autocrine or paracrine in pituitary and extrapituitary tissues; its induction of local rather than hepatic IGFs causally mediates some of its developmental effects in target tissues. These local effects may be particularly relevant in the embryo, since GH and GH receptors are expressed in the embryo prior to the differentiation of pituitary somatotrophs. The expression of different GH isoforms also differs between tissues and during ontogeny and these distinct patterns may result in tissue-specific and developmentally-specific GH actions. GH is thus a chameleon, in that the isoforms produced and the consequent cellular response can be modified to suit the needs of different tissues and different developmental stages. The heterogeneity of GH production and action during development is thus the focus of this brief review.

Pituitary GH

Growth hormone (GH) is a pituitary protein that induces cellular proliferation and differentiation and stimulates the growth of soft and skeletal tissue. These actions partly result from growth hormone regulation of protein, carbohydrate, lipid, and nucleic acid metabolism and from its ability to induce angiogenesis. These actions may be directly induced by GH or be indirectly mediated by local growth factors such as insulin-like growth factor I (44).

Ontogeny in the embryo

The development of GH function in the embryo represents an ontogenic shift between exclusively local GH actions in the early embryo and endocrine as well as local actions in the late embryo and post-hatch chick. The timing and regulation of this transition remains an active and contentious research area. For instance, as shown in Table 1, the first appearance of GH immunoreactivity in the pituitary gland (or somatotroph precursor cells) has been reported as early as embryonic day (ED) 4.5 (50) or as late as ED 18 (51). These widely differing estimates are likely to reflect methodological differences, including different dating methods, antibodies, detection systems, and fixation methods. Indeed, Murphy *et al.* (37) was able to detect GH-IR in Rathke's pouch of ED7 embryos fixed in paraformaldehyde but not in Carnoy's fixative. Paraformaldehyde, a cross-linking fixative, fixes smaller proteins than does Carnoy's; thus, the GH-IR at ED 7 may reflect the preferential expression of smaller GH moieities. Since Pit-1 expression has been detected as early as ED2 (19) or ED5 (52), this early GH-IR may reflect transcription and translation of the GH gene. The immunoreactive cells are also morphologically distinct from somatotrophs, since they are small and round with abundant nuclear GH-IR (50). Nuclear immunoreactivity for proteins that are traditionally secreted is characteristic of rapidly-dividing tissues in which proteins escape packaging into secretory vesicles . These early GH-containing cells may thus contain a GH isoform that is not secreted; rather, it may act in an intracrine fashion to alter cell development.

Table 1: Events in somatotroph maturation

ED	Event	Reference
2-5	pituitary *pit-1* mRNA detected	(19, 52)
9	pituitary GH mRNA detected	(52)
11	[↑]serum thyroid hormone	(29)
4.5-12	pituitary GH-IR detected	(40, 50)
12.5	structural maturation, hypothalamo-pituitary axis	(49)
14	[↑] serum corticosterone	(29)
14	pituitary TRH detected	(15)
16	*in vitro* pituitary GRF/TRH sensitivity	(14)
16	*in vitro* GH secretion	(39)
17	plasma GH detected	(21)
18	hypothalamic TRH secreted	(15)

These GH-containing cells appear to differentiate into GH-secreting somatotrophs by ED16, since GH secretors are only occasionally detected by the reverse hemolytic plaque assay (RHPA) in cultures of younger pituitary cells (39). These *in vitro* data are supported by the finding of Harvey *et al.* (21), in which plasma GH levels were not detectable until ED17. Prior to somatotroph differentiation, the number (beginning at ED10) and size (ED 13-15) of GH-containing cells begin to increase, as does pituitary content of GH-IR (ED12-14) and GH mRNA (ED 14-16), particularly in the caudal lobe (25, 33, 50; Table 1). Secretory granules, conversely, do not assume their adult size and configuration until approximately ED18 (50). Thus, although immunoreactive GH is present very early in embryogenesis, the differentiation and proliferation of cells that are morphologically and functionally identifiable as somatotrophs is not initiated until ED16. Late embryonic development may therefore represent a critical period in which pituitary GH shifts identity, from an intracrine to an endocrine hormone.

The conversion of GH-containing somatotroph precursors into mature somatotrophs reflects a number of structural and functional changes in the hypothalamus, pituitary, and other endocrine axes. Since adult GH secretion is largely dependent on hypothalamic TRH, maturation of the hypothalamo-pituitary axis may be a pivotal event. As shown by Thommes & Rousseau (49), TRH –containing fibers appear in the median eminence by ED8 and the portal vasculature is developed by ED12.5. The hypothalamus is not, however, secreting significant amounts of TRH until ED 19 (15); early somatotroph differentiation may thus be dependent on pituitary TRH instead, which does appear at ED 16 (15). Somatotroph responsiveness to both TRH and GHRH also develops at approximately ED16 *in vitro* and ED18 *in vivo* and increases (particularly for TRH) until ED20 (13, 14, 27).

The coordinate acquisition of secretagogue responsiveness and the ability to secrete GH would suggest that TRH and GHRH themselves do not induce somatotroph differentiation. Secretagogue sensitivity may be a sign of somatotroph maturation rather than an inducing event. Porter and colleagues hypothesize that corticosterone may be the initiating factor, since the differentiation of ED12 pituitary cells into somatotrophs can be induced by incubation with corticosterone (40) and plasma glucocorticoid levels rise by ED14 (29). GHRH potentiates the effect of glucocorticoids but is ineffective alone (40). Thyroid hormones, which are present in serum at ED9.5 and increase significantly between ED11 and 13, also act synergistically with corticosterone to increase somatotroph differentiation (29). Thus, a series of events are involved in the transition between a small cell population expressing, but not secreting, GH to a large population of morphologically and functionally identifiable somatotrophs. Pit-1 may drive the initial production of GH in the immature cells, whereas corticosterone, pituitary and eventually hypothalamic TRH, GHRH, and thyroid hormones may coordinately induce differentiation and proliferation.

Heterogeneity

It is now accepted that pituitary GH is composed of a family of homologous, but not identical, proteins produced by mechanisms such as gene duplication, alternative splicing, posttranslational modification and post- secretion modification. The resulting size and charge GH variants have been shown to differ in respect to bioactivity and regulation. Thus it has been proposed that the functional diversity of GH may be partially explained by its molecular heterogeneity.

Using the chicken as experimental model, it has been shown that GH variants (or isoforms) in the pituitary gland and plasma result from posttranslational modifications of monomeric (22 kDa) GH (4, 35). These modifications include glycosylation (26 kDa)(9, 11, 12), phosphorylation (8), oligomerization (44, 66, 88, 110 kDa) (5) and proteolytic processing (15 kDa)(3). Also, three charge variants of chicken GH (Rf= 0.23, 0.30 and 0.35) have been shown by native PAGE and up to ten isoforms (pIs ranging from 6.8-8.4) by IEF (46).

Functional analysis has revealed that these structural modifications alter the bioactivity of the GH isoform (Table 2). For instance, two of the monomeric charge variants (0.23 and 0.30) are similarly somatogenic in Snell dwarf mice (45) but are either lipolytic (0.23) or anti-lipolytic (0.3) in chicken adipose tissue (2, 7). Growth hormone has also been shown to exert an angiogenic effect on chicken corioallantoid membrane (16) and preliminary results suggest that this resides on a 15 kDa amino terminal fragment of GH (3). This GH isoform is also able to inhibit type III deiodinase in chicken embryo hepatocytes in culture. Recently, glycosylated cGH (G-cGH), but not its unglycosylated 22 kDa isoform, has been shown to stimulate Nb2 cell proliferation (12). Together, these data suggest that some bioactivities may be shared by several GH variants whereas others can be ascribed to a specific isoform.

If GH molecular variants are able to express different bioactivities, then it would be expected that their synthesis and release might be differentially regulated (Table 2). Indeed, recent data has shown that GHRH (and/or secretagogue analogs 8Br-cAMP and phorbol ester) stimulates the secretion of glycosylated and phosphorylated cGH isoforms to a greater extent than the 22 kDa GH monomer (8, 34). TRH, conversely, increases the proportion of monomeric and dimeric GH moieties in plasma at the expense of oligomers (1, 6). Thus, the production of different GH isoforms, and thus the type of GH bioactivity, can be controlled by judicious manipulation of different GH secretagogues.

Table 2. Chicken growth hormone variants

Isoform type	MW*	Characteristics	Reference
GH fragment	15 kDa	[↑] in chick embryos	(5)
	(NRC, RC)	Low RRA activity	(5)
		Angiogenesis (+)	(3)
		Lipolysis (+)	(3)
		Type III deiodination (–)	(3)
GH monomer	22 kDa (NRC)	Most activities related to	(4)
	26 kDa (RC)	GH	(3, 5)
		Most abundant in pituitary	(5, 31)
Charge variant	Rf= 0.23	[↑] in young chicks	(2, 7)
		Somatotrophic (+)	(2, 7)
Charge variant	Rf= 0.30	Lipolysis (+)	(2, 7)
		Somatotrophic (+)	(2, 7)
		Lipolysis (–)	(2, 7)
Charge variant	Rf= 0.35	Antilipolysis (+)	
		Uncharacterized	
P-cGH		released by GHRH and	(8)
		PMA	(8)
		phosphorylated by PKA	
Cleaved GH	25 kDa (NRC)	Low RRA activity	(5)
G-cGH	26 kDa (NRC)	Type III deiodination (+)	(11)
	29 kDa (RC)	20% lower clearance rate	(11)
		lower affinity for GHR	(12)
		preferential release by	(34)
		GHRH	(12)
		Nb2 proliferation activity	
		(+)	
GH dimer	44 kDa (NRC)	Low RRA activity	(5)
GH oligomers	66, 88, 110	Low RRA activity	(5)
	kDa	[↑] in aging chickens	(5)
	(NRC)		

*as determined by SDS-PAGE under non-reducing conditions (NRC) or reducing conditions (RC). Rf: relative mobility in native PAGE. P-cGH: phosphorylated chicken growth hormone; G-cGH: glycosylated chicken growth hormone. RRA: radioreceptor assay in chicken liver membranes. GHR: growth hormone receptor; GHRH growth hormone releasing hormone; PMA: phorbol myristate acetate; PKA: protein kinase A.

Ontogenic patterns of heterogeneity

The changes in plasma GH levels during embryonic and post-embryonic development have been well-documented (18). However, GH is a heterogeneous family of related proteins expressed in embryonic pituitary precursors as well as in the embryonic and post-hatch pituitary gland. Plasma GH levels may not, therefore, accurately reflect GH bioactivity, since the pattern of expression of different GH isoforms is also developmentally-specific. For instance, embryonic (ED18) plasma contains significantly greater proportions of submonomeric (15 kDa) and dimeric (44 kDa) GH but less 22 kDa GH than the plasma of newly hatched chicks (2). Since only the 22 kDa form is active in radioreceptor assays (2), conventional determinations of embryonic GH concentrations (radioimmunoassay, immunocytochemistry, RHPA) may overestimate the actual somatogenic bioactivity. It is also possible that the distinct pattern of GH heterogeneity in the embryo indicates that submonomeric and perhaps even dimeric GH are affecting distinct functions by binding to a different receptor.

Extrapituitary GH

GH was thought to contribute to post-natal, but not pre-natal, growth, since plasma GH is detectable only late in embryonic development and growth is not markedly impaired in hypophysectomized chicks (reviewed in 23). Extrapituitary GH may nevertheless play an important role in embryonic development. GH cDNA can be detected in headless chick bodies by RT-PCR as early as ED2, long before the differentiation of the pituitary gland and the initiation of pituitary GH gene transcription on ED9-10 (19, 22). GH immunoreactivity is virtually ubiquitous in paraformaldehyde-fixed early (ED3) embryos, but is subsequently confined to specific cells in each tissue by ED5 and to specific tissues (i.e. not the heart) by ED8 (22). Conversely, GH-immunoreactivity in ED7 embryos fixed with Carnoy's is much less widespread, since it is largely confined to neural tissues, the crop, intracostal muscles, the Mullerian duct, and hepatocytes (36). Since Carnoy's preferentially fixes larger moieties, these data would suggest that larger GH isoforms are particularly abundant in neural tissues, whereas smaller (perhaps sub-monomeric) GH moieties detected in paraformaldehyde-fixed embryos are preferentially expressed in peripheral tissues. This hypothesis is supported by the detection of submonomeric GH moieties in reproductive and lymphoid tissues (30, 31), but not the brain or hypothalamus (43), as determined by Western analysis. The distinct biological activities of sub-monomeric GH, which may be mediated by a novel GH receptor, may thus be particularly relevant to early embryos, particularly in non-neural tissues.

GH is also present in numerous extrapituitary tissues of older embryos and post-hatch birds, including the nervous system, reproductive system, and immune system (24). Ontogenic and tissue-specific patterns of GH isoform expression may enable GH to induce developmentally- and tissue-specific actions.

Heterogeneity in the nervous system

GH immunoreactivity is abundantly present in neurons of the hypothalamus and the hippocampus and in the medial suprachiasmatic nucleus of the turkey (41). In the chicken, neural GH immunoreactivity is thought to reflect translation of a 689 bp GH transcript identical in sequence to that obtained from the pituitary (43). Western analysis has revealed the presence of monomeric and dimeric GH moieties in chicken brain and hypothalamic extracts (43).

The neural retina, conversely, may be a particularly rich source of submonomeric GH moieities. A study by Baudet *et al.* (10) revealed that GH immunoreactivity in the embryonic neural retina is mainly associated with proteins of 15 and 16 kDa, whereas only small amounts of 22 kDa GH are present. The smallest moiety may be preferentially secreted, since a 15 kDa protein is the most abundant GH-like moiety in the vitreous humor. The submonomeric, 15 and 16 kDa GH-like proteins persist in the neural retina after hatch, but they are no longer observed after 42 d of age. These moieties may reflect transcriptional heterogeneity, since Takeuchi et al., (48) has reported a novel truncated GH mRNA transcript encoding a 16.5 kDa GH protein in the retina. Baudet *et al.* (10), conversely, has detected full-length GH transcripts homologous to the product of the pituitary GH gene. The neural retina is also a site of GH action, since cGH significantly increases (4- to 5-fold) the content of IGF-1 mRNA in 48-h cultured ED8 neural retinas (10).

Heterogeneity in the reproductive system

Growth hormone immunoreactivity (GH-IR) has been found by ELISA in several male (testis, vas deferens) and female (oviduct, follicles, follicular membranes) reproductive tissues in the chicken, although their GH concentration was generally <1-2% of that found in the pituitary gland (32). The immunoassayable GH concentration in the testis was relatively unchanged between 4 and 66 weeks of age, but approximately 10-fold higher than at 1-week of age and 25-fold higher than that in 1-day-old and perinatal (18dE) embryos. This GH-IR was associated, as in the pituitary, with several proteins of different molecular size (14,17,23,26,29,32,35,40,45,52 kDa) when analyzed by SDS-PAGE under reducing conditions. However, while most of the GH-IR in the pituitary (40 and 15%, respectively) is associated with monomer (26 kDa) or dimer (52 kDa) GH isoforms GH-IR in the testis is primarily (30-50%) associated with a 17 kDa moiety. GH bands between 32 and 45 kDa are also relatively more abundant in the testis than in the pituitary. During ontogeny the relative abundance of the 17 and 45 kDa moieties in the testis increased with advancing age (from 29% and 5% in the 18dE embryo, to 47% and 12% in the sexually mature bird, respectively); whereas the relative abundance of the 14 kDa GH and 40 kDa GH moieties significantly declined (from 4 and 11% in the embryo to 0.2 and 1% in the adult, respectively). The monomeric 26 kDa form however, showed no significant changes in its relative proportion during development of the chicken testis. In contrast, in the pituitary the 26 kDa variant and the glycosylated 29 kDa isoform showed important changes during ontogeny (31). In adult birds, intense GH-IR in the seminiferous tubules was shown by immunohistochemistry. It was abundantly present in secondary spermatocytes and spermatids in the luminal compartments of the tubules as well as in some Leydig cells and surrounding myocytes, although it was absent in the basal compartment of Sertoli cells and in the spermatogonia and primary spermatocytes (31).

Using RT-PCR and oligonucleotide primers for pituitary GH cDNA, GH mRNA was shown to be present in the testes and vas deferens of adult chickens. Although testicular GH mRNA was of low abundance, a 690 bp fragment of the amplified testicular GH cDNA was cloned and had a nucleotide sequence 99.6% homologous with pituitary GH cDNA. GH mRNA was localized by *in situ* hybridization in spermatogonia and primary spermatocytes of the seminiferous tubules, but unlike testicular GH-immunoreactivity, GH mRNA was not present in secondary spermatocytes, spermatids nor spermatozoa (20). The stage-specific presence of GH and GH mRNA in the chicken testis suggests GH gene expression is correlated with the onset of gametogenesis and that the induction of GH in the haploid cells may have a functional significance. It is interesting then that several actions of GH in gametogenesis have been described (17, 38) and that GH deficiency is thought to be causally linked to male infertility (47). The presence of GH immunoreactivity in the interstitial cells between tubules may indicate that testicular GH also has roles in gonadal steroidogenesis, as exogenous GH has been shown to have steroidogenic actions (26). All these findings support the hypothesis that GH production in the testis may play local autocrine/paracrine roles in reproductive function, particularly in the gametogenesis of sperm.

Heterogeneity in the immune system

The presence of GH has been described in mammalian leukocytes (53) and lymphoid cells (28). In the chicken GH-IR was detected in the thymus, spleen and bursa of Fabricius (30, 42) with a concentration <0.5% in comparison of that found in the pituitary at 4 weeks of age. Again, the GH-IR in these tissues was associated with several variants with molecular size ranging between 14-52 kDa when analyzed by SDS-PAGE under reducing conditions, and the relative proportions of the different GH moieties in the immune tissues were quite different than those found in the pituitary. Once more, the most abundant GH isoform was a 17 kDa protein, which accounted for 27% of the total GH-IR in the bursa; 29% in the spleen and 36% in the thymus, respectively. The heterogeneity and the relative abundance of bursal GH-immunoreactive moieties showed a changing pattern during development between ED13 and 9 weeks of age. In each stage, the 17 kDa variant was the most abundant form (30). Bursal GH mRNA was studied by RT-PCR and the resultant GH cDNA sequence was shown to be identical to that of the pituitary gland (42).

Table 3. Total GH content in different organs in the chicken

	Pituitary			Testes			Bursa		
Age (wk)	Weight (mg)	mg GH/mg protein	Content mg GH/organ	Weight (mg)	mg GH/mg protein	Content mg GH/organ	Weight (mg)	mg GH/mg protein	Content mg GH/organ
1	2.3	101.5	11.6	-	-	-	276	0.474	3.6
4	5.1	60.2	28.9	180	0.173	0.52	635	0.55	22
9	10.4	47.5	48.9	2674	0.193	12.4	460	0.346	7.3
25	15.8	11.9	15.8	9700	0.329	145.8	-	-	-

Conclusion

It is now clear that the GH gene can be expressed in extrapituitary tissues and that the production of different structural variants is regulated in a tissue-specific and developmentally-specific manner. Since the GH receptor (GHR) mRNA and protein are co-localized with GH in many of those tissues, it is likely that extrapituitary GH acts primarily as a paracrine/autocrine. Moreover, since GH variants in the chicken pituitary can differ in biological activity and potency (3, 7), tissue- and age-related differences in the relative proportion of the GH moieties in extrapituitary tissues might provide a viable mechanism to modulate the action of GH. This phenomenon contributes to the complex physiology of GH. Although the local concentration of GH in extrapituitary tissues is much lower than that found in the pituitary (as expressed by milligram of protein), the total production of "peripheral" GH might be equivalent to (or even higher than, under certain conditions) the total amount synthesized in the pituitary, taking into account the size of the organs (Table 3). The statement "Everything is produced everywhere and does everything" is thus particularly relevant to GH, which is produced in many diverse tissues and exerts a wide range of cellular effects. GH may, however, assume different hues in order to initiate particular responses, depending on the developmental stage and the specific tissue.

Acknowlegments

The authors would like to acknowledge the support of PAPIIT-DGAPA, UNAM and CONACYT, México, NSERC, Canada, and the Senate Research Committee, Bishop's University.

References

1 Arámburo C, Berghman L & Scanes CG (1994) *17th Conference of European Comparative Endocrinologists, September, Spain.*:233 (Abstr.)

2 Arámburo C, Campbell RM & Scanes CG (1989) Heterogeneity of chicken growth hormone (cGH). Identification of lipolytic and non-lipolytic variants. *Life Sciences* **45** 2201-7

3 Arámburo C, Carranza M, Reyes M, Luna M, Martinez-Coria H, Berumen L,& Scanes CG (2001) Characterization of a bioactive 15 kDa fragment produced by proteolytic cleavage of chicken growth hormone. *Endocrine* **15** 231-40

4 Arámburo C, Carranza M, Sanchez R & Perera G (1989) Partial biochemical and biological characterization of purified chicken growth hormone (cGH). Isolation of cGH charge variants and evidence that cGH is phosphorylated. *General and Comparative Endocrinology* **76** 330-9

5 Arámburo C, Luna M, Carranza M, Reyes M, Martinez-Coria H & Scanes CG (2000) Growth hormone size variants: changes in the pituitary during development of the chicken. *Proceedings of the Society for Experimental Biology and Medicine* **223** 67-74

6 Arámburo C, Martínez-Coria H, Luna M, Carranza M, McCann-Levorse LM, Berghman L & Scanes CG (2004) Differential stimulation of the release (in vitro and in vivo) of chicken growth hormone isoforms: Effects of Thyrotropin Releasing Hormone (TRH) and Growth Hormone Releasing Hormone (GHRH). *Endocrinology (in preparation)*

7 Arámburo C, Montiel JL, Perera G, Navarrete S & Sanchez R (1990) Molecular isoforms of chicken growth hormone (cGH): different bioactivities of cGH charge variants. *General and Comparative Endocrinology* **80** 59-67

8 Arámburo C, Montiel JL, Proudman JA, Berghman LR & Scanes CG (1992) Phosphorylation of prolactin and growth hormone. *Journal of Molecular Endocrinology* **8** 183-91

9 Arámburo C, Navarrette S, Montiel JL, Sanchez R & Berghman LR (1991) Purification and electrophoretic analysis of glycosylated chicken growth hormone (G-cGH): evidence of G-cGH isoforms. *General and Comparative Endocrinology* **84** 135-46

10 Baudet ML, Sanders EJ & Harvey S (2003) Retinal growth hormone in the chick embryo. *Endocrinology* **144** 5459-5468

11 Berghman LR, Buyse J, Huybrechts LM, Darras VM, Vandesande F, Kuhn ER, Decuypere E, & Scanes CG (1994) Disappearance rate of glycosylated and non-glycosylated chicken growth hormone: influence on biological activity. *Comparative Biochemistry and Physiology: Pharmacology, Toxicology and Endocrinology* **108** 161-9

12 Berumen L, Luna M, Carranza M, Martínez-Coria H, Cárabez A, & Arámburo C (2004) Chicken growth hormone: further characterization and

ontogenic changes of an N-glycosylated isoform in the anterior pituitary gland. *General and Comparative Endocrinology* **139** 113-23
13 Darras VM, Finné MF, Berghman LR & Kühn ER (1994) Ontogeny of the sensitivity of the somatotrophes to thyrotrophin releasing hormone (TRH) and growth hormone releasing factor (GRF) in the embryonic and post-hatch chick. *Annals of Endocrinology* **55** 255-60
14 Dean CE, Piper M & Porter TE (1997) Differential responsiveness of somatotrophs to growth hormone-releasing hormone and thyrotropin-releasing hormone during chicken embryonic development. *Molecular and Cellular Endocrinology* **132** 33-41
15 Geris KL, Berghman LR, Kühn ER & Darras VM (1998) Pre- and posthatch developmental changes in hypothalamic thyrotropin-releasing hormone and somatostatin concentrations and in circulating growth hormone and thyrotropin levels in the chicken. *Journal of Endocrinology* **159** 219-25
16 Gould J, Arámburo C, Capdevielle M, & Scanes CG (1995) Angiogenic activity of anterior pituitary tissue and growth hormone on the chick embryo chorio-allantoic membrane: a novel action of GH. *Life Sciences* **56** 587-94
17 Gravance CG, Breier BH, Vickers MH & Casey PJ (1997) Impaired sperm characteristics in postpubertal growth-hormone-deficient dwarf (dw/dw) rats. *Animal Reproduction Science* **49** 71-6
18 Harvey S (1993) Hypophysiotrophic control of GH secretion and its feedback regulation: an update. In *Avian Endocrinology*, pp. 11-29. Ed P Sharp. Journal of Endocrinology, Ltd: Bristol
19 Harvey S, Azumaya Y& Hull KL (2000) Pituitary and extrapituitary growth hormone: Pit-1 dependence? *Canadian Journal of Physiology and Pharmacology* **78** 1013-28
20 Harvey S, Baudet ML, Murphy AE, Luna M, Hull KL& Arámburo C (2004) Testicular growth hormone (GH): GH expression in spermatogonia and primary spermatocytes. *General and Comparative Endocrinology* **139** 158-67
21 Harvey S, Davison TF & Chadwick A (1979) Ontogeny of growth hormone and prolactin secretion in the domestic fowl (*Gallus domesticus*). *General and Comparative Endocrinology* **39** 270-3
22 Harvey S, Johnson CD & Sanders EJ (2000) Extra-pituitary growth hormone in peripheral tissues of early chick embryos. *Journal of Endocrinology* **166** 489-502
23 Harvey S, Sharma P, Johnson CDM, Sanders EJ & Hull KL (1998) Growth hormone: an embryonic growth factor? *Comparative* Biochemistry *and Pharmacology* **119** 305-15
24 Hull KL, Janssens WC & Harvey S (1997) Growth hormone: actions in sites of synthesis. In *Perspectives in Avian Endocrinology*, pp. 375-387. Eds S Harvey, RJ Etches. Journal of Endocrinology, Ltd.: Bristol
25 Kansaku N, Shimada K, Terada O & Saito N (1994) Prolactin, growth hormone, and luteinizing hormoneβ subunit gene expression in the cephalic and caudal lobes of the anterior pituitary gland during embryogenesis and

different reproductive stages in the chicken. *General and Comparative Endocrinology* **96** 197-205

26 Kanzaki M & Morris PL (1999) Growth hormone regulates steroidogenic acute regulatory protein expression and steroidogenesis in Leydig cell progenitors. *Endocrinology* **140** 1681-6

27 Kuhn ER, Vanderpooten A, Huybrechts LM, Decuypere E, Darras V & Sharp PJ (1988) Hypothalamic hormones that release growth hormone stimulate hepatic 5'-monodeiodination activity in the chick embryo. *Journal of Endocrinology* **118** 233-6

28 Lantinga V, Teske E, van G & Mol JA (2001) Growth hormone gene expression in normal lymph nodes and lymphomas of the dog. *Anticancer Research* **20** 2371-6

29 Liu L, Porter TE (2004) Endogenous thyroid hormones modulate pituitary somatotroph differentiation during chicken embryonic development. *Journal of Endocrinology* **180** 45-53

30 Luna M, Barraza N, Berumen L, Pedernera E, Harvey S & Arámburo C (2004) Immune growth hormone (GH): heterogeneity in lymphoid tissues and changes during ontogeny in domestic fowl. *General and Comparative Endocrinology (submitted)*

31 Luna M, Huerta L, Berumen L, Martinez-Coria H, Harvey S & Arámburo C (2004) Growth hormone in the male reproductive tract of the chicken: heterogeneity and changes during ontogeny and maturation. *General and Comparative Endocrinology* **137** 37-49

32 Luna M, Martínez-Coria H, Carranza M, Huerta L, Harvey S & Arámburo C. 2000. *20th Conference of European Comparative Endocrinologists, Faro, Portugal.*:33 (Abstr.)

33 Malamed S, Gibney JA, Cain LD, Perez FM & Scanes CG (1993) Immunocytochemical studies of chicken somatotrophs and somatotroph granules before and after hatching. *Cell and Tissue Research* **272** 369-74

34 Martinez-Coria H, Lopez-Rosales LJ, Carranza M, Berumen L, Luna M & Arámburo C (2002) Differential secretion of chicken growth hormone variants after growth hormone-releasing hormone stimulation in vitro. *Endocrine* **17** 91-102

35 Montiel JL, Berghman LR & Arámburo C (1992) Identification of growth hormone molecular variants in chicken serum. *General and Comparative Endocrinology* **88** 298-306

36 Murphy AE & Harvey S (2001) Extrapituitary beta TSH and GH in early chick embryos. *Molecular and Cellular Endocrinology* **185** 161-71

37 Murphy AE, Peek H, Baudet ML & Harvey S (2003) Extrapituitary GH in the *chicken*: underestimation of immunohistochemical staining by Carnoy's fixation. *Journal of Endocrinology* **177** 223-34

38 Ovesen P, Jorgensen JO, Ingerslev J, Orskov H & Christiansen JS (1998) *Growth* hormone treatment of men with reduced sperm quality. *Ugeskr. Laeger* **160** 176-80

39 Porter TE, Couger GS, Dean CE & Hargis BM (1995) Ontogeny of growth *hormone* (GH) secreting cells during chicken embryonic development: Initial somatotrophs are responsive to GH-releasing hormone. *Endocrinology* **136** 1850-6

40 Porter TE, Dean CE, Piper MM, Medvedev KL, Ghavam S, & Sandor J (2001) Somatotroph recruitment by glucocorticoids involves induction of growth hormone gene expression and secretagogue responsiveness. *Journal of Endocrinology* **169** 499-509

41 Ramesh R, Kuenzel WJ, Buntin JD & Proudman JA (2000) Identification of growth-hormone- and prolactin-containing neurons within the avian brain. *Cell and Tissue Research* **299** 371-83

42 Render CL, Hull KL & Harvey S (1995) Expression of the growth hormone gene in immune tissues. *Endocrine* **3** 729-35

43 Render CL, Hull KL & Harvey S (1995) Neural expression of the pituitary GH gene. *Journal of Endocrinology* **147** 413-22

44 Sanders EJ, Harvey S. 2004. *Anatomy and Embryology*. In press

45 Scanes CG, Arámburo C, Campbell RM (1990) Hormonal involvement in avian growth and development: growth hormone and insulin-like growth factor. In *Endocrinology of Birds - Molecular to Behavioural*, pp. 93-110. Ed M Wada. Springer-Verlag: Berlin

46 Scanes CG, Arámburo C, Campbell RM, Kopchick J & Radecki S (1993) Chemistry and physiology of poultry growth hormone. In *Avian Endocrinology*, pp. 93-110. Ed PJ Sharp. Journal of Endocrinology: Bristol

47 Shimonovitz S, Zacut D, Benchetrit A & Ron M (1993) Growth hormone status in patients with maturation arrest of spermatogenesis. *Human Reproduction* **8** 919-21

48 Takeuchi S, Haneda M, Teshigawara K & Takahashi S (2001) Identification of a novel GH isoform: a possible link between GH and melanocortin systems in the developing chicken eye. *Endocrinology* **142** 5158-66

49 Thommes RC & Rousseau RP (1959) Vasculogenesis in the adenohypophysis of the developing chick embryo. *Growth* **23** 205-19

50 Thommes RC, Umprowicz DM, Leung FC & Woods JE (1987) Ontogenesis of immunocytochemically demonstrable somatotrophs in the adenohypophyseal pars distalis of the developing chick embryo. *General and Comparative Endocrinology* **67** 390-8

51 Tixier-Vidal A (1954) Etude histophysiologique de l'hypophyse anterior de l'embryon de poulet. *Archives D Anatomie Microscopique et de Morphologie Experimentale* **43** 463-86

52 Van As P, Buys N, Onagbesan OM & Decuypere E (2000) Complementary DNA cloning and ontogenic expression of pituitary-specific transcription factor of chickens (*Gallus domesticus*) from the pituitary gland. *General and Comparative Endocrinology* **120** 127-36

53 Yang Y, Guo L, Ma L & Liu X (1999) Expression of growth hormone and insulin-like growth factor in the immune system of children. *Hormone and Metabolic Research* **31** 380-4

Functional Avian Endocrinology
A Dawson & P J Sharp (Eds)

Ocular pituitary hormones in the chick embryo

S Harvey, D Rattray, M-L Baudet and E J Sanders

Department of Physiology, University of Alberta, Edmonton, Alberta, T6G 2H7, Canada

Introduction

The extrapituitary expression of pituitary hormones is now well established. Growth hormone (GH) (21), prolactin (39), adreno-corticotropin (ACTH) (12), luteinizing hormone (LH) (43), and thyrotropin (TSH) (33), have all been shown to be abundantly present in tissues outside the avian adenohypophysis, particularly in neural tissues (21, 33, 39) and in early embryos (21, 33, 43). As recent studies have demonstrated the presence of pituitary hormones in the neural retina and other ocular tissues of chick embryos, these studies are the focus of this brief review.

Ocular GH: Presence in the eye

GH immunoreactivity is present in the optic cup of chick embryos from embryonic day (ED) 2 of the 21 day incubation period (18). This immunoreactivity was detectable by Harvey *et al* (18) using 3 different antibodies raised against chicken pituitary GH and its specificity was shown by its complete loss following the preabsorption of the antibodies with recombinant chicken GH. This immunoreactivity was widespread in the inner and outer neural retina and, at ED 7, was specifically localized in immunologically identified retinal ganglion cells (RGCs) (3, 41). It was also present in the underlying retinal pigmented epithelium (RPE) (16) and, in lesser amounts, in the choroid layer, the ciliary body and cornea and in the fibre cells and the epithelium of the lens capsule (16, 17). It is also present in the optic nerve (17), derived from the axons of the RGCs. Takeuchi *et al* (45) also found GH immunoreactivity in the embryonic chick eye, using an antibody raised against rat GH, although they only found GH immunoreactivity in the RPE and only after ED 10 of the incubation period. The discrepancy between these studies may reflect differences in the chicken strains used (White Leghorns vs broilers), differences in the fixatives used (paraformaldehyde vs Bouins solution) or differences in the specificity of the primary antibodies, since rat GH antibodies underestimate GH abundance in chickens (23) and may not detect all the GH variants in the eye (3, 41).

Harvey *et al* (3, 17) showed the GH immunoreactivity in the neural retina of the ED 7-ED 9 chick embryo was primarily associated with proteins of 15 kDa and 16 kDa. In contrast, these moieties were only present in trace amounts in the pituitary gland, in which the GH monomer (22 kDa under non-reducing conditions, 25 kDa under reducing electrophoresis) was the most abundant GH species. These submonomer proteins were detectable in the chick retina until at least postnatal day 12, although they were not present by 42 days of age. Interestingly, Takeuchi *et al* (45) also found the GH immunoreactivity that they detected in the embryonic RPE was not present in 1- or 7-day old neonates.

In addition to being in the neural retina, the GH immunoreactivity detected by Baudet *et al* (3) was secreted by retinal explants *in vitro*. However, the only GH moiety in the incubation media was the 15 kDa protein, which suggests the 16 kDa moiety is not a secretory form. The finding of 15 kDa GH in the vitreous humor (3) supports this view. The high concentration of GH immunoreactivity in the vitreous (comparable to that in the pituitary gland, 3), suggests it may act as an extracellular store, especially as GH immunoreactivity is still abundantly present in the vitreous of 42 day-old chickens when it is no longer present in retinal tissue (3). Within the vitreous some of the 15 kDa GH moiety is bound to a 45 kDa protoglycan, opticin (41), representing a unique, perhaps tissue-specific, GH-binding protein. The binding of 15 kDa GH to opticin is discernible when vitreous is electrophoresed under non-reducing condition and this binding may prevent its degradation in the vitreous body.

This 15 kDa GH moiety is thought to be a fragment (2) produced by the proteolytic cleavage of the full-length GH molecular (between amino acids Ang_{133}-Gly_{134}), rather than the product of a truncated transcript. The trace amounts of the 25 kDa GH monomer in the eye and neural retina of ED 7-ED 8 chicks (3) supports this view, as does the generation of a 15 kDa fragment when recombinant (25 kDa) pituitary GH is incubated with retinal tissue *in vitro* (unpublished observations).

The GH immunoreactivity in the neural retina, as in other neural tissues (22) appeared to be in both nuclear and cytoplasmic compartments (18). However, ultra-structural studies showed that it was primarily in the cytoplasmic compartments of ED 9 neural retina cells and RPE cells (16). Although not associated with secretory granules, GH in the RPE was particularly associated with the membranes of its melanin granules.

Ocular GH: Synthesis in the eye

The occurrence of GH immunoreactivity in ocular tissue of the chick embryo occurs prior to the differentiation of pituitary somatotrophs (in mid-late incubation, 20) and the appearance of GH in peripheral plasma (at ED 17, 20, 36). It cannot, therefore, reflect the sequestration of pituitary GH from plasma. Indeed, the retinal-blood barrier is functional by ED 14 (7, 31) and this prevents the entry of systemic GH into the eye (26). The high concentrations of GH in the vitreous of neonatal and juvenile chickens must therefore result from its local synthesis. This possibility is supported by the finding of *Pit-1* and *Pit -1* mRNA in the neural retina of ED 7 embryos (17), since the expression of the GH gene in the pituitary gland is dependent upon this transcription factor. GH mRNA is, furthermore, also present in the eyes of embryonic chicks (3, 16, 17).

A 690 bp cDNA was generated from reverse-transcribed ED 7 RPE RNA when amplified by the polymerase chain reaction with oligonucleotide primers designed to span the full-length (exons 1-5) pituitary GH transcript (17). A 466 bp cDNA was similarly generated from the neural retina RNA and RPE RNA of ED 9 embryos, after its amplification with primers coding from a portion of exon 3 and spanning exons 4 and 5 (16). Full-length GH cDNA from the neural retina of ED 7 embryos has also been cloned and sequenced (3) and found to be 99.9% homologous (689/690 bases) to the published sequence for chicken pituitary GH cDNA (the only difference being a base substitution at position 588 (623 of the cDNA fragment, an A for G substitution). The amino acid sequence of the predicted retinal protein would, however, be 100% identical with pituitary GH, as the base substitution would not change the coded amino acid (lysine). This substitution is, moreover, merely a gene polymorphism, since it is also present in GH mRNA in the testis and pituitary of this chicken strain (19).

This full-length GH transcript was widespread throughout the neural retina of ED 7 embryos but it was particularly intense in large cells in the RGC layer (3). In these cells the staining to a DIG-labeled riboprobe was cytoplasmic, although it was not present in the optic fibre layer derived from the RGC axons. This GH transcript was not, however, detected by Takeuchi *et al* (45), who nevertheless cloned a novel transcript that coded for a smaller GH isoform (s-cGH). This truncated transcript is comprised of two exons: the first corresponds to a part of intron 3 and exon 4 of the pituitary GH gene, the second exon corresponds to exon 5 of pituitary GH mRNA. The protein coded by this cDNA has a deduced molecular size of 16.5 kDa and 140 amino acid residues, but it lacks the signal peptide and the N-terminal 71 amino acids of monomer pituitary GH. Its N-terminus would have a unique sequence of 20 amino acids, whereas the C-terminal 120 amino acid (positions 97-216) would be identical to 22 kDa pituitary GH and contain helixes 3 and 4 that may bind to the GHR. It would also have the conserved four cysteine residues in its C-terminus that would permit the formation of intramolecular disulfide bridges and the secondary (coil) structure that is characteristic of 22 kDa GH. It is therefore possible that it could therefore bind to and activate GH receptors (GHRs). However, as its N-terminus would be hydrophilic it would be unlikely to act as a signal peptide and s-cGH would probably not be secreted by cells. If it has biological activity, it may thus act as an intracrine, especially as it has leucine residues in its third helical domain that are arranged every 7 amino acids, giving it the characteristic structure of the leucine-zipper transcription factors involved in gene regulation. Although Takeuchi *et al* (45) did not detect s-cGH in the embryonic eye, our finding of a 16 kDa protein in the neural retina that was not secreted *in vitro* nor present in the vitreous *in vivo* may indicate translation of the s-cGH transcript.

Ocular GH: Actions in the eye

The GHR gene is expressed in the neural retina of embryonic chicks (17). cDNA fragments for its extracellular and intracellular domains have been generated by RT-PCR using specific oligonucleotide primers for the GHR gene. The presence of predictable cleavage sites in these fragments for endonuclease digestion further demonstrates sequence homology with the GHR mRNA. Immunoreactivity for the chicken GHR has also been demonstrated by the Western blotting of proteins extracted from the neural retina of ED 8 chicks (17). The neural retina is therefore likely to be an autocrine/paracrine target site for GH produced locally.

Within the retina endogenous GH may participate in neurogenesis by stimulating retinal cell survival. Exogenous chicken GH markedly reduces the number of TUNEL positive, apoptotic cells when incubated for 24-48h with ED 5 or ED 6 retinal explants *in vitro* (4). Conversely, immunoneutralization of endogenous GH increases the number of apoptotic cells in cultured retinal cells and intravitreal administration of GH antisera to the eyes of ED 3 embryos also increases the number of apoptotic retinal cells in ED 5 embryos (4, and unpublished observations). The antiapoptotic action of exogenous GH *in vitro* is accompanied by a decrease in caspase 3 mRNA and a decrease in AIF mRNA (4), consistent with its neuroprotective action in the CNS (1, 22, 42) and its antiapoptotic actions in other tissues (22). Since chicken GH also induces a five-fold increase in the production of IGF-I mRNA in 48h cultured chick retinal explants (3), its antiapoptotic action may be direct or mediated through IGF-I, which is also neuroprotective in chick retina (11, 25), possibly through a caspase 3 dependent mechanism (28).

In addition to neurogenesis, GH may act in the embryonic eye to regulate angiogenesis. Human GH binds to retinal blood vessels (30) and GH excess is implicated in the etiology of retinal vascularization and retinopathy in humans (6, 15), whereas impaired retinal vascularization accompanies a clinical GH deficiency (24). The angiogenic actions of GH and its cleavage fragments on the blood vessels of the chick chorioallantoic membrane (14, 44) supports this view.

Ocular prolactin

Prolactin mRNA is expressed in the retinal blood vessels (and leukocytes) of humans with retinopathy of prematurity (ROP) (8, 38). A 16 kDa prolactin fragment of the full-length (23 kDa) protein is also present in the retrolental fibrovascular membranes of these patients, in which 16 kDa prolactin is present in the aqueous humor and sub retinal fluid. This fragment is generated by proteases in the subretinal fluid and its concentration in ocular fluids is unrelated to that in serum, although it is directly related to the amount of prolactin mRNA in the ocular blood vessels (8). Rat retinal endothelial cells also express prolactin mRNA and full-length 23 kDa prolactin, which is present in the cytoplasm of these cells and released from them *in vitro* (35). The concentration of prolactin in these retinal blood vessels is 300 fold higher than that in other blood vessels, suggesting important roles in ocular function. Receptors for prolactin have been localized in the retina and the cornea, in which exogenous 16 kDa prolactin has anti-angiogenic activity (9).

In the ED 7 chick embryo, preliminary studies show that presence of prolactin immunoreactivity (using an antibody raised against chicken prolactin, kindly provided by Dr. J.A. Proudman) in the cytoplasm of retinal ganglion cells, in the choroid and in the endothelial cells of choroidal blood vessels (40, unpublished observations and Fig1)

Ocular LH

While LH is present in the central nervous system (5, 27, 34), its possible presence in the eye has yet to be assessed. Preliminary studies have, however, shown the presence of LH immunoreactivity in the cytoplasm of large cells in the RCG layer of the ED7 neural retina (40, unpublished observations and Fig 1). Western blotting with the same antisera (raised against chicken LH) indicated that this immunoreactivity was associated with a 25 kDa protein (unpublished observations), as similarly detected in the lungs of embryonic chicks (43). In addition, to being a site of LH production or LH sequestration, the neural retina may be a site of LH action, since LH receptors are present in the bovine neural retina (48). This may indicate that LH has hitherto unsuspected roles in retinal function.

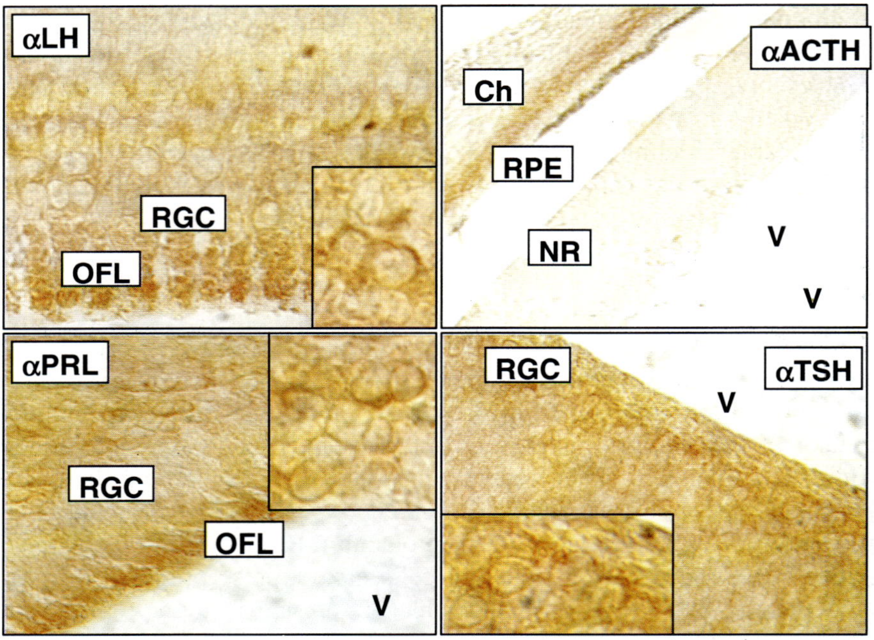

Fig. 1 Immunoreactivity for pituitary hormones in the eyes of embryonic day (ED)7 chick embryos. *Upper left panel*: Immunoreactivity against an antibody raised in rabbits against chicken luteinizing hormone (LH). Diffuse staining is seen throughout the neural retina, but specific staining is confined to the cytoplasm of large, nucleated cells in the retinal ganglion cell (RGC) layer, that have the morphological appearance and location of RGC's (as shown in the insert). Intense staining is also present in fascicles of the optic fibre layer (OFL), comprised of RGC axons. Magnification x 400. *Upper right panel*: Immunoreactivity against an antibody, raised in rabbits, against human adrenocorticotropin (ACTH). Intense staining is seen in a discrete layer of the choroid (Ch) but no staining is seen in the neural retina (NR). Abbreviations: RPE, retinal pigmented epithelium, V, vitreous. Magnification x 40. *Lower left panel*: Immunoreactivity against an antibody, raised in rabbits against chicken prolactin (PRL). Specific staining is seen in the cytoplasm of cells in the RGC layer (as shown in the insert) and in the fascicles of the OFL. Magnification x 400. *Lower right panel*: Immunoreactivity against an antibody, raised in rabbits, against rat β-thyrotropin (TSH). Specific staining is seen in the cytoplasm of cells in the RGC layer (as shown in the insert), but not in the OFL. Magnification x 400.

Avidin-biotin-peroxidase immunocytochemistry was performed on 8 μm sections of eyes fixed in Carnoy's fixative, using diaminobenzidine as a stain. Final antibody concentrations were 1:5000 for the ACTH and LH antibodies and 1:1000 for the TSH and PRL antibodies. No staining was seen when the primary antibodies were replaced with normal rabbit serum.

Ocular TSH

TSH-immunoreactivity has been reported in the ganglion cells and perivascular fibres of the human retina (10). Immunoreactive TSH (detected using a human TSH antibody) has also been observed in the cytoplasm of RGC's and their axons in ED7 chick embryos (37), prior to its presence in Rathke's pouch (33), the pituitary progenitor (Fig 1). This finding was recently confirmed using an antibody raised against the β subunit of bovine TSH (40, unpublished observations). TSH immunoreactivity is also present in the inner and outer plexiform layer of the retina at ED 17, in cells with the appearance a location of amacrine and horizontal cells (37). In the post hatching period numerous TSH-immunopositive ganglion and amacrine cells are present in the retina of day-old chicks, although no TSH staining is present in the plexiform layers (37). The role of TSH in the neural retina is unknown, although Prada *et a*l (37) suggested it was involved in the differentiation and migration of the different retinal cell-types. An ocular site of TSH action has been established by the ability of exogenous TSH to rapidly increase cAMP production in intact cultured RPE cells (29).

Ocular ACTH

ACTH is a melanocortin, derived from the proopiomelanocortin (POMC) precuror. This precursor has been shown to be expressed in the rat retina, particularly in the RGC's that have melanocortin receptors and respond to exogenous melanocortins with increased neurite outgrowth and survival (32). The presence of ACTH in rat ocular tissues has, however, yet to be established, although ACTH immunoreactivity (reacting against an antibody raised against human ACTH) has been demonstrated in ocular tissues of ED 7 chick embryos (40, unpublished observations and Fig 1)), prior to its detection in the pituitary gland (36). Intense ACTH staining is present in the cornea, choroid layer and the lens capsule, although no ACTH staining is present in the RGC's or the neural retina, in maked contrast to the staining for GH, LH, TSH and prolactin (40), nor in the RPE, in contrast with the staining for GH and MSH (16, 45). The intensity of the ACTH staining suggests it has physiological significance.

Ocular MSH

Immunoreactivity for α-melanocyte stimulating hormone (MSH) has been detected in the ganglion cell layer of the adult rat retina (32). As these cells abundantly express melanocortin (MC_3-MC_5) receptors (32), this immunoreactivity may reflect the sequestration of α-MSH in these cells, rather than the synthesis of α-MSH in them. The actions of α-MSH in these cells include an induction of cell survival and neurite outgrowth. Within the eye, α-MSH also stimulates the release of retinal dopamine and γ-aminobutyric acid, increases the permeability of the blood-aqueous barrier and stimulates the production of eicosanoids by the RPE (32).

In birds, a combination of RT-PCR, *in situ* hybridization and immunocytochemistry has demonstrated that the RPE cells in the chick embryo express α-MSH in a developmentally regulated way (47), coincident with the temporal production of s-cGH (45). α-MSH immunoreactivity in the cells of the chick RPE is initially detected at ED10 and increases in intensity as development proceeds. As the prohormone convertases responsible for the production of α-MSH from POMC are also found in the RPE (47), this is likely to be the source of α-MSH production in the embryonic eye. However, immunoreactivity for these covertases is not present in the RPE of newly hatched birds, which also lack POMC mRNA. In newly hatched birds, α-MSH immunoreactivity is present in the RPE and in the photoreceptor cone cells, in which POMC mRNA is abundantly expressed. As the POMC convertases are not present in the cone cells during embryogenesis, the RPE and cone cells would appear to be the embryonic and neonatal sources, respectively, of the α-MSH in the chicken eye (47).

Although the roles of endogenous α-MSH in ocular function are still uncertain, MC_1, MC_4 and MC_5 receptors are expressed in layers of the chick choroid and neural retina (46), adjacent to the RPE. α-MSH from the RPE or the photoperiodic cone cells is therefore thought to regulate ocular development in a paracrine fashion, possibly through s-cGH interactions (45,46).

In addition to the RPE and the cone cells of the retina, α-MSH has also been observed in retinal cultures of Muller cells, devoid of neurons (13). These astroglial cultures also contain β- MSH immunoreactivity (13). The roles of α- or β- MSH in these cells is, however, still uncertain.

Conclusion

All of the hormones that are primarily produced in the anterior pituitary gland neonatally are present in ocular tissues (particularly retinal tissues) during early chick embryogenesis, prior to the differentiation of the pituitary gland. The embryonic chick eye is thus likely to be an extrapituitary site of adenohypophyseal hormone production, although some of these 'ocular pituitary hormones' appear to be developmentally regulated and absent in adults, in which the blood-ocular barriers would exclude the entry of adenohypophyseal-derived hormones from peripheral plasma into the eye. The presence of receptors for most of these 'pituitary' hormones in the embryonic eye suggests they have ocular roles during early development, although these are, as yet, still largely unknown. These embryonic roles would, however, appear to be autocrine or paracrine in nature, in contrast to the endocrine actions of these hormones in later development.

Acknowledgements

This work was supported by grants from the Canadian Institutes for Health Research and the Natural Sciences and Engineering Research Council of Canada. Marie-Laure Baudet was in receipt of a University of Alberta F.S.Chia Studentship and a studentship from the Alberta Heritage Foundation for Medical Research.

References

1 Ajo R, Cacicdeo L, Navarro C, & Sanchez-Franco F (2003) Growth hormone action on proliferation and differentiation of cerebral cortical cells from fetal rat. *Endocrinology* **144** 1086-1097

2 Aramburo C, Carranza M, Reyes M, Luna M, Martinez-Coria H, Berumen L, & Scanes CG (2001b) Characterization of a bioactive 15 kDa fragment produced by proteolytic cleavage of chicken growth hormone. *Endocrine* **15** 231-240

3 Baudet M-L, Sanders EJ, & Harvey S (2003) Retinal growth hormone in the chick embryo. *Endocrinology* **144** 5459-5468

4 Baudet M-L, RattrayD Sanders EJ & Harvey S (2004) Growth hormone : a survival agent in the developing chick retina. *GH and IGF Research* **14** 105

5 Bowen RL, Smith MA, Harris PL, Kubat Z, Martins RN, Castellani RJ, Perry G *et al* (2002) Elevated luteinizing hormone expression colocalizes with neurons vulnerable to Alzheimer's disease pathology. *Journal of Neuroscience Research* **70** 514-518

6 Chantelau E (2002) Effect of growth hormone receptor antagonist on proliferative diabetic retinopathy. *Ophthalmology* **109** 2187-2188

7 Cunha-Vaz JG (1997) The blood-ocular barriers: past, present, and future. *Doc Opthalmology* **93** 149-157

8 Duenas Z, Rivera JC, Quiroz-Mercado H, Aranda J, Macotela Y, Montes de Oca P, Lopez-Barrera F, Nava G, Guerrero JL, Suarez A, De Regil M, Martinez de la Escalera G, Clapp C (2004) Prolactin in eyes of patients with retinopathy of prematurity: implications of vascular regression. *Invesigative Ophthalmology and Visual Science* **45** 2049-2055

9 Duenas Z, Torner L, Corbacho AM, Gutierrez-Ospina G, Lopez-Barrera F, Barrios FA, Berger P, Martinez de la Escalera G, & Clapp C (1999) Inhibition of rat corneal angiogenesis by 16-kDa prolactin and by endogenous prolactin-like molecules. *Investigative Ophthalmology and Visual Science* **40** 2498-2505

10 Fdez-Trujillo FJ, Prada A & Verastegui C (1996) Thyrotopin –like immunoreactivity in human retina: immunoreactive co-localization in ganglion cells and perivascular fibres. *Neurchemistry International* **28** 381-384

11 Frade JM, Marti E, Bovolenta P, Rodriguez-Penez-Garcia D, Rohrer H, Edgar D, & Rodriguez-Tebar A (1996) Insulin-like growth factor-1

stimulates neurogenesis in chick retina by regulating expression of the alpha 6 integrin subunit. *Development* **122** 2497-2506

12 Franchini A & Ottaviani E (1999) Immunoreactive POMC- derived peptides and cytokines in the chicken thymus and bursa of Fabricius microenvironments: age-related changes. *Journal of Neuroendocrinolgy* **11** 685-692

13 Girard A, Cristanti-Combes P, Faivre-Bauman A, Dubois MP & Pessac B (1981) Demonstration of neuropeptides in cultured neuroretinal cells of chick embryo. *Comptes Rendes Seances Academie Sciences* **293** 291-295

14 Gould J, Aramburo C, Capdevielle M, & Scanes CG (1995) Angiogenic activity of anterior pituitary tissue and growth hormone on the chick embryo chorio-allantoic membrane: a novel action of GH. *Life Sciences* **56** 587-594

15 Hansen R, Koller EA, & Malozowski S (2000) Full remission of growth hormone (GH)-induced retinopathy after GH treatment discontinuation: long-term follow-up. *Journal of Clinical Endocrinology & Metabolism* **85** 2627

16 Harvey S, Kakebeeke M, & Sanders EJ (2003) Growth hormone localization in the neural retina and retinal pigmented epithelium of embryonic chick. *Journal of Molecular Neurosciences* **22** 139-145

17 Harvey S, Kakebeeke M, Murphy AE, & Sanders EJ (2003) Growth hormone in the nervous system: autocrine or paracrine roles in retinal function? *Canadian Journal of Physiology and Pharmacology* **81** 371-384

18 Harvey S, Johnson CDM, & Sanders EJ (2001) Growth hormone in neural tissues of the chick embryo. *Journal of Endocrinology* **169** 487-498

19 Harvey S, Baudet ML, Murphy AE, Luna M, Hull KL & Aramburo C (2004) Testicular growth hormone (GH): GH expression in spermatogonia and primary spermatocytes. *General and Comparative Endocrinology* **139** 158-167

20 Harvey S, Johnson, CDM, Sharma P, Sanders EJ, & Hull KL (1998) Growth hormone: a paracrine growth factor in embryonic development. *Comparative Biochemistry and Physiology* **119** 305-315

21 Harvey S & Hull KL (1997) Growth hormone: a paracrine growth factor? *Endocrine* **7**: 267-279

22 Harvey S & Hull KL (2003) Neural growth hormone: an update. *Journal of Molecular Neuroscience* **20** 1-14

23 Harvey S, Hoshino S, & Suzuki M (1987) Plasma immunoreactive-growth hormone in domestic fowl: measurement by homologous and heterologous radioimmunoassays.*General and Comparative Endocrinology* **65** 92-98

24 Hellstrom A, Svensson E, Carlsson B, Niklasson A, & Albertsson-Wikland K (1999) Reduced retinal vascularization in children with GH deficiency. *Journal of Clinical Endocrinology and Metabolism* **84** 795-798

25. Hernandez-Sanchez D, Lopez-Carranza A, Alarcon D, de La Rosa EJ, & de Pablo F (1995) Autocrine/paracrine role of insulin-related growth factors in neurogenesis: local expression and effects on cell proliferation and differentiation in retina *Proceedings of the National Academy of Sciences USA* **92** 9834-9838

26 James CRH, & Cotlier E (1983) Fate of insulin in the retina: an autoradiographic study. *British Journal of Opthalmology* **67** 80-88

27 Hojvat S, Emanuele N, Baker G, Connick E, Kirsteins L, & Lawrence AM (1982) Growth hormone (GH), thyroid-stimulating hormone (TSH), and luteinizing hormone (LH)-like peptides in the rodent brain: non-parallel ontogenetic development with pituitary coutnterparts. *Brain Research* **256** 427-434

28 Kermer P, Anderhold R, Klocker N, Krajewski S, Reed JC, & Bahr M (2000a) Caspase-9: involvement in secondary death of axotomized rat retinal ganglion cells *in vivo. Brain Research: Molecular Brain Research* **85** 144-150

29 Koh SW & Chader GJ (1984) Agonist effects on the intracellular cyclic AMP concentration of retinal pigment epithelial cells in culture. *Journal of Neurochemistry* **42** 287-289

30 Lewis UJ, Sinha YN, & Lewis GP (2000) Structure and properties of members of the hGH family in a review. *Endocrine Journal* **47** 51-58

31 Liebner S, Berhardt H, & Wolburg H (1997) Maturation of the blood-retina barrier in the developing oculi of the chicken. *Brain Research: Developmental Brain* Research **100** 205-219

32 Lindqvist N, Napankangas U, Lindblom J & Halbook F (2003) Proopiomelanocortin and melanocortin receptors in the adult rat retino-tectal system and their regulation after optic nerve transection. *European Journal of Pharmacology* **482** 85-94

33 Murphy AE & Harvey S (2001) Extrapituitary beta TSH and GH in early chick embryos. *Molecular and Cellular Endocrinology* **185** 161-171

34 Noteborn HP, de Koning J, de Jong FH, Ebels I, & Salemink CA (1992) Identification of luteinizing hormone-like proteins in the ovine pineal gland. *Journal of Pineal Research* **12** 118-127

35 Ochoa A, Montes de Oca P, Rivera JC, Duenas Z, Nava G, de La Escalera GM, & Clapp C (2001) Expression of prolactin gene and secretion of prolactin by rat retinal caspase endothelial cells. *Investigative Ophthalmology and Visual Science* **42** 1639-1645

36 Porter TE (1997) Regulation of somatotroph differentiation during chicken embryonic development: a review. In *Perspectives in Avian Endocrinology*, pp 47-56. Eds RJ Etches and S Harvey. Society of Endocrinology, Bristol, UK

37 Prada A, Verastegui C, Perez-Rios N, Gonzalez-Moreno M & Fdez-Trujillo FJ (2000) Thyrotropin-like immunoreactivity in the developing chicken retina. *European Journal of Morphometry* **38** 34-40

38 Quiroz H, Duenas Z & Lopez-Barrera F (2000) Detection of prolactin and prolactin mRNA in the eye of patients with retinopathy orf prematurity *Investigative Opthalmology and Visual Science* 41 S12

39 Ramesh R, Kuenzel WJ Buntin JD & Proudman JA (2000) Identification of growth-hormone- and prolactin-containing neurons within the avian brain. *Cell and Tissue Research* **299** 371-383

40 Rattray D, Baudet M-L & Harvey S (2003) Ocular pituitary hormones in the developing eye of embryonic chicks. *Proceedings of the 11th Western Perinatal Research Society, Banff Canada* p63

41 Sanders EJ, Walter MA, Parker E, Aramburo C, & Harvey S (2003) Opticin binds retinal growth hormone in the embryonic vitreous. *Investigative Opthalmology and Visual Science* **44** 5404-5409
42 Scheepens A, Sirimanne ES, Breier BH, Clark RG, Gluckman PD, & Williams CE (2001) Growth hormone as a neuronal rescue factor during recovery from CNS injury. *Neuroscience* **104** 677-687
43 Shirasawa N, Shiino M, Shimizu Y, Nogami H and Ishii S (1996) Immunoreactive luteinizing hormone (ir-LH) cells in the lung and stomach of chick embryos. *Cell and Tissue Research* **283** 19-27
44 Struman I, Bentzien F, Lee H, Mainfroid V, D'Angelo G, Goffin V, Weiner RI, & Martial JA (1999) Opposing actions of intact and N-terminal fragments of the human prolactin/growth hormone family members on angiogenesis: an efficient mechanisms for the regulation of angiogenesis. *Proceedings of the National Academy of Sciences USA* **96** 1246-1251
45 Takeuchi S, Haneda M, Tshigawara K, & Takahashi S (2001) Identification of a novel GH isoform: a possible link between GH and melanocortin systems in the developing chicken eye. *Endocrinology* **142** 5158-5166
46 Takeuchi S, Takahashi S, Okimoto R, Scioth HB & Boswell T (2003) Avian melanocortin system; alpha - MSH may act as an autocrine/paracrine hormone : a minireview. *Annals of the New York Academy* of *Sciences* **994** 366-372
47 Teshigawara K, Takahashi S, Boswell T, Li Q, Tanaka S, & Takeuchi S (2001) Identification of avian alpha-melanocyte-stimulating hormone in the eye: temporal and spatial regulation of expression of the developing chicken. *Journal of Endocrinology* **168** 527-537
48 Thompson DA, Othman MI, Lei Z, Li X, Huang ZH, Eadie DM, & Rao CV (1998) Localization of receptors for luteinizing hormone/chorionic gonadotropic neural retina. *Life Sciences* **63** 1057-1064

Functional Avian Endocrinology
A Dawson & P J Sharp (Eds)

CRH control of thyroid function in the chicken

B De Groef, K L Geris, K Vandenborne, V M Darras and E R Kühn

Laboratory of Comparative Endocrinology, Department of Biology, Catholic University of Leuven, B3000 Leuven, Belgium

Components of the avian thyroidal axis

In birds, as in other vertebrates, three structural components make up the so-called hypothalamo-pituitary-thyroidal axis (HPT). The thyroid gland releases thyroid hormones (THs), mainly 3,5,3',5'-tetraiodothyronine (T_4) or thyroxine. T_4 is activated by outer ring monodeiodination in peripheral tissues, producing the active thyroid hormone 3,5,3'-triiodothyronine (T_3). The secretory activity of the thyroid gland is stimulated by the pituitary gland, secreting thyroid-stimulating hormone (TSH) or thyrotropin in the systemic circulation. In turn, the pituitary is controlled by the hypothalamus, releasing both stimulatory and inhibitory hormones in a portal blood vessel system connecting both structures.

A central position within the HPT is taken by the thyrotropes, *i.e.* the TSH-producing pituitary cells. In a recent study, the development of thyrotropes was investigated by the immunohistochemical detection of TSH protein using a homologous anti-chicken TSH antibody, as well as the detection of TSH-β mRNA by RT-PCR and *in situ* hybridization (28). TSH-β mRNA-expressing cells were found from embryonic day 9 (E9) and immunoreactive TSH-β was present from E10 onwards. Based on these results, the development of the thyrotropes is believed to start around E9. On E20, thyrotrope density was almost four times greater than on E10. As previous studies using heterologous anti-TSH-β antibodies had already indicated (27), thyrotropes are confined to the cephalic lobe of the chicken pars distalis and never occur in the caudal lobe (7, 21, 28).

By analogy with its function in mammals, the hypothalamic tripeptide thyrotropin-releasing hormone (TRH) is often considered to be the main regulator of the secretory activity of the thyrotropes. Upon TRH challenge, a 3-fold increase in calculated plasma TSH index was recorded from three day post- hatch chicken (C3) pituitaries *in vitro* (1). TRH also increased TSH-β mRNA by 60% on average in cultured E19 anterior pituitary cells (18). Increasing evidence, however, suggests an important role for corticotropin-releasing hormone as a thyrotropin-releasing factor.

Thyrotropic activity of corticotropin-releasing hormone

Corticotropin-releasing hormone (CRH) is a 41-residue peptide characterized in mammals for its adrenocorticotropin (ACTH)-releasing action (33). Ovine CRH (oCRH) induces ACTH release from dispersed chicken pituitary cells in a concentration-dependent manner (2). When injected in E18 chicken embryos, oCRH causes not only the expected rise in plasma corticosterone levels, but also a marked increase in plasma T_4 one and two hours after injection (22, 26). The involvement of TSH in this effect was later confirmed (13, 14). Ovine CRH injected into E19 chicken embryos increases the plasma concentration of TSH within 15 min and for at least 2 hours (Fig. 1). Perifusion of pituitaries and thyroid glands consecutively, shows that oCRH induces an increase in the secretion of T_4 from the thyroid glands. Ovine CRH does not influence the *in vitro* thyroidal secretion of T_4 directly, but when applied to pituitaries *in vitro,* it increases TSH released into the perifusate. Porcine ACTH influences neither the *in vitro* secretion of TSH from pituitaries, nor T_4 from thyroid glands from C1 chicks (14). These experiments indicate that CRH is a potent TSH-releasing factor in the chicken, and that it exerts this function directly at the level of the pituitary. On an equimolar basis (100 nM) oCRH is twice as potent as TRH in inducing TSH release from C1 pituitaries *in vitro* (Table 1) (K.L. Geris, unpublished results).

The ontogeny of the *in vitro* TSH-releasing activity of oCRH has also been investigated in the chicken (16). Several embryonic stages (E14, E16, and E18) as well as posthatch stages (C1, C8, C22, and adult) were included in this study. Both oCRH concentrations tested (*i.e.* 10 and 100 nM) increased TSH secretion in all ages studied, demonstrating that thyrotropes are responsive to CRH before and after hatch. Comparison of the TSH response for both doses, showed that the lowest responses were at E14, and increased with age. However, the relative increase in TSH secretion induced by oCRH is much lower in adult birds due to a markedly higher basal TSH secretion. These ontogenetic changes in efficacy of CRH to stimulate TSH release in embryonic, growing, and adult chickens corresponds with the reported different sensitivities to exogenous TRH in the early period of chicken embryonic development, as well as before, during and after hatching (16).

It is important to note that the experiments described above all made use of ovine CRH, because the endogenous chicken CRH has been characterized only very recently (34). The chicken proCRH gene

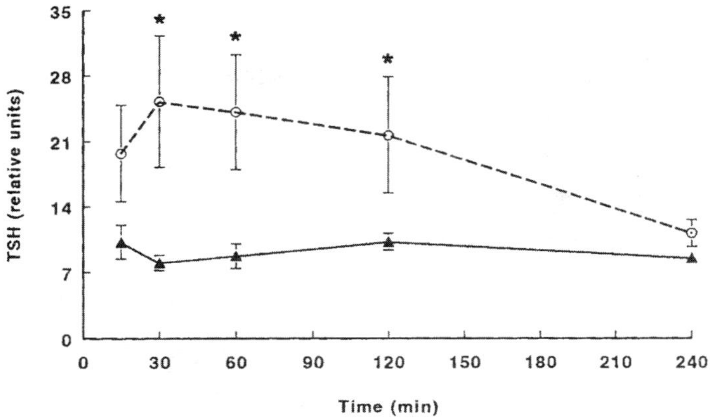

Fig. 1 Effect of a single injection of oCRH (2 μg) on the circulating concentration of TSH in day 19 chick embryos. Values represent means ± SEM (n = 5-7). Asterisks indicate differences between the two experimental groups within one time point ($P < 0.05$). ▲ control, ○ oCRH. (data from 14)

Table 1 Effect of 100 nM oCRH, TRH or both on *in vitro* TSH secretion in 1-day-old chicks. Data are mean stimulation factors ± SEM (n = 4).

Treatment	TSH
oCRH	18.0 ± 5.1
TRH	9.2 ± 1.2
oCRH + TRH	35.6 ± 8.2

consists of two exons separated by an intron. Exon 1 encodes the 5'-untranslated region, whereas exon 2 encodes proCRH and the 3'-untranslated region. The mature chicken CRH peptide is 100% identical to CRH in human, rat, and mouse (hCRH) (Fig. 2). During the last week of embryonic development, CRH content in the median eminence parallels plasma corticosterone values, indicative of an ACTH-releasing action (34). In a static incubation, hCRH induced a 6-fold increase in α subunit secretion from C3 chicken pituitaries, suggesting a thyrotropin-releasing function for endogenous chicken CRH as well (B. De Groef, unpublished results).

Although the regulation of ACTH secretion by CRH is commonly considered to be the primary hypophysiotropic role for this neuropeptide, TSH-releasing effects similar to those in the chicken have been found in other vertebrates, including the coho salmon (*Oncorhynchus kisutch*) (25), frogs (*Rana pipiens, Xenopus laevis*, and *Hyla regilla*) (10), and

```
chicken  MKLQPLVCAGILLLALLPCHECRALSKS---PGA-----ARGALQQPDFFPQQQQQQCQQ  52
Xenopus  MKFQLWVSTGILLVSLLPCHECRAFIKS---PAS-----SPGAL-----LPALSNSQ---  44
human    MRLPLLVSAGVLLVALLPCPPCRALLSRGPVPGARQAPQHPQPLDFFQPPPQSEQPQQPQ  60
sheep    MRLPLLVSVGVLLVALLPSPPCRALLSRGPIPGARQASQHPQPLSFFQPLPQ---PQEPQ  57
tilapia  MKLNLFGTTVILLVAFLPRYECRAIESP----GS--------TLR--VPALQTQNSQCQQ  46

chicken  TLPVLLRMGEEYFLRLGHLTKRPAGPFSASSGG-------HLRPEASAE----------  94
Xenopus  --PFLLRMGEEYFLRLGNLHKHSPGSFPEASAGNFVRAVQQLQAQQWSS----------  91
human    ARPVLLRMGEEYFLRLGNLNKSPAAPLSPASSLLAGGSGSRPSPEQATANFFRVLLQQLL 120
sheep    ALPTLLRVGEEYFLRLGNLDETRAAPLSPAASPLASRSSSRLSPDKVAANFFRALLQ--- 114
tilapia  SGPILERLGEEYFIRLGNGDSNSFPSTSMYPGGSPSIFNRALQLQLTRR----------  95

chicken  -----LLRAAAAQLQGSGSPEGDEGAGEAVEREKRSEEPPISLDLTFHLLREVLEMARAE 149
Xenopus  -----QPGMRAASLDGADSPYSAQ--EDPTEKAKRAEEPPISLDLTFHLLREVLEMARAE 144
human    LPRRSLDSPAALAERGARNALGGH-QEAPE-RERRSEEPPISLDLTFHLLREVLEMARAE 178
sheep    -PRRPLDSPAGPAKRGTENALGSR-QEAPAARKRRSQEPPISLDLTFHLLREVLEMTKAD 172
tilapia  -----LLQGKVGNIRALISGFGDR-GDDSMERGRRSEDPPISLDLTFHLLREMMEMSRAE 149

chicken  QLAQQAHSNRKLMEIIGK 167
Xenopus  QIAQQAHSNRKLMDIIGK 162
human    QLAQQAHSNRKLMEIIGK 196
sheep    QLAQQAHSNRKLLDIAGK 190
tilapia  QLAQQAQNNRRMMELFGK 167
```

Fig. 2 Comparison of the deduced amino acid sequence of chicken preproCRH with its orthologs in human, sheep, the African clawed frog *Xenopus laevis*, and the Nile tilapia *Oreochromis niloticus*. Identical amino acids have black backgrounds; similar amino acids have gray backgrounds. The solid black line indicates the mature CRH peptide and the dotted black line indicates the signal peptide-encoding region (data from 34)

hatching turtles (*Trachemys scripta*) (12). *In vitro* studies using pituitaries from all these species show that the stimulation of TSH release by CRH occurs through a direct action on the pituitary. Specific radioimmuno-assays for TSH in the salmon and the turtle have verified that the thyrotropic activity released in response to CRH is authentic TSH. It appears that a primitive role for CRH may have been to regulate both the thyroidal and adrenal (or interrenal) axes (11), whereas TRH may have obtained its TSH-releasing capacity only later in vertebrate evolution. This hypothesis is supported by the observation that CRH potently releases TSH from cultured salmon pituitaries, while TRH is not able to induce TSH secretion at all (25). Similarly, TRH stimulates TSH release in adult amphibians (3), but the bulk of evidence suggests that TRH does not influence TSH release in tadpoles (11). In the chicken, TRH is also involved in the control of growth hormone (GH) secretion, as TRH is a potent GH-releasing factor in birds (19). The sensitivity of the chicken somatotropes to GH-releasing hormone (GHRH) and TRH increases between E16 and E20, and continues to increase during the first days posthatch (4, 6). Injection of TRH in *adult* chickens increases plasma T_3 levels, but circulating T_4 decreased or remained unaffected, indicating that TRH is mainly somatotropic in adult chickens as GH is known to increase plasma T_3 (23). Both chicken thyrotropes and somatotropes express a type 1 TRH receptor (7).

Cellular distribution of CRH receptors in the chicken pituitary

Since CRH exerts its TSH-releasing function directly at the level of the pituitary, the presence of a CRH receptor in the pituitary has been investigated. At the inception of the investigation, only a type 1 CRH receptor (CRH-R1) was known in the chicken, whereas two isoforms had been identified in mammals and *Xenopus laevis* (20, 35). The coding sequence of chicken CRH-R1 consists of a 1260-bp open reading frame encoding a 450-amino acid G protein-coupled receptor (35). The amino acid sequence is 86.9% and 87.7% identical to human and murine CRH-R1s, respectively. The chicken CRH-R1 shows a higher binding affinity for urotensin I, a CRH-like peptide isolated from fish urophysis, than for human CRH. It shares this property with a second mammalian CRH-R subtype that was named type 2 (CRH-R2). However, the cloned chicken CRH-R1 has a higher sequence homology to the mammalian CRH-R1 than to the mammalian CRH-R2, hence its name. A major difference between the chicken and mouse CRH-R1 is five additional amino acids in the receptor's amino terminus just after its putative signal peptide, followed by a cluster of sequence divergence. Another cluster of sequence divergence is located in the region proximal to the first transmembrane domain. The species difference in amino acid sequence in the extracellular domain of the receptor may therefore be responsible for the high binding affinity for urotensin I (35).

Cellular localization studies using a combination of *in situ* hybridization and immunohistochemical staining of the thyrotropes shows that the chicken TSH-producing cells express mRNA for type 1 TRH receptors (TRH-R1) but are devoid of CRH-R1 mRNA (7). CRH-R1 mRNA could only be detected on corticotropes, suggesting that CRH-induced ACTH release is the result of a direct stimulation of corticotropes, whereas CRH-stimulated TSH release is not directly mediated by the chicken CRH-R1. To elucidate the mechanism underlying the thyrotropic activity of CRH, two plausible scenarios were tested: CRH could stimulate the thyrotropes either directly or in a paracrine fashion, for instance through the release of a corticotrope-derived factor (8). To test the latter hypothesis, the effect of melanocortins on the *in vitro* TSH release was investigated. Neither ACTH, α-melanotropin (α-MSH), nor its powerful agonist NDP-MSH could mimic the TSH-releasing effect of CRH, and a nonselective melanocortin receptor blocker had no effect on CRH-induced TSH secretion whatsoever (8). However, we discovered and cloned a partial cDNA for a type 2 CRH receptor (CRH-R2) and showed that this receptor is expressed in thyrotropes, but not in corticotropes (8). The involvement of CRH-R2 in CRH-induced TSH release was further demonstrated by perifusion experiments using a CRH-R2-specific agonist and antagonist. The agonist, human urocortin III, elicited a modest increase in TSH secretion from chicken pituitaries, and the CRH-R2-specific antagonist, antisauvagine-30, completely abolished the TSH-releasing capacity of oCRH (8). These results suggest that the hypophysiotropic actions of CRH in the chicken are mediated through different CRH receptors: corticotropes are stimulated to release ACTH after activation of CRH-R1, whereas CRH-induced TSH secretion is mediated by CRH-R2 on thyrotropes. This mechanism allows for the fine-tuning of the hypophysiotropic actions of CRH; according to the changing needs, CRH can preferentially release ACTH, TSH, or both by differentially regulated receptors (Fig. 3).

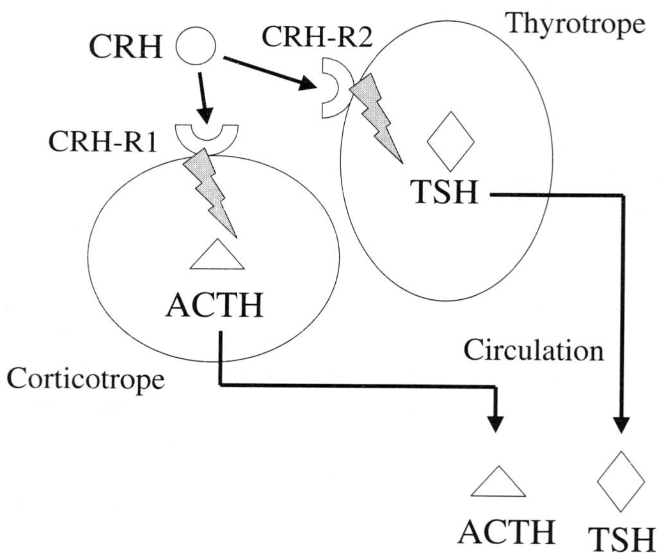

Fig. 3 Diagram to illustrate the finding in the chicken, that the stimulatory effects of corticotropin releasing hormone (CRH) on corticotropes and thyrotropes are mediated by different isoforms of the corticotropin-releasing hormone receptor (CRH-R1 and CRH-R2 respectively).

Because of the importance of CRH-R2 in CRH-induced TSH release, we cloned the complete coding sequence of this receptor isoform by rapid amplification of cDNA ends (RACE) (9). The chicken CRH-R2 is a 412-

amino acid 7-transmembrane G protein-coupled receptor, showing 87% identity to the *Xenopus laevis* and *Oncorhynchus keta* CRH-R2s, and 78-80% to mammalian CRH-R2s. As determined by RT-PCR analysis, both CRH-R1 and CRH-R2 mRNA are expressed throughout the chicken brain, as well as in the pituitary. In peripheral organs, CRH-R1 mRNA shows a more restricted distribution, whereas CRH-R2 mRNA is expressed in every tissue investigated, indicating that a number of actions of CRH and/or CRH-like peptides remain to be discovered in the chicken as well as in other vertebrates (9). The pharmacological binding properties of the chicken CRH-R2 remain to be investigated.

CRH-induced TSH release is inhibited by thyroid hormones and somatostatin, but not by corticosterone

The hypothalamic hormone somatostatin (SRIH) inhibits TRH-stimulated GH release (19) as well as TRH-stimulated TSH release in the chicken (17). SRIH was also found to inhibit CRH-induced TSH secretion (16). Pretreatment of pituitaries with SRIH had no effect on basal TSH secretion, but it clearly diminished the *in vitro* TSH-releasing capacity of oCRH both in embryonic and posthatch chickens (Fig. 4). Although experiments performed by other authors suggest a direct inhibitory effect of SRIH on the thyroid gland (24), our results show that SRIH exerts its inhibitory control of the HPT at least partially through a direct action at the level of the pituitary. In the chicken, this action is already present during embryonic development, though less pronounced than in posthatch chicks. In rats on the other hand, no endocrine function could be demonstrated for SRIH during fetal development; although after birth, SRIH is involved in the control of concentrations of plasma GH and TSH (29, 30).

Five SRIH receptors (SSTR1-5) have been identified in mammals. To study the specific function of each receptor subtype, researchers have synthesized specific nonpeptidyl agonists (ag1-5) (31). Using two *in vitro* protocols (static incubation and a perifusion system), the SSTRs controlling the secretion of GH and TSH from the chicken pituitary have been identified (17). SRIH actions at the level of the pituitary were found to be regulated through specific receptor subtypes. In both the static and perifusion systems ag2 (L-779,976) lowered the GH response to TRH, whereas TRH-stimulated TSH release was diminished by both ag2 and a g5 (L-817,818). These data indicate that SRIH inhibits chicken thyrotrope function through SSTR2 and SSTR5. Recent experiments have shown that these same two receptors are involved in the inhibition of CRH-induced TSH secretion by SRIH (B. De Groef, unpublished). Chicken thyrotropes were found to express SSTR2 mRNA, which is the only (partially) characterized SSTR mRNA to date (7).

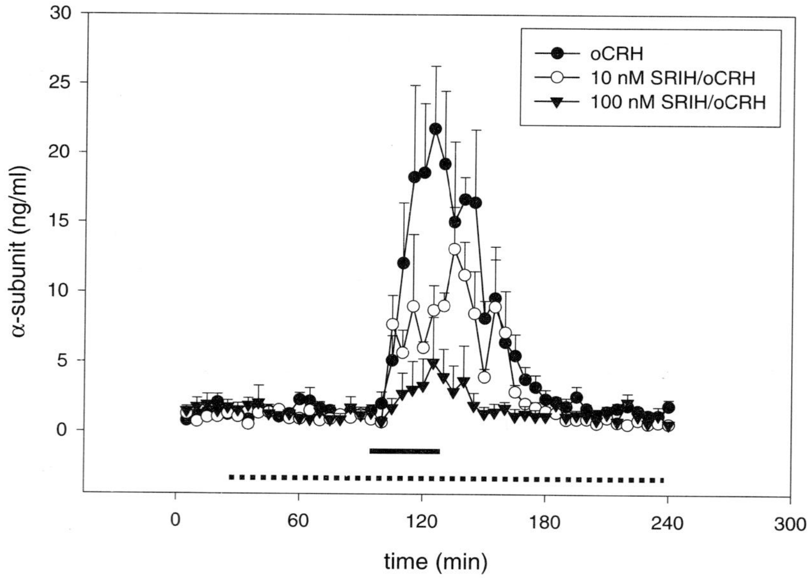

Fig. 4 The effect of two doses of SRIH (10 and 100 nM) on the *in vitro* α subunit (*i.e.* TSH)- releasing activity of oCRH (100 nM) in 19-day-old chicken embryos. The continuous bar indicates the presence of oCRH in the medium, whereas the dotted bar represents the period during which SRIH was added to the medium (data from 16).

In addition to the hypothalamic control through CRH, TRH, and SRIH, the secretory activity of the thyrotropes is also regulated by peripheral hormones through negative feedback mechanisms. *In vitro* perifusion of isolated E19 and C8 chicken pituitaries shows that THs blunt the TSH response to CRH at the level of the pituitary (15). T_3 causes a dose-dependent decrease in the oCRH-induced TSH release *in vitro* at both ages (see Table 2 for E19). A single injection of TH suppresses expression of both pituitary CRH-R2 and TRH-R1 mRNA as determined by RT-PCR and *in situ* hybridization (B. De Groef, unpublished results). SSTR2 mRNA on the other hand, is strongly upregulated in the pituitary of E19 chicken embryos after intravenous injection of T_3 or T_4. Taken together, the negative feedback effect of TH on stimulated TSH release might be – at least partially – due to an increased sensitivity of the thyrotropes to hypothalamic inhibition by SRIH, combined with a decreased responsiveness to hypothalamic stimulation by CRH and TRH (B. De Groef, unpublished results).

In vitro preincubation of rat anterior pituitary tissue with the synthetic glucocorticoid dexamethasone significantly reduced the release of immunoreactive TSH evoked by TRH (32). In the chicken the endogenous glucocorticoid corticosterone (B) does not suppress secretagogue-induced TSH release at the level of the pituitary as in mammals. Administration of B 30 minutes before an oCRH challenge diminished the *in vivo* sensitivity of thyrotropes to oCRH in E19 embryos, but not in C8 chicks (15). *In vitro* on the other hand, B did not alter the TSH response to oCRH at both ages studied (Table 2). This observation suggests that any observed effect *in vivo* is likely to be a secondary result of B-mediated effects, for instance an altered peripheral TH metabolism. Indeed, B is known to provoke a rise in plasma T_3 in embryonic but not in post-hatch chicks (5), which in turn could exert a negative feedback on CRH-induced TSH release *in vivo*.

Table 2 Effect of T_3 and corticosterone (B) *in vitro* on the TSH-releasing activity of 100 nM oCRH using pituitary glands from19-day-old chicken embryos. (data from 15)

Treatment	TSH
oCRH	16.18 ± 1.86^{a}
0.01 μM T_3 + oCRH	6.89 ± 1.79^{b}
1 μM T_3 + oCRH	1.81 ± 0.74^{b}
oCRH	12.91 ± 3.52^{a}
0.01 μM B + oCRH	9.28 ± 1.23^{a}
1 μM B + oCRH	8.73 ± 1.78^{a}

Data shown are mean stimulation factors ± SEM (n = 4). Within one treatment (T_3 or B) data with a common letter are not significantly different (Bonferroni; $P < 0.05$).

Conclusion

Contrary to the original view, not one but several hypothalamic factors are involved in the control of TSH secretion in birds. TSH release is stimulated by TRH and inhibited by SRIH. CRH also appears to be a potent stimulator of TSH release throughout life. CRH exerts its control over TSH and ACTH release through different receptors: CRH-R2 is expressed by thyrotropes, whereas CRH-R1 is expressed by corticotropes. As more and more evidence emerges that a similar mechanism exists in other vertebrate species, the chicken seems a good animal model to study this interaction between the thyroidal and adrenal axes.

Acknowledgements

BDG is a Postdoctoral Fellow of the Fund for Scientific Research – Flanders (FWO-Vlaanderen). KV was supported by the Institute for the Promotion of Innovation by Science and Technology in Flanders (IWT-Vlaanderen).

References

1 Berghman LR, Darras VM, Chiasson RB, Decuypere E, Kühn ER, Buyse J & Vandesande F (1993) Immunocytochemical demonstration of chicken hypophyseal thyrotropes and development of a radioimmunological indicator for chicken TSH using monoclonal and polyclonal homologous antibodies in a subtractive strategy. *General and Comparative Endocrinology* **92** 189-200

2 Carsia RV, Weber H & Perez FM Jr (1986) Corticotropin-releasing factor stimulates the release of adrenocorticotropin from domestic fowl pituitary cells. *Endocrinology* **118** 143-148

3 Darras VM & Kühn ER (1982) Increased plasma levels of thyroid hormones in a frog *Rana ridibunda* following intravenous administration of TRH. *General and Comparative Endocrinology* **48** 469-475

4 Darras VM, Finné M-F, Berghman LR & Kühn ER (1994) Ontogeny of the sensitivity of the somatotrophes to thyrotrophin releasing hormone (TRH) and growth hormone releasing factor (GRF) in the embryonic and posthatch chick. *Annales d'Endocrinologie (Paris)* **55** 255-260

5 Darras VM, Kotanen SP, Geris KL, Berghman LR & Kühn ER (1996) Plasma thyroid hormone levels and iodothyronine deiodinase activity following an acute glucocorticoid challenge in embryonic compared with posthatch chickens. *General and Comparative Endocrinology* **104** 203-212

6 Dean CE, Piper M & Porter TE (1997) Differential responsiveness of somatotrophs to growth hormone-releasing hormone and thyrotropin-releasing hormone during chicken embryonic development. *Molecular and Cellular Endocrinology* **132** 33-41

7 De Groef B, Geris KL, Manzano J, Bernal J, Millar RP, Abou-Samra AB, Porter TE, Iwasawa A, Kühn ER & Darras VM (2003) Involvement of thyrotropin-releasing hormone receptor, somatostatin receptor subtype 2 and corticotropin-releasing hormone receptor type 1 in the control of chicken thyrotropin secretion. *Molecular and Cellular Endocrinology* **203** 33-39

8 De Groef B, Goris N, Arckens L, Kühn ER & Darras VM (2003) Corticotropin-releasing hormone (CRH)-induced thyrotropin release is directly mediated through CRH receptor type 2 on thyrotropes. *Endocrinology* **144** 5537-5544

9 De Groef B, Grommen SVH, Mertens I, Schoofs L, Kühn ER & Darras VM (2004) Cloning and tissue distribution of the chicken type 2 corticotropin-releasing hormone receptor. *General and Comparative Endocrinology* **138** 89-95

10 Denver RJ (1988) Several hypothalamic peptides stimulate *in vitro* thyrotropin secretion by pituitaries of anuran amphibians. *General and Comparative Endocrinology* **72** 383-393

11 Denver RJ (1999) Evolution of the corticotropin-releasing hormone signaling system and its role in stress-induced phenotypic plasticity. *Annals of the New York Academy of Sciences* **897** 46-53

12 Denver RJ & Licht P (1989) Neuropeptides influencing pituitary hormone secretion in hatching turtles. *Journal of Experimental Zoology* **251** 306-315

13 Geris KL, Darras VM, Berghman LR & Kühn ER (1995) Influence of corticotropin-releasing factor on the *in vitro* thyroxine and thyrotropin secretion in newly hatched fowl. *Belgian Journal of Zoology* **125** 143-156
14 Geris KL, Kotanen SP, Berghman LR, Kühn ER & Darras VM (1996) Evidence of a thyrotropin-releasing activity of ovine corticotropin-releasing factor in the domestic fowl (*Gallus domesticus*). *General and Comparative Endocrinology* **104** 139-146
15 Geris KL, Laheye A, Berghman LR, Kühn ER & Darras VM (1999) Adrenal inhibition of corticotropin-releasing hormone-induced thyrotropin release: a comparative study in pre- and posthatch chicks. *Journal of Experimental Zoology* **284** 776-782
16 Geris KL, De Groef B, Kühn ER & Darras VM (2003) *In vitro* study of corticotropin-releasing hormone-induced thyrotropin secretion: ontogeny and inhibition by somatostatin. *General and Comparative Endocrinology* **132** 272-277
17 Geris KL, De Groef B, Rohrer SP, Geelissen S, Kühn ER & Darras VM (2003) Identification of somatostatin receptors controlling growth hormone and thyrotropin secretion in the chicken using receptor subtype specific agonists. *Journal of Endocrinology* **177** 279-286
18 Gregory CC, Dean CE & Porter TE (1998) Expression of chicken thyroid-stimulating hormone beta-subunit messenger ribonucleic acid during embryonic and neonatal development. *Endocrinology* **139** 474-479
19 Harvey S, Scanes CG, Chadwick A & Bolton NJ (1978) The effect of thyrotropin-releasing hormone (TRH) and somatostatin (GHRIH) on growth hormone and prolactin secretion *in vitro* and *in vivo* in the domestic fowl (*Gallus domesticus*). *Neuroendocrinology* **26** 249-260
20 Hillhouse EW, Randeva H, Ladds G & Grammatopoulos D (2002) Corticotropin-releasing hormone receptors. *Biochemical Society Transactions* **30** 428-432
21 Iwasawa A, Nakayama H, Yamamoto Y, Doi O, Nakamura T & Kato Y (2002) Specific anti-peptide antibody against beta subunit of chicken thyrotropin: production, characterization and application. *Journal of Reproduction and Development* **48** 197-204
22 Kühn ER, Huybrechts LM, Darras VM, Meeuwis R & Decuypere E (1990) Impaired peripheral T_3 production but normal induced thyroid hormone secretion in the sex-linked dwarf chick embryo. *Reproduction, Nutrition and Development* **30** 193-201
23 Kühn ER, Herremans M, Dewil E, Vanderpooten A, Rudas P, Bartha T, Verheyen G, Berghman L & Decuypere E (1991) Thyrotropin-releasing hormone (TRH) is not thyrotropic but somatotropic in fed and starved adult chickens. *Reproduction, Nutrition and Development* **31** 431-439
24 Lam S-K, Harvey S & Scanes CG (1986) Somatostatin inhibits thyroid function in fowl. *General and Comparative Endocrinology* **63** 179-185
25 Larsen DA, Swanson P, Dickey JT, Rivier J & Dickhoff WW (1998) *In vitro* thyrotropin-releasing activity of corticotropin-releasing hormone-family peptides in coho salmon, *Oncorhynchus kisutch*. *General and Comparative Endocrinology* **109** 276-285

26 Meeuwis R, Michielsen R, Decuypere E & Kühn ER (1989) Thyrotropic activity of the ovine corticotropin-releasing factor in the chick embryo. *General and Comparative Endocrinology* **76** 357-363
27 Mikami S & Takahashi H (1987) Immunocytochemical studies on the cytodifferentiation of the adenohypophysis of the domestic fowl. *Japanese Journal of Veterinary Science* **49** 601-611
28 Nakamura K, Iwasawa A, Kidokoro H, Komoda M, Zheng J, Maseki Y, Inoue K & Sakai T (2004) Development of thyroid-stimulating hormone beta subunit-producing cells in the chicken embryonic pituitary gland. *Cells Tissues Organs* **177** 21-28
29 Oliver C, Griaud P, Lissitzky JC, Cote J, Boudouresque F, Gillioz P & Conte-Devolx B (1982) Influence of endogenous somatostatin on growth hormone and thyrotropin secretion in neonatal rats. *Endocrinology* **110** 1018-1022
30 Rieutort M (1981) Ontogenetic development of the inhibition of growth hormone release by somatostatin in the rat: in-vivo and in-vitro (perifusion) study. *Journal of Endocrinology* **89** 355-363
31 Rohrer SP, Birzin ET, Mosley RT, Berk SC, Hutchins SM, Shen D-M, Xiong Y, Hayes E, Parmar RM, Foor F, Mitra SW, Degrado SJ, Shu M, Klopp JM, Cai S-J, Blake AD, Chan WWS, Pasternak A, Yang L, Patchett AA, Smith RG, Chapman KT & Schaeffer JM (1998) Rapid identification of subtype-selective agonists of the somatostatin receptor through combinatorial chemistry. *Science* **282** 737-740
32 Taylor AD, Flower RJ & Buckingham JC (1995) Dexamethasone inhibits the release of TSH from the rat anterior pituitary gland *in vitro* by mechanisms dependent on *de novo* protein synthesis and lipocortin 1. *Journal of Endocrinology* **147** 533-544
33 Vale W, Spiess J, Rivier C & Rivier J (1981) Characterization of a 41-residue ovine hypothalamic peptide that stimulates secretion of corticotropin and β-endorphin. *Science* **213** 1394-1397
34 Vandenborne K, De Groef B, Geelissen SME, Denver RJ, Boorse GC, Kühn ER, Darras VM & Van der Geyten S (2004) Molecular cloning and developmental expression of corticotropin-releasing factor in the chicken. *Endocrinology* **146** 301-308
35 Yu J, Xie LY & Abou-Samra AB (1996) Molecular cloning of a type A chicken corticotropin-releasing factor receptor with high affinity for urotensin I. *Endocrinology* **137** 192-197

Functional Avian Endocrinology
A Dawson & P J Sharp (Eds)

Differential regulation of intracellular thyroid hormone availability in developing chicken brain and peripheral tissues

S Van der Geyten, C H J Verhoelst, G E Reyns, E R Kühn, E Decuypere[1] and V M Darras

Laboratory of Comparative Endocrinology, Zoological Institute, KU Leuven, Naamsestraat 61, 3000 Leuven, Belgium
[1]Laboratory for Physiology of Domestic Animals, Department of Animal Production, KU Leuven, Kardinaal Mercierlaan 92, 3000 Leuven, Belgium

Introduction

The thyroid secretes predominantly thyroxine (T_4), a relative inactive prohormone that has to be converted into 3,5,3'-triiodothyronine (T_3) to exert its full biological activity. This activation, through outer ring deiodination (ORD) of T_4, occurs in the peripheral tissues by way of a family of selenocysteine containing oxidoreductases: the iodothyronine deiodinases (Fig. 1). Thyroid hormones can also be inactivated by iodothyronine deiodinases through inner ring deiodination (IRD) of T_4 or T_3 into inactive 3,3',5'-triiodothyronine (rT_3) or 3,3'-T_2 respectively. In birds, as in mammals, so far three types of deiodinases have been identified (1, 4).

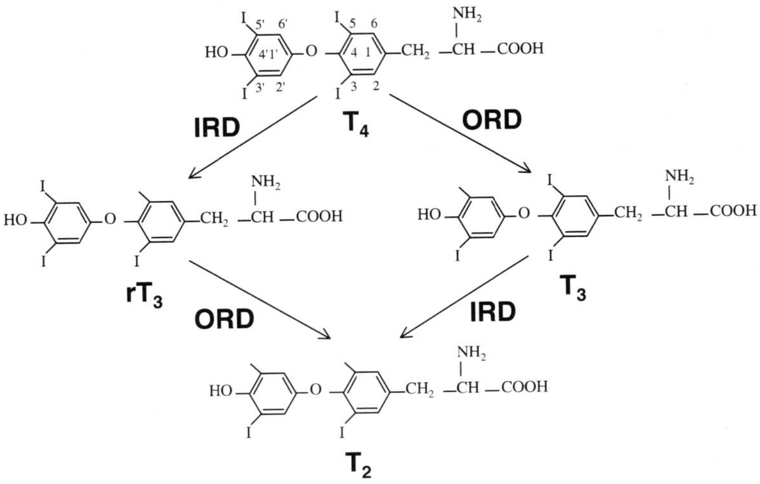

Fig. 1 Stepwise deiodination of T_4

Type I deiodinase (D1), which is capable of both ORD and IRD, is a high K_m (µM range for rT_3), high V_{max} enzyme highly sensitive to inhibition by 6-n-propylthiouracyl (PTU) (18). Type II deiodinase (D2), which is only capable of ORD, is a low K_m (nM range for T_4), low V_{max} enzyme that is not inhibited by PTU (10). Also the type III deiodinase (D3) is a selective low K_m (nM range for the T_3), low V_{max} enzyme that is only capable of IRD and is in turn insensitive to PTU (17).

The type III iodothyronine deiodinase plays a crucial role in regulating T_3 availability during development

Thyroid hormones play an important role in many crucial developmental events in all vertebrate classes. In birds, T_3 is essential for yolk sac retraction, functional maturation of the lungs, pipping and hatching (8). These phenomena are related to a steep increase in circulating T_3 levels (Fig. 2a), that reaches its maximum at pipping, when the embryo switches from allantoic to lung respiration (5, 6). Initially it was assumed that this rise in plasma T_3 was the result of increased hepatic T_3 production *via* D1 (13). However, more recent studies demonstrated that reduced T_3 breakdown *via* a reduction in hepatic D3 expression (Fig. 2b) is also essential for the increase in T_3 at the end of incubation (2, 18). Since growth hormone (GH) as well as corticosterone were known to affect plasma T_3 levels in the chicken, it was hypothesized that these hormones were also important in regulating the changes observed in peripheral deiodination during embryonic development (7, 14).

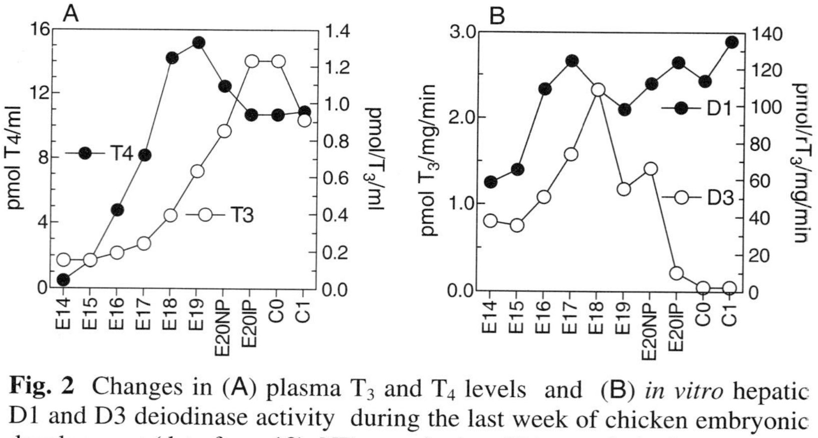

Fig. 2 Changes in (A) plasma T_3 and T_4 levels and (B) *in vitro* hepatic D1 and D3 deiodinase activity during the last week of chicken embryonic development (data from 18). NP non-pipping, IP internal pipping

Because most of the data published in this respect in recent years are concerned with the effects of glucocorticoids, we will focus for the remainder of this chapter on the role of glucocorticoids in the regulation of thyroid hormone function during chick development.

Regulation of hepatic deiodination by glucocorticoids during embryonic development

Intravenous administration of corticosterone as well as its long acting synthetic analogue dexamethasone (DEX) to 18-day-old chicken embryos (E18), results in significantly increased plasma T_3 levels within 30 min of administration (Fig. 3a). In contrast, plasma T_4 levels decreases within the same time frame (Fig. 3b), thereby indicating that an increased thyroidal T_4 production is not causative for the observed increase in circulating T_3 (19).

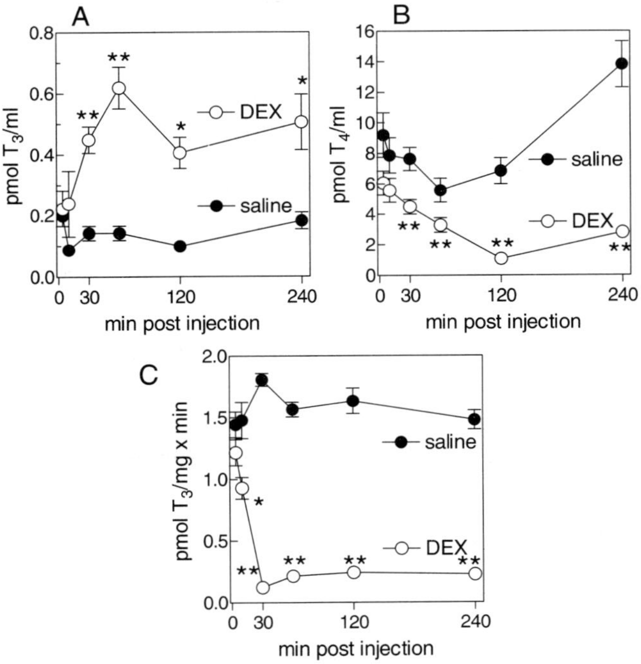

Fig. 3 Effects of *iv* administration of 50 μg dexamethazone (DEX) or saline on (A) plasma T_3, (B) plasma T_4 and (C) *in vitro* hepatic D3 activity in E18 chicken embryos (data from 19). Data are presented as mean ± SEM (n=10). Significant differences between the control and the DEX treated animals are indicated as * $P < 0.01$; ** $P < 0.001$.

An increased hepatic T_4 uptake, and hence T_4 to T_3 conversion, can be excluded as being causative since T_4 concentrations in the liver decrease after DEX treatment (16), whereas D1 mRNA (Fig. 4a, b) and D1 protein levels (Fig. 4c, d) remain unaffected (19, 23). *In vitro* hepatic D3 activity as well as D3 mRNA expression, however, decrease very rapidly after DEX injection (Fig. 3c), thereby demonstrating the essential role of hepatic T_3 degradation in the regulation of circulating T_3 levels under these circumstances (19).

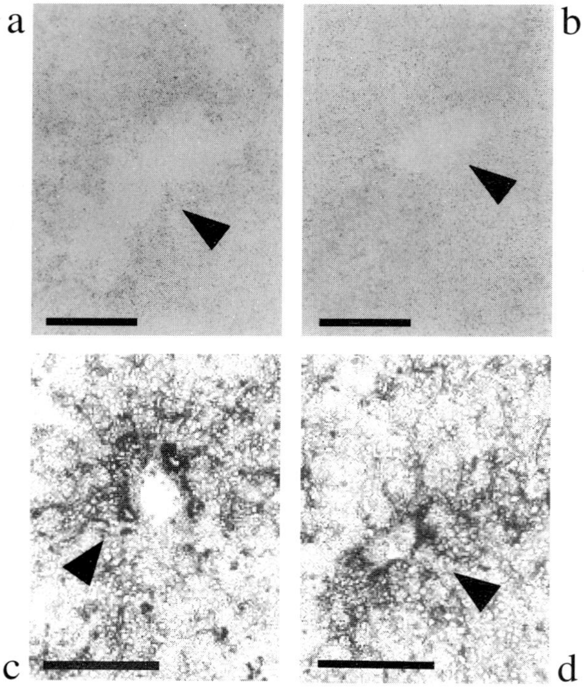

Fig 4. *In situ* hybridization on chicken liver cryosections using (a) a specific D1 cRNA probe of control or (b) dexamethazone (DEX) treated E18 embryos; and (c) alkaline phosphatase immunolabelling of hepatic paraffin sections with an anti-D1 antiserum of control or (d) DEX treated E18 embryos. Arrowheads indicate blood vessels. The black bars represent 100 μm.

With the use of specific D3 antibodies it could later be demonstrated that after glucocorticoid treatment the D3 protein is restricted to the hepatocytes circumscribing the blood vessels, whereas under control conditions D3 protein expression is present throughout the entire liver (Fig. 5) (23). Although the very rapid downregulation of hepatic D3 activity after glucocorticoid treatment seems to favor the involvement of post-translational enzyme inactivation, a series of experiments including blocking gene transcription using actinomycin D, or protein synthesis using cycloheximide showed that DEX directly inhibits hepatic D3 gene transcription (19, 20). The same studies also demonstrated that both the D3 mRNA and protein have very short half-lives (± 33 min and ± 24 min, respectively). This is in contrast to the much longer half-life for D1 mRNA and protein (> 8h). The short half-life of both the D3 mRNA and the D3 protein also explains why the blocking of D3 gene transcription by glucocorticoids can very rapidly decrease D3 enzyme availability.

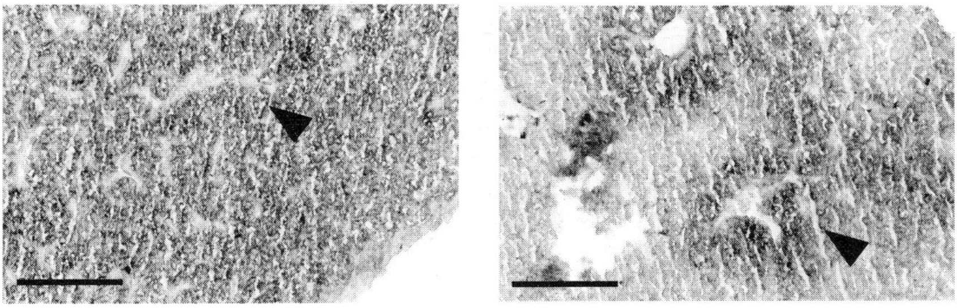

Fig. 5 Alkaline phosphatase immunolabelling using an anti-D3 serum of hepatic paraffin-embedded sections from (a) control or (b) DEX treated E18 embryos. Arrowheads indicate blood vessels. The black bars represent 100 μm.

Regulation of deiodination in the central nervous system by glucocorticoids

In contrast to hepatic D3, however, *in vitro* brain D3 activity (Fig. 6a) was not affected by DEX treatment (20). Although this lack of effect could be explained by assuming the existence of a blood-brain barrier for DEX, this explanation is highly unlikely since in mammals DEX is able to pass from the blood to the cerebrospinal fluid (15). Since renal D3 was also not affected by corticosterone or DEX administration (3, 20), it appears that the downregulation of D3 transcription by glucocorticoids in chicken embryos is tissue specific. This hypothesis finds additional support in the observation that although DEX has no effect on cerebral D3, glucocorticoids stimulate D2 expression in this tissue at the D2 activity (Fig. 6b), mRNA (Fig. 6c, d) and protein levels (Fig 6e, f) (20). This increase in overall cerebral D2 activity suggests that glucocorticoids increase T_3 availability in the brain, and by doing so stimulate brain maturation. However, the measurement of intracellular thyroid hormone concentrations showed that while T_3 levels were indeed increased in the telencephalon, diencephalon and cerebellum 2-6 h after DEX injection this was not the case in other brain areas (*e.g.* optic lobes and brainstem), and this despite the clear increase in D2 expression in these areas. This observation suggests a fine-tuning mechanism in the brain that enables it to maintain T_3 concentrations within narrow limits (16).

T_3 supply in the avian brain

The development in our lab, of specific antisera for D1, D2 and D3 allowed a detailed cellular localization of the different iodothyronine deiodinase proteins in the avian brain (22, 24, 25). One of the most surprising results from these studies was the demonstration of D1 expression in both the external germinal layer and the internal granular layer of the cerebellum (25), since it had always been assumed that D1

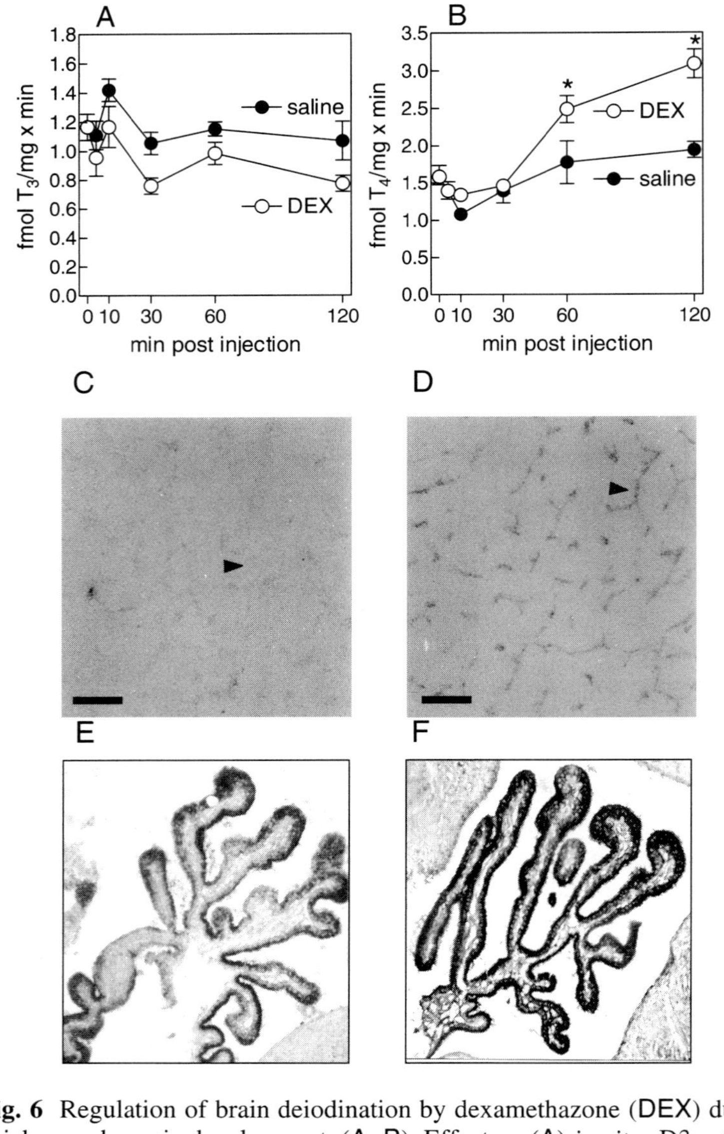

Fig. 6 Regulation of brain deiodination by dexamethazone (DEX) during chicken embryonic development. (A, B). Effect on (A) *in vitro* D3 activity and (B) *in vitro* D2 activity (data from 20). Data are presented as the mean ± SEM (n=10). Significant differences between the saline control and the DEX treated animals are indicated as * $P < 0.01$; ** $P < 0.001$. (C, D). *In situ* hybridization to chicken brain cryosections (telencephalon) using a D2 cRNA probe of (C) control or (D) DEX treated E18 embryos. Arrowheads indicate elongated cell clusters. The black bars represent 100 μm. (E, F). Alkaline phosphatase immunolabelling of chicken choroid plexus with an anti-D2 serum of (E) control or (F) DEX treated E18 embryos (data from 24).

was not expressed in the chicken central nervous system (21). D3 protein was detectable throughout the brain, with the highest levels of expression localized in the neuronal Purkinje cells of the cerebellum (22). Similarly, D2 protein was present in all brain regions, thereby confirming *in situ* hybridization data (11). The use of a specific anti-D2 serum demonstrated D2 protein expression in the cuboidal epithelial cells of the embryonic choroid plexus (Fig. 7a), whereas neither D1 (Fig. 7b) nor D3 (Fig. 7c) was detected in this structure (24).

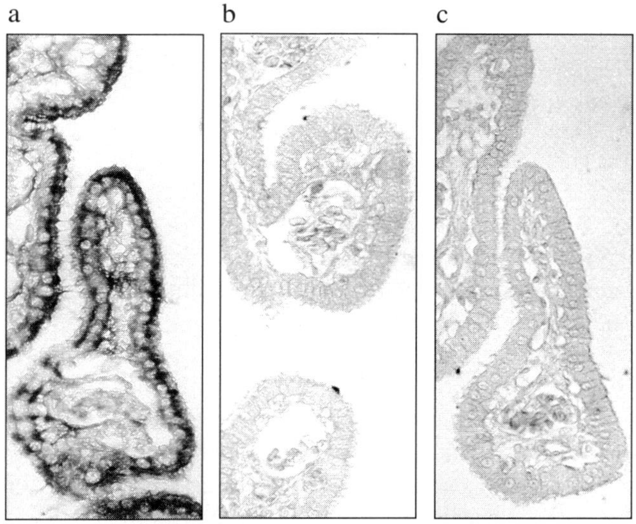

Fig. 7 Alkaline phosphatase immunolabelling of chicken choroid plexus (E18) with (a) anti-D2 serum, (b) anti-D1iserum and (c) anti-D3 serum (data from 24).

These observations are differ from the rat, where no D2 mRNA was detected in the choroid plexus using *in situ* hybridization. This led to the hypothesis that T_4 entering the mammalian brain through the choroid plexus binds to transthyretin (TTR) and is subsequently transported to the astroglial cells (9). These astroglial cells express D2, and can therefore convert the T_4 into T_3, which in turn is delivered to the surrounding neurons (12). Because the chicken choroid plexus expresses D2, this allows for an additional mechanism for the supply of T_3 to neurons (Fig. 8). The D2 in the choroid plexus deiodinates T_4 that enters *via* the basolateral membrane, thereby forming T_3. This T_3 then binds to TTR, which in birds has a higher affinity for T_3 than for T_4, and is subsequently transported directly to the neurons. So, in contrast to the mammals, in birds astrocytes appear not to be the sole source of bioactive T_3 and both the choroid plexus and the astroglial cells seem to be involved in T_3 supply to neurons.

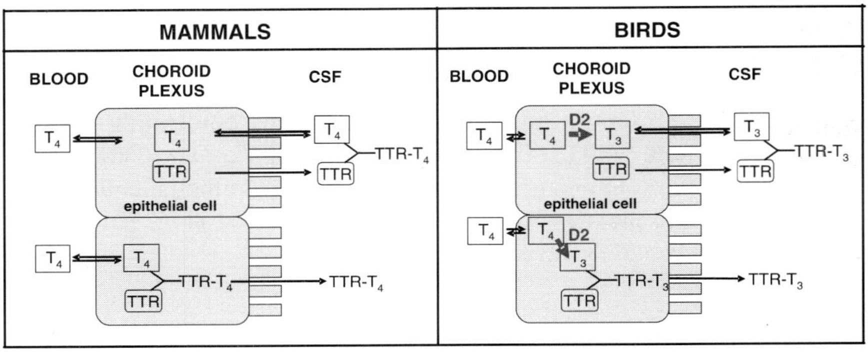

Fig. 8 Schematic diagram of the putative mechanism for thyroid hormone transport along the blood-brain-barrier in birds, compared to the existing scheme in mammals (from 24). CSF = cerebrospinal fluid; TTR = transthyretin

Conclusions

In the chicken, the developmentally crucial increase in circulating T_3 levels that occurs around hatching is predominantly regulated through a decrease in hepatic D3 expression, thereby strongly reducing the livers' T_3 degrading capacity. This decrease in D3 expression in the liver is predominantly induced by the increase in both GH and corticosterone secretion near the end of incubation. These effects are tissue and enzyme subtype specific since glucocorticoid administration has no effects on either brain or renal D3, whereas overall brain D2 is significantly stimulated by corticosterone as well as DEX administration. Although these changes in D2 favor increased thyroid hormone activation in the brain, this does not result in increased intracellular T_3 concentrations in all brain areas. These observations indicate that the avian brain possesses a very efficient fine-tuning mechanism, besides deiodination, to keep T_3 concentrations within narrow limits. Also, the discovery of D2 in the embryonic chicken choroid plexus suggests a new hypothesis concerning the supply of T3 to the brain. Thyroxine enters the brain *via* the choroid plexus and (a) binds to transthyretin and is transported to the astroglial cells were it is converted into T_3 by D2, and from there delivered to the surrounding neurons, or (b) is converted into T_3 by the D2 expressed in the choroidal epithelial cells, after which T_3 binds to transthyretin and is transported directly from the choroidal epithelium to the neurons.

Acknowledgements

Serge van der Geyten was supported by the Fund for Scientific Research – Flanders. This work was financed by the K.U.Leuven Research Council (OT/97/22, OT/01/23), the Fund for Scientific Research –

Flanders (G.0295.99, G.0272.04) and European Union Grant QLG3-CT2000-00930.

References

1 Bianco AC, Salvatore D, Gereben B, Berry MJ & Larsen PR (2002) Biochemistry, cellular and molecular biology and physiological roles of the iodothyronine selenodeiodinases. *Endocrine Reviews* **23** 38-89

2 Darras VM, Visser TJ, Berghman LR & Kühn ER (1992) Ontogeny of type I and type III deiodinase activities in embryonic and posthatch chicks: relationship with changes in plasma triiodothyronine and growth hormone levels. *Comparative Biochemistry and Physiology A* **103** 131-136

3 Darras VM, Kotanen SP, Geris KL, Berghman LR & Kühn ER (1996) Plasma thyroid hormone levels and iodothyronine deiodinase activity following an acute glucocorticoid challenge in embryonic compared with posthatch chickens. *General and Comparative Endocrinology* **104** 203-212

4 Darras VM, Verhoelst CHJ, Reyns GE, Kühn ER & Van der Geyten S (2004) Thyroid hormone deiodination in birds. *Thyroid* in press

5 Decuypere E, Nouwen EJ, Kühn ER, Geers R & Michels H (1979) Differences in the serum iodohormone concentration between chick embryos with and without the bill in the air chamber at different incubation temperatures. *General and Comparative Endocrinology* **37** 264-267

6 Decuypere E, Kühn ER, Clijmans B, Nouwen EJ & Michels H (1982) Prenatal peripheral monodeiodination in the chick embryo. *General and Comparative Endocrinology* **47** 15-17

7 Decuypere E, Scanes CG & Kühn ER (1983) Effects of glucocorticoids on circulating concentrations of thyroxine (T_4) and triiodothyronine (T_3) and on peripheral monodeiodination in pre- and post-hatching chickens. *Hormone and Metabolic Research* **15** 233-236

8 Decuypere E, Dewil E & Kühn ER (1990) The hatching process and the role of hormones. In *Avian Incubation*. pp. 239-256. Ed S.C. Tullet.Butterworths: London

9 Dickson PW, Aldred AR, Menting JGT, Marley PD, Sawyer WH & Schreiber G (1987) Thyroxine transport in choroid plexus. *The Journal of Biological Chemistry* **262** 13907-13915

10 Gereben B, Bartha T, Tu HM, Harney JW, Rudas P & Larsen PR (1999) Cloning and expression of the chicken type 2 iodothyronine 5'-deiodinase. *Journal of Biological Chemistry* **274** 13768-13776

11 Gereben B, Pachucki J, Kollar A, Liposits Z & Fekete C (2004) Ontogenic redistribution of type 2 deiodinase mRNA in the brain of chicken. *Endocrinology* **145** 3619-3625

12 Guadaño-Ferraz A, Obregón MJ, St Germain DL & Bernal J (1997) The type 2 iodothyronine deiodinase is expressed primarily in glial cells in the neonatal rat brain. *Proceedings of the National Academy of Sciences of the United States of America* **94** 10391-10396

13 Hughes TE & McNabb FMA (1986) Avian hepatic T_3 production by two pathways of 5'-monodeiodination: effects of fasting and patterns during development. *Journal of Experimental Zoology* **238** 393-399

14 Kühn ER, Verheyen G, Chiasson RB, Huts C, Huybrechts L, Van den Steen P & Decuypere E (1987) Growth hormone stimulates peripheral conversion of thyroxine into triiodothyronine by increasing the liver 5'-monodeiodinase activity in the fasted and normal fed chicken. *Hormone and Metabolic Research* **19** 304-308

15 Meijer OC, de Lange EC, Breimer DD, de Boer AG, Wortel JO, & de Kloet ER (1998) Penetration of dexamethasone into brain glucocorticoid targets is enhanced in mdr1a P-glycoprotein knockout mice. *Endocrinology* **139** 1789-1793

16 Reyns GE, Verhoelst CHJ, Kühn ER, Darras VM & Van der Geyten S (2004) Regulation of thyroid hormone availability in liver and brain by glucocorticoids. *General and Comparative Endocrinology* in press

17 Valverde CR, Aceves C & Reyes EZ (1993) Ontogenesis of iodothyronine deiodinase activities in brain and liver of the chick embryo. *Endocrinology* **132** 867-872

18 Van der Geyten S, Sanders JP, Kaptein E, Darras VM, Kühn ER, Leonard JL & Visser TJ (1997) Expression of chicken hepatic type I and type III iodothyronine deiodinases during embryonic development. *Endocrinology* **138** 5144-5152

19 Van der Geyten S, Buys N, Sanders JP, Decuypere E, Visser TJ, Kühn ER & Darras VM (1999) Acute pretranslational regulation of type III iodothyronine deiodinase by growth hormone and dexamethasone in chicken embryos. *Molecular and Cellular Endocrinology* **147** 49-56

20 Van der Geyten S, Segers I, Gereben B, Bartha T, Rudas P, Larsen PR, Kühn ER & Darras VM (2001) Transcriptional regulation of iodothyronine deiodinases during embryonic development. *Molecular and Cellular Endocrinology* **183** 1-9

21 Van der Geyten S, Van den Eynde I, Segers IB, Kühn ER & Darras VM (2002) Differential expression of iodothyronine deiodinases in chicken tissues during the last week of embryonic development. *General and Comparative Endocrinology* **128** 65-73

22 Verhoelst CHJ, Vandenborne K, Severi T,Bakker O, Zandieh Doulabi B, Leonard JL, Kühn ER, Van der Geyten S & Darras VM (2002) Specific detection of type III iodothyronine deiodinase protein in chicken cerebellar Purkinje cells. *Endocrinology* **143** 2700-2707

23 Verhoelst CHJ, Van der Geyten S & Darras VM (2004) Renal and hepatic distribution of type I and type III iodothyronine deiodinase protein in chicken. *Journal of Endocrinology* **181** 85-90

24 Verhoelst CHJ, Darras VM, Roelens SA, Artykbaeva GM & Van der Geyten S (2004) Type II iodothyronine deiodinase protein in chicken choroid plexus: additional perspectives on T_3 supply in the avian brain. *Journal of Endocrinology* **183** 235-241

25 Verhoelst CHJ, Darras VM, Zandieh Doulabi B, Reyns G, Kühn ER & Van der Geyten S (2004) Type I iodothyronine deiodinase in euthyroid and hypothyroid chicken cerebellum. *Molecular and Cellular Endocrinology* **214** 97-105

Author Index

Subject Index